W9-ADM-404

GENETICS OF THE EVOLUTIONARY PROCESS

THEODOSIUS DOBZHANSKY

Genetics OF THE *Evolutionary Process*

COLUMBIA UNIVERSITY PRESS

New York

Library of Congress Catalog Card Number: 72-127363

Printed in the United States of America
ISBN: 0-231-02837-7 *Clothbound*
ISBN: 0-231-08306-8 *Paperback*

9 8 7 6 5

To the Memory of My Mother

PREFACE

This book was started as a fourth edition of my "Genetics and the Origin of Species" (three previous editions: 1937, 1941, and 1951). It soon became apparent that so much has happened in evolutionary biology during the years since 1951 that no revision of the old book can be satisfactory. This does not exactly mean that everything in it is wrong. On the contrary, most of it is valid. Something more interesting has occurred: new problems have replaced the old at the forefront of our attention. Many things that had to be argued and documented in 1937, and even in 1951, now seem almost trite. For example, does one need nowadays to convince the reader that mutants are not mere laboratory products but occur as well in natural populations? Or that the differences between subspecies of wild animals and plants are mostly genetic? Or that species differences are compounded of the same building blocks as intraspecific and individual differences? On the other hand, it is now more clearly realized than it was in the past that natural selection is a common name for a complex of processes of rather diverse kinds and different biological significance. The discoveries of molecular geneticists have advanced our understanding of the origin of the evolutionary raw materials, and are throwing new light on the dynamics of the evolutionary process itself.

Some troublesome (and even painful) decisions had to be made in selecting the topics to be included in the book. The plain fact is that the relevant literature is now overwhelmingly vast. In the nineteen twenties, at the beginning of my career as a geneticist, I could truthfully claim to have at least glanced through a majority of the articles on genetics published until then in major European languages. I doubt whether anybody could make this claim at present, and I certainly cannot do so. An attempt to summarize all the available literature (even if this could be done) would not only make this book unduly long, but would also transform it into a kind of annotated bibliography. My intention is, rather, to present ideas with necessary examples, and not a miscellany of literature references. To this end, I have selected some illustrative references and, of necessity, left the rest aside. The choice could hardly be other than arbitrary. Many excellent contributions are not mentioned, and I can only beg forgiveness of their authors.

A whole field of mathematical genetics has developed in recent years. I am not at home in this field and must forbear from explaining it in detail. Yet no evolutionist can ignore the achievements of genetical mathematicians. My only recourse is to try to understand their assumptions and conclusions, and to hope that what comes in between is valid. Although much of the current research in genetics employs microorganisms as experimental materials, higher organisms get more attention in the book than lower ones, diploids more than haploids, sexually reproducing more than asexual ones. Drosophila is no longer the queen of genetics, as at one time it was. It remains, however, probably the best material for studies on evolutionary and population genetics, and I happen to be most familiar with it. Hence it will figure prominently in the following pages.

Evolutionary genetics is at present in an exciting period of development. Discoveries currently being made upset some classical theories that have acquired a status almost of dogmata. Clashes of opinion and polemics abound; new theories are being put forward that may or may not gain acceptance. No matter how much one tries to keep a stance of objectivity, some persons will find their views not given sufficient prominence, or presented in ways not conforming to their tastes. Moreover, a book attempting to describe a rapidly developing field will inevitably soon be out of date. The bibliography was completed in November 1969, which means that only a few more recent papers with which I was familiar in manuscript are referred to. I regret most of all the non-inclusion of the collection of essays edited by Drs. M. K. Hecht and W. C. Steere (Appleton-Century-Crofts, New York, 1970), since all of these essays are most relevant to the topics discussed in this book.

When there is a choice between discussing older or newer works dealing with the same topic, the newer ones are generally chosen. Of course, I have not done this because new works are always better than the old. My rationale is rather that recent articles usually contain references to the older ones, while the converse is not true. For the same reason, reviews and secondary sources are sometimes cited, even when I am familiar with the primary ones. Problems of priority have often been given short shrift. And yet, to trace the evolution of ideas about evolution one often has to go back to Morgan, Mendel, Darwin, and even Aristotle and Plato.

In writing this book the help of several colleagues and friends was invaluable. It is a great pleasure to acknowledge my obligations to Drs. F. Ayala and Lee Ehrman, who read the entire text; and to V. C. Allfrey, Verne Grant, I. M. Lerner, R. C. Lewontin, R. C. Richmond, Bruce Wallace, and L. J. Wangh, each of whom read one or more chapters. None of them is responsible for mistakes which will doubtless be found in the book, especially since their advice was not always heeded. But who can produce a book entirely free of mistakes? Finally, I am grateful to Mr. Andrea Palestrina, who patiently typed and retyped the entire book.

Theodosius Dobzhansky

July 20, 1970
New York City

CONTENTS

GENETICS OF THE EVOLUTIONARY PROCESS

CHAPTER ONE

THE UNITY AND DIVERSITY OF LIFE

Life and Evolution

A man consists of some seven octillion (7×10^{27}) atoms, grouped in about ten trillion (10^{13}) cells. This agglomeration of cells and atoms has some astounding properties; it is alive, feels joy and suffering, discriminates between beauty and ugliness, and distinguishes good from evil. There are many other living agglomerations of atoms, belonging to at least two million, possibly twice that many, biological species. What is most remarkable is that the individuals of every one of these species are so designed that they are able to live and reproduce in some existing environments. In other words, each species is adapted to a certain way of life. How has this come about? How can agglomerations of atoms accomplish any of these things?

Two kinds of answers have been proposed. Vitalists assume that living bodies are formed through the intervention of occult forces, variously called entelechy (Aristotle, Driesch), *vis essentialis* (C. F. Wolff), psyche, or inherent directiveness (Sinnott). Mechanists, on the other hand, claim that all biological structures and processes are only highly elaborate patterns of physical and chemical phenomena. Life can be understood without recourse to the assumption of any transcendental powers.

Vitalism goes often, though not always, together with creationism, that is, with the belief that the world as a whole, and the living species in particular, were created a few thousand years ago and have remained essentially unchanged since then. Lamarck, Darwin, and others after them expounded a different view: the living world that we observe at present has been shaped during billions of years of evolutionary history. The organisms now living have evolved gradually from ancestors that were generally more and more different from their descendants as one looks progressively farther back in time. Some vitalists do not deny evolution, but claim that it is guided toward predestined ends by inscrutable forces. Darwin thought otherwise. He posited natural selection as a process that impels and directs evolutionary changes. Subsequent research has on the whole vindicated his view.

Natural selection is a strictly biological phenomenon, in the sense that it is a sequel to life, and it exists solely in the living world. The essence of natural selection is the differential reproduction of the carriers of different hereditary endowments (see Chapters 4-7 for further discussion). Reproduction is the most important, or at any rate one of the most important, functions of life. Therefore, natural selection could not have begun before life appeared. (See, however, Spiegelman's work, discussed in Chapter 3.) Speaking of prebiological natural selection represents either a loose metaphor or a misapplication of a fundamental biological concept. On the other hand, reproduction, in conjunction with hereditary variation, makes natural selection inevitable. Natural selection is a pattern of physical events, and a pattern that is contingent upon and is restricted to life. Living bodies that have evolved under the control of natural selection can be described as machines, though machines of a very special sort. (For a contrary view, see Polanyi 1968.)

Cartesian and Darwinian Biology

The mechanist hypothesis was enunciated by Descartes in the seventeenth century. "The body of a living man," he wrote, "differs from that of a dead man just as does a watch or other automation (i.e., a machine that moves of itself), when it is wound up and contains in itself the corporeal principle of those movements for which it is designed, along with all that is requisite for its action, from the same watch or other machine when it is broken and when the principle of its movement ceases to act."

Any living organism, even the "simplest," is a machine of prodigious complexity. To understand how a machine works one should examine it in two different but complementary ways. It is necessary both to know the components of the machine and to comprehend how they fit together. Also, the more complex is a machine, the more critical become its structure and the interrelationships of its parts. The Cartesian reductionist method calls for a description of life in terms of the chemical and physical components and processes in living bodies. Some of the major triumphs of biology have been achieved with the aid of the Cartesian method. The discovery of the role of nucleic acids in the transmission and realization of heredity is an outstanding recent example.

The Cartesian approach has as its goal the reduction of biology to the status of a specialized branch of chemistry and physics. Some "molecular biologists" proclaim aggressively that the only worthwhile way to study life is in terms of its chemical and physical components and processes. They are confident that, in the fullness of time, when this study has advanced far enough, all the more complex biological phenomena will be seen clearly, or even predicted, as patterns of simple physicochemical ones. The interest and importance of molecular biology are, of course, unquestionable. The self-sufficiency of the reductionist program is, however, quite another matter. Reduction is "the explanation of a theory or a set of experimental laws established in one area of inquiry, by a theory usually though not invariably formulated in some other domain" (Nagel 1961).

It may well be doubted that all biological explanations and laws can be deduced from the more "powerful" physicochemical laws. This does not imply any kind of attenuated vitalism. Biological laws, such as Mendel's laws, deal with particular patterns of physical and chemical processes that occur only in living bodies. They are simply irrelevant, not contrary, to physics and chemistry. Nagel has written:

> There are sectors of biological inquiry in which physicochemical explanations play little or no role at present, and a number of biological theories have been successfully exploited which are not physicochemical in character. . . . Thus, there is a genuine alternative in biology to both vitalism and mechanism—namely, the development of systems of explanation that employ concepts and assert relations neither defined in nor derived from the physical sciences.

Biology is faced with several hierarchically superimposed levels of integration of structures and functions: molecular, cellular, individual, populational, ecosystemic. Discoveries have certainly been made on all these levels. Far more often, however, findings on higher levels point to the need of investigations of underlying levels than vice versa. Mendelian inheritance was not deduced from studies on the chemistry of DNA; these studies derive their significance in part from their bearing on Mendelian inheritance. Embryonic development was studied before the processes of transcription and translation in the DNA-RNA-protein codes were discovered. These findings suggest research strategies for further advances of embryology.

It is convenient to distinguish the molecular level from the organismic (cellular + individual + populational + ecosystemic) levels of integration. Explanations of the type termed by Simpson (1964a) as compositionist are important in biology, particularly on organismic levels. In this connection Simpson wrote:

In biology, then, a second kind of explanation must be added to the first or reductionist explanation made in terms of physical, chemical, and mechanical principles. This second form of explanation, which can be called compositionist in contrast to reductionist, is in terms of adaptive usefulness of structures and processes to the whole organism and to the species of which it is a part, and still further, in terms of ecological function in the communities in which the species occurs.

Adaptedness

Explanations in terms of adaptedness or teleology are not only appropriate but indeed necessary in biology, whereas they are meaningless in the nonliving world. The often misunderstood concept of teleology has been analyzed by Ayala (1968). The structures and functions of living bodies are said to exhibit adaptedness, or end-directedness, when they are shown to contribute to individual survival or to reproduction, which makes possible the survival of the species. Now, end-directedness may be due to external teleology, imposed by man for his purposes. Man-made tools, machines, and regulatory mechanisms are examples of such external teleology. The end-directedness of living bodies is of a different kind; it is internal teleology. The purposefulness of the components and of their configurations in living bodies that constitutes this internal teleology is a fact of observation. Such adaptedness makes survival and reproduction possible.

To understand and explain this internal teleology is the foremost, or at any rate one of the foremost, problems of compositionist, or Darwinian, biology. Satisfactory explanation of internal teleology eluded philosophers and biologists for a long time. Creationists believe that living bodies are engineered by a wise Creator; according to this view, organisms, like human contrivances, exhibit external teleology. To vitalists, teleology is an elemental property immanent in life. It cannot and need not be analyzed any further. To most biologists since Darwin, internal teleology is a product of evolutionary development. The adaptedness is neither devised nor planned by any external conscious agent; it is not guaranteed by a providential ability of living matter to act purposefully. Rather, it has evolved and is being maintained and often improved by natural selection.

Cartesian biology deals with organisms as they are today; Darwinian biology asks also how they got to be what they are. The "arrow of time" is most important in Darwinian biology. An individual organism develops from a fertilized egg or from a bud to an embryo, a juvenile, an adult, a senescent, a corpse. The development is directional and, except for minor details, irreversible. The directionality of individual development has always been recognized; the development of the living world as a whole, and of the Universe as a whole, is a relatively recent discovery.

The world images of Oriental philosophers and the philosophers of classical Greece were dominated by static or cyclic conceptions of time. Parmenides thought that changes which one observes happening in the world are illusions of the senses. Others believed that changes are cyclic and repeat themselves without end, as day becomes night and night changes to day. The most subtle synthesis of these speculations was given in the fourth century B.C. by Plato. What we observe in the world are only shadows of the eternal archetypes or ideas (*Eidos*). An individual man is only a distorted image of the unimaginably perfect and beautiful Man. All horses are warped replicas of the ideal Horse, and all dogs of the ideal Dog. The earthly replicas can perhaps be improved to reflect their archetypes more faithfully, but they cannot equal, much less exceed, the perfection of their celestial models. In what follows we shall have several occasions to criticize the application of this typological mode of thought to biological problems.

The evolutionary world view assumes a linear, instead of a cyclic, concept of time. Things, especially living things, were different in the past and will be different in the future. History is not an illusion, not a tedious return of past states. It is evolution, which has brought about the present state and will usher in the future states of the world. Because of an egregious miscomprehension, some Christians have fought Darwin's evolutionary interpretation of the living world, and eventually of the universe as a whole, though a linear concept of time is basic to Christian religious thought. At any rate, in biology nothing makes sense except in the light of evolution. It is possible to describe living beings without asking questions about their origins. The descriptions acquire meaning and coherence, however, only when viewed in the perspective of evolutionary development.

Chemical Composition

Living bodies are composed of the same chemical elements as are found also in the inorganic world. Data on the composition of living matter are summarized in Table 1.1. The three most abundant elements, oxygen, carbon, and hydrogen, constitute about 98.5 percent of organic matter. Oxygen and hydrogen are also among the elements most abundant in nonliving things. Carbon, on the other hand, is conspicuously more abundant in living bodies or in their remains and derivatives than elsewhere.

More than a century ago (in 1848 and 1860) Pasteur discovered another characteristic of living matter that proved to be remarkably general. Many chemical substances exist in two dissymmetric molecular forms, the right (D) and the left (L) isomers. This leads to optical activity, manifested in a rotation of the plane of polarized light either to the right or to the left. Compounds synthesized in the laboratory, on the other hand, are usually racemic, that is, optically inactive. The

TABLE 1.1

Average composition of living matter in percentages of the chemical elements by weight (After Vernadsky 1965)

Element	%	Element	%	Element	%
O	70	Fe	0.01	Cr	2×10^{-4}
C	18	Al	0.005	Br	1.5×10^{-4}
H	10.5	Ba	0.003	Ge	1×10^{-4}
Ca	0.5	Sr	0.002	Ni	5×10^{-5}
K	0.3	Mn	0.001	Pb	5×10^{-5}
N	0.3	B	0.001	St	5×10^{-5}
Si	0.2	Th	0.001	As	3×10^{-5}
Mg	0.04	Ti	8×10^{-4}	Co	2×10^{-5}
P	0.04	F	5×10^{-4}	Li	1×10^{-5}
S	0.05	Zn	5×10^{-4}	Mo	1×10^{-5}
Na	0.02	Rb	5×10^{-4}	Y	1×10^{-5}
Cl	0.02	Cu	2×10^{-4}	Cs	1×10^{-5}
		V	2×10^{-4}		

famous experiments of Pasteur demonstrated that such artificially prepared racemic substances consist of equal amounts of D and L isomers and that the isomers can be artificially separated.

What is notable is that in most organisms the principal constituents of protoplasm, particularly the amino acids, are represented exclusively by L isomers. Why this should be so is conjectural at present. This phenomenon may be evidence of life having arisen monophyletically (from a common stem), with L optically active constituents (Gause 1941 and Lederberg 1965). Certain bacteria contain also D amino acids, but these are mostly components of antimetabolites, toxic to competing organisms, whereas the essential proteins of the bacteria still contain L amino acids. In addition, D isomers are found in some relatively advanced forms, among earthworms and insects (Cloud 1968 and Corrigan 1969). This should not be taken as evidence that life is polyphyletic, having arisen independently two or more times. The many similarities in the key constituents of most diverse organisms (e.g., the two varieties of nucleic acids, DNA and RNA, and energy-storing compounds such as adenosine triphosphate) argue against a polyphyletic origin.

Much greater diversity of chemical compounds is found in living bodies than in nonliving ones. Nevertheless the distinction between organic and inorganic chemistry is not sharp. In 1829, Wöhler synthesized in the laboratory an organic compound, namely, urea. So many organic substances have subsequently been synthesized that additions to the list are important only if some particularly interesting or useful substance is involved. Although the progress of biochemistry has been impressive, it hardly warrants flights of fancy like that of Pollard (1965), who proposes, apparently in all seriousness, to construct not only the chemical components but actually a living bacterial cell!

How did life originate in the first place? It has been one of the biological certitudes since the days of Redi (seventeenth century), Spallanzani (eighteenth), and Pasteur (nineteenth) that life arises only from other life—*omne vivum ex vivo.* An impressive amount of research and an even greater amount of speculation have been devoted to this problem in recent years. It would be out of place here to attempt to review this large topic in detail, especially since technical as well as popular accounts are available (e.g., Oparin 1964, Sullivan 1964, and Bernal 1967).

Very briefly, the following scheme is proposed. Some chemical compounds were formed under the conditions of the primitive, lifeless earth that are now obtained only, or at least mainly, from living bodies. Experiments are therefore contrived to simulate the conditions supposed to have existed on the primitive earth. Several amino acids and even peptides and nucleotides have been thus obtained. Primordial oceans may, then, have contained a very dilute "broth" of organic compounds. But even if such a "broth" existed, the origin in it of the first self-reproducing, and hence living, systems remains an unsolved problem. Some writers have bravely, but not very convincingly, declared that, given the basic chemical properties of matter, the origin of life was inevitable. Since life has in fact appeared, its origin was indeed inevitable. Yet the problem cannot be resolved by a fiat; nobody really knows just how great an order of improbability this "inevitable" event involved. Hence we cannot be sure that, if there exist a million million planets more or less resembling the earth, any of them had extra-terrestrial life originate and evolve.

The Carriers of Genetic Information

Perhaps the most impressive demonstration of the unity of life is that in all organisms the genetic information is coded in two related groups of substances—the deoxyribonucleic (DNA) and ribonucleic (RNA) acids. Yet this method of coding is so versatile that the number of possible genetic "messages" is virtually infinite. Here, then, is the basis of the diversity as well as the unity of life.

A new era in understanding the physical basis of heredity, and hence of evolution, began in 1953 with the publication by Watson and Crick of the double-helix model of the structure of DNA. In the unbelievably short period since then, the model not only was shown to correspond to reality but also led to novel insights into the processes of gene and chromosome replication, gene mutation, and gene action in protein synthesis, and to a beginning of an understanding of the regulation of gene action in development. This explosive growth of biochemical and biological knowledge resulted from both independent and joint efforts of thousands of investigators. Although a detailed account of this success story, the greatest in the history of biology, cannot be given in this book, many accounts, ranging all the way from highly technical to popular ones (Watson 1965 and G. W. Beadle and M. Beadle 1966, to name only two examples from opposite ends of the spectrum), are now available. The brief description that follows is intended merely as a base line for the references to molecular genetics found elsewhere in this book. Only a sampling of the bibliography is given.

All nucleic acids are basically uniform in structure. Can the various genes in the same and in different organisms be so similar in composition? Most biologists held this impossible until Watson and Crick proposed their double-helix model of DNA. Nucleic acids are chainlike molecules; DNA molecules are paired chains, wound helically around a common axis. The backbone of each chain is made up of alternating deoxyribose sugars and phosphate groups; the links between the chains involve the nucleotide bases. Four nucleotide bases are commonly found—two purines (adenine, A, and guanine, C) and two pyrimidines (thymine, T, and cytosine, C). The paired chains are held together by weak hydrogen bonds between the nucleotides, in such a way that an A in one chain is always linked to a T in the other, and vice versa. Similarly, each G is linked with a C, and vice versa. As a consequence, the DNA's of quite diverse organisms have as many A as T bases, and as many C's and G's (in other words, the A : T and C : G ratios equal unity). By contrast, the proportions of (A + T) : (C + G) bases vary rather widely.

The important variables are, however, the sequences of the bases in the chains. A particular sequence of specific letters of the English alphabet can make up any word of the English language; it can also convey information or a message. So can the sequence of the four genetic "letters" in DNA; genetic "messages" are composed of different linear sequences of the genetic letters. Four letters taken two at a time can give 16, and taken three at a time, 64 different combinations. If a gene is a genetic message consisting of n letters, 4^n variant messages are possible. The numbers of nucleotide pairs in a gene may range in the hundreds or even the thousands. The possible variety of genes is enormous.

The double-helix model makes it possible to envisage how the genes direct their own precise replication. One has to suppose that the hydrogen bonds between the nucleotides in the paired chains break, and the chains unwind; each A attracts a T, each T an A, each C a C, and each G a C. The new bases are linked by a new sugar-phosphate backbone. The outcome is a pair of double helices similar to each other and to the parental one.

Meselson and Stahl (1958) submitted this scheme of the replication of DNA to a test. *Escherichia coli* were grown first in a medium containing N^{15}, a heavy isotope of nitrogen, and subsequently on ordinary nitrogen, N^{14}. The bacterium has no nucleus but a single circular chromosome; when grown on N^{15}, both strands of the DNA helix have this isotope. After the transfer to N^{14} and a single division, one strand of the helix has N^{15} and the other N^{14}. After two divisions one-half of the cells have no heavy nitrogen, and the other half have one heavy and one light strand. The replication of the chromosomes is therefore "semiconservative;" both strands of the DNA in a helix serve as templates for the synthesis of new strands, but the integrity of the old and the new strands is conserved. The replication of the chromosomes in higher organisms may also be semiconservative (Taylor 1969).

The near universality of the four genetic letters, A, T, C, and G, indicates that evolution of life has taken place by means of the composition of ever new "messages," not of new "letters." The existence of some exceptions must nevertheless be noted. Methyl cytosine, hydroxy methyl cytosine, and methyl adenine replace cytosine and adenine in some bacteria and bacteriophages (Jukes 1966 and Vanyushin, Belozersky, et al. 1968). It is conceivable that among the primordial forms of life there may have existed a greater diversity of genetic "letters," and that the four now used proved more convenient and hence replaced the others. Sinsheimer (1959) showed that the tiny bacteriophage ϕ X 174 has a molecule of single-stranded, instead of double-stranded, DNA as its hereditary material. It probably becomes double-stranded while replicating, and reverts to the single-stranded condition later.

Some plant viruses, of which the tobacco mosaic virus is the best-known representative, and also some bacteriophages have RNA instead of DNA as the carrier of genetic information. However, RNA is present in all organisms, since it is involved in the translation of the information stored in the DNA into protein structure (see below). RNA differs from DNA in having a ribose, instead of a deoxyribose, sugar and in having thymine (T) replaced by uracil (U). A tobacco mosaic virus particle has a core of a single strand of RNA containing about 6000 nucleotides, surrounded by a protein coat. When the virus infects a cell of a tobacco plant, it sheds its protein coat, and its RNA subverts the metabolic machinery of the host cell to form virus proteins instead of tobacco proteins. Ingenious experiments of Schramm (1956) and of Fraenkel-Conrat and Singer (1957) demonstrated that it is, indeed, the RNA and not the protein that directs the formation of the virus. They separated the RNA of the virus from the protein, and showed that new complete virus particles are formed on infection of tobacco leaves by the RNA alone. An analogous feat was achieved by Guthrie and Sinsheimer (1960), who infected the protoplasts (i.e., cells stripped of the surface cuticle) of *Escherichia coli* with purified DNA obtained from the ϕ X 174 bacteriophage.

The brilliant work of Kornberg and his students has demonstrated that DNA may serve as a template for self-replication *in vitro* as well as in a living cell (Kornberg 1962 and other works). The key to their success was the discovery of an enzyme, DNA polymerase, which catalyzes the synthesis (or repair) of DNA chains replicating those of a given "primer" DNA. A cell-free system is made containing this enzyme isolated from *Escherichia coli*, the four deoxynucleoside triphosphates, A, T, G, and C, and magnesium ions. To such a system are added as primers small amounts of DNA isolated from *E. coli* or from some other organism. Not only is new DNA synthesized, but also —and more significant—this new DNA resembles in composition that of the organism furnishing the primer, rather than that furnishing the enzyme. If self-replication is considered the fundamental manifestation of life, in these experiments we come close to a reproduction of a life process. This is true even though the materials involved are isolated chemically from other living beings.

The Genetic Code

Presumably every gene has at least two functions. It makes more of itself, serving as a template for the production of its facsimiles. It also forwards the genetic information stored in its DNA to direct the metabolic processes in the cell, and eventually in the body that carries it. The double-helix model has not only successfully explained the first function but also pointed the way toward elucidation of the second. S. Brenner, F. H. C. Crick, H. G. Khorana, M. Nirenberg, S. Ochoa, C. Yanofsky, and others have made major contributions in this field.

Although the basic understanding of heredity has come from studies on higher organisms, most of the current work uses prokaryotes (organisms lacking discrete nuclei) as materials. Bacteria, especially *Escherichia coli,* and bacteriophages are the predilect objects of molecular genetics. Caution is called for in extrapolation of the findings to the eukaryotes (the higher organisms having organized nuclei) and indeed to the entire living world. Unfortunately, such caution has not always been exercised.

Many, perhaps most, genes are structural genes, specifying the sequences of amino acids in proteins. Like nucleic acid, proteins are chain molecules. The backbone of a protein consists of so-called peptide linkages between successive amino acids. Many proteins (e.g., hemoglobins) are composed of two or more chains, coded by different genes, bent and folded in various ways. Although at least 170 different amino acids are known, only 20 are the common constituents of proteins. These 20 "letters" of the protein "alphabet" stand in a remarkable relationship to the 5 letters of the DNA and RNA alphabets.

Two processes, transcription and translation, intervene between the gene and the protein. At least three different kinds of RNA are involved—messenger, transfer, and ribosomal. The specificity of a gene resides in the order of the genetic letters in a certain section of the DNA helix. This order is transcribed, with the aid of the enzyme RNA polymerase, in a corresponding sequence of letters in a single-stranded messenger RNA. The latter is then translated into a specific sequence of amino acids in a polypeptide chain of a protein. Thus the DNA of a chromosome serves as a template, the sequence of the genetic letters in it dictating that in the messenger RNA (see, however, Bell 1969). The process is similar in principle to the synthesis of new DNA strands, except that in RNA uridine (U) corresponds to thymidine (T) in DNA, and that, for reasons not yet understood, only one of the two strands of the DNA helix is transcribed.

The translation is an even more complex process, which takes place on the cytoplasmic organelles known as ribosomes. There probably exist at least 60 different varieties of transfer RNA, corresponding to the triplets of the genetic code (see the next paragraph), and at least 20 activating enzymes that attach the different amino acids to the transfer RNA's. The ribosomes contain additional RNA's peculiar to them. Messenger RNA becomes associated with the ribosomes. The various amino acids, "activated" by the attachment to their transfer RNA's, are then added one by one to the growing polypeptide chain, in the order specified by the sequence of the code letters in the messenger. The translation of the nucleotide sequence in the messenger begins at a fixed end of the RNA strand and proceeds to the opposite end. Every three consecutive letters specify a certain amino acid.

The genetic code is a triplet and nonoverlapping code. This means that the translation proceeds by a "reading frame," which moves by three nucleotides in the messenger RNA and by a single amino acid in the protein at a time. One of the splendid achievements started in 1961 by the work of Nirenberg and Matthei was "breaking" the genetic code. Table 1.2 shows the correspondence between the 64 possible triplets, the 20 amino acids, and the protein chain initiation and termination.

The genetic code is said to be "degenerate." This refers to the fact that the same amino acid can be coded, at least *in vitro*, by more than a single triplet. Indeed, there are six triplets each coding for leucine, arginine, and serine. When a given amino acid is coded by different triplets in the same or in different organisms, these different triplets and the corresponding transfer RNA's need not be equally abundant in all cells. We shall return to this problem in Chapter 8. Three triplets, namely, UAA, UAG, and UGA, code for no known amino acids. They have been dubbed "nonsense" codons, a name hardly deserved since they appear to serve the important function of terminating the translation of the RN codons into polypeptide chains.

Nirenberg and Matthei (1961) have demonstrated that the same triplet code is "recognized" in protein synthesis in quite diverse organisms. For example, RNA of the tobacco mosaic virus induces protein synthesis in cell-free extracts that contain ribosomes of *Escherichia coli.* More recently Marshall, Caskey, and Nirenberg (1967) obtained even more remarkable results. They worked with cell-free systems including ribosomes, the necessary enzymes, messenger RNA of *E. coli,* and transfer RNA of either *E. coli,* a toad (Xenopus), or a mammal (guinea pig). As many as 50 codons were "recognized" in these systems by the transfer RNA's of the very different organisms just named, but considerable differences in the efficiency of certain codons were also noted. Hence, although these data are usually interpreted to mean that the genetic code is universal (or nearly so) in the living world, some skeptics (e.g., Commoner 1964) are impressed by the differences as well as by the similarities.

Conclusive evidence of the "colinearity" of the sequences of the nucleotides in a gene, and of the amino acids in a protein that this

TABLE 1.2

The genetic code of RNA nucleotide triplets, specifying the amino acids in protein chains: U, uracil; A, adenine; C, cytosine; G, guanine

Triplet	Amino Acid	Triplet	Amino Acid
UUA, UUG, CUU, CUC, CUA, CUG	Leucine (Leu)	ACU, ACC, ACA, ACG	Threonine (Thr)
UCU, UCC, UCA, UCG, AGU, AGC	Serine (Ser)	GGU, GGC, GGA, GGG	Glycine (Gly)
CGU, CGC, CGA, CGG, AGA, AGG	Arginine (Arg)	AUU, AUC, AUA	Isoleucine (Ile)
GUU, GUC, GUA, GUG	Valine (Val)	UUU, UUC	Phenylalanine (Phe)
CCU, CCC, CCA, CCG	Proline (Pro)	UGU, UGC	Cysteine (Cys)
GCU, GCC, GCA, GCG	Alanine (Ala)	UAU, UAC	Tyrosine (Tyr)
UAA, UAG, UGA	Chain termination	CAU, CAC	Histidine (His)
AUG	Methionine (Met)	CAA, CAG	Glutamine (Gln)
		AAU, AAC	Asparagine (Asn)
		AAA, AAG	Lysine (Lys)
		GAU, GAC	Aspartic acid (Asp)
		GAA, GAG	Glutamic acid (Glu)
		UGG	Tryptophan (Trp)

gene makes, has been provided by Yanofsky and his colleagues (Yanofsky, Drapeau, et al. 1967 and other works). A collection of mutants of *Escherichia coli* is available with different variant forms of the enzyme tryptophan synthetase. The sequences of the amino acids in these variant enzymes have been determined, pin-pointing the amino acids that have been substituted because of the mutations (see Chapter 2). The locations of the mutational changes in the chromosome of the bacterium were determined by genetic recombination studies. The linear arrangements of the changes in the chromosome and in the protein are similar.

Organic Diversity

The apparent simplicity, uniformity, and universality of the genetic code make even more impressive the prodigious diversity of the organisms found on earth. The virus of the foot-and-mouth disease is an approximately spherical body 8–12 millimicrons (a millimicron is 10^{-6} millimeter) in diameter. This is the order of magnitude of large protein molecules. Tobacco mosaic virus is a rod some 15 millimicrons in diameter and 300 millimicrons long. The psittacosis virus is a sphere with a diameter of about 450 millimicrons. Perhaps the smallest forms of cellular life are the pleuropneumonia-like organisms, 100–250 millimicrons in diameter. The pigmy shrew, weighing about 2.3 grams, is the smallest mammal. At the opposite extreme, the blue whale (*Balaenoptera musculus*) reaches 100 feet in length and 150 tons in weight. The weight of the largest *Sequoia gigantea* tree is estimated at 6167 tons.

The variability in the duration of an individual's life is also impressive. Colon bacteria divide under favorable conditions once in about 20 minutes. A bacteriophage particle infecting a bacterium gives 200–300 new particles within about 30 minutes. The oldest being now living is a bristlecone pine, *Pinus aristata,* about 4900 years old (Ferguson 1968). There are, however, many species of trees that can perpetuate themselves by stump-sprouting; among them is the redwood, *Sequoia sempervirens.* The life of an individual of this type is limited only by climatic and other environmental changes that exceed the tolerance of the species concerned.

Some algae, such as *Sphaerella nivalis,* grow and multiply on the surface of alpine snows, that is, at temperatures close to 0°C, while *Bacillus megaterium* and the fungus Sporobolomyces grow in Antarctica in saline pools at −23°C, the lowest temperature at which active life has been recorded. Some algae living in the outflow from the hot springs in Yellowstone Park have an optimal temperature range of 50–55°C but can still grow at 73–75°; some bacteria grow there at 80–85°C and can tolerate a temperature of 91°C, which is only 2° below the boiling point of water at the elevation where they live. The bacterium *Thiobacillus thiooxidans* grows in strongly acid media, whereas the alga *Plectonema nostrocorum* tolerates an alkalinity of pH 13 (Skinner 1968). A species of Delphinium (larkspur) grows on Mt. Everest at 20,340 feet, and some spiders even exist at 22,000 feet, feeding apparently on springtails and other insects, also living there or blown in by winds.

The ways of obtaining nutrition, and hence sources of energy for life, are likewise diversified. Autotrophic organisms are independent of external supplies of organic materials. By far the most widespread and successful autotrophs are green plants, which need only water, carbon dioxide and some inorganic substances, plus energy derived from solar radiation. Chemosynthetic bacteria derive their energy not from sunlight but from the oxidation of certain inorganic compounds. Thus, sulfur bacteria (Thiobacteriaceae) oxidize hydrogen sulfide (H_2S) and certain other sulfur compounds to sulfates (SO_4). Nitrifying bacteria (Nitrobacteriaceae) oxidize ammonia (NH_3) to nitrites (NO_2), and nitrites to nitrates (NO_3). Iron bacteria (Siderocapsaceae) use the oxidation of ferrous to ferric ions as the source of energy in the assimilation of carbon dioxide. Heterotrophic organisms utilize as energy sources a wide range of organic compounds, derived ultimately from autotrophs.

Some heterotrophs are specialized to an astonishingly narrow degree (monophagous), while others feed on many substances (polyphagous). Thus, at least 78 species of food plants are used by the gypsy moth (*Lymantria dispar*), whereas some other moth and butterfly species occur on a single kind of food plant only. Some parasites occur on just one host species, while others feed on many hosts. Perhaps the most extraordinary food source, or at least feeding place, is that of the petroleum fly, *Psilopa petrolei* (Thorpe 1930 and Oldroyd 1964). Its larvae live in pools of crude oil in California oil fields and, as far as is known, nowhere else. Its food consists of corpses of other insects caught and killed by the oil. The larva of *Psilopa petrolei* is the only known insect that can live in the oil, and even the adults of the same species can walk on the surface unharmed as long as only the tarsi of their legs are in contact with the oil. If any other part of the body touches the oil, the fly is trapped, killed, and devoured by the larvae.

The diversity of organisms must be somehow inscribed in the DNA of their cell nuclei. Yet the differences most readily detectable in the various DNA's are merely quantitative. One kind of difference concerns the relative prevalence of the four kinds of nucleotides. As stated previously, each A(denine) in one chain of the double helix corresponds to a T(hymine) in the other, and each G(uanine) is linked to a C(ytosine). The two purines together (A + G) are, then, equally as numerous as the two pyrimidines (T + C). The ratios of (A + T) : (C + G) are, however, variable, especially in microorganisms. In higher organisms, both animals and plants, the proportions of C + G are usually above 30 but less than 50 percent. In the bacteria *Micrococcus lysodeikticus* and *Streptomyces griseus* the amount of C + G reaches between 70 and 80 percent, whereas in *Clostridium perfringens* and *C. tetani* it is only 30–32 percent (cited after Jukes 1966). The diversity of living beings is evidently based not on the proportions but on the arrangements of the genetic "letters."

The amounts of DNA per cell are, as a rule, uniform in different tissues and individuals of the same species. Sex cells carry one-half as much DNA as do body cells. The amounts vary, however, in different organisms, as shown in Table 1.3. More complex organisms generally have more DNA per cell than do simpler ones, but this rule has conspicuous exceptions. Man is nowhere near the top of the list, being exceeded by Amphiuma (an amphibian), Protopterus (a lungfish), and

TABLE 1.3

Estimated amounts of DNA (in 10^{-12} gram) per haploid chromosome complement (After Mirsky and Ris 1951 and other sources)

Amphiuma	84	Duck	1.3
Protopterus	50	Chicken	1.3
Frog	7.5	Sea urchin	0.90
Toad	3.7	Snail	0.67
Man	3.2	Yeast	0.07
Cattle	2.8	Colon bacteria	0.004,7
Green turtle	2.6	Bacteriophage T2	0.000,2
Carp	1.6	Bacteriophage ϕ X 174	0.000,003,6

even ordinary frogs and toads. Why this should be so has long been a puzzle. It seems unreasonable that Amphiuma needs twenty-six times as many genes as man does. The amounts of DNA in *Escherichia coli,* in man, and in Amphiuma correspond to some 4.5×10^6, 2.9×10^9, and 8×10^{10} nucleotide pairs, respectively. If it is assumed that a protein coded by one gene has on the average 200 amino acids, or 600 nucleotides, *E. coli* has enough DNA for 7500, and man for 5,000,000, genes. Yet the variety of enzymes and proteins in higher organisms does not seem to be greater by three orders of magnitude than that in the lower ones.

A lead to the solution of this puzzle has perhaps been found by Britten and Kohne (1968). Heating DNA causes its "denaturation," since the paired strands of the double helix come apart; lowering the temperature leads to "renaturation," that is, to reassociation of the strands with complementary nucleotide sequences. For purposes of experiment, the DNA is broken by shearing into fragments some 400–500 nucleotides in length, heated to dissociate the paired strands, and cooled again to observe the rate at which the strand fragments are reassociated. To become reassociated, the fragments with similar nucleotide sequences must by chance come into contact; the more different fragments there are in a solution, the slower will be the reassociation of the fragments present in only two or another small number of copies. Conversely, if there are many similar fragments, their reassociation will be rapid.

Britten and Kohne found a most interesting difference between the lower and the higher organisms. A large fraction of the DNA in eukaryotes consists of segments with similar or even identical sequences, some of them repeated thousands and even as many as a million times. This redundancy of genetic materials is absent or at least is inconspicuous, however, in the prokaryotes, such as *Escherichia coli.* The following results were obtained with mouse DNA: about 10 percent of it consists of nucleotide sequences repeated close to one million times, some 20 percent of sequences repeated 1000–100,000 times, and 70 percent of unique sequences. About 40 percent of calf DNA consists of sequences repeated 10,000–100,000 times; a small percentage shows a slight degree of repetition, and 60 percent is made up of unique sequences. Redundancy has been found even in some unicellular algae (Euglena); whether or not redundancy increases systematically from the less complex to the more complex organisms remains to be seen.

The Discontinuity of Individuals

The diversity of living matter as it exists on earth shows two fundamental properties: the diversity is discontinuous, and the discontinuity is hierarchically organized. Some pioneer biologists (Bonnet) and philosophers (Leibnitz) thought that the "order of nature" cannot be incomplete, and consequently missing links between all existing living creatures must eventually be discovered. This may have been true of the past but is certainly not of the present. Provided that the evolutionary process occurred by gradual modifications, rather than by sudden jumps, different organisms now living have evolved from common ancestors, and the lines of descent had no major gaps or interruptions. But such gaps are certainly found between existing organisms.

Life occurs in discrete quanta, in individuals. The boundaries between individuals are as a rule evident. Multicellular individuals may, in one aspect, be viewed as colonies of cells. The cells of a body are, however, so thoroughly integrated and interdependent that they are constituent parts of individuals rather than autonomous individuals themselves. If some cells excised from my body are propagated in a tissue culture, they will no longer be parts of me. Individuality becomes seriously ambiguous only in some colonial forms, such as siphonophores, certain hydroids, corals, and tapeworms. Here the colony acts as a functionally effective individual, while its components may be specialized in nutrition (gastrozooids), reproduction (gonozooids), perception of external stimuli (dactylozooids), etc. Boundaries between individuals may also be lost in plants that reproduce asexually by runners, stolons, bulbs, or sprouts. Interesting situations present themselves among colonial insects, such as ants and termites. The bodily separateness of the individual members of a colony is here unmistakable; nevertheless, a vast majority of these individuals are sterile workers, and the functionally and genetically effective unit is a colony that includes individuals of the reproductive caste.

The bodily discontinuity of individuals may or may not be accompanied by genetic diversification, depending on the reproductive biology of a given form of life. Among higher organisms, especially animals, including man, sexual reproduction and outbreeding are prevalent. Every individual is then likely to possess a genetic endowment, a genotype, that is unique and different from the genotypes of all other individuals who live now, who lived in the past, and, probably, who will live in the future. This genetic uniqueness is a corollary to the Mendelian mechanism of inheritance. The matter is basically simple. A diploid individual heterozygous for n genes has a potentiality of producing 2^n kinds of sex cells with different gene complements. Two parents heterozygous for the same n genes may produce 3^n kinds of progeny with different gene constellations. With two parents each heterozygous for n different genes the potentiality rises to 4^n. In Chapter 8 evidence will be presented which suggests that, among higher animals and plants, n is of the order of hundreds or thousands.

The Mendelian mechanism of inheritance appears to be capable of generating a genetic diversity of individuals vastly greater than can actually be realized, because the number of individuals of any biological species is minuscule in relation to the diversity potentially possible. For example, the world population of human beings consists of somewhat more than 3 billion individuals, that is, between 2^{31} and 2^{32}, or 3×10^9. With only 100 variable gene loci in the human species (a patent underestimate), a vast majority of the possible genotypes will not be materialized. Similarly, Williams (1960) estimates that the world population of all species of insects combined is of the order of 10^{18} living individuals. He assumes that there may be as many as 3 million living species of insects, and estimates the median number of individuals per species as about 1.2×10^8. One or two most abundant insect species may consist of between 3×10^{16} and 5×10^{16} living individuals. Even these figures are diminutive compared to the genetic variety potentially possible. Most of this variety has never been realized and probably never will be.

The numbers of individuals of microorganisms are certainly greater than those of insects. No estimates seem to be available, and it happens that such estimates would be of limited interest in the present connection. The reason is that asexual reproduction is much more prevalent among the simpler than among the more complex forms of life, and fission and other modes of asexual reproduction, as well as certain kinds of parthenogenesis or apogamy (see Chapter 12), yield progenies that consist of individuals genetically similar to the mother and to each other. Unless mutations intervene, there may arise aggregations, clones, of genotypically identical individuals. Stebbins (1950) gives examples of clones of several species of flowering plants that have perpetuated themselves apparently without change for thousands of years, and that are represented by numerous individuals growing over extensive territories.

In many normally asexual species the asexual reproduction may from time to time be interrupted by a sexual generation or by other forms of gene exchange (Chapter 11). Conversely, in many normally sexual forms, including man, there occur from time to time monozygotic twins and other multiple births. Monozygotic co-twins are members of a clone, and they are, barring mutation, genetically identical. The biological meaning of individuality is, thus, not the same in sexual and in asexual organisms. In the former, the genetic uniqueness of every individual makes him a separate "experiment," testing a novel kind of being. In the latter, what is being tested is not a unique individual, but an array of individuals with similar or identical genotypes. Johannsen (1909) termed such arrays biotypes. In contrast to Johannsen's genotype and phenotype concepts, which have played important and constructive roles in the development of genetics (Chapter 2), the concept of biotypes has sometimes been misconstrued. It is not really useful in outbreeding sexual species; every man or fly or corn plant represents a biotype of its own. In asexual species a biotype is usually (barring mutation) equivalent to a clone.

The Discontinuity of Arrays of Individuals

Suppose that we make a fairly large collection, say some 10,000 specimens, of birds or butterflies or flowering plants in a small territory, perhaps 100 square kilometers. No two individuals will be exactly alike. Let us, however, consider the entire collection. The variations that we find in size, in color, or in other traits among our specimens do not form continuous distributions. Instead, arrays of discrete distributions are found. The distributions are separated by gaps, that is, by the absence of specimens with intermediate characteristics. We soon learn to distinguish the arrays of specimens to which the vernacular names English sparrow, chickadee, bluejay, blackbird, cardinal, and the like, are applied. Our collection of specimens, like the living world at large, is not a single array in which any two variants are connected by a series of intergrading variants. It is instead an array of discontinuous arrays, intermediates between which are absent. The small arrays, like those named above, are clustered into larger ones—passerine birds, the crow family, hummingbirds, birds of prey, etc. Still larger arrays are birds, mammals, reptiles, and fishes.

Biologists have exploited the discontinuity of variation as an aid in the construction of a classification of the living world. The hierarchical nature of the clusters and of the discontinuities lends itself admirably to this purpose. The small clusters are designated species, larger ones genera, still larger ones subfamilies, families, and orders. The classification thus obtained is natural, inasmuch as it reflects the objectively ascertainable discontinuities and the hierarchical order of the organic variations. The dividing lines between the species, genera, subfamilies, etc., are drawn to correspond to the gaps between the discrete clusters of the living forms. Biological classification is a man-made system of pigeonholes, serving the pragmatic purpose of recording observations in a convenient manner; it is also a reflection and an acknowledgment of the ubiquity of discontinuities in the living world.

The classification is also artificial, to the extent that it is a matter of convenience and convention which clusters are to be called genera or subfamilies or families. There have always been skeptics who contend that it is equally arbitrary which clusters we choose to call species. A great majority of biologists are convinced, however, that there are clusters which have, in at least the sexually reproducing organisms, biologically significant qualities which other clusters do not have, and that it is convenient to call these particular clusters species. Moreover, the vernacular names of animals and plants are most frequently attached to the clusters that biologists call species. The reality of species has apparently been perceived by biologically untrained people. This problem will be discussed in Chapter 11. Since we shall use the numbers of described species of organisms as measures of organic diversity, we may consider a single example as an illustration of what biological species really are.

Any two cats are individually distinguishable, as probably are any two lions. No individual has ever been seen about which there could be a doubt as to whether he was a cat or a lion. The species of cats (*Felis domestica*) and of lions (*Felis leo*) are discrete, because intermediates between them are absent. Difficulties that may arise in defining the two species will not be due to any artificiality of these species; however, the words cat and lion are sometimes used to refer neither to individual animals nor to all existing representatives of these species, but to some modal, average cat and lion. This is not far from Platonic "ideas" of these species. No matter how difficult it may be to define such ideal cats and lions, the discreteness of the species is not thereby impaired.

In sexually reproducing organisms, the existence of species could, in principle, be demonstrated without reference to the discontinuities in their bodily structures, by observing the pairing and procreation of the creatures concerned. Species are more or less discrete reproductive communities. Members of these communities are united by the bonds of sexual unions, of common descent, and of common parenthood. It will easily be discovered that there is a reproductive community of cats, and a separate reproductive community of lions. No lion cub was ever born to a pair of cats, or any kitten to a pair of lions. Members of different reproductive communities usually show no sexual interest in each other, or if they do, the results of their matings are inviable, sterile, or less fit than the progenies of mating within a community. A species, like a race or a genus or a family, is a group concept and a category of classification. A species is, however, also something else: a supraindividual biological system, the perpetuation of which from generation to generation depends on the reproductive bonds between its members.

The total number of species now living on earth is not known with precision. One reason is that many species have not yet been found, studied, described, and named. Aristotle knew approximately 500 species of animals, and Theophrastus 450 species of plants. Linnaeus listed 4235 animal species in 1758, and 5250 and 7000 species of plants in 1753 and 1774, respectively (see Zavadsky 1968). The available modern estimates, shown in Table 1.4, add up to some 1,594,565 species. Moreover, this number will quite probably be doubled or tripled by future studies. Another, though less important, reason for the inadequacy of the information concerning the numbers of species is that not all species are so clearly discrete as the cat and the lion. Darwin's crucial argument in favor of the origin of species by evolution was that all intergradations between distinct species and "varieties" (we would say at present races or subspecies) can be observed. This

TABLE 1.4

Estimated numbers of described species (After Grant 1963, Zavadsky 1968, and Mayr 1969)

Animals		Plants	
Vertebrates	41,700	Flowering Plants	286,000
Tunicata and Prochordata	1,300	Gymnosperms	640
Echinoderms	6,000	Ferns and allies	10,000
Molluscs	107,000	Bryophytes	23,000
Arthropods	838,000	Green algae	5,275
Annelids	8,500	Red algae	2,500
Bryozoans	3,750	Brown algae	900
Nematodes	11,000	Fungi	40,000
Rotifers	1,500	Slime molds	400
Nemertines	800		
Flatworms	12,700	Total plants	368,715
Coelenterates	5,300	Monerans	
Sponges	4,800	Blue-green algae	1,400
Minor phyla	800	Bacteria	1,630
Protozoans	28,350	Viruses	200
Total animals	1,071,500	Total monerans	3,230

argument fully conserves its validity. Moreover, the nature of the species differences varies in organisms with different reproductive biologies. This will be discussed in more detail in Chapter 11.

It is evident that the numbers of species in the different phyla of the animal and plant kingdoms listed in Table 1.4 are grossly unequal. The arthropods have about 82 percent of all animal species; among the arthropods some 92 percent of the species are insects; among the insects about 40 percent of the species are beetles, 16 percent moths and butterflies, 15 percent hymenopterans, 12 percent flies, etc. (Sabrosky 1952). By contrast, the very ancient and distinctive phylum Brachiopoda (lamp shells), abundantly represented in Paleozoic seas, is reduced to at most 260 species, and the phylum Phoronida to a mere 15 species. Inequalities almost as striking are found among the plant phyla in Table 1.4.

Adaptive Peaks

Organic diversity is impressive, wonderful, fascinating, or exasperating, according to one's tastes and temperament. Does it have some biological function and meaning? Is it merely a product of the creative exuberance of a deity or of some forces of nature? Or does it fulfill specific needs? The answer was already implicit in Darwin's and even in Lamarck's writings. It was fully articulated by the creators of the modern biological theory of evolution, S. Tshetverikov (1926), R. A. Fisher (1930), J. B. S. Haldane (1932), and Sewall Wright (1931). Organic diversity is a response of living matter to the diversity of environments, and of opportunities for different modes of life on our planet.

One can imagine a planet having just one kind of living beings, capable of surviving in some specially protected and lenient environment. Perhaps primordial life was thus uniform and sheltered. Life has, however, a propensity ever to expand in numbers and in mass, and to spread, invade, and assimilate ever new environments. We may envisage two strategies with the aid of which such expansion could be achieved. One is a strategy of environmental adaptability; an extraordinarily versatile genotype would evolve, the carriers of which could survive and perpetuate themselves in all kinds of environments. The same form of life could then live everywhere. The second is a strategy of diversification and environmental specialization, whereby multitudes of genotypes, each of them ideally suited to live in one and only one environment, would be formed. These two strategies are the conceivable limiting cases. In reality both strategies have been used in the evolution of the living world, but in different proportions in different lines of descent.

There are no fewer than two million species, and possibly twice or more than twice that number. Here, then, is the most general result of the application of the strategy of diversification. On the other hand, no species is specialized to live in just a single environment. Since any environment constantly changes, such a species would not endure for long. Really successful species are masters of a variety of environments, including some that occur over large parts of our planet. This is an accomplishment of the first strategy, environmental versatility. The human species, mankind, is perhaps most adept at this strategy: it is capable not only of choosing but also of artificially contriving environments to suit its needs and tastes.

Reference has already been made to the prodigious powers of the Mendelian mechanism of inheritance to generate ever-new gene constellations. Suppose there are only 1000 kinds of genes in the world, each gene existing in 10 different variants or alleles. Both figures are patent underestimates. Even so, the number of gametes with different combinations of genes potentially possible with these alleles would be 10^{1000}. This is fantastic, since the number of subatomic particles in the universe is estimated as a mere 10^{78}. The question that presents itself is whether it is a matter of chance which of these potentially possible gene combinations are realized. Clearly, it is not chance alone that operates here. Some gene combinations, indeed a vast majority of them, would be discordant and inviable in any environment. Other combinations, perhaps a tiny minority of the potentially possible ones, are suitable for life in some environments.

This situation can be envisaged with the aid of a symbolic picture of adaptive "peaks" and "valleys" first contrived by Wright (1932). With two gene loci each having 10 alleles, 100 combinations are possible. They can be diagramed as a two-dimensional grid, on which each combination will be represented by a point. Now, some of the combinations will have a higher and others a lower Darwinian fitness or adaptive value (see Chapter 4). The fitness may be depicted as a third dimension, giving the diagram the appearance of a topographic map (Fig. 1.1). This representation is evidently an oversimplification; with n variable genes a complete picture would require every gene combination and its fitness to be represented by a point in an $(n + 1)$. dimensional space. Nevertheless, Wright's diagram is very helpful. The contours in the figure symbolize the adaptive values of the gene combinations. Groups of related gene combinations that make their possessors able to inhabit certain environments are represented by adaptive peaks (plus signs in Fig. 1.1). Unfavorable gene combinations that make their carriers unfit to live in these environments are symbolized by adaptive valleys (minus signs in Fig. 1.1).

The diversity of living forms may then be envisaged as a multitude of adaptive peaks, corresponding to the multitude of ways of living that are possible on our planet. The variety of these possible ways of living—ecological niches—is, however, not only great; it is also discontinuous. One species of insect may feed, for example, on oak leaves, and another species on pine needles; an insect that required food intermediate between oak and pine would probably starve to death. Hence, the living world is not a formless mass of randomly combining genes and traits, but a great array of families of related gene combinations, which are clustered on a large but finite number of adaptive peaks. Each living species may be thought of as occupying one of the

FIGURE 1.1

"Adaptive peaks" and "adaptive valleys" in the field of gene combinations. The contour lines symbolize the adaptive values of the carriers of various genotypes. (After Wright)

available peaks in the field of gene combinations. The adaptive valleys are deserted and empty.

Furthermore, the adaptive peaks and valleys are not interspersed at random. "Adjacent" adaptive peaks are arranged in groups, which may be likened to mountain ranges in which the separate pinnacles are divided by relatively shallow notches. Thus, the ecological niche occupied by the species lion is relatively much closer to the niches occupied by tiger, puma, and leopard than to those occupied by wolf, coyote, and jackal. The feline adaptive peaks form a group different from the group of canine peaks. But the feline, canine, ursine, musteline, and certain other groups of peaks form together the adaptive "range" of carnivores, which is separated by deep adaptive valleys from the ranges of rodents, bats, ungulates, primates, and others. In turn, these ranges are again members of the adaptive system of mammals, which are ecologically and biologically segregated, as a group, from the adaptive systems of birds, reptiles, etc. The hierarchic nature of biological classification reflects the objectively ascertainable discontinuity of adaptive niches—in other words, the discontinuity of ways and means by which organisms that inhabit the world derive their livelihood from their environments.

Evolution

The classical theory of evolution, as formulated by Darwin and his immediate followers, can be summed up in four assertions: (1) the beings now living have descended from very different beings that lived in the past; (2) the evolutionary changes were more or less gradual, so that if all the past as well as present inhabitants of the earth could be assembled, a fairly continuous array of forms would emerge; (3) some changes were divergent, and many species now living are descended from fewer and fewer ancestral species as one goes farther and farther back in the past; (4) all the changes were products of causes that now continue in operation and therefore can be studied experimentally.

Evolutionists of the nineteenth century were interested primarily in demonstrating that evolution had in fact taken place. They succeeded; evolution as a process that has occurred in the history of the earth is no longer questioned either by scientists or by the informed public. As the study of evolution proceeded, two main approaches were employed. The first concentrated on unraveling actual evolutionary histories, that is, phylogenies of various groups of animals and plants. The methods used are those of systematics, comparative anatomy, comparative embryology, and especially paleontology, the study of fossils. The second approach emphasized studies of the mechanisms that bring evolution about, of causal rather than historical aspects. Genetics, especially population genetics and ecological genetics, has supplied the basic concepts and the experimental as well as observational methods.

In the nineteen twenties and thirties, the foundations of the modern biological theory of evolution were laid by Tshetverikov, Fisher, Wright, and Haldane, as stated above. These four pioneers were brilliant theoreticians; they leaned heavily on mathematical deduction from a few fundamental postulates, especially those of Mendelian inheritance. In the late nineteen thirties, the forties, and the fifties, a number of biologists of various specialities, working in different countries, showed that the theory made good sense in their respective fields. The resulting synthesis is, therefore, truly a biological, not only a genetical or ecological or paleontological, theory.

This book presents an outline of the biological theory of evolution, with special emphasis on its genetic aspects. The general principles of the biological theory are widely, but not universally, accepted among present-day biologists. A stridently dissenting voice is, for example, that of Koestler (1967). Though not himself a biologist, this author lists, as the first of the "monumental superstitions" on which "the citadel of orthodoxy" in modern science is built, the view "that biological evolution is the result of random mutations preserved by natural selection." I shall endeavor to show that Koestler's view and similar ones result from a monumental miscomprehension of what the biological theory of evolution really is.

CHAPTER TWO

GENETIC CONTINUITY AND CHANGE

The Conservatism of Heredity

Heredity is a conservative force: the genes function as templates for the production of their exact copies; by making the offspring resemble their parents, heredity confers stability upon biological systems. Evolution is the antithesis of permanence; the most general definition of evolution maintains, "The current state of a system is the result of a more or less continual change from its original state" (Lewontin 1968). If heredity were always perfectly exact, evolution could not occur.

The precision of heredity is tempered, however, by occasional instability. The counterpart of heredity is variation. Variation has two aspects, a static and a dynamic one. Variation as a status (variability) is the observable diversity between individuals or groups within a species, or the diversity between species. Variation as a process means that the development of individuals may be modified by environmental influences, and the hereditary endowment may be changed by gene recombination or mutation (Philiptschenko 1927). Changes of the latter type are the ultimate source of all genetic variability. If all life is monophyletic, derived from a single kind of primordial life, then all organic diversity must be the outcome of the accumulation and ordering of mutational changes. Not all mutations are conserved, however; as we shall see in Chapter 3, most of them are cast out by natural selection.

All living beings grow and reproduce. Growth occurs by the assimilation of materials taken up from the environment, and their transformation into body constitutents. The organism reproduces itself, in its progeny, from the food that it consumes. The processes whereby self-reproduction is accomplished are the essence of heredity. The basic discovery of genetics is that the units of self-replication are molecular-level systems called genes. The self-replication normally occurs within larger systems—chromosomes, nuclei, cells, organisms. Nevertheless, some components of the body are self-reproducing, whereas others are not. The former constitute both what early precursors of genetics (e.g., Weismann) called the germplasm or idioplasm and what we call genes, irrespective of whether they are located in the chromosomes (chromosomal genes) or in some cytoplasmic organelles, such as plastids or mitochondria (plasmagenes).

The distinction between self-reproducing and non-self-reproducing systems has proved incomprehensible to some people, such as Lysenko in Russia. The evidence of the actuality of self-reproduction seems, however, secure enough at present. A single example will be sufficient here.

The now classical work of Kornberg, mentioned in Chapter 1 (a review in Kornberg 1962), on the *in vitro* synthesis of DNA became possible because of the isolation of an enzyme DNA polymerase from *Escherichia coli*. A mixture is prepared containing this enzyme, the four kinds of nucleotides (in the form of deoxyribonucleoside triphosphates), and magnesium ions. After several hours of incubation, short chains of nucleotides are formed. The reaction is greatly accelerated by adding to the system "templates," that is, small amounts of DNA extracted from some organisms. Not only are DNA chains synthesized, but also—and this is crucial—the DNA that is formed is like the primer in composition. Kornberg and his colleagues used as primers DNA's from organisms as different as cattle, the bacteria *E. coli* and *Mycobacterium phlei*, and bacteriophage T2. The DNA's from the two bacteria differ in the proportions of the genetic "letters"; in *E. coli* the percentages of A, T, G, and C are 25, 24, 24, and 26, and in *M. phlei* 16, 16, 34, and 34, respectively. When the primers act as templates for the *in vitro* synthesis, the resulting DNA's have percentages of A, T, G, and C of about 26, 25, 24, and 25 with the *E. coli* templates, and of about 16, 16, 33, and 34 with those of *M. phlei* (the small differences are expected experimental imprecisions).

Phenotype and Genotype

The concepts of phenotype and genotype were introduced by Johanssen (1909) and they remain basic for clear thinking about genetic and evolutionary problems. They can at present be defined as follows. The phenotype of an individual is what is perceived by observation: the organism's structures and functions—in short, what a living being appears to be to our sense organs, unaided or assisted by various devices. The genotype is the sum total of the hereditary materials received by an individual from its parents and other ancestors. The phenotype of an individual changes continuously from birth to death. Barring somatic mutation, the genotype, however, remains stable. This stability is due to the genes reproducing themselves, not to the genes being chemically inert materials or being somehow isolated from the environment.

The error of the Lamarckian belief in the inheritance of acquired characters lay in its failure to recognize that the phenotype is a by-product of gene activity, while the genes reproduce by serving as templates in the copying process. A very brief description of the present status of the problem of the translation of genetic information encoded in the DNA of genes into the sequences of amino acids in the proteins was given in Chapter 1. As pointed out by many authors (e.g., Crick 1967), the process is unidirectional: DNA → RNA → protein. The replication of the genetic material, DNA, requires, of course, the presence of certain enzymes, which are proteins. The protein, the composition of which is specified by a given gene and hence by a certain section of the DNA chain, does not, however, serve as a template for the synthesis of a new gene. In other words, genetic information is transferred from a section of DNA to the corresponding protein, but not in the reverse direction, that is, not from the protein to the DNA.

This conclusion is sometimes referred to as the "central dogma" of molecular genetics. How felicitous this designation is may well be questioned; at any rate, the basic idea was clearly present in the minds of those, beginning with Weismann, who discounted the hypothesis of inheritance of acquired traits. Consider such an acquired trait as big muscles strengthened by exercise. Its inheritance would require that some product secreted by the muscles changed the nucleotide sequence or number in the DNA chains of some genes. Such changes are unknown and seem quite improbable.

The statement that the genotype does not change during an individual's lifetime must be clarified to stave off ambiguity. The amount of DNA in a cell is doubled in the interval between successive cell divisions. New DNA chains are synthesized on the old ones. Does an adult person have the same genes that he had as an infant, an embryo, and a fertilized egg cell? The answer is that he has true copies of these genes. Even more liable to misunderstanding is the statement that genes go without change through many generations of individuals. An individual has copies of some of the genes of his ancestors. The lack of change means only that the gene-copying process is as a rule scrupulously exact.

Genetic Programming of Potential Phenotypes

A gene, or a cistron according to Benzer's (1957) terminology, is a functional unit; it usually corresponds to a section of the DNA chain coding for an amino acid sequence in a protein. If this description of the action of structural genes is valid, then any gene can yield one and only one primary product, if it functions at all. Actually, however, the process of development is more complex. Between the genes and their messenger RNA's at one end, and the adult phenotypes at the other, there intervenes a set of developmental processes, which, particularly in complex organisms, may be exceedingly long and elaborate. This leaves ample opportunity for the uniform primary action of a gene to yield a variety of manifestations in the developing phenotypes. The genotype does not, therefore, determine the phenotype; it determines a range of potentially possible phenotypes. The range of phenotypes that can develop with a given genotype is technically known as the norm of reaction of that genotype. Which potentialities of the norm of reaction will in fact be realized in a given individual at a certain stage of his development is decided by the sequence of the environments in which the development takes place.

It used to be regarded as not implausible that gene action is a by-product of gene synthesis, and that all genes act hand in hand with cell division. The evidence is now overwhelmingly against these possibilities. Transcription of the genetic information in the DNA may occur independently of its replication, for example, in nondividing cells such as neurones. Cells in different tissues of a multicellular body carry probably identical sets of genes. Nevertheless, different batteries of genes are active in different tissues. It has been shown, particularly by Mirsky and his colleagues (Allfrey, Littau, and Mirsky 1963 and other publications), that a majority of genes are silent in cells of a given tissue at a given stage of body development. Only a minority of the genes are being transcribed in messenger RNA and are translated into proteins. Cell and tissue differentiation during the development of multicellular organisms can be envisaged as a succession of reactions between the genes in different cells of the embryo and the processes going on in neighboring cells. Some genes are switched on and others off in different tissues and at different stages of development (Bonner 1965 and Davidson 1968).

The processes that control the gene action in different cells are not fully understood, despite much research in this important and fascinating field. The already classical work of Jacob and Monod (1961) on the enzyme beta-galactosidase in *Escherichia coli* has led to the recognition of functional groups of genes, constituting operons. An operon is composed of several structural genes and of an operator gene, located in the chromosome in close proximity to the structural genes it controls. Other kinds of controlling elements are the regulator genes, which may have their chromosomal locations far from the operons they control. When the bacteria are grown on a medium lacking a lactose sugar, the enzyme is usually absent (although there are strains in which the enzyme is "constitutive," i.e., present regardless of whether the medium contains lactose). In the absence of lactose, the regulator produces a repressor substance, which combines with the operator and prevents transcription of the structural genes in the operon. Lactose acts as an inducer by rendering the repressor inactive and thus permitting the operon to produce its messenger RNA; the latter initiates the formation of the enzyme.

Evidence is rapidly accumulating that regulator genes are very important in the developmental processes in higher organisms (see Zuckerkandl 1964, Welshons 1965, and Britten and Davidson 1969). We saw in Chapter 1 that the amounts of DNA vary greatly in the nuclei of different organisms. It is a plausible hypothesis that these variations are due to the fact that higher organisms possess greater numbers of regulator genes than of structural genes. According to Britten and Davidson, "The principal difference between a poriferan and a mammal could be in the degree of integrated cellular activity, and thus in vastly increased complexity of regulation rather than a vastly increased number of producer [structural] genes." A much larger quantity of DNA may therefore represent the regulators rather than the structural genes. The evolutionary importance of regulator genes was stressed first by Wallace (1963a) and by Stebbins (1969).

Genes and Characters

A hypothesis once widely accepted among geneticists postulated that each gene is responsible for the production of one and only one enzyme. This one-gene-one-enzyme hypothesis is now modified to state that each structural gene is transcribed into a single messenger RNA, and the latter is translated into a single polypeptide chain of a protein. It by no means follows that every gene produces just a single character or trait. This is a misconception refuted by the striking manifold (pleiotropic) effects of many genes (see below). The process of the development of an organism should not be misinterpreted as a gradual accumulation and superposition of independent contributions of its genes. Actually, the development is a complex network of processes. In these processes the gene products play, of course, the leading roles; taken all together, these processes are, however, integrated into harmonious systems capable of being alive. In other words, the genetic materials are aggregations of particulate or atomistic genes; the development is a unified network of interrelated events.

The phenotypic manifestation of a gene varies, depending on which other genes it is associated with—in other words, on the internal genetic environment. Textbook examples of such "epistatic" interactions are the arrays of genes controlling the coat colors of various mammals. A cat or a rabbit homozygous for the albino allele is an albino regardless of what other color genes it contains; in the presence of the pigment allele the animal can be yellow, black, tortoise-shell, tabby (in cats), or agouti (in rodents) according to its genotypic constitution. Perhaps the most dramatic epistatic interactions cause the synthetic lethals, abundantly represented in natural populations of some Drosophila (Chapter 3). Either one of two genes, a or b, permits the viability to be normal, but the combination $a + b$ is a synthetic lethal that causes death.

The phenotypic manifestation of the genes can be modified by environmental agencies of various sorts, physical, chemical, and biological. Such modifications may spell the difference between life and death. Many mutants, especially in microorganisms, are inviable on culture media on which the ancestral forms grow easily. This inviability is often caused by the lack in the mutants of certain enzymes needed to catalyze essential metabolic reactions; however, additions to the culture medium of the substances that these reactions produce enable the

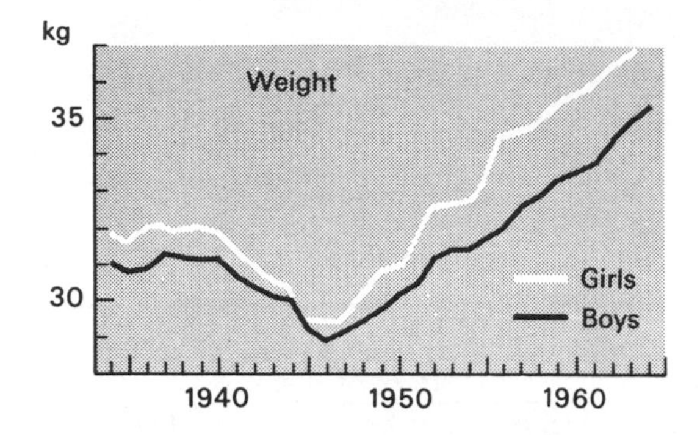

FIGURE 2.1

Changes in height and weight of 12-year-old children in Sendai, Japan. (After Takahashi)

mutants to survive and to grow vigorously. Quite similar situations are known in higher organisms. According to Walles (1963) and Boynton (1966), mutants unable to synthesize their own vitamin B_1 are known in the molds *Penicillium notatum, Neurospora sitophila,* and *N. crassa*; in the bacteria *Escherichia coli, Aerobacter aerogenes,* and *Bacillus subtilis*; in the algae *Chlamydomonas moewusei* and *Ch. reinhardi*; and in the flowering plants *Arabidopsis thaliana,* barley, and tomato. The mutants are lethals, except when vitamin B_1 is supplied in the experimental environment.

Modification, Morphoses, Homeostasis, and Canalization

A phenotype is a biological system constructed by successive interactions of the individual's genotype with the environments in which the development takes place. What the genotype determines is the norm of reaction of the organism to its environments. The word norm does not imply, in this context, that some reactions are intrinsically normal and others abnormal. The norm of reaction is the entire range, the whole repertoire, of the variant paths of development that may occur in the carriers of a given genotype in all environments, favorable and unfavorable, natural and artificial. It is fully known for no genotype. To experiment with the carriers of a genotype by exposing

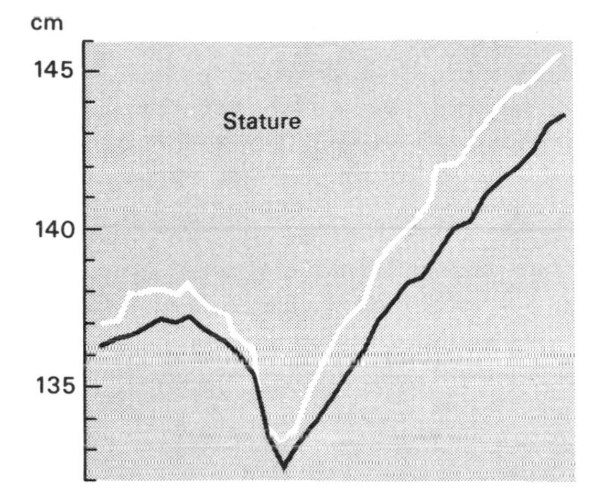

them to every existing or obtainable environment is patently impossible, since the number of environments is virtually infinite. And yet it is of the greatest importance to acquire information as complete as possible about the norms of reaction of human genotypes, and about those of agricultural animals and plants. Medical and educational studies, and much agronomic research, are directed toward this end.

Biological evolution is genetic change; environmental modifications of phenotypes do not constitute evolution. Increases of average stature over time have been recorded in diverse human populations (Fig. 2.1), wherever relevant statistical data are available (for an introduction to the extensive literature dealing with these increases, see Tanner 1962, Takahashi 1966). One would hesitate to call this evolutionary change if, as seems probable, most of the increases in stature are caused by nutritional and general hygienic improvements, rather than by genetic changes. This statement implies no underestimate of the evolutionary importance of the phenotype. Some paths of development bring about phenotypes that are adaptive in some environments and not in others. These phenotypes decide survival or death, the leaving or the not leaving of progeny. By and large, phenotypes that develop in wild species, in response to environmental stimuli that recur regularly in the habitats of these species, are conducive to survival and to reproductive success. Conversely, responses to unusual or artificially created environments are adaptively ambiguous and often even unfavorable. Schmalhausen (1949) has termed the former modifications and the latter morphoses. Modifications are forged in the evolutionary history of a species; morphoses are "new reactions which have not yet attained a historical basis."

Modifications take two outwardly different but closely related forms: physiological homeostasis or the "wisdom of the body" (Cannon 1932 and Dubos 1965), and developmental homeostasis or phenotypic flexibility (Thoday 1953 and Lerner 1954). Familiar examples of physiological homeostasis in man are the maintenance of constant body temperature despite fluctuating temperatures in the environment; constant ionic, glucose, and water concentrations in the blood; and antibody response to infections. Physiological homeostasis maintains constant the *milieu interieur,* to use Claude Bernard's famous phrase. This should not be misconstrued, however, as a kind of biological stubbornness. What is maintained constant is a complex of functions essential for the continuation of life, though this constancy is achieved by means of changes in the operation of physiological mechanisms that serve as regulatory devices. For example, constancy of blood composition is achieved through changes in the work done by the kidneys and the liver.

When adaptive changes are conspicuous and not easily reversible within an individual's lifetime, we are dealing with developmental rather than physiological homeostasis. An obviously intermediate case is the tanning reaction of human skin. This protection against sunburn requires some time to develop or to regress. Striking examples of developmental homeostasis are found among plants (Bradshaw 1965) and in lower animals. If the food available remains within a certain critical amount, starving larvae of Drosophila and many other insects do not die but metamorphose into adults of dwarf size. The coloration of some insects is modifiable to match that of the background. On the other hand, there is good evidence (Ergene 1951) that some grasshoppers choose, if available, surroundings matching their own coloration, thus achieving a protective camouflage by behavioral means. The rotifer *Brachionus calyciflorus* develops no spines unless its predator *Asplanchna sieboldi* lives in the same medium. The spines have a protective function; the most remarkable fact is that the effects of the Asplanchna-conditioned medium are exerted on the Brachionus eggs and seem to be irreversible thereafter (Gilbert 1966).

In higher animals, including man, the most interesting phenotypic flexibility occurs in the development of behavioral characteristics. The old "nature-nurture" problem is dead, in the sense that no serious investigator tries to dichotomize human or animal characteristics into those due to "nature" and to "nurture." All traits are products of the interactions of heredity with environment. Nevertheless, heritability studies, partitioning the observed variance into genetic and environmental components, are needed for many traits, especially those of social, medical, or economic interest. An account of the present status of this field is outside the purview of this book. A point worthy of emphasis is, however, the overwhelming importance of the early environment and experience in the formation of many behavioral characteristics (studies on man reviewed in Berelson and Steiner, 1964, Bloom 1964, and Edwards and Cauley 1964; on animals in Eibesfeldt 1965, Thorpe 1963, and Marler and Hamilton 1966). The adult phenotype, though never absolutely fixed, is nevertheless powerfully circumscribed by what happened in infancy.

Most fascinating are experimental studies of the socialization and training of dogs (Scott and Fuller 1965), wolves (Woolpy and Ginsburg 1967), and chimpanzees (Hayes 1951). A wolf cub can be brought up not only to have no fear of man but in fact to seek human companionship; a chimpanzee child can be taught many behavior patterns of a human child, with the notable exception of spoken symbolic language. Indeed, these experiments have revealed hitherto quite unsuspected potentialities of the norms of reaction of such "wild" species as wolves and chimpanzees.

A still different kind of phenotypic flexibility is found in some insects, particularly locusts and other orthopterans (Uvarov 1928, Uvarov and Thomas 1942, Gunn and Hunter-Jones 1952, and Key 1950, 1957). Some species of locusts occur in two "phases," so distinct in appearance and behavior that they were originally believed to be different species. One phase is solitary and the other gregarious in its habits. In some years the gregarious phase makes spectacular outbreaks; multitudes of these insects engage in distant migrations and become "plagues," recorded even in the Bible. Although important details of the causation of these phenomena are in dispute, it appears that the species involved have bimodal or even plurimodal norms of reaction. Their development turns toward either the solitary or the gregarious phase, as dictated by the degrees of crowding and other environmental conditions. Of course, the possibility that the populations of locusts have also a variety of genotypes, some of which can be turned toward the gregarious phase more easily than others, is not ruled out.

Rigidity or flexibility of development in the carriers of a genotype is a factor of major importance in determining the adaptive value of the genotype. A theoretically imaginable ideal genotype would so stabilize the course of development that vitally important organs and functions would appear in all environments, while the modifications arising in other characters would always be adaptive in the environments that evoked them. The reality is, in varying degrees, short of the ideal. However, especially in higher organisms, physiological and developmental homeostasis provides a buffering of the developmental path against environmental shocks. According to Waddington (1957), the development is "canalized." Under most or all conditions that the species encounters regularly in its habitats, and with most genotypes comprising the adaptive norm of the species population, development leads to generally similar results. Hence canalization can be symbolized by a landscape, in which the topography is such that the waters flow usually into the same valley or canal.

The development of the female and male sexes provides an excellent example of canalization. The female and male genotypes, usually determined by the X and Y chromosomes, give, under most circumstances, either functional females or males. Intersexes that are sterile or suffer various impairments occur rarely if at all in natural populations. They can, however, be produced by drastic environmental or genetic changes. Brust and Horsfall (1965) find that thermal stresses applied to larvae of the mosquito *Aedes communis* suppress the development of masculine traits in genetic males, but leave genetic females unaffected. The sexual traits in triploid intersexes of *Drosophila melanogaster* are greatly modified by temperature and other environmental variations that are without effect on normal diploid females and males. Sexual development of females and males is evidently canalized, whereas that of intersexes is not (Laugé 1966).

Phenocopies

The concept of phenocopy was developed by Goldschmidt (1938). His assumption was that any alteration of the phenotype caused by a genetic change, a mutation, should be reproducible as a purely phenotypic modification if a suitable environmental agency is discovered. By employing temperature shocks or treatment with ether and other chemicals, Goldschmidt (1938), Rapoport (1939), Schatz (1951), Hadorn (1961), and others have obtained phenocopies mimicking some well-known wing, thorax, and pigmentation mutants in Drosophila. The treatments are administered during sensitive stages of development, specific for each phenocopy. Extensive and detailed studies of phenocopy induction in poultry have been made by Landauer and his colleagues (Landauer 1948, and a review in Hadorn 1961).

The degree of similarity between a phenocopy and the genetically determined change that it resembles varies from case to case. The curly-haired and blond-haired phenocopies induced by beauticians are usually recognized as such. In any case, the untreated progeny of a phenocopy are normal, whereas at least some of the progeny of a mutant inherit the mutant genes or chromosomes.

The phenocopy concept is also applicable in reverse—genetic mutants made by environmental treatments to resemble the ancestral type. Under this rubric belongs the "cure" of hereditary diseases and malformations by environmental treatments. Mutant plants and microorganisms unable to synthesize vitamin B_1 have been mentioned previously; they are lethal without treatment but viable when the vitamin is supplied. Some of the inborn errors of metabolism in man can be at least partly corrected by excluding from the diet substances that the affected individuals cannot metabolize, or by including substances that are needed. To the first category belong restrictions of phenylalanine intake in phenylketonuria, of galactose in galactosemia, and of phytanic acid in Refsum's disease; the second category is represented by the supplying of vitamins in hereditary vitamin deficiencies (Scriver 1967 and Larson 1961) and of insulin to diabetics. Learning how to induce phenocopies of vigorous health in carriers of deleterious genes is one of the central problems of human engineering (Neel, Fajans, et al. 1965).

A Brief History of the Mutation Concept

The ultimate source of organic diversity is mutation. The mutation concept has had a tortuous history. The paleontologist Waagen (1869) designated as mutations the smallest perceptible changes in the temporal series of forms in a species of ammonites. Around the turn of the century, Bateson (1894) in England and Korzhinsky (1899) in Russia stressed the importance of the sudden origin of discontinuous variations as a source of evolutionary change. But it was de Vries (1901) who wrote, "As the theory of mutation I designate the statement that the properties of organisms are built from sharply distinct units. . . . Intergrades, which are so numerous between the external forms of plants and animals, exist between these units no more than between the molecules of chemistry." A mutation is, then, a change in one or more genes. Thus far, de Vries's statements have a decidedly modern ring. De Vries contrasts, however, his mutation theory and Darwin's selectionism:

> The latter [selectionism] assumes that the usual or the so-called individual variability is the starting point of the origin of new species. According to the mutation theory the two [individual and mutational variabilities] are completely independent. As I hope to show, the usual variability cannot lead to a real overstepping of the species limits even with a most intense steady selection. . . .

On the other hand, each mutation "sharply and completely separates the new form, as an independent species, from the species from which it arose."

De Vries's "species" are evidently not identical with the usual Linnaean ones. Attempts were made to introduce the terms elementary species or Jordanons or biotypes for the former. These attempts met with little sympathy; the word species was here used in a new sense, and in sexually reproducing organisms one would have to consider almost every individual an elementary species. Individual variability, in so far as it is hereditary at all, is due to the populations of most species being mixtures of genotypes differing from each other in one gene or in several. Finally, the mutants obtained by de Vries in his classical investigations with Oenothera proved to be an assemblage of diverse changes, including gene alterations, segregation products due to hybridity of the initial material, and chromosomal aberrations.

Studies by Morgan and his colleagues of mutability in the fly *Drosophila melanogaster* (Morgan 1911, 1919, and Morgan, Sturtevant, et al. 1915) showed that the distinction which de Vries attempted to draw between individual and mutational variability was spurious. The changes produced by mutations range all the way from ones so drastic that they are lethal, through moderate, to barely detectable. Mutants that are easily recognizable to an untrained eye are most useful for genetic experimentation; they are preserved, while the slight ones are generally discarded. This has created a false impression that all Drosophila mutants show strikingly visible alterations, although Morgan (1919) repeatedly emphasized that slight mutants commonly occur in Drosophila. Slight mutants were observed as early as 1909 by Johannsen in beans, and Baur (1924) found them to be very common in the snapdragon (*Antirrhinum majus*). In the nineteen forties, fifties, and sixties, molds, bacteria, and viruses became favorite materials for mutation studies. The Drosophila story was re-enacted: clear-cut mutants were selected for experimental convenience, but ones with minor effects proved to be quite common.

Another turning point in mutation studies came with Muller's discoveries (1927, 1928a,b). Before this time mutations were said to arise spontaneously. A natural phenomenon is "spontaneous" when it is not brought about by man-made causes, and usually when we are, in fact, ignorant of its causes. Working with Drosophila, Muller showed that the frequencies of certain classes of mutations can be accurately measured. He also demonstrated that these frequencies are greatly increased in the progeny of parents treated with X-rays. This work opened the doors to later discoveries of other mutagenic radiations and of chemical mutagens. Of course, some mutations still arise spontaneously, in cultures not known to be treated with any mutagen. It is possible, however, to increase the mutation frequencies at will, and in some instances to promote the occurrence of particular kinds of mutations (see below). The most recent breakthrough followed the elucidation of the chemical structure of nucleic acids. It became possible, at least in some favorable materials and situations, to determine the chemical nature and mechanisms of the changes that on the biological level manifest themselves as mutations.

A Classification of Mutations

The term *mutation* subsumes a variety of phenomena. In the inclusive sense, any change in the genotype not due to gene recombination is a mutation. Chromosomes, as well as self-reproducing cytoplasmic organelles, undergo mutational changes. Gene mutations are caused by alterations within genetic materials; chromosomal aberrations involve loss, multiplication, or rearrangement of genes in the chromosomes. A synopsis of the kinds of mutations may be as follows:

I. *Gene Mutations,* or point mutations in older genetic literature—changes caused by substitution, addition, or deletion of nucleotides within a section of the DNA or the RNA of a gene.

II. *Structural Chromosomal Changes,* affecting the arrangement of genes in the chromosomes.

A. Changes due to loss or reduplication of some of the genes.

a. *Deficiency* (deletion). A section containing one gene or a block of genes is lost from one of the chromosomes. If a normal chromosome carries genes *ABCDEFG,* the deficient chromosome may contain only *ABEFG.*

b. *Duplication.* A section of a chromosome may be present at its normal location in addition to being present elsewhere. If a normal chromosome has genes *ABCDEFG,* the duplication may be *ABCDCDEFG* or the equivalent. Studies of chromosomes in the salivary gland cells of certain flies have shown that in the "normal" chromosomes certain sections are represented two or more times in the haploid set. Such "repeats" are duplications that have become established in the phylogeny of these flies.

B. Changes due to an alteration in the arrangement of the genes.

a. *Translocation.* Two chromosomes, with genes *ABCDEFG* and *HIJK,* may exchange parts, giving rise to "new" chromosomes having *ABCDJK* and *HIEFG.*

b. *Inversion.* The location of a block of genes within a chromosome may be changed by rotation through 180°. The resulting chromosome carries the same genes as the original one, but their arrangement is modified, for instance, from *ABCDEFG* to *AEDCBFG.*

c. *Transposition.* A block of genes is moved to a new position within a chromosome, for instance, *ABCDEFG* to *ADEFBCG.*

III. *Numerical Changes,* affecting the number of chromosomes.

A. *Aneuploidy.* One or more chromosomes of the normal set may be lacking (monosomics, nullosomics) or present in excess (trisomics, tetrasomics, etc.).

B. *Haploidy.* Higher organisms are mostly diploid during a major part of the life cycle, that is, they possess two chromosomes of each kind in the nuclei of most cells. Gametes, as well as gametophytes in plants, are haploid and carry one chromosome of each kind. Under experimental conditions some diploid organisms have produced haploid aberrants, which have a single set of chromosomes in the tissues that are normally diploid.

C. *Polyploidy.* Normally diploid organisms may give rise to forms with more than two sets of homologous chromosomes. Such forms are known as polyploids.

Discrimination between point mutations and chromosomal aberrations is often difficult. Numerical chromosomal changes are most readily detectable by cytological examination. Only in more favorable materials, particularly in the giant chromosomes of the larval salivary glands of Drosophila and other flies, can structural changes, down to those involving very small portions of the chromosome, be seen under the microscope. Very small structural changes may conceivably be overlooked even under the most favorable circumstances. Stadler argued in several closely reasoned papers (see Stadler 1954 for further references) that what we call point mutations are merely the residue of mutational changes for which no structural basis in the chromosomes is detectable under the microscope. Goldschmidt (1938, 1940) went even further—all mutations, he said, are due to rearrangements in the chromosomes, whether or not one can see these rearrangements. The development of molecular genetics has made these views obsolete.

The Molecular Basis of Gene Mutations

The effects of genetic changes are most often detected in higher organisms through the morphological or physiological alterations that they produce in the adults. Developmental geneticists used to believe that, if one could patiently unravel the development of a mutational difference backward from the adult to the early stages, it might be possible eventually to discover the nature of the alteration that took place in the gene structure. Things did not go as these early workers

Human Beta	VAL	HIS	LEU	THR	PRO	GLU	GLU	LYS	SER	ALA	VAL
Horse Beta	VAL	GLN	LEU	SER	GLY	GLU	GLU	LYS	ALA	ALA	VAL
Human Alpha	VAL	—	LEU	SER	PRO	ALA	ASP	LYS	THR	ASN	VAL
Whale Myoglobin	VAL	—	LEU	SER	GLU	GLY	GLU	TRY	GLN	LEU	VAL

Human Beta	GLY	GLY	GLU	ALA	LEU	GLY	ARG	LEU	LEU	VAL	VAL
Horse Beta	GLY	GLY	GLU	ALA	LEU	GLY	ARG	LEU	LEU	VAL	VAL
Human Alpha	GLY	ALA	GLU	ALA	LEU	GLU	ARG	MET	PHE	LEU	SER
Whale Myoglobin	GLY	GLN	ASP	ILEU	LEU	ILEU	ARG	LEU	PHE	LYS	SER

Human Beta	SER	THR	PRO	ASP	ALA	VAL	MET	GLY	ASN	PRO	LYS
Horse Beta	SER	ASP	PRO	GLY	ALA	VAL	MET	GLY	ASN	PRO	LYS
Human Alpha	SER	HIS	GLY	SER	—	—	—	—	—	ALA	GLN
Whale Myoglobin	LYS	THR	GLU	ALA	GLU	MET	LYS	ALA	SER	GLU	ASP

Human Beta	GLY	LEU	ALA	HIS	LEU	ASP	ASN	LEU	LYS	GLY	THR
Horse Beta	GLY	VAL	HIS	HIS	LEU	ASP	ASP	LEU	LYS	GLY	THR
Human Alpha	ALA	VAL	ALA	HIS	VAL	ASP	ASP	MET	PRO	ASN	ALA
Whale Myoglobin	ILEU	LEU	LYS	LYS	LYS	GLY	HIS	HIS	GLU	ALA	GLU

Human Beta	ASP	PRO	GLU	ASN	PHE	ARG	LEU	LEU	GLY	ASN	VAL
Horse Beta	ASP	PRO	GLU	ASN	PHE	ARG	LEU	LEU	GLY	ASN	VAL
Human Alpha	ASP	PRO	VAL	ASN	PHE	LYS	LEU	LEU	SER	HIS	CYS
Whale Myoglobin	PRO	ILEU	LYS	TYR	LEU	GLU	PHE	ILEU	SER	GLU	ALA

Human Beta	PRO	PRO	VAL	GLN	ALA	ALA	TYR	GLN	LYS	VAL	VAL
Horse Beta	PRO	GLU	LEU	GLN	ALA	SER	TYR	GLN	LYS	VAL	VAL
Human Alpha	PRO	ALA	VAL	HIS	ALA	SER	LEU	ASP	LYS	PHE	LEU
Whale Myoglobin	ALA	ASP	ALA	GLN	GLY	ALA	MET	ASN	LYS	ALA	LEU

Indicates amino acid common to the four proteins

Indicates amino acid common to the three hemoglobins

Indicates amino acid common to the two beta hemoglobins

THR	ALA	LEU	TRY	GLY	LYS	VAL	ASN	—	—	VAL	ASP	GLU	VAL
LEU	ALA	LEU	TRY	ASP	LYS	VAL	ASN	—	—	GLU	GLU	GLU	VAL
LYS	ALA	ALA	TRY	GLY	LYS	VAL	GLY	ALA	HIS	ALA	GLY	GLU	TYR
LEU	HIS	VAL	TRY	ALA	LYS	VAL	GLU	ALA	ASP	VAL	ALA	GLY	HIS

TYR	PRO	TRY	THR	GLN	ARG	PHE	PHE	GLU	SER	PHE	GLY	ASP	LEU
TYR	PRO	TRY	THR	GLN	ARG	PHE	PHE	ASP	SER	PHE	GLY	ASP	LEU
PHE	PRO	THR	THR	LYS	THR	TYR	PHE	PRO	HIS	PHE	—	ASP	LEU
HIS	PRO	GLU	THR	LEU	GLU	LYS	PHE	ASP	ARG	PHE	LYS	HIS	LEU

VAL	LYS	ALA	HIS	GLY	LYS	LYS	VAL	LEU	GLY	ALA	PHE	SER	ASP
VAL	LYS	ALA	HIS	GLY	LYS	LYS	VAL	LEU	HIS	SER	PHE	GLY	GLU
VAL	LYS	GLY	HIS	GLY	LYS	LYS	VAL	ALA	ASP	ALA	LEU	THR	ASN
LEU	LYS	LYS	HIS	GLY	VAL	THR	VAL	LEU	THR	ALA	LEU	GLY	ALA

PHE	ALA	THR	LEU	SER	GLU	LEU	HIS	CYS	ASP	LYS	LEU	HIS	VAL
PHE	ALA	ALA	LEU	SER	GLU	LEU	HIS	CYS	ASP	LYS	LEU	HIS	VAL
LEU	SER	ALA	LEU	SER	ASP	LEU	HIS	ALA	HIS	LYS	LEU	ARG	VAL
LEU	LYS	PRO	LEU	ALA	GLN	SER	HIS	ALA	THR	LYS	HIS	LYS	ILEU

LEU	VAL	CYS	VAL	LEU	ALA	HIS	HIS	PHE	GLY	LYS	GLU	PHE	THR
LEU	ALA	VAL	VAL	LEU	ALA	ARG	HIS	PHE	GLY	LYS	ASP	PHE	THR
LEU	LEU	VAL	THR	LEU	ALA	ALA	HIS	LEU	PRO	ALA	GLU	PHE	THR
ILEU	ILEU	HIS	VAL	LEU	HIS	SER	ARG	HIS	PRO	GLY	ASN	PHE	GLY

ALA	GLY	VAL	ALA	ASN	ALA	LEU	ALA	HIS	LYS	TYR	HIS	—	—
ALA	GLY	VAL	ALA	ASN	ALA	LEU	ALA	HIS	LYS	TYR	HIS	—	—
ALA	SER	VAL	SER	THR	VAL	LEU	THR	SER	LYS	TYR	ARG	—	—
GLU	LEU	PHE	ARG	LYS	ASP	ILEU	ALA	ALA	LYS	TYR	LYS	GLU	LEU

—	—	—	—
—	—	—	—
—	—	—	—
GLY	TYR	GLN	GLY

FIGURE 2.2 The sequences of amino acids in alpha and beta chains of the human adult hemoglobins, in the beta chain of horse hemoglobin, and in whale myoglobin. The sequences are arranged from left to right across both pages, starting from the upper left corner (position No. 1, valine) to the lower right (No. 153, Glycine, in the myoglobin). Empty spaces are gaps in the sequences in some of the chains. (After Dayhoff and Eck)

expected; the gene structure and primary gene action are at present understood far better than development and differentiation.

More and more mutational changes are being shown to produce specific changes in single proteins. For example, Pauling, Itano, et al. (1949) found sickle-cell anemia to be a "molecular disease." This ailment, widespread in some human populations native to Africa, is inherited as a single Mendelian recessive gene. The homozygote dies, usually before adolescence, of severe anemia; the heterozygote survives and in point of fact enjoys an advantage over the normal homozygote, being relatively resistant to certain malarial fevers (see Chapter 6). Pauling and his colleagues discovered that the hemoglobins of normal, heterozygous, and anemic individuals are distinguishable by electrophoresis, because these hemoglobins move at different rates in an electric field. Moreover, while normal homozygotes have a hemoglobin called A, and anemic homozygotes have hemoglobin S, heterozygotes have both A and S in approximately equal amounts.

A great amount of work has been done in recent years on the structure of normal hemoglobin and of its genetic variants in man (excellent reviews in Ingram 1963, Baglioni 1967, and Dayhoff and Eck 1968). A hemoglobin molecule consists of four protein (polypeptide) chains, two of them called alpha and two beta chains, and two iron-containing heme groups. An alpha chain has 141 amino acids and a beta chain 146, the sequences of which are shown in Fig. 2.2. Although the alpha and beta chains are encoded by different genes, their amino acid sequences are more similar than could reasonably be ascribed to chance (the similar portions are outlined in Fig. 2.2). The genes responsible for the formation of alpha and beta chains are probably the modified descendants of a single ancestral gene that underwent a duplication and a gradual divergence in evolution.

Hemoglobin S, found in carriers of the sickle-cell gene, has proved to differ from ordinary hemoglobin A in the substitution of a single amino acid: it has a valine (Val) instead of the normal glutamic acid (Glu) at position 6 in the beta chain. This change could be effected by substitution of a single letter of the genetic alphabet in the DNA and RNA produced at the gene coding for the beta chain of hemoglobin. Reference to Table 1.2 shows that the triplets for glutamic acid are GAA and GAG; substitution of U for A gives the triplets GUA and GUG for valine. Here, then, is a change in a single molecule, which is so amplified in the process of development that the individual who inherits this molecule from both parents dies of sickle-cell anemia.

Table 2.1 lists some of the variants of alpha and beta chains of human hemoglobins which have been found to differ in substitutions of single amino acids. Mutation evidently hits many spots in the nucleotide sequences of the genes coding for these chains. Only a few of these mutant hemoglobins (especially S and C in the beta chain) are common in populations of certain geographic regions. Others have been encountered mostly in single families or persons. Some of these "private" hemoglobins have, however, been discovered by different investigators in places remote from each other; this almost certainly means that similar mutations occurred independently in different populations (Konigsberg, Huntsman, et al. 1965). The carriers of variant hemoglobins are often, though not always, found to be suffering from diverse forms of ill health. Some have appeared, however, to be healthy; taken at face value, this seems to mean that certain changes in the hemoglobin molecule make it neither more nor less serviceable. The decided predominance of "normal" hemoglobin in man suggests, however, that this particular molecular species has proved itself advantageous.

The variant hemoglobins listed in Table 2.1 differ from the common one in single amino acid replacements, and the corresponding genes presumably by single nucleotide replacements. Most of the nucleotide replacements appear to be transitions, that is, replacements of one purine by another or of one pyrimidine by another (thymine by cytosine or vice versa, and adenine by guanine or vice versa). Less frequent are transversions, that is, replacements of a purine by a pyrimidine or vice versa. Hemoglobin C Harlem differs from the common one not by a single replacement but by two amino acids—substitution of valine for glutamic at position 6, and of asparagine for aspartic at position 73 in the beta chain. Still more complex are the changes in Lepore hemoglobins; here two normal alpha chains are combined with chains each of which consists of a part of the normal beta and a part of the delta chain. The latter is a normal constituent of hemoglobin A_2, which is present in small amounts in the blood of normal persons. The mutation that produced the Lepore hemoglobin chain probably involved crossing over between the partially homologous genes coding for the beta and the delta chains.

The mechanics of the origin of rearrangements of genic materials within chromosomes are far from completely understood. Deletions, duplications, inversions, and translocations of chromosome sections involve one or two chromosomes being broken in one or more places, and the resulting chromosome fragments being recombined in new ways. The basic event in all these changes is, then, chromosome fracture. This may be a process analogous to gene mutation, though it disrupts the nucleic acid backbone in one or more places in the chromosomes, instead of substituting one nucleotide "letter" for another. The fracture points in the chromosomes are said to be "sticky,' because

TABLE 2.1

Variant human hemoglobins, caused by point mutations substituting single amino acids at certain positions (After Dayhoff and Eck 1968)

Variant	Position	Amino Acid	
		Old	New
Alpha Chain			
J Toronto	5	Ala	Asp
J Oxford, I Interlaken	15	Gly	Asp
I, I Texas	16	Lys	Glu
J Medellin	22	Gly	Asp
G Audhali	23	Glu	Val
G Honolulu, G Singapore, G Hong Kong	30	Glu	Gln
Umi, Kokura, L Ferrara	47	Asp	Gly
Sealy, Hasharan	47	Asp	His
Russ	51	Gly	Arg
Shimonoseki	54	Gln	Arg
Mexico	54	Gln	Glu
Norfolk, G Ibadan	57	Gly	Asp
M Boston, M Osaka	58	His	Tyr
G Philadelphia, G Bristol, G ST 1, D alpha St. Louis, Stanleyville 1	68	Asn	Lys
Stanleyville 2	78	Asn	Lys
Etobicoke	84	Ser	Arg
M Kankakee, M Shibata, M Iwate	87	His	Tyr
Chesapeake	92	Arg	Leu
J Cape Town	92	Arg	Gln
J Tongariki	115	Ala	Asp
O Indonesia	116	Glu	Lys

TABLE 2.1 (continued)

Variant	Position	Amino Acid Old	Amino Acid New
Beta Chain			
Tokuchi	2	His	Tyr
S, X	6	Glu	Val
C	6	Glu	Lys
C Georgetown, Siriraj	7	Glu	Lys
G	7	Glu	Gly
Porto Alegre	9	Ser	Cys
Sogst	14	Leu	Arg
D Bushman	16	Gly	Arg
J Baltimore, J Trinidad, J Ireland, J New Haven	16	Gly	Asp
G Saskatoon	22	Glu	Lys
E	26	Glu	Lys
Genova	28	Leu	Pro
Hammersmith	42	Phe	Ser
G Galveston, G Texas, G Port Arthur	43	Glu	Ala
K Ibadan	46	Gly	Glu
G Copenhagen	47	Asp	Asn
J Bangkok. J Meinung, J Korat	56	Gly	Asp
Hikari	61	Lys	Asn
M Saskatoon, M Emory, M Kurume, M Chicago, M Hamburg	63	His	Tyr
Zurich	63	His	Arg
M Milwaukee 1	67	Val	Glu
Sydney	67	Val	Ala
J Cambridge	69	Gly	Asp
J Iran	77	His	Asp
G Accra	79	Asp	Asn
D Ibadan	87	Thr	Lys
Agenogi	90	Glu	Lys
Oak Ridge	94	Asp	Asn
Hopkins, N Jenkins, N Baltimore	95	Lys	Glu
Köln	98	Val	Met
Kansas	102	Asn	Thr
New York	113	Val	Glu
D Los Angeles, D Punjab, D Cyprus D Conley, D Chicago, D Portugal	121	Glu	Glu
O Arabia	121	Glu	Lys
K Woolwich	132	Lys	Gln
Hope	136	Gly	Asp
Kenwood	143	His	Asp

they tend to rejoin with other fractures. The reunion may either restore the original gene arrangements or create new ones.

Mutagenic Radiations and Temperature

Reference has been made to Muller's discovery (1927, 1928a,b) that the frequencies of both point mutations and chromosomal changes are increased in the offspring of X-rayed Drosophila. The large and still rapidly growing literature on radiation genetics has been reviewed in the books by D. E. Lea (1955), Purdom (1963), and Dubinin (1964). Muller's findings were soon confirmed and extended by Stadler, Timofeeff-Ressovsky, Oliver, Demerec, Gowen, and others, working with animals and plants and with simple and complex organisms. All so-called ionizing radiations, from soft X-rays to gamma rays, neutron beams, and presumably cosmic rays, are mutagenic. Very important is the fact that the frequencies of induced gene mutations are directly proportional to the amounts of radiation administered, as measured by the ionizations produced in the living tissue (r or rem units). This direct proportionality rules out the possibility that "spontaneous" mutations might be accounted for by the small amounts of ionizing radiations omnipresent in nature; the radiation "background" is far too weak to account for the observed spontaneous mutation frequency. At the same time, no amount of radiation exposure can be regarded as safe or innocuous, since it will inevitably produce some genetic damage. This fact is fundamental for a reasoned evaluation of possible radiation damage to human populations.

Ultraviolet light is also mutagenic. Its action is different, however, from that of ionizing radiations (Witkin 1966 and references therein): its penetration and absorption in living tissue depend on the wavelength employed; the numbers of mutations induced are not in general directly proportional to the amounts used in treatment; ultraviolet light induces mainly point mutations and few or no chromosomal aberrations. It is probably of little consequence as a natural mutagen, in at least the higher animals and plants, because their reproductive cells are generally too well protected for the ultraviolet part of the light spectrum to penetrate.

The effects of temperature on mutation rates were studied by Muller (1928b), Timofeeff-Ressovsky (1935), and others. Timofeeff-Ressovsky gives the following data for the frequencies of origin of sex-linked lethal mutations in *Drosophila melanogaster* at three temperatures:

Temperature, °C	Number of Chromosomes Examined	Number of Lethals Found	Percentage of Lethals
14	6871	6	0.087 ± 0.035
22	3708	7	0.188 ± 0.071
28	6158	20	0.325 ± 0.072

Notwithstanding the large experimental errors, it seems certain that the mutation rate is doubled or trebled with a 10°C rise in temperature. Timofeeff-Ressovsky points out, however, that the development of the fly is more rapid at high than at low temperatures; he also finds that mutation is proportional to time, since the frequency of sex-linked lethals is higher in the spermatozoa of old males than in that of young ones. Taking this factor into consideration, he estimates that the temperature coefficient of the mutation process (i.e., the ratio of increase per 10°C) is in the neighborhood of 5.

Chemical Mutagens

Over the years, many geneticists have been inclined to believe that mutational changes in the genes are produced by chemical agencies. Nevertheless, for a long time attempts to find chemical mutagens met with little success, and mutations were artificially induced first by radiations. In the nineteen thirties, a group of Russian investigators obtained suggestive results in *Drosophila melanogaster*, increasing the mutation rates by treatments with iodine and potassium iodide (Sacharov 1936, Samjatina and Popova 1934, and Kondakova 1935), with copper sulfate (Law 1938 and Magrzhikovskaja 1938), with ammonia (Lobashov and Smirnov 1934), with potassium permangate (Naumenko 1936), with sublimate (Kosiupa 1936), with lead salts (Ponomarev 1937–1938), and with asphyxia (Lobashov 1935). Later Rapoport (1946), Kaplan (1948), Herskowitz (1949b), and Vogt (1950) increased the mutation frequency by feeding Drosophila larvae on media with sublethal doses of formalin or urethane.

It remained, however, for Auerbach and her collaborators (Auerbach

1949, 1965; see Auerbach and Ramsay 1968 for references) to discover the first really powerful chemical mutagen, namely, mustard gas $(Cl \cdot CH_2 \cdot CH_2)_2S$. Under the most favorable conditions, about as high a proportion (up to 25 percent) of the X chromosomes of *Drosophila melanogaster* acquire sex-linked lethals when treated with mustard gas as is observed after treatments with the highest doses of X-rays that the insect can stand without being completely sterilized. Hadorn and his colleagues (Hadorn and Niggli 1946, and Hadorn, Rosin, and Bertani 1949) treated ovaries of *D. melanogaster in vitro* with phenol solutions. In some experiments striking increases of mutation rates were obtained, whereas other experiments were negative.

Development of the genetics of microorganisms and of molecular genetics led to rapid advances in the studies of chemical mutagenesis. As materials for such studies microorganisms have an evident advantage—one can experiment with numbers of individuals several orders of magnitude greater than is possible with higher animals or plants. Even rare mutations are thus detected without undue amount of work.

In respect to the modes of action of chemical mutagens, some apparently operate by substitution of genetic "letters," the bases in the DNA chains; others cause deletion of some of the letters normally present, or insertion of additional letters (Brenner, Barnett, et al. 1961, Strauss 1964, and Freese 1965). Base substitutions may occur through formation of so-called tautomeric shifts in the nucleotides of DNA or RNA. These are occasional shifts in the positions of the hydrogen atoms responsible for bonding the purines, adenine and guanine, to the pyrimidines, thymine and cytosine, respectively. The tautomeric shifts cause alterations in the process of replication of the DNA strands. A modified adenine may now pair with a cytosine instead of with a thymine, or vice versa; and a modified guanine with a thymine, or vice versa. The tautomeric shifts are unstable, and at the next chromosome replication the normal bonding is restored. The nucleotide introduced into the altered chain of the helix will, however, be linked to its proper partner, thus making a permanent alteration of one genetic letter in the sequence. This may result in a triplet that contains the modified letter coding for a different amino acid, and hence in a modified protein.

Mutational changes that replace one purine in the sequence of genetic letters by another, or one pyrimidine by another, are called transitions. Replacements of a purine by a pyrimidine, or vice versa, are transversions. Transitions seem to occur more often than transversions; the frequency of transitions is greatly increased in bacteria, bacteriophages, and yeasts by treatments with 5-bromouracil, 2-aminopurine, and nitrous acid. The first two substances are chemical analogues of the nucleotide bases and presumably act by being occasionally incorporated in the DNA in place of some normal bases, thus increasing the frequency of mispairing to a value above the spontaneous rate. Nitrous acid is believed to act in a different manner—by replacing the amino groups in some of the bases with hydroxyl groups. The mutations induced by treatments with these substances are reversible, and the frequency of the reverse mutations is increased by treatments with the same substances.

Ethyl ethanesulfonate, nitrogen mustard, and other so-called alkylating agents act as powerful mutagens. Carlson, Sederoff, and Cogan (1967) and Jenkins (1967) have studied the induction of mutants in Drosophila by ethyl ethanesulfonate. Close to one-third of the mutants produced by this substance in the bacteriophage T4 appear to be transversions.

A different kind of mutation is induced by the acridine dyes proflavin and acridine orange. These substances are believed to become inserted between the normal bases in the DNA chains, pushing these bases apart. This results in addition or deletion of one or several bases during the process of DNA replication, and in the appearance of the so-called frameshift mutations (Freese 1965, Brenner, Barnett, et al. 1961, and Streisinger, Okada, et al. 1966). Briefly, their nature is as follows. The genetic code is a nonoverlapping triplet code (cf. Chapter 1). The sequence of three succeeding nucleotides is transcribed and translated into one of the twenty amino acids in a protein chain. The transcription process begins at a fixed point in the sequence of the nucleotides and proceeds in a definite direction, as though a "reading frame" were applied to transcribe the consecutive triplets of the nucleotides in the sequence. Insertion into or deletion from the sequence of one or two nucleotides causes a more drastic change in the gene product than substitution of one nucleotide for another, or than insertion or deletion of three nucleotides constituting a triplet. Indeed, substitutions, insertions, or deletions of single triplets permit the remainder of the series of triplets to be transcribed and translated correctly. On the contrary, in frameshift mutations, the "reading frame" will now be transcribing triplets different from those it transcribed in the ancestral gene. Moreover, there may be formed triplets UAA, UAG, or UGA, which, it will be recalled, are "nonsense" or chain-termination triplets (see Table 1.2). If a frameshift mutation produces such triplets, the result may be premature termination of a protein chain (Garen 1968).

The Quest for Directed Mutation

Geneticists have always been on the lookout for means to control and direct the mutation process. Theoretically, it should be possible to change chosen genes in a predetermined direction. Thus far the quest has proved elusive. Ionizing radiations increase the frequencies of all kinds of mutations. Ultraviolet light, as pointed out above, induces relatively more point mutations than chromosome breaks. Of course, the frequencies of mutations that arise depend not only on the mutagen used but also on the organism treated. Thus, Witkin (1947, 1966) and Greenberg (1964) selected strains of colon bacteria both with increased and with decreased resistance to ultraviolet and to X-rays. Glass (1955) and Thomas and Roberts (1966) found that chromosomal aberrations induced by similar X-ray exposures are more frequent in male than in female sex cells of Drosophila. There seems to be little if any difference in the frequencies of radiation-induced gene mutations in female and in male sex cells, although more data on this subject are to be desired.

A greater differentiation of the mutation spectra is found with chemical mutagens. Working with bacteria, Demerec (1955) compared the mutabilities of certain genes after treatments with manganese salts and with X-rays; some of these genes responded more strongly to the manganese and others to the radiation. As mentioned above, some chemical mutagens induce base-pair substitutions and others frameshift mutations. De Serres (1964) found in the mold *Neurospora crassa* clear-cut differences between the mutation spectra induced by X-rays, nitrous acid, 2-aminopurine, and the spontaneous mutants.

Transformation is a genetic phenomenon really distinct from mutation. It leads, however, to a directed genetic change and should logically be discussed in this chapter (it will be mentioned in another context in Chapter 12). It was discovered and studied by Griffith, Dawson, Avery, MacLeod, McCarthy, Hotchkiss, and others. An account of this classical work can be found in any modern textbook on genetics; very briefly, the story is as follows. Some forms of pneumonia in man and in animals are caused by bacteria belonging to the species *Diplococcus pneumoniae*. The ability of the bacteria to cause infection (their virulence) depends on the presence on their cell surfaces of an envelope composed of polysaccharides. When grown on a laboratory medium, virulent pneumococci give colonies with a smooth, glistening surface. If, however, they are maintained for a long time by repeated transfers on laboratory media, the bacteria undergo a characteristic change. They lose their polysaccharide envelopes and their virulence; the colonies become small with rough outlines. The change from smooth to rough is reversible. Griffith showed in 1928 that mice inoculated with a mixture of living rough pneumococci and of heat-killed smooth ones became infected. Moreover, pneumococci that form smooth colonies when cultured on laboratory media can be isolated from such infected mice. The change from rough to smooth is conditioned by the presence of dead smooth bacteria.

The changes from smooth to rough, and vice versa, may be due to selection of spontaneous mutations adapted to one or the other of these conditions. The smooth phase is able to invade susceptible hosts, while the rough one is superior to the smooth on laboratory culture media. One or the other genotype is selected by the environment in which the strain is placed. Such transformations, involving differential survival of spontaneous mutants, are well known in many microorganisms. Dawson found, however, that the transformation from rough to smooth can be induced in a test tube by killed smooth cells. Avery, MacLeod, and McCarthy proved that the material responsible for the transformation in DNA was derived from dead smooth cells.

Furthermore, the transformation is remarkably specific. There are many variants or "types" of pneumococci, distinguishable by immunological tests and also by the kinds of polysaccharides in their envelopes. Suppose now that a smooth strain of type I loses its envelope and becomes a rough strain. Suppose further that it is transformed back to smooth with the aid of killed cells of type III. Will the new smooth strain belong to type I or to type III? It belongs to the latter; in other words the transformation depends on the strain that furnishes the DNA "transforming principle." Most or all known types of pneumococci can be transformed into other variants by using the dead cells of the desired types. Moreover, the transformed strains retain the induced-type characteristics after cultivation on laboratory media or after passage through animal hosts. They have acquired not merely a temporary polysaccharide envelope of a kind different from that which their ancestors had, but also the ability to synthesize the new polysaccharide indefinitely.

The "transforming principle" isolated by Avery and his associates is a viscous colorless substance, which proves to be a highly polymerized deoxyribonucleic acid with little or no impurities. The transforming power of this substance is so great that it is capable of causing transformation in a dilution of 1 : 600,000,000. Subsequent studies by Hotchkiss and others have revealed important details of the transformation process. The DNA of the donor type penetrates into the recipient cell; sections of this transforming DNA are then incorporated into the chromosome of the recipient cell. The genes located near each other in the transforming DNA are acquired by the transformed cells more often than are genes located far apart.

Genetic transformations first discovered in the pneumococci were subsequently achieved in several other genera of bacteria. Can they occur also in higher organisms? More than a decade ago, Benoit, Leroy et al. (1960) reported that they had induced transformations in the progeny of Pekin ducks injected with DNA extracted from ducks of the Khaki-Campbell breed. This claim has not been confirmed. Fox and Yoon (1966) reported carefully conducted experiments, treating dechorionated eggs of *Drosophila melanogaster* with DNA prepared from several classical mutants (yellow body, white eyes, singed bristles, etc.). When the adult flies obtained from the treated eggs were inspected, some of them showed patches of tissue seemingly altered in the direction of the mutants that were the DNA donors. These results are a suggestive and perhaps promising beginning, but as yet no more than that.

The potential importance of transformation, if it could be achieved in higher organisms, especially man, is enormous. It would open possibilities for genetic engineering, or "genetic surgery" as Muller (1965, 1967) called it, that stagger the imagination. One could envisage such operations as implanting desired genes or removing undesirable ones. Almost unlimited potentialities for eugenic betterment, as well as the improvement of domesticated animals and plants, would be within grasp. For the time being, however, one can say only that transformation may be an important evolutionary force in some microorganisms, especially those in which true sexual union and gene recombination does not occur (see Chapter 12).

Mutational Changes in the Cytoplasm

Correns, one of the rediscoverers of Mendel's laws, studied the inheritance of a green and white variegation in *Mirabilis jalapa* and in some other plants as early as 1909. The inheritance was simple—seeds formed in the flowers on green branches gave green progeny those on white branches gave plants without chlorophyll, and those on variegated branches gave variegated, green or white plants. Crosses showed that the inheritance was through the mother only, the color of the progeny depending on the ovule and not on the pollen. The obvious explanation was that the difference between the green and the white tissues depends in this case not on nuclear genes but on some self-reproducing entities in the cytoplasm, most likely the chloroplasts, which are transmitted only through the female line. Subsequently such extra-chromosomal or cytoplasmic inheritance was found in several species of higher plants and in microorganisms (Paramecium, Neurospora, yeasts, Chlamydomonas; there is a review in Jinks 1964). It appears to be rare in animals, although a case has been established in mosquitoes (Laven 1967).

In recent years interest in cytoplasmic inheritance has been enhanced by the utilization of cytoplasmic male sterility in agricultural practice, especially in the production of hybrid corn and hybrid wheat (Kihara 1967). Moreover, both chloroplasts and mitochondria have been shown to contain DNA and thus to possess an unqualified genetic continuity. This does not mean that the chromosomal and the extrachromosomal carriers of genetic information are independent of each other. One of the commonest kinds of Mendelian recessive mutant in higher plants shows reduction or total absence of chlorophyll in the chloroplasts. On the other hand, Rhoades (1946) described a nuclear gene in corn, which induces irreversible mutational changes in the chloroplasts, subsequently inherited according to the classical Correns scheme just described. Very little is known about the mutational origin of other cytoplasmic variants, although chemical induction of such variants in yeasts has been recorded. Maly (1951) induced plastid variants by irradiation in fern prothallia.

Pleiotropism or Manifold Effects

Early theorists of genetics, particularly Weismann, envisaged the germ plasm as a mosaic of particles, each representing an anatomically defined part of the body. This idea harks back to the preformistic notions that led some early microscopists to imagine a tiny image of man, a homunculus, in the head of the human spermatozoon. Darwin's provisional hypothesis of pangenesis followed the same tradition, since it postulated that the hereditary materials in the sex cells are compounded of particles, gemmules, secreted by the cells of each part of the body. Preformistic ideas became, largely by indirection, attached to the gene concept as well. Pioneers of Mendelian genetics in the early years of the current century imagined the organism to be a mosaic of "unit characters," each determined by a special gene.

The method of naming mutant genes and, by extension, their ancestral gene alleles reinforces these ideas. In Drosophila, mutations of the gene white turn the eye color from red to white, the mutation vestigial produces vestigial wings, etc. The names are convenient, but they can be misleading if taken for descriptions of the total range of effects of a particular gene. In point of fact, the mutation white changes not only the color of the eyes but also that of the testicular sheath, the shape of the spermatheca, the longevity, and the viability. Vestigial not only reduces wing size but also modifies the balancers, makes certain bristles erect instead of horizontal, and changes wing muscles, spermatheca shape, development rate, longevity, fecundity, and viability. Under favorable environmental conditions vestigial decreases the number of ovarioles relative to the long-winged form; under unfavorable culture conditions the wild type has fewer ovarioles than does the vestigial. The mutant split in *Drosophila melanogaster* makes the eye surface rough and the bristles on the thorax split or doubled. Stern and Tokunaga (1968) studied mosaic flies that have most of the body wild type, but have spots showing the mutant trait. Such spots show the roughness if they include parts of the eyes, and changed bristles if they occur on the thorax. The two manifestations of the gene split are autonomous in development.

Genes that change more than one character are said to be pleiotropic or to have manifold effects. The more detailed a comparison of a mutant with the ancestral form, the more differences are detected. For instance, most classical mutants in Drosophila reduce the viability of their carriers, in addition to whatever changes they produce in the appearance of the insects. The reduced viability may be evident in crowded but not in uncrowded cultures. Dobzhansky (1927) attempted to estimate the prevalence of manifold effects by examining a sample of mutants in *Drosophila melanogaster* for changes in an arbitrarily selected organ, namely, the spermatheca. In ten of the twelve mutants, differences in the spermatecae were detected, although these mutants were known as eye-color, body-color, and wing-shape mutants and were not suspected to differ in the internal anatomy in general or in spermatheca shape in particular. It remained uncertain whether the differences in spermatheca shape were due to polygenes lying in the chromosome in the vicinity of the gene loci responsible for the externally visible changes, or to the latter genes themselves (Schwab 1940).

In a subsequent study Dobzhansky and Holz (1943) obtained several mutations of the genes white and yellow in an inbred strain of *Drosophila melanogaster,* and showed that the mutants differed from the parent strain in spermatheca shape as well as in the colors of the eyes and of the body. It is unlikely that mutations of the genes modifying the shape of the spermatheca arise by chance every time the genes white and yellow undergo mutation; the changes in spermatheca shape must be ascribed to pleiotropic effects of these genes. Parsons and Green (1959) have likewise shown that the decreased fitness of vermilion-eyed flies is caused by the vermilion locus and not by associated linked genes.

Because of the importance of pleiotropism for understanding evolutionary mechanisms, we shall examine the following random examples that show how widespread it is. Varieties of garden onions may have white, cream-colored, red, or purple bulbs. The color variations are determined chiefly by alleles of a single Mendelian gene. Jones, Walker, et al. (1946) and Walker (1951) found that this gene also determines the resistance to the smudge fungus, Colletotrichum. White bulbs are easily infected with the fungus; cream-colored bulbs are slightly susceptible; red or purple bulbs are resistant. The resistance is due to the presence in the colored bulbs of catechol and protocatechuic acids, which are poisonous to the spores of the fungus. Ali (1950) found a parallel case in varieties of beans, some of which are susceptible and others resistant to the bean mosaic virus.

Hereditary diseases in man and higher animals are often complicated "syndromes," composed of changes in many body parts, organ systems, and physiological functions. A mutation in the rat causes thickened ribs, narrowed lumen of the trachea, emphysema of the lungs, hypertrophy of the heart, blocked nostrils, blunt snout, and low viability. Grüneberg (1938) found that the whole syndrome stems from a single primary change, an anomaly of the cartilage.

Homozygotes for the gene lozenge-clawless in *Drosophila melanogaster* differ from the ancestral form in the following syndrome of traits: size and shape of the eyes, structure of the ommatidia, distribution of the pigment in the eyes, reduction in size of the third segment of the antenna, reduction of the basiconic sensilia on the antennae and the palpi, reduction of the tarsal claws, lack of the spermathecae and parovaria in the female reproductive organs, tendency of the sperm to congeal in a mass in the vagina, failure of most of the eggs deposited by these females to hatch (Anders 1955). Anders believes that this syndrome may also be reduced to a single primary change.

Hadorn (1956 and other works) studied what he describes as biochemical pleiotropy in the fly Drosophila and the moth Ephestia. Figure 2.3 shows the relative amounts of the substances detected by

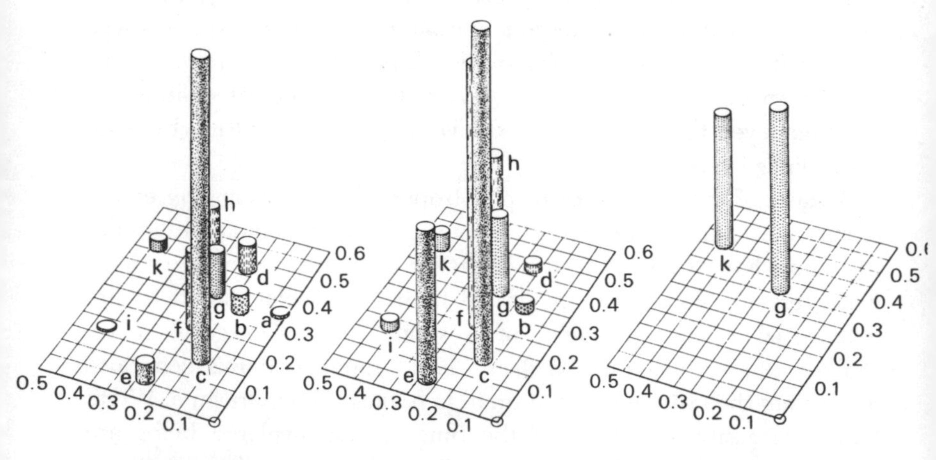

FIGURE 2.3

Relative amounts of the fluorescent substances detected in normal (left) and mutant (center and right) moths Ephestia kühniella. The heights of the columns symbolize the amounts of the substances, and their positions on the grids show their locations on two-dimensional chromatograms. (After Hadorn and Kühn)

ultraviolet fluorescence on chromatograms of the wild-type and two mutants of *Ephestia kühniella.*

Rothe (1951) and Stebbins and Yagil (1966) have described complex syndromes produced by mutants in the snapdragon and in barley, respectively. These are due to single genes that alter "the course of development at an early primordial stage, and initiate an entirely new epigenetic sequence of development."

Particularly interesting are pleiotropic syndromes that combine morphological traits with specific changes in behavior. The mutant yellow in *Drosophila melanogaster* changes the body color and also the courtship pattern; yellow males are less successful than normal ones in mating with normal females, and Bastock (1956) found that the difference is not due to discrimination based on the yellow coloration. Cotter (1967) found the following effects of a single locus mutation in the moth *Ephestia kühniella*: change in eye color, reduced mean rate of development, decreased oxygen consumption, higher preadult mortality, increased variance in rate of development, and changes in the antennae; in addition, mutant males initiate the courting of females earlier but are less successful in copulation than wild-type males, and mutant females have increased mean progeny production. Belyaev and Evsikov (1967) have ascertained that the recessive genes modifying the coat color in mink (*Lutreola vison*) reduce the fertility of the animals when homozygous, though some of them increase the fertility when heterozygous. Rather less well established is the association claimed by Cattell, Young, and Hindleby (1964) between the classical ABO blood system in man, intelligence, and tender-mindedness versus tough-mindedness. The same judgment can be applied to the alleged association between eye and hair colors and susceptibility to poliomyelitis (D. E. Lea 1955).

The strikingly pleiotropic and apparently nonpleiotropic genes are certainly not two different classes of genes. Suppose that an ancestral gene, *A*, produces a certain phenotype in cooperation with all the other genes that the organism has; a mutant gene, *a*, gives rise to a different phenotype, but again in cooperation with the same residual genotype. The differences between the phenotypes of the ancestral type and the mutant are indicative of the effects of the change $A \rightarrow a$, not of the sum total of the effects of either *A* or *a*. More information about the total effects of a gene can be obtained by observation of what a physical removal, a deletion, of this gene does to the organism. In nonpolyploid organisms, such as Drosophila, most homozygous deficiencies act as lethals. As Demerec had already shown in 1936, most deficiencies in Drosophila are cell lethals, that is, the absence of a gene is fatal not only to the whole organism but also to a patch of cells surrounded by tissues in which this gene is present.

Furthermore, the gene may appear pleiotropic or nonpleiotropic, according to the level on which its action is studied. A "dogma" of molecular genetics is that each structural gene specifies one and only one polypeptide chain in a protein. On the molecular level, then, we would find no "genuine pleiotropism"; all pleiotropisms are "spurious pleiotropisms" in Grüneberg's (1943) terminology. This in no way diminishes their biological interest and evolutionary significance. In immunogenetics and enzyme genetics the phenomenon of codominance is most frequently observed: a heterozygote for two alleles coding for two variant enzymes or antigens shows both of these enzymes or antigens present. Formation of "hybrid substances" or interaction products is rather an exception.

As the traits studied are further and further removed from the primary gene action, the possibilities of epistatic interactions of different genes, as well as modifications due to environmental influence, increase. Contrary to the views of early geneticists, the organism is not an aggregate of "unit" traits or characters or qualities. Traits, characters, and qualities are not biological units; they are abstractions, words, semantic devices that a student needs in order to describe and communicate the results of his observations. A trait has no adaptive significance in isolation from the whole developmental pattern that an organism exhibits at a certain stage of its life cycle; one may define a trait only as an aspect of the path of development of the organism (Dobzhansky 1956a). Talking about traits as though they were independent entities is responsible for much confusion in biological, and particularly in evolutionary, thought.

CHAPTER THREE

MUTATION AND GENETIC VARIABILITY

The Building Blocks of Evolution

Reference was made in Chapter 2 to the distinction drawn by de Vries between mutations creating new species and Darwinian "fluctuating" variability. This distinction is invalid. In point of fact, only chromosome doubling in interspecific hybrids (allopolyploidy) is a special kind of mutation that may lead directly to the emergence of new species (see Chapter 11). The process of mutation supplies only the building blocks, the raw materials, from which evolutionary changes, including species differences, are compounded by natural selection. Mutation is, then, the ultimate source of evolution, but there is more to evolution than mutation. It will be shown in the concluding pages of the present chapter that mutation is a random process with respect to the adaptive needs of the species. Therefore, mutation alone, uncontrolled by natural selection, would result in the breakdown and eventual extinction of life, not in adaptive or progressive evolution.

All genetic changes, except those due to gene recombination, are mutations by definition. Only the phenomena of transformation (see Chapters 2 and 12) are, in a sense, bridging the gap between mutational and recombinational variability. The synopsis on pp. 44-45 shows that a collection of diverse phenomena is subsumed under the name mutation. Attempts have been made repeatedly to hypothesize that different kinds of mutations produce changes of different taxonomic value. Goldschmidt (1940) contended that not only new species but also new genera, families, and orders arise by means of special "systemic" mutations. Singleton (1951) supposed that the "corn grass" mutation, derived from cultivated maize, may be a "macromutation" of possible significance as an "ancestral type," but neither Singleton himself nor Goldschmidt claimed it to be a systemic mutation. Lamprecht argued in a series of papers (summary in Lamprecht 1964) that there are two categories of genes and of mutations, some distinguishing species and others only varieties. Böcher (1951) believed that there are two kinds of mutations, some responsible for adaptation to the environment and others for progressive evolution. These views have very few adherents at present.

The Types of Changes Produced by Mutation

Mutations change all sorts of characteristics—structural, physiological, biochemical, and behavioral. The classical mutants in *Drosophila melanogaster,* it will be recalled from Chapter 2, were hand picked for unambiguous recognition by inspection of the external characteristics of the fly. These "visibles" alter the eye and body colors; the numbers and shapes of the bristles; the size, shape, and venation of the wings; the manner in which the fly holds its wings in relation to the body; the antennae and the legs, etc. Mutants that can be distinguished from wild-type flies under some but not under all environments in which the flies are cultivated in laboratories were considered inferior, and usually discarded.

As early as 1912, the first recessive sex-linked lethal gene, which caused no visible changes in heterozygous females but killed hemizygous males, was identified. Recessive lethals, sex-linked and autosomal, eventually proved to be considerably more frequent than conveniently usable visibles; therefore, lethals rather than visibles were used for quantitative studies of the mutation process, with and without radiation, starting with the classical work of Muller (1928a,b). The nature of the changes that cause the lethality is largely unknown, although several studies have been directed specifically at detecting the manner of action of lethals (Hadorn 1951, Rizki 1952, Lindsley, Edington, and von Halle 1960, and summary in Lindsley and Grell 1968).

Lethals may cause death at any stage of the development, from early to late embryogenesis, any one of three larval instars, early or late pupa, or newly emerged adult. Hadorn and Rizki found indications that certain developmental stages are more sensitive than others. Different lethals harm different organ systems—digestive tract, Malpighian tubes, gonads, fat bodies, mouth parts, tracheal tubes, musculature, etc. The imaginal discs may fail to form, be underdeveloped, or fail to evert. Tumors, though mostly benign and rarely invasive, may develop. Physiological and biochemical changes were described in some lethal genotypes.

Mutants that modify the sex-determination mechanisms are also known. A dominant mutant in *Drosophila pseudoobscura* transforms females into intersexes without changing either the sexual traits or the fertility of males (Dobzhansky and Spassky 1941). The recessive gene double sex in *D. melanogaster* changes both females and males into almost similar intersexes (Hildreth 1965). Another recessive gene makes the females intersexual but does not change the males. A mutant strain in which the males evinced homosexual behavior was found in *D. melanogaster* by Gill (1963) and in the stickleback fish (*Pygosteus pungitius*) by Morris (1952).

Homeotic mutants produce quite spectacular transformations of some organs into others. A single pair of wings, a pair of balancers, and antennae and mouth parts built in certain ways are diagnostic characteristics of the order Diptera (flies). The mutant aristapedia has antennae replaced by leglike organs; in proboscipedia the proboscis becomes antennalike or leglike; bithorax and tetraptera transform the balancers into a second pair of wings; hexaptera adds a pair of winglike appendages on the prothorax, etc. (Herskowitz 1949a). Of course, all these mutants continue to belong not only to the order of flies but also to the genus and species *Drosophila melanogaster.*

The kinds of mutants found in a given organism depend in part on the methods used for their detection. The classical mutants in Drosophila, as mentioned previously, were discovered by visual inspection. Mutants in bacteria and other microorganisms change mostly biochemical traits, such as the ability or inability to metabolize certain substances, or to synthesize some vitamins and amino acids. However, a considerable number and variety of morphologically visible mutants, in addition to the biochemical type, are known in the bread mold *Neurospora crassa* (Garnjobst and Tatum 1967). Studies in higher organisms have in recent years disclosed a tremendous amount of genetic variability in enzyme, hemoglobin, and serum polymorphisms.

The Numbers of Genes and Mutations

The existence of genes was inferred by Mendel from observations on the distribution of traits in progenies of hybrids between varieties of peas differing in clear-cut characteristics. If all members of a species were genetically alike, Mendelian genetics would be thwarted. One discovers the genes in man that control the blood groups, the eye colors, and other traits because people vary in these respects. Mendelian segregation of blood groups and of eye colors can be observed in families and in pedigrees. Only genes that have mutated and are represented by two or several allelic forms in the same species can be discovered by Mendelian methods. This is why geneticists are always on the lookout for genetic diversity. Genetically, highly variable species are predilect materials for investigation.

The number of genes is not known with precision in any species. The 56 genes in the bacteriophage T4 (Watson 1967) constitute probably the greatest fraction of the genes present in any organism that have been detected or studied. Since DNA is the carrier of genetic information, attempts have been made to estimate the numbers of genes through measurement of the DNA contents of cell nuclei. We saw in Chapter 1 that the haploid chromosome set in man contains approximately 3.2×10^{-12} gram of DNA, which corresponds to some 2.9×10^{9} nucleotide pairs. On the assumption of 600 nucleotides per gene, this is enough for more than 5 million genes. This number seems altogether excessive, and McKusick (1966) gives 100,000 as a reasonable estimate of the number of different genes in man. The amount of DNA in a haploid chromosome set in Drosophila suffices for at least 300,000 genes. This is again at least one order of magnitude greater than the number estimated by the pioneers of genetics. Their estimates were based on the assumption (which was admitted to be an oversimplification) that all genes mutate equally frequently, and that the genes which have been observed to mutate are a random sample of all the genes in a given species.

Several possible explanations for the above discrepancies can be offered. Perhaps not the whole mass of DNA consists of genes; a great majority of the genes have never been observed to mutate; mutations in most genes may be altogether inviable; most genes may be regulators rather than structural genes; mutations in most genes may produce slight and not easily detectable changes in the phenotypes of their carriers. The recent discovery of high redundancy of some genetic materials in higher organisms (Britten and Kohne 1968; see Chapter 2) affords at least a partial explanation. Unless the redundant gene loci are periodically derived from a single gene (which is unlikely), mutations will not be easily detected. They may be mutationally "silent" genes.

Drosophila melanogaster and man are the two species in which the greatest numbers of genes are known. A painstaking review of the mutants in the first of these species has been published by Lindsley and Grell (1968). They list 483 gene loci in the X chromosome, 279 in the second, 214 in the third, and 16 in the fourth—a total of 992 genes located with varying degrees of precision. McKusick (1966) catalogues 793 presumed autosomal dominants, 629 presumed autosomal recessive, and 123 sex-linked traits in man. However, among these only some 344 autosomal dominants, 280 autosomal recessives, and 68 sex-linked traits have their inheritances reasonably well established. Drosophila and man are higher organisms best studied genetically. Among microorganisms, bacteriophages T4 and lambda are also genetically well explored; about 56 genes in the former and 20 in the latter are known.

The Mutation Rates of Different Genes

The difficulty of obtaining accurate data on the mutation pressure is apparent. Either one tries to determine the total frequency of mutations for all the genes that the organism possesses, or else a particular gene is selected and its mutability measured. In the former case, mutations that produce slight changes present an obstacle, for no known experimental procedure permits the detection of all such mutations, and yet they probably constitute the most frequent class. On the other hand, if a single gene is selected, the mutation frequency is usually so low that accumulation of accurate data on higher organisms is difficult, slight mutations may be overlooked, and there is no assurance that all mutations of the gene in question (because of its manifold effects) produce changes in the same character. Conversely, the same character may be modified in similar ways by mutations in different genes ("mimics"), so that the mutation rate ascribed to a single gene locus may be an overestimate. The techniques of estimation of the mutation rates in human genes and the sources of error encountered in such estimation are well described in Stern (1960).

A sampling of the recorded spontaneous mutation rates is shown in Table 3.1. The so-called unstable or mutable genes, some of which cause piebald, mottled, or dotted color patterns in animals and plants,

TABLE 3.1

Spontaneous mutation rates of specific genes in various organisms (After Strickberger 1968, modified)

Species and Traits	Mutations per 100,000 Cells or Gametes
Escherichia coli (K12)	
Streptomycin resistance	0.00004
Resistance to phage T1	0.003
Leucine independence	0.00007
Arginine independence	0.0004
Tryptophan independence	0.006
Salmonella typhimurium	
Tryptophan independence	0.005
Diplococcus pneumoniae	
Penicillin resistance	0.01
Neurospora crassa	
Adenine independence	0.0008-0.029
Inositol independence	0.001-0.010
Drosophila melanogaster	
Yellow body	12
Brown eyes	3
Ebony body	2
Eyeless	6
Zea mays	
Shrunken seed	0.12
Colorless	0.23
Sugary seed	0.24
Pr to pr	1.10
I to i	10.60
R^r to r^r	
Mus musculus	
Brown	0.85
Pinkeye	0.85
Piebald	1.70
Dilute	3.40
Homo sapiens	
Epiloia	0.4-0.8
Retinoblastoma	1.2-2.3
Aniridia	0.5
Achondroplasia	4.2-14.3
Pelger's anomaly	1.7-2.7
Neurofibromatosis	13.0-25.0
Microphthalmos-anophthalmos	0.5
Huntington's chorea	0.5

are not included, because they may represent phenomena of a different nature. Even so, calculated in frequencies per generation (or cell division in unicellular organisms), the rates range over several orders of magnitude (10^{-10} to 10^{-4}). These differences would, of course, be much less pronounced if the mutation rates were calculated per unit time, since the generation lengths in the organisms mentioned in Table 3.1 differ by factors up to 10^{6}. It should be noted, however, that genes in the same species may have mutabilities differing by at least two orders of magnitude, and data like those in Table 3.1 are necessarily a biased sample, since very rare mutations are likely not to be observed at all.

The mutation rates in varieties of a single species, and presumably in different species as well, are under genetic control. It has been known for at least two decades that some strains of *Drosophila melanogaster* have higher mutation rates than others, and that these differences may be caused by mutability enhancer genes, which can be located in one or another linkage group (see Ives 1950 and Thompson 1962 for further references). Abundant data have been accumulated on the frequency of origin of recessive lethal mutations in the X chromosomes of *D. melanogaster*. Dubinin (1966) has summarized the data on the frequencies of such mutations in 71 strains from different parts of the world. The total 385,207 chromosomes analyzed had 719 lethal mutants; this means that 0.187 ± 0.007 percent of X chromosomes acquire such mutants per generation. The mutation rates in different strains varied, however, from 0.05 to 1.09 ± 0.15 percent per generation. Similar variations are found also in the second chromosome, although here the information is considerably less extensive.

Table 3.2 summarizes the results of a comparative study of muta-

TABLE 3.2

Frequencies (in percentages) of recessive lethal and semilethal mutations in homologous chromosomes of four species of Drosophila (CL = 95 percent confidence limits)

Species and Population	Frequency	CL
Pesudoobscura	0.999	0.733–1.340
Persimilis	1.783	1.403–2.264
Prosaltans	0.638	0.467–0.869
Willistoni, São Paulo	1.714	1.203–2.438
Willistoni, São Paulo	0.762	0.472–1.231
Willistoni, Belem	0.568	0.317–1.016
Willistoni, Total	0.894	0.678–1.179

tion rates in four species of Drosophila (Dobzhansky, B. Spassky, and N. Spassky 1952, 1954). All experiments were made at the same temperature, 24°C. *Drosophila persimilis,* a species living in cool and humid parts of the western United States, has a mutation rate higher than that of the more widespread *D. pseudoobscura. Drosophila willistoni* is a common and widespread, and *D. prosaltans* a rare and specialized, tropical species. The last-named has the lowest mutation rate of the four species. Among the three populations of *D. willistoni* studied, the one from equatorial Brazil (Belem) has the lowest mutation rate; the populations from southern Brazil (São Paulo) have higher rates. The mutation rates seem to be adjusted to the climatic and ecological conditions of the habitats in which these insects live.

Many, and perhaps all, genes may be changed in various ways and may produce series of multiple alleles. The frequencies of different kinds of mutations depend on the structure of the gene itself and on the genotype as a whole. The classical work of Timofeeff-Ressovsky (1937 and earlier) showed that the gene *W* (for the normal red eye color) in *Drosophila melanogaster* changes to the extreme allele *w* (for white eye color) more frequently than to intermediate alleles, such as w^e (eosin color) or w^a (apricot color). The "normal" *W*, allele of this gene is, however, different in different strains. A strain of American origin, and another of Russian, were given identical X-ray treatments. In the former, 55 mutations at the white locus were observed among 59,200 chromosomes; in the latter, 40 mutations among 75,000 chromosomes. The "Russian" allele changed mostly to white, and the "American" one to white and to intermediates (eosin) with about equal frequency. Through further experiments Timofeeff-Ressovsky proved that the difference in the behavior of the Russian and the American strains was due to different mutabilities of the white gene itself, and not to modifying genes in other chromosomes.

Gene mutation is, in principle, a reversible process. Most mutations are caused by substitution of a single amino acid in a protein, and of a single nucleotide in the DNA chain coding for this protein. With such mutations, the reverse substitutions may also occur. Unfortunately, the experimental approach to the problem of mutation reversibility meets with complications. Suppose that a mutation changes the eye color in Drosophila from red to vermilion; a reverse mutation should change vermilion back to red. Study of such reversals shows, however, that some of them are due, not to mutations in the vermilion gene, but to mutations in other genes that suppress the phenotypic manifestation of vermilion. Be that as it may, the observable frequencies of "forward" and "reverse" mutations are often sharply unequal.

Schlager and Dickie (1966) summarize the data on spontaneous mutations in five genes in the house mouse, based on a study of approximately 1.5 million mice. The frequencies of forward and back mutations for 1 million gametes are as follows:

Gene Locus	*Forward*	*Reverse*
a	71	4.7
b	0	0
o	9.7	0
d	19.2	0.4
ln	15.1	0

An even more detailed study of forward and reverse mutations was made by Yanofsky and his colleagues (Yanofsky, Berger, and Brammar 1969, and other works) in *Escherichia coli*. Mutants that changed the gene coding for the enzyme tryptophan synthetase were collected, and the amino acids substituted in the different mutants were determined. Reverse mutations, which restored the original phenotype, that is, the normal physiological function of the enzyme, were also observed. Most of the reverse mutations did not, however, restore the original amino acids (and, hence, the original nucleotide sequence in the gene); in point of fact, some of the phenotypic reversions differed from the original enzyme in having more amino acid substitutions than did the mutants.

Lethal and Subvital Mutations Induced by Radiation

Mutations form a spectrum, ranging from drastic changes that cause death, or dramatic alterations of the external appearance, to barely perceptible, and perhaps even imperceptible, modifications of body structures or of viability (subvital mutants). Drastic or easily visible changes are, of course, advantageous for study. In organisms such as Drosophila, the detection of lethals is easy and accurate. In contrast, some observers detect small changes in the external traits of the fly that are overlooked by others. With lethals such "personal equation" is unimportant. And yet it is possible that slight mutants outnumber drastic ones. Moreover, the drastic and spectacular mutants may really be only pathological by-products of the evolutionary changes, which are compounded of many small mutations.

Timofeeff-Ressovsky (1935) and Kerkis (1938) did the pioneering work in comparing the frequencies of large and small mutations in *Drosophila melanogaster*. Males from an inbred strain were treated with X-rays, and crossed to ClB females which previously had been crossed repeatedly to the same inbred strain. ClB females have in one of their X chromosomes a gene marker, Bar (*B*), making narrow eyes, a recessive lethal gene (*l*), and an inversion (*C*) which suppresses the recombination between the ClB chromosome and the other X chromosome present in the same female. In the F_1 generation, females with narrow eyes, which carry the ClB chromosomes, were selected and outcrossed to untreated wild-type males. In the progeny of such matings, some of the sons receive the ClB chromosome and die of the lethal contained in it; the sons receiving the other X chromosome survive, provided no lethal mutation has been induced in this chromosome by the treatment. The expected sex ratio is, therefore, 2 females : 1 male. If a lethal mutation is induced, the offspring are females only.

If a mutation that is not lethal but decreases the viability arises in the X chromosome, the resulting sex ratio falls between 2♀ : 1♂ and 2♀ : 0♂, depending on the degree of the deleterious effect produced by the mutation. For technical reasons it is preferable to record only the daughters that do *not* carry the X chromosome (ClB); they form about half of all females and can be recognized by round instead of Bar eyes. The frequencies of such females and males turn out to be 1♀ : 0.95♂ if no mutation has been induced in the treated chromosome, and 1♀ : 0♂ if a lethal mutation has been induced. Table 3.3 gives the sex ratios produced by individual females. The control series shows the ratios obtained in the progeny of males that have not been treated with X-rays.

TABLE 3.3

Percentages of cultures giving various sex ratios obtained by Timofeeff-Ressovsky (1935) in his experiments on mutations affecting viability

Culture	Ratio												
	1: 1.15	1: 1.05	1: 0.95	1: 0.85	1: 0.75	1: 0.65	1: 0.55	1: 0.45	1: 0.35	1: 0.25	1: 0.15	1: 0.05	1: 0.00
Control	2.1	14.1	77.1	5.5	0.7	...	...	0.5	...	...	...	...	...
Treated	0.7	10.1	44.9	8.8	7.2	5.3	4.2	1.8	1.1	0.7	1.4	0.9	13.0

In the treated series, 13 percent of the cultures had no males at all, and about 3 percent produced less than one-third as many males as females; these cultures contained newly arisen lethal and semilethal mutants. Many cultures—no fewer than 20 percent of the total—gave appreciably fewer males than were observed in most cultures of the control experiment. These male-deficient cultures contained subvital mutants, which cause viability losses not drastic enough to be classed as semilethal. Although it is difficult to identify every culture that carries a subvital mutant, both Timofeeff-Ressovsky and Kerkis concluded that subvital mutants are more frequent than lethal and semilethal ones, and by extrapolation that mutations with small effects are generally more frequent than those with large effects. This was questioned by Käfer (1952), Bonnier and Jonsson (1957), Paxman (1957), and Friedman (1964). These authors found lethals the most numerous class of mutations, at least in the progenies of X-ray-treated flies. However, the well-designed and large-scale experiments of Mukai (1964, 1969) established conclusively the prevalence of small mutations.

The Techniques of Chromosome Assay and of Accumulation of Mutants

At this point we must review certain techniques developed for Drosophila, which are being used in many experiments. There is a laboratory strain of *Drosophila melanogaster* that carries the dominant genes Curly wing (*Cy*) and Lobe eye (*L*) in one of its second chromosomes, and Plum-colored eyes (*Pm*) in the other second chromosome. Both chromosomes have also inverted sections to suppress the recombination in the second chromosomes. Analogous strains, with different mutant markers, exist in other Drosophila species.

Suppose that we wish to determine the proportion of the sex cells that contain newly arisen second-chromosome mutants, or the proportion of the second chromosomes in a natural population of *Drosophila melanogaster* carrying recessive lethal, semilethal, subvital, or any other recessive mutant genes. Following the scheme in Fig. 3.1, we cross a fly to be tested to *Cy L/Pm* flies. A single Curly-Lobe male is taken in the F_1 generation, and crossed further as shown in the figure. In the third generation, one quarter, i.e., 25 percent of the flies, carry in duplicate (in homozygous condition) the chromosomes $+_2$ or $+_3$,

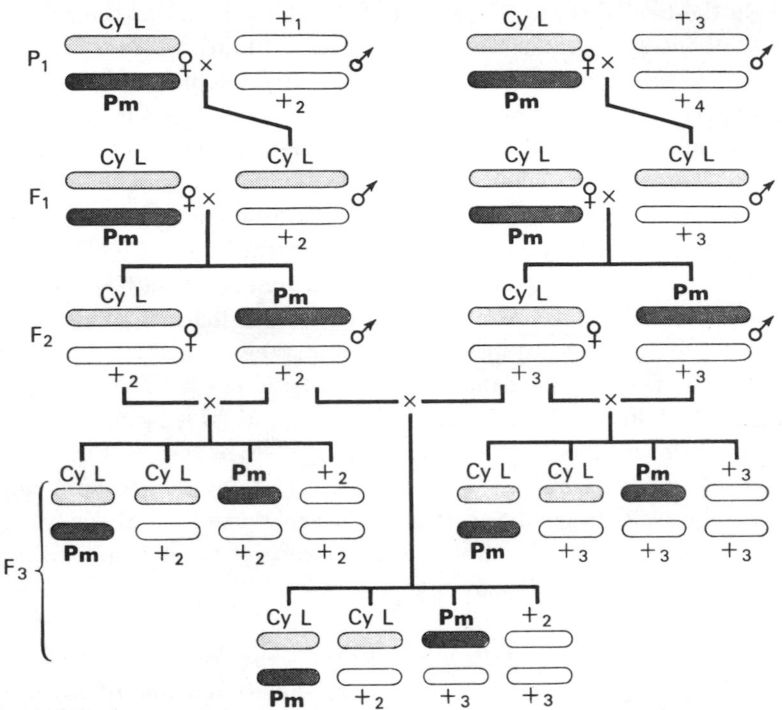

FIGURE 3.1

Technique of chromosome assay for recessive genetic variants in Drosophila melanogaster. The chromosomes with the dominant marker genes Cy, L, and Pm are shown shaded; wild-type chromosomes ($+_1$, $+_2$, $+_3$, and $+_4$) are white.

the effects of which in double dose we wish to test. These flies are recognizable by inspection, because they do not carry the marker genes Curly, Lobe, or Plum. Suppose now that the tested chromosome contains a recessive lethal gene or genes; the wild-type flies will then be absent in the test culture. A recessive condition that gives a proportion of wild-type flies more than zero but less than 12.5 percent is a semilethal; a subvital will make the wild-type class frequency more than 12.5 but less than 25 percent, and a supervital significantly more than 25 percent.

Another variant of the same technique involves intercrossing in the F_2 generation (Fig. 3.1) females and males that carry one Curly-Lobe and one wild-type chromosome ($Cy\ L/+_1$ or $Cy\ L/+_2$). One quarter of the zygotes in the F_2 will now die on account of homozygosis for the lethal genes Curly and Lobe. Two-thirds of the surviving zygotes carry one Curly-Lobe chromosome, and a chromosome the genetic contents of which are to be tested, represented as white in Fig. 3.1; one-third of the zygotes will have the chromosome to be tested in double dose (in homozygous condition). Now, if these homozygotes are equal in viability to the Curly-Lobe heterozygotes, the frequencies of the two classes will be 33.3 percent nonmutant (wild-type) and 66.7 percent Curly-Lobe flies, respectively. Suppose, however, that the tested chromosome contains a recessive lethal gene; no wild-type flies will then be found in the culture. A recessive condition that gives a proportion of wild-type flies of more than zero but less than 16 or 17 percent is conventionally called semilethal; a subvital will make the wild-type class more frequent than 17 but less than 33 percent. A recessive sterility gene will make the homozygous flies sterile, a recessive visible mutant will render them morphologically abnormal, etc. If large enough numbers of the flies are counted, the method is sufficiently sensitive to detect slight mutational changes.

Mukai's experiment (1964) was started by crossing a single male, carrying one wild-type second chromosome known to give homozygotes of good viability, and also a Plum second chromosome, to *Cy L/Pm* females (see Fig. 3.1). In the progeny, 104 lines were established, by outcrossing in every generation a single *Pm*/wild-type male to *Cy L/Pm* females. The purpose of this technique is to "accumulate" mutations in the 104 wild-type second chromosomes transmitted generation after generation from the *Pm*/wild-type fathers to their sons. Any completely recessive mutant will be protected from elimination by natural selection in the heterozygous state, even if it is completely lethal when homozygous. However, in the 10th, 15th, 20th, and 25th generations, tests were made, by means of crosses like those diagramed in Fig. 3.1, to examine the viability of flies homozygous for each of the 104 second chromosomes. By the 25th generation, 15 chromosomes had acquired recessive lethal, and 2 chromosomes semilethal, mutants. The remaining lines were "quasi-normal" when homozygous, but their mean viability gradually declined because in some of them recessive subvital mutations accumulated. This was manifested by the average percentage of the wild-type class declining from the theoretical value of 33.3 percent to 31.60 in the 10th generation, 30.93 in the 20th, and 28.35 in the 25th.

Hand in hand with the decline in average viability, the variance

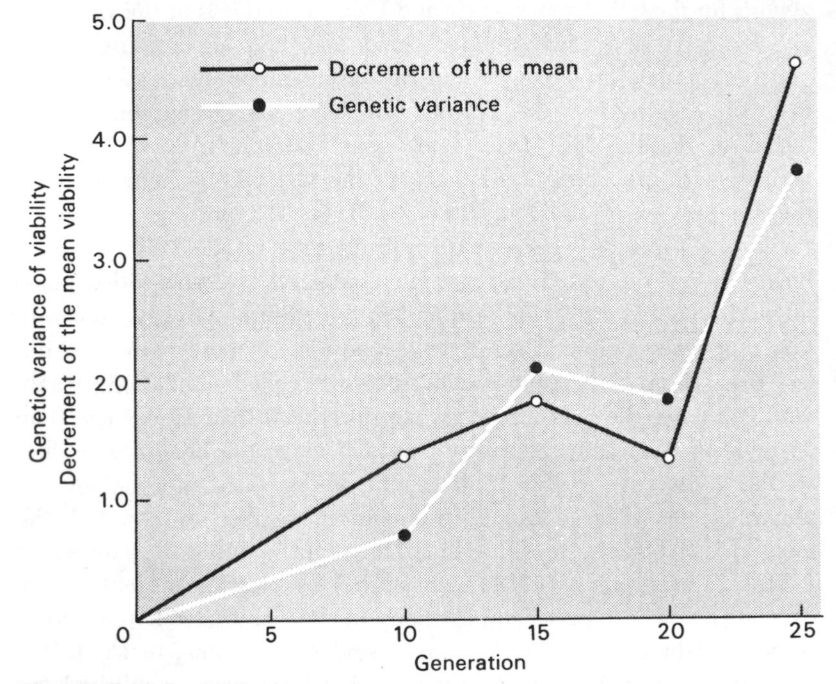

FIGURE 3.2

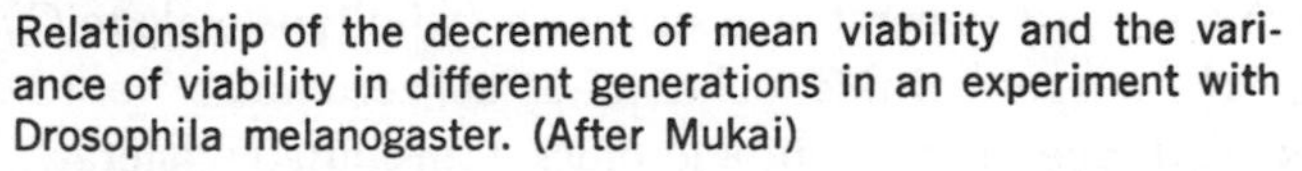
Relationship of the decrement of mean viability and the variance of viability in different generations in an experiment with Drosophila melanogaster. (After Mukai)

in the viabilities of the quasi-normal chromosomes increased, as shown in Fig. 3.2. The chromosomes that were originally similar descendants of the chromosome with which the experiments were started became more and more diversified, because many of them acquired mutations changing (usually decreasing) their viability. To analyze the data, Mukai has made a reasonable assumption: that spontaneous mutations occur among the chromosomes by chance, following a Poisson distribution. This permits computation of the average proportion of the chromosomes that acquire new mutations per generation. It can be shown that this average should be approximately equal to the squared regression coefficient of the decrement of the mean viability, divided by the regression coefficient of the increment of variance. The value found by Mukai was close to 0.15. In other words, about 15 percent of the second chromosomes of *Drosophila melanogaster* acquire new spontaneous mutants in each generation. Only about one-twentieth of the mutations are lethal or semilethal; a majority involve small, subvital changes.

In classical genetics, mutation was believed to be a rare phenomenon. And so it is, if changes of individual genes are considered. However, when the whole genotype is taken into account, mutations are seen to be quite frequent. Since the second chromosome of *Drosophila melanogaster* contains about two-fifths of the total chromosomal materials, one may conservatively estimate that at least 30 percent of the sex cells transport one or several newly arisen mutations affecting the viability in every generation. High rates of occurrence of small mutations are suggested also by observations on the genetic divergence of inbred strains in mice (Bailey 1959; see also Wallace 1965), and by the studies of Sprague, Russell, and Penny (1960) on maize. These findings have important practical as well as theoretical consequences. Inbred strains of experimental animals and plants are widely used in medical, physiological, and other research, in the belief that individuals of such strains are all alike genetically. This confidence is often misplaced because of high mutation frequencies. The decided predominance of small mutations over drastic ones is a fact of major significance to the evolutionist; some of the classical models of the evolutionary process were built on the implied assumptions that mutations are rather infrequent, but that those which do occur may be individually recorded and studied. A multitude of small mutations requires different approaches.

Mutation and Adaptedness to External Environments

Most biologists were skeptical, and justifiably so, of the mutation theory of de Vries, who claimed that new species arise by sudden mutations. Likewise, when Morgan and his associates described mutant Drosophilae, many biologists remained skeptical, because these mutants looked like a collection of freaks rather than changes fit to serve as raw materials of evolution. Some biologists continue, though no longer justifiably, to be skeptical. The reason is that most mutations, large as well as small, are more or less deleterious to their carriers. Mutation appears to be a destructive, rather than a constructive, process. One should not forget, however, that a mutation is neither useful nor harmful in the abstract; it can be so only in some environment. If the environment is not specified, the statement that a mutation is useful or harmful is meaningless. A mutant that is harmful when its carrier is placed in one environment may be neutral in another, and useful in still other environments. Furthermore, a mutant gene does not exert its effects on adaptedness regardless of what other genes an individual carries; a changed gene may be harmful on some genetic backgrounds but useful on others.

One must be careful, of course, not to overstate the argument just presented. It does not mean that some environment, external or genetic, can be found to make every mutant gene useful. To be sure, it is rash to declare that a given mutant can never be useful. For example, scaleless chickens may appear to be hopelessly inferior, and yet Abbott, Asmundson, and Shortridge (1962) found that these mutants can grow with less methionine and cysteine in their diet than can normal chickens. Yet the fact remains that mutational changes are genetic accidents. They can be compared to random knocks on some delicate mechanism, such as a watch; such knocks will very rarely improve the functioning of the mechanism. If most mutations are substitutions of single nucleotides in the DNA of a gene, and of single amino acids in the protein that this gene produces, the chances of improvement are still small. The genes and proteins that a species has are products of selection in the evolutionary process; at least the frequently occurring changes have had time to be incorporated into the genotype. Another way of saying the same thing is that the level of adaptedness of every existing species is fairly close to the maximum achievable in its present environment. Therefore it is not surprising that most mutations are not clear-cut improvements in external and genetic environments in which the species normally exists.

To maximize the chances of observing favorable mutations, the experimental organisms should be placed in novel environments. Examples are plentiful, particularly in microbiology. It was known for a long time that, confronted with adverse environmental conditions or unusual food sources, bacterial cultures may quickly give rise to new genetically stable lines able to cope with these conditions successfully. In point of fact, it was precisely the ease with which such "adaptation" or "training" takes place that inclined bacteriologists to believe that these changes do not arise by mutation regardless of their usefulness, but are directly induced by specific environmental needs.

It took the brilliant work of Luria and Delbrück (1943) and Demerec and Fano (1945) to show that spontaneous mutation is involved. The bacteria *Escherichia coli* are attacked by bacteriophages, which reproduce in the bacterial cells and cause their breakdown, or lysis. If bacteriophages are added to a culture of bacteria, the latter are destroyed within minutes or hours, whereupon the medium contains great numbers of bacteriophages capable of infecting other bacterial cells. However, a few cells may survive and form colonies of bacteria that are henceforward resistant to the bacteriophage strain in the presence of which they appeared. Luria and Delbrück showed that this resistance arises by mutation, at the rate of about 2 × 10 [illegible] cell generation. The bacteriophage does not produce resistant bacteria; its role is that of a selective agent which destroys all nonmutant cells. In the presence of bacteriophages normal bacteria are killed and only the mutants survive and reproduce.

Several bacteriophage strains are known, and a bacterial strain that is resistant to one bacteriophage may not be resistant to others: the resistance is specific. Demerec and Fano (1945) found in *Escherichia coli* at least eight different kinds of mutants, each resistant to one or more of the seven bacteriophage strains used by these authors. If the bacteria are exposed to the proper bacteriophage, each kind of mutant can easily be obtained. Hence, by exposing bacteria to a succession of bacteriophages, it is possible to build up strains resistant to several or even to all phage strains. The rates of mutation to resistance to a given phage strain are independent of what other mutations have occurred previously.

The discovery and the widespread use of chemotherapeutic drugs and antibiotics were followed by the appearance and spread of bacterial mutants resistant to these drugs and antibiotics (a review, no longer up to date, in Wolstenholme and O'Connor 1957). Some diagramatically clear examples of adaptive mutations have been encountered. Demerec (1948) exposed cultures of *Escherichia coli* to streptomycin, which kills all the bacteria except streptomycin-resistant mutants. Some of these mutants are, however, not only genetically resistant to streptomycin but also streptomycin-dependent, that is, unable to grow without the presence in the nutrient medium of this substance, fatal to their ancestors. Reverse mutations, which remove the streptomycin dependence, also occur; by placing a large inoculum of streptomycin-dependent bacteria on streptomycin-free medium, independent strains of the bacteria can be selected. Some of these strains continue to be resistant to streptomycin, whereas others lose the resistance.

The question then presents itself: Since streptomycin-resistant mutants arise from time to time in cultures not exposed to this antibiotic, why have all the bacteria in nature not become resistant? This is easily answered for mutants that are streptomycin-resistant as well as dependent: they evidently could not exist in streptomycin-free environments. The answer is not so obvious for the resistant but independent mutants. One must suppose that they too are for some reason at a disadvantage in media free of streptomycin. The mutation to resistance, as well as the reverse mutation, is favored and discriminated against in different environments.

Domestic animals and cultivated plants are, next to microorganisms, the most favorable materials for the detection of useful mutants. The environments provided for them by cultivators, though permissive and sheltered in many ways, are quite different from the environments of their wild ancestors and hence demand genetic adaptation. The pioneer experiments of Gustafsson and his colleagues (Gustafsson 1941, 1963a,b, Gustafsson and Nybom 1950, and Gustafsson, Nybom, and Wettstein 1950) showed that even genetically so "destructive" an agent as X-radiation may induce some mutants in barley that are agriculturally useful. This lead has been followed by numerous other workers seeking useful mutants (e.g., Bandlow 1951, in wheat; Scholz and Lehmann 1962, in barley; and Stubbe 1950, in snapdragon). The Food and Agricultural Organization (FAO) of the United Nations has held a symposium at which many successful experiments were reported (Anonymous 1965 and Russian works reviewed in Shkvarnikov 1966).

Extracellular Selection of Mutants

As stated in Chapter 1, natural selection is a function of differential reproduction; it could not have begun before there was life. Yet we have seen that DNA and RNA extracted from living organisms can "reproduce" *in vitro,* that is, act as templates for the synthesis of their replicas. Spiegelman and his collaborators (Spiegelman, Pace, et al. 1969, Levisohn and Spiegelman 1969, and references therein) have utilized this fact to make most elegant "extracellular Darwinian experiments with replicating RNA molecules."

They have isolated from *Escherichia coli* infected with RNA bacteriophages MS-2 and Q-beta replicase enzymes, which catalyze the *in vitro* synthesis of RNA chains. A fact of importance is that these replicases are specific in their action; the replicase extracted from the bacteria infected with MS-2 "recognizes" the RNA from that bacteriophage, while the replicase coming from Q-beta-infected bacteria is active only with RNA templates from that phage. A standard reaction mixture is used which contains the replicase enzyme, the four triphosphate nucleotides (A, U, G, and C), magnesium ions, and an RNA template. If the replicase and the template are of the same origin (i.e., both MS-2 or both Q-beta), new RNA chains are synthesized by copying the template. The new RNA can in turn serve as a template, and by making serial transfers to fresh reaction mixtures this reproduction can continue, presumably indefinitely.

Spiegelman et al. utilized a temperature-sensitive mutant of the Q-beta phage; the mutant grows poorly at 41°C, whereas the original phage thrives at this temperature. The RNA of the temperature-sensitive variant has replicated its mutant characteristic *in vitro*, that is, the new RNA synthesized is also temperature sensitive. Clones descended from single RNA strands, which reproduce their kind, can be obtained.

Spiegelman and his colleagues have now turned to selection for mutant RNA's arising *in vitro*, in their reaction mixtures. When multiplying in living bacterial hosts, the RNA of the bacteriophage must not only replicate itself but also induce the synthesis of the protein coat of the phage particle. The latter requirement is no longer binding on RNA's that replicate *in vitro*. Their V-1 mutant is a chain 83 percent shorter than the original or wild-type RNA chain. Yet the loss of the now dispensable parts permits the mutant to replicate appreciably faster in the standard mixture. Levisohn and Spiegelman (1969) selected also what they call "nutritional" mutants V-2, V-4 and V-6. These RNA templates can replicate more and more successfully in mixtures with less cytosine (cytidylic acid) than the standard mixture. Another mutant is resistant to tubercidine, an analogue of adenosine, which inhibits the replication of the original form. All this is quite properly described as Darwinian natural selection, though selection on a postbiological rather than a prebiological level.

Mutation and Adaptedness to Genetic Environments

As early as 1934, Timofeeff-Ressovsky showed that the viabilities of some mutants of *Drosophila funebris* depend on both external and genetic environments (Table 3.4). Thus, the mutant eversae is, at 15–16°C and 28–30°C, inferior to the wild type, whereas at 24–25°C it is superior. The viability of venae abnormes and miniature mutants is only slightly inferior to that of the wild type at 15–16° but is much inferior at 28–30°C. On the contrary, the viability of the bobbed mutant is low at 15–16°C but approaches normal at 28–30°C. Overpopulation of the cultures decreases relative viability in the mutations eversae, venae abnormes, and miniature, but has an opposite effect on bobbed. Combination of venae abnormes and lozenge, each of which decreases viability, produces a summation of the individual deleterious effects; combination of miniature and bobbed gives a compound that is more viable than either mutation by itself.

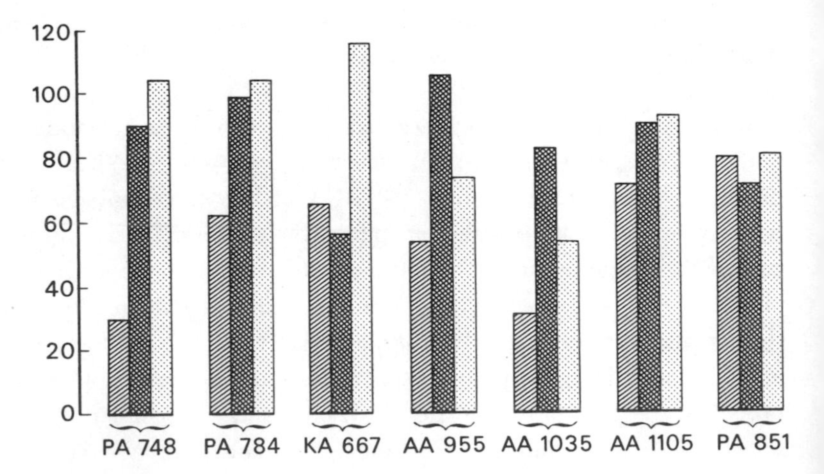

FIGURE 3.3

Fifty generations of natural selection for viability in laboratory cultures of seven strains of Drosophila pseudoobscura. Ordinates—percents of "normal" viability; left (crosshatched)—the viability at the beginning of the experiment; middle and right (light and dark stippled)—after fifty generations with and without exposure to low doses of X-rays. (After Dobzhansky and Spassky)

TABLE 3.4

Viabilities of some mutations and their combinations in Drosophila funebris, expressed in percentages of the viability of the wild type (After Timofeeff-Ressovsky)

Mutation	Temperature, °C			Combination	Temperature, 24–25 °C
	15–16	24–25	28–30		
Eversae	98.3	104.0	98.5	Eversae signed	103.1
Singed		79.0		Eversae abnormes	83.7
Abnormes	96.2	88.9	80.7	Eversae bobbed	85.5
Miniature	91.3	69.0	63.7	Singed abnormes	76.6
Bobbed	75.3	85.1	93.7	Singed miniature	67.1
Lozenge		73.8		Abnormes miniature	82.7
				Abnormes lozenge	59.3
				Abnormes bobbed	78.7
				Miniature bobbed	96.6
				Lozenge bobbed	69.2

Dobzhansky and Spassky (1947) took a different approach. Many chromosomes in natural populations of *Drosophila pseudoobscura* and other species carry gene complexes that make them lethal, semilethal, or subvital in double dose (for more details, see Chapter 4). Three second and four fourth chromosomes were chosen which, as shown by the cross-hatched columns in Fig. 3.3, reduced the viability of their homozygous carrier to between 30 and 80 percent of the normal value. The average viability of individuals carrying two chromosomes of each pair, taken at random from a given population, is defined as "normal" (see Chapter 4). Populations of flies homozygous for each of the seven chromosomes were bred in cultures that were deliberately made grossly overpopulated. The experiment extended for fifty consecutive generations. The rationale of this experiment was that mutations might arise which would compensate for the loss of viability caused by the chromosomes subvital or semilethal in the homozygous condition. In a population living in a stringent environment, caused by crowding, such favorable mutants should survive and gradually enhance the adaptedness of the whole population. Furthermore, each population was subdivided into two sections, one of which received in every generation 1000 r of X-rays. It was hoped that the irradiation would increase the frequencies of mutations, including the favorable ones.

From time to time during the course of the experiments, as well as after the fifty generations were completed, the viability of the homozygotes for each of the seven chromosomes was tested by a series of crosses analogous to those diagramed in Fig. 3.1. The light-stippled columns in Fig. 3.3 show the viabilities of the populations not treated with X-rays, and the dark-stippled columns the viabilities of the irradiated ones. Every one of the chromosomes, except that designated PA-851, showed improvement of the viability either in the X-ray-treated or in the untreated population, or in both lines. It can be seen in Fig. 3.3 that at least six of the fourteen experimental lines attained fully "normal" viabilities, within the limits of experimental errors. The irradiated and the unirradiated lines did not differ significantly in the viability levels attained.

The results of the experiments in Fig. 3.3 do not show that all, or even most, mutations arising in the cultures were favorable. In addition to the cultures homozygous for each of the seven experimental chromosomes, there were kept also cultures in which these chromosomes were transmitted, from one generation to the next, always in the heterozygous condition. This was effected by means of a technique analogous to that previously described in connection with Mukai's experiments (see above). Lines treated with X-rays and untreated ones were kept. When, after fifty generations, the chromosomes were finally tested for viability in the homozygous condition, the results contrasted sharply with those shown in Fig. 3.3. Of the fourteen lines, seven became lethal when homozygous; in five there were no changes or slight deteriorations; and only in two cases were ostensible improvements observed. The irradiated and the unirradiated lines were not conspicuously different. It may be concluded that when mutations are allowed to accumulate indiscriminately they result in deterioration of the viability; there also appear, however, some favorable mutants, which improve the viability when conditions are propitious for their selection.

In the experiments of Ayala (1966, 1967, 1969b), irradiated populations of two closely related species, *Drosophila serrata* and *D. birchii*, outstripped the unirradiated ones in the pace of genetic adaptation to the experimental environments. The male progenitors of the experimental populations were treated with 2000 r of X-rays for three generations, and crossed to unirradiated sisters in uncrowded cultures. Experiments were then started with large numbers of flies, faced with rigorous competition for food and space in crowded cultures. Control populations were set up in the same way, except that the flies were not irradiated. At the beginning of the experiments the average number of flies per culture was smaller in the populations with radiation histories than in the unirradiated controls for several generations. Some deleterious mutants were evidently induced, and it took time to purge the populations of the excess of these mutants. Thereafter, however, the irradiated populations became more flourishing than the controls. Over a period of 1 year, the average sizes of the two irradiated populations of *D. serrata* were 493 and 972 flies greater than those of the control, and in *D. birchii* the irradiated cultures averaged 621 and 587 flies more than the unirradiated populations. The conclusion is inevitable that, although the average effect of newly induced mutations was deleterious, a minority of the induced mutations proved favorable, in the sense that they permitted the populations to exploit more efficiently their stringently limited environments.

Genetic Techniques of Pest Control

Although geneticists and evolutionists are most interested in detecting the minority of mutants that are beneficial, in some situations the more numerous deleterious mutations can also be utilized. More than 40 years ago Muller (1927) observed the high mortalities of eggs deposited by untreated Drosophila females mated to males subjected to X-rays. This unusual mortality is due to dominant lethal mutations, many of them chromosome breaks that cause inviability, induced in the spermatozoa by the treatment. After a heavy dose of irradiation, a male may be as active as an unirradiated male in mating with females, but all eggs fertilized by his spermatozoa will be inviable. Bushland and Hopkins (1951) and Knipling (1955) saw here a possible method of combating insect pests, by releasing large numbers of sterilized males to compete with their normal counterparts for the available females, thus destroying the progeny of the latter. Much work has been done to develop this method to the point of practical applicability (reviews and bibliographies in Knipling 1967 and LaChance 1967).

Suppose that the population of some insect consists of 1,000,000 fertile females and 1,000,000 males, and that we release an additional 9,000,000 males sterilized by irradiation or by a chemical mutagen (of which several suitable for this purpose have been found). Fertile and sterile males will then be present in the ratio 1 : 9. If they are equally active, and each female mates only once, only 100,000 females will produce progeny. Suppose further that a fertile female can leave on the average 5 daughters and 5 sons. In the next generation the population will consist of 500,000 females and 500,000 fertile males. If then another 9,000,000 sterile males are released, they will outnumber the fertile ones in the ratio 18 : 1. Only some 26,316 females will produce fertile progeny, which, granting the same assumptions as above, will number about 131,625 females and males. Two additional releases of 9,000,000 sterile males will, theoretically, reduce the fertile populations to 9535, 50, and 0.

The most successful application of this method was made on the 170-square-mile island of Curaçao in the Caribbean, to control the population of the screw-worm fly (*Cochliomyia hominivorax*), the larvae of which develop in abscesses in the flesh of living mammals. A technique has been devised, however, to grow them on mass scale in shallow pans with a mixture of ground horse meat, water, blood, and a small amount of formaldehyde. The female of the species is monogamous. Releases of sterile males have destroyed the population on the island. A campaign to control the screw-worm fly in the southeastern United States in 1958 and 1959 was also effective, but another in Texas and New Mexico met with only partial success. Programs for the control of other insect pests, among them the true fruit flies (*Dacus dorsalis, Anastrepha ludens, Ceratitis capitata*) and tsetse flies (Glossina), are in the experimental stage.

Recombination and Interaction of Genes

As mentioned above and discussed in detail in the next chapter, many chromosomes found in natural populations are lethal or otherwise deleterious in double dose. But some chromosomes, often only a minority, give homozygotes equal, or even superior (supervital), in viability to the natural population average. The question naturally arises, Are these the elusive "normal," typical, optimal chromosomes, unspoiled by mutation? One way to approach this problem is to find out whether these apparent "normals" are alike in their gene contents, carrying only the optimal gene alleles.

Dobzhansky (1946) in *Drosophila pseudoobscura* and Wallace, King, et al. (1953) in *D. melanogaster* have shown that, when pairs of chromosomes, both of which produce normally viable homozygotes, are allowed to undergo crossing over, some of the recombination products are semilethal or lethal when homozygous. Such lethals arisen by recombination are called synthetic lethals. A considerable literature has accumulated on synthetic lethals, chiefly in species of Drosophila but also in Tribolium (Sokoloff 1964). Most of the synthetic lethals are recessive, but Gibson and Thoday (1962) found also a dominant synthetic lethal in *D. melanogaster.*

It is hardly surprising that not all chromosomes, and not all populations, produce synthetic lethals (Hildreth 1956 and Spiess and Allen 1961). A systematic search for them has been carried out in *Drosophila pseudoobscura* (B. Spassky, N. Spassky, et al. 1958), *D. persimilis* (Spiess 1959), *D. prosaltans* (Dobzhansky, Levene, et al. 1959) and *D. willistoni* (Krimbas 1961). In each case samples of 10 chromosomes were chosen from one or two natural populations of a species, by methods analogous to that shown in Fig. 3.1. The chromosomes selected were those that produced homozygotes of nearly normal, and sometimes even superior, viability. Females containing all possible combinations of the 10 chromosomes (45 combinations) were obtained, and 10 chromosomes from each progeny (i.e., 450 chromosomes) were tested, again with the aid of techniques analogous to that represented in Fig. 3.1, for viability in the homozygous condition. Since crossing over occurs in Drosophila females, at least a part of the 10 chromosomes derived from each cross represented recombination products of the chromosomes that the mother carried. An example of the results obtained is as follows. Eleven of the 45 combinations of second chromosomes of *D. pseudoobscura* from a Texas population produced from 1 to 4 synthetic lethals in the groups of 10 chromosomes tested; 8 of the 45 combinations of the chromosomes from California gave from 1 to 7 lethals; and 19 of the 100 combinations in which 1 chromosome was of Texas and the other of California origin gave from 1 to 5 lethals. In sum, 77 out of 1900 recombination chromosomes tested, or about 4 percent, were lethal.

It should be noted that the chromosomes yielding lethal recombination products were often among the most viable ones in the homozygous condition. They evidently carried internally well-balanced gene complexes (Mather 1941), the components of which interacted favorably to produce viable homozygotes. Recombination, however, caused breakdowns of the internal balance; the epistatic interactions of the new combinations of the genes then resulted in lethality (Lucchesi 1968). Changes in traits other than viability may evidently arise by similar mechanisms; Krimbas (1960) obtained in *Drosophila willistoni* some recombination products of chromosomes making the flies fertile that caused "synthetic" sterility.

How frequent synthetic lethals are in natural populations of Drosophila and of other organisms remains to be ascertained. Magalhães and his colleagues (Magalhães, de Toledo, and da Cunha 1965 and Magalhães, da Cunha, et al. 1965) found that in experimental populations of *Drosophila willistoni* more than 20 percent of the second chromosomes were lethal in double dose after seven generations of breeding. Conversely, chromosomes previously known to be lethal had this effect expunged by "suppressors" found in natural populations. Evidently, synthesis and suppression of lethals by epistatic interactions with other genes are the two sides of the same coin. A lethal can be "synthesized" by combining in a chromosome genes that are not lethal separately, but that interact to produce a lethal in double dose; the lethal can be "suppressed" or "desynthesized" by breaking up this combination. A proof of this has been given by Dobzhansky and Spassky (1960); they obtained some synthetic lethals in *D. pseudoobscura* and then broke them up by subjecting the lethal chromosomes to crossing over.

Synthetic lethals are evidently the most spectacular products of gene recombination and of epistasis. More important still is the fact that crossing over between ostensibly similar chromosomes generates an enormous amount of genetic variation, ranging all the way from lethality and semilethality through subvitality and ostensible "normality" to supervitality. Dobzhansky, Levene, et al. (1959) measured this release of genetic variability by comparing the variance of the homozygous viabilities of the chromosomes found in natural populations of a given species (see Chapter 4 for more detail), with the variance observed among the recombination products of similar chromosomes. The results are collated in Table 3.5. It can be seen that the recombination in a single generation produces between 24 percent (*Drosophila persimilis*) and 43 percent (*D. pseudoobscura*) of the total variance in viability in the homozygous condition. If lethal chromosomes (both natural and synthetic) are disregarded, the remaining quasi-normal chromosomes

TABLE 3.5

Mean viabilities and variances of homozygotes for chromosomes of three species. "Natural" refers to mean viabilities and variances for samples of wild chromosomes obtained directly from natural populations. "Recombination" refers to mean viabilities and variances for chromosomes that were recombination products of pairs of quasi-normal original chromosomes obtained from nature.

Species and Chromosomes	Mean Viability: All Chromosomes	Mean Viability: Quasi-normal	Variance: All Chromosomes	Variance: Quasi-normal	Recombination Variance % / Natural Variance: All Chromosomes	Recombination Variance % / Natural Variance: Quasi-normal
D. pseudoobscura						
Natural	20.29	24.26	140	65	43	74
Recombination	22.92	23.93	60	48		
D. persimilis						
Natural	23.54	28.00	110	60	24	27
Recombination	28.16	28.76	26	16		
D. prosaltans						
Natural	21.17	25.98	200	85	25	28
Recombination	28.14	29.72	50	24		

show at least as great a recombination effect—from 27 percent in *D. persimilis* to 74 percent in *D. pseudoobscura.*

These observations throw some light on one of the basic problems of evolutionary genetics, that of the maintenance of the genetic diversity found in natural populations of sexually reproducing species. The simplistic view, adhered to by some geneticists, is that most genes are identical in most individuals comprising a population, and whatever genetic diversity is observed is due to the presence of rather recently arisen deleterious mutants, not yet eliminated by natural selection. This view is hardly tenable any longer. Overt genetic diversity is evidently far exceeded by concealed or potential variability. The latter is stored in the gene combinations found in naturally occurring chromosomes, and it can be released by recombination (Mather 1953 and Thoday, Gibson, and Spickett 1964).

The variance released in one generation by recombination of parts of chromosomes selected for a relative uniformity amounts to at least one-quarter of the total expressed variance. The fact that some of the recombination products are semilethal or lethal when homozygous is particularly illuminating. The frequency of such synthetic lethals will be maintained in a population not by newly arising mutations, but rather by an equilibrium between the frequencies of their being "synthesized" and "desynthesized." Genetic diversity is maintained primarily not by new mutants, but by the advantages of heterozygosis for gene alleles and gene complexes that are kept up by natural selection, and also by environmental fluctuations in space and in time that alter the signs and the magnitudes of selective advantages and disadvantages. With a population structure of this sort, a total suppression of the mutation process would probably fail to change the evolutionary plasticity of the species for many generations.

The Randomness of Mutations

Random mutations are the raw materials of the evolutionary process. Natural selection orders them in functionally coherent, adaptive systems. Mutations are often described as accidental, random, undirected, chance events. Just what do these epithets mean? Mutations are accidents, because the transmission of hereditary information normally involves precise copying. A mutant gene is, then, an imperfect copy of the ancestral gene. It would be absurd, however, to say that human genes are only distorted copies of the primeval genetic materials. The serviceability of human genes, or of those of any existing species, has been validated by natural selection. Mutations are undirected with respect to the adaptive needs of the species. They arise regardless of their actual or potential usefulness. It may seem a deplorable imperfection of nature that mutability is not restricted to changes that enhance the adaptedness of their carriers. However, only a vitalist Pangloss could imagine that the genes know how and when it is good for them to mutate.

The frequencies of spontaneous mutations of some genes have been ascertained in genetically well-studied organisms. Yet where, when, and in which individual a particular mutation will appear is unpredictable. Even the rather more specific chemical mutagens discussed in Chapter 2 rarely change 100 percent of the genes exposed to treatment. The mutational repertoire of the gene is great but not infinite; it is limited by the composition of the gene. Consider only mutations due to single nucleotide pair substitution in a gene that codes for a protein chain 150 amino acids long, and has 150 triplet codons or

450 nucleotide base pairs. Since each base in a triplet can be substituted in three ways, a codon can be changed in nine ways. For 150 codons this means 1350 possible mutational changes. Because of the so-called degeneracy of the genetic code, about one-fourth of the single nucleotide substitutions will be to synonymous codons, that is, those coding for the same amino acid (see Chapter 1).

Approaching the matter from another angle, we note that there are 61 different codons specifying amino acids (Table 1.2); this gives $61 \times 9 = 549$ possible substitutions, of which 134 will be to synonymous codons (King and Jukes 1969). These figures do not, of course, include frameshift mutants and mutants substituting more than a single nucleotide. Frameshifts are likely, however, to give "nonsense" mutations, many of which may be lethal, and multiple substitutions are probably rare. The variety of possible mutations in a gene is impressively large.

Mutations cannot be said to change the development in random directions because a single nucleotide substitution is rarely if ever sufficient for a gene to change its function radically. The hundred or more mutant hemoglobins known in man (Table 2.1, and Perutz and Lehman 1968) produce variant hemoglobins rather than entirely different proteins. Although some of these variants are more or less deleterious to their carriers, others seem to be functionally equivalent to the ancestral condition. Accumulation of mutational changes may, of course, change a gene more radically. Wittmann-Liebold and Wittmann (1963) found that chemically induced mutants in the tobacco mosaic virus usually differ from the original kind in single, and rarely in two, amino acid replacements. Naturally occurring strains of the virus, however, differ from each other in as many as 30 of the 158 amino acids. The successive mutational gene changes acquire a direction because natural selection controls the fitness of the resulting phenotypes and thus indirectly imposes a restriction on the randomness of the mutational events. If it is assumed that life on earth arose only once, all existing genes have a common descent, and they are now different because of this long series of nucleotide substitutions that have taken place.

Hemoglobins and myoglobins are animal proteins of interest in connection with this discussion. There are good reasons to think that the genes coding for hemoglobins and myoglobins are modified descendants of a common ancestral gene. Human alpha differs from human beta hemoglobin in 84 amino acids (out of 141 and 146, respectively). Myoglobin (of sperm whale) has 153 amino acid residues; it differs in 115 amino acids from human alpha, and in 117 from beta, hemoglobins (Dayhoff and Eck 1968). Transformation of the gene coding for hemoglobin into that coding for myoglobin, or of the hemoglobin alpha gene into the hemoglobin beta gene or vice versa, has a zero probability of occurring by a single mutation. Yet these transformations have in fact taken place in evolutionary history, by way of a sequence of mutations, presumably controlled by natural selection. While each mutation in this sequence was, if considered on the molecular level, an accident, the sequence as a whole is in no sense accidental or random. Although we do not know the physiological functions served by the proteins intermediate between the modern hemoglobins and myoglobins, the alterations through which they passed were far from random.

The following illustration of the nonhaphazard nature of mutations is flippant but apt. Suppose that one wishes to transform, by selective breeding, the human race into a race of angels. We can be virtually certain that it would be much easier to breed for angelic disposition than for a pair of wings because there is available in human populations a variance in disposition, and it is not implausible to suppose that part of this variance is genetic. If so, selection may accentuate the dispositions that can be called angelic, and more variance may arise by mutation. There is much less chance of encountering variants on the basis of which the development of wings may be started, and to expect mutations providing such a basis seems rather farfetched. And yet birds and mammals, or bats and primates, have had a common, albeit remote, ancestry. There is no possibility, however, of reversing and repeating the evolutionary process that gave rise to these winged and wingless creatures.

CHAPTER FOUR

NORMALIZING NATURAL SELECTION

A Historical Sketch

Charles Darwin relates that the idea of evolutionary changes being brought about by natural selection came to him in a flash of insight as he was riding in a carriage on a country road. The same idea came to A. R. Wallace with equal suddenness during a paroxysm of malarial fever. Empedocles in ancient Greece, Lucretius in ancient Rome, Maupertuis and Buffon in the Age of Enlightenment, and Erasmus Darwin, grandfather of Charles, all had germs of the same idea. That so many people arrived at this same idea independently is not strange; the idea is quite simple and even obvious. The most poignant discovery in science comes when one suddenly sees a truth that was open to view all the time.

Eiseley (1959) brought to light the forgotten writings of Blyth, published between 1835 and 1837, in which natural selection (of course, not so named) was quite clearly discussed. Darwin was probably familiar with Blyth's articles, yet nowhere acknowledged this. The reason is, I believe, quite simple. Blyth argued that natural selection keeps species constant; Darwin declared that it changes them and forms new species. Darwin was an evolutionist; Blyth, an antievolutionist.

That natural selection is a common name for several cognate but distinct processes has been realized only within relatively recent years. Schmalhausen (1949) distinguished dynamic (directional) and stabilizing selection. The first changes the adaptive norm of the population; the second tends to keep it constant. A natural population is adapted to a certain range of environments. Environmental change is likely to cause a decline in this adaptedness; some formerly favorable genetic variants become disadvantageous and are replaced by new ones. Conversely, in a population that has achieved a high degree of adaptedness in a certain range of environments, the genetic endowment is advantageous, and deviations from it are inopportune. Stabilizing selection eliminates such deviations and promotes, in Schmalhausen's words, "more stable mechanisms of normal morphogenesis."

Waddington (1957) distinguishes two kinds of stabilizing selection, normalizing and canalizing. The former protects the adaptive norm by the elimination of harmful mutants, malformations, and weaknesses of various sorts. It was normalizing selection that Blyth wrote about. Canalizing selection, Waddington states, favors "genotypes which control developmental systems which are highly canalized and therefore not very responsive either to abnormalities in the environment or to new gene mutations of minor character." Canalization refers to the "limited responsiveness of a developing system." Human development is a familiar example. It is so canalized that all human beings are fundamentally similar and recognizably human, despite manifold variations in environments and in individual genotypes. Only major mutations and drastic environmental stresses deflect the development from its regular course (or creode, in Waddington's terminology); the result is usually death or teratological changes (morphoses in Schmalhausen's terminology; see Chapter 2).

Several kinds of selective processes are subsumed under the name of balancing selection (Dobzhansky 1964b). Their common feature is that they maintain genetic heterogeneity or polymorphisms, that is, the continued presence in a population of two or several alleles of some genes or of variant chromosome structures, the frequencies of which are more or less fixed by selection.

Directional selection enhances the frequencies of some gene alleles or gene combinations, and depresses the frequencies of others. Usually it acts in response to environmental changes. If novel and favorable gene mutations or gene combinations appear, however, directional selection may come into action without environmental change. In either case it is an agency of innovation, with which Darwin was mainly concerned (Lerner 1958 and Dobzhansky 1964a,b). The fact that Darwin's natural selection was adumbrated by several predecessors does not in the least diminish his greatness. No one else has stated this as clearly as Wilkie (1959):

The theory of evolution must be considered as a scientific theory, a theory, that is, proposed to explain or systematize a set of facts, and no one has any claim to be considered as a serious rival of Darwin in the "discovery" of this theory who did not conduct his evolutionary studies upon a reasonably wide basis of fact. To have ideas, *aperçus*, is not enough, and it is the overvaluation of such clever but uncontrolled guesses which is apt to produce the ludicrous . . . fallacy of combination, in which fragments of the final theory are collected from widely scattered sources and are combined in such a way as to impugn the originality of him who was the first to see how such a synthesis is possible.

Selection, Struggle, Mortality, and Fitness

Recognition of the diversity of selectional processes should not obscure the fact that all forms of natural and artificial selection have a common denominator. Lerner (1958) defines selection as nonrandom differential reproduction of genotypes; this can also be paraphrased as differential perpetuation of gene alleles and gene complexes.

Darwin and Wallace deduced natural selection from Malthus' principle that any population tends to increase in numbers in a geometric progression, and consequently will sooner or later collide with the limited resources on which it subsists. Only a part, often a small part, of the progeny survive; a majority die out. The carriers of better-adapted genotypes survive more often than do progeny inferior in this respect. The incidence of the former will increase and of the latter decrease from generation to generation. Yet natural selection may occasionally occur even if all the progeny survive in some generations. Consider a species that gives several generations per year; if food and other resources are not limiting during the warm season, the population may increase geometrically with no mortality among the progeny. Selection will nevertheless take place if the carriers of different genotypes have produced different numbers of offspring or have developed to sexual maturity at varying rates. A winter season may reduce the numbers of individuals drastically but unselectively; exponential growth and selection may be resumed the following year. Differential fecundity is, in principle, as powerful a selective agent as differential survival or mortality. This does not mean that the modern version of the theory of natural selection substitutes fecundity for survival as the principal method. Living beings must survive to reproduce, and must reproduce to survive in the following generation.

Darwin accepted, not without hesitation, Herbert Spencer's slogan "survival of the fittest in the struggle for life" as an alternative description of natural selection. At present this term seems inappropriate. Not only the fittest but also the tolerably fit survive and reproduce. Under flexible or soft selection (see Chapter 7) all the carriers of the same array of genotypes may survive when the population is increasing in size, and may be eliminated when it is shrinking. Moreover, as will be explained in more detail below, the fit are not necessarily the brawniest, the toughest, or the most aggressive. They are simply those who produce the largest numbers of viable and fertile progeny.

Selection does not create the materials that it selects. It operates effectively only as long as the population contains two or more genotypes, which perpetuate themselves at different rates (with natural selection), or among which a breeder can choose the parents of the next generation (with artificial selection). The crucial problem is, then, the source of the genetic raw materials, of the genetic diversity with which selection works. Darwin realized the importance of this problem, as well as the lack in his time of data adequate for its solution.

Early in the current century, Johannsen was making experiments, as simple as they were ingenious, on pure lines of beans. Beans are normally self-pollinating, that is, the seeds are produced by fertilization of the ovules by the pollen of the same plant. Commercial "varieties" usually contain mixtures of genotypes. By keeping the progenies of individual large-seeded and small-seeded plants separate, Johannsen easily isolated from these mixtures lines with large and with small seeds. On the contrary, selection failed to work in pure lines obtained by self-pollination from single individuals. The progenies of large and of small seeds from the same pure line were similar on the average.

In retrospect it seems strange that Johannsen's experiments, in the opinion of many of his contemporaries, demolished Darwin's theory of evolution by natural selection. Approximately between 1900 and 1925, the reputation of the selection theory was at its lowest ebb. The turn of the tide started with the work of T. H. Morgan and his school on mutations in Drosophila. It became evident that the process of mutation supplies the genetic raw materials on which selection may work. It remained, however, for Tshetverikov (1926), Fisher (1930), Wright (1931), and Haldane (1932) to lay the foundations of a modern version of the theory of natural selection. In the following decades, this theory expanded by inclusion of the results of other biological disciplines; it became the biological or synthetic theory of evolution (see Chapter 1).

The Hardy-Weinberg Law

The difference between the concept of heredity on which Darwin had to rely, and the one that serves as a basis of modern views, is the antithesis between the blending and the particulate theories of inheritance. The corollaries of the two theories are strikingly different. If the heredities of the parents blend in the progeny as a water-soluble dye commingles with water, the genetic variance in a sexually reproducing, Mendelian population will be halved in every generation. No matter how large a genetic diversity may be present at the start, there will be a rapid decay of the variability, and eventual homogeneity. No such difficulty is encountered with gene heredity. The different alleles of a gene present in a heterozygote do not contaminate each other; at meiosis they segregate, uninfluenced by their temporary sojourn in the same body and the same cell.

Suppose that two strains of a sexual and cross-fertilizing species are introduced into a previously unoccupied territory, in which they are equally adapted to live. Suppose further that they differ in a single gene, one strain being *AA* and the other *aa*, interbreed at random, and are introduced in the proportions p of *AA* individuals and $q = (1 - p)$ of *aa* individuals. We assume that the individuals composing the population contribute equal numbers of gametes, some carrying the gene *A* and others its allele *a*, to the gene pool of this population. What, then, will be the frequencies of *A* and of *a* in the gene pool, and what will be the proportions of the homozygotes, *AA* and *aa*, and of the heterozygotes, *Aa*, in this Mendelian population in the next generation and in the following ones? The solution of this problem was given by Hardy and by Weinberg independently in the same year, 1908. The frequencies p and q in the gene pool will remain constant generation after generation. The distribution of the genotypes among the zygotes will be:

$$p^2\,AA + 2pq\,Aa + q^2\,aa = 1$$

This expression, known as the Hardy-Weinberg equilibrium, describes the situation in a randomly breeding Mendelian population. There may, however, be assortative mating, such as some preference for mating of likes (homogamy) or of unlikes (heterogamy), or inbreeding or self-pollination (in plants). The relative frequencies of the homozygotes (*AA* and *aa*) and the heterozygotes (*Aa*) will then be modified, but the frequencies, p and q, in the gene pool will remain constant.

Maintenance of constant gene frequencies is evidently a conservative factor. Evolution is change in the frequencies of some genes. We shall proceed now to discuss the agencies that modify gene frequencies. It is interesting that each of these agencies is counteracted by another, opposite in sign, which tends to conserve the status quo. Hence a living population is constantly under the stress of opposing forces; evolution results when one group gains the upper hand over the other.

Mutation and Genetic Equilibrium

The value of q, the frequency of a gene or a chromosome structure in a population, can be modified by mutation pressure, that is, by gene mutations and chromosomal changes. If the change from A to a takes place, the frequencies p and q must also change. Let the mutation in the direction $A \rightarrow a$ have a rate of u per generation; the change in the frequency of A in the population per generation will be $\Delta p = -up$, where p is the frequency of A. If the mutation in the direction of $A \rightarrow a$ is unopposed by any other factor, the population will eventually reach uniformity for a. If the frequency of gene A in a certain generation is p_o, its frequency n generations later, p_n, will be:

$$p_n = p_o(1 - u)^n$$

Since the mutation rates, u, of most genes are small, of the orders of 10^{-4} to 10^{-7}, many generations are required to bring about considerable changes in gene frequencies in a gene pool by mutation pressure alone.

When the mutation is reversible, the change in the direction $A \rightarrow a$ is opposed by the change $a \rightarrow A$. With the rate of reverse mutation equal to v, the frequency of A will change as $\Delta p = -up + vq$. An equilibrium will be reached when the change per generation is $\Delta p = 0$.

The equilibrium value of p determined by the two mutually opposed mutation rates is therefore $p = v / (u + v)$. If, for example, the rate of mutation $A \rightarrow a$ is taken to be 1 in a million gametes per generation ($u = 0.000,001$), and the rate of mutation $a \rightarrow A$ to be $v = 0.000,000,5$, the equilibrium value for p will be 0.33, which means that 33 percent of the gametes in the gene pool will carry gene A and 67 percent gene a. If the mutation rates to and from a given allele are alike ($v = u$), the equilibrium values for q and p will be equal, 0.5.

Starting from an initially homogenous population, the mutation pressure will tend to increase the variability until the equilibrium values determined by the opposing mutation rates are reached. The process of mutation may, accordingly, make the population polymorphic, that is, composed of a variety of genotypes occurring with fixed frequencies in the same population. Whether polymorphisms due to opposing mutation rates are in fact widespread in nature is an open question. This is not improbable in microorganisms, where the generation time is short; in higher organisms, however, the numbers of generations and the time intervals required are so great that the adaptive values of the genotypes are not likely to remain equal. In this case, changes in gene frequencies will be determined more by selection than by mutation rates.

Darwinian Fitness and the Selection Coefficient

The Hardy-Weinberg formula assumes that the carriers of genotypes *AA*, *Aa*, and *aa* contribute, on the average, equal numbers of gametes to the gene pool of the next generation. The carriers of some genotypes may, however, be more viable and may reach the reproductive stage more often than the carriers of other genotypes. Or the carriers of different genotypes may vary in fecundity (numbers of eggs or seeds produced), in sexual activity, or in time of maturity or cessation of the reproductive period. All these variables influence the contribution that the carriers of a given genotype make to the gene pool. This contribution, relative to the contributions of other genotypes in the same population, is a measure of the Darwinian fitness of a given genotype. ("Adaptive value" and "selective value" are terms often used synonymously with "Darwinian fitness.") Thus, if the carriers of genotypes *AA* and *Aa* contribute on the average 100 gametes to the gene pool of the next generation, whereas the *aa* carriers contribute only 90 gametes, the Darwinian fitness of the former is $W_1 = 1$, and of the latter $W_2 = (1 - s) = 0.9$. The value of s is the selection coefficient. It is merely conventional to give the value of unity to the fittest genotype; one might just as well make the values of W_1 and W_2 equal $(1 + s)$ and 1, respectively.

If the Darwinian fitness of genotypes *AA* and *Aa* is 1, and that of *aa* is $(1 - s)$, the frequencies of these genotypes in the population before and after selection will be as follows:

Genotypes	*AA*	*Aa*	*aa*	Total Population
Darwinian fitness	1	1	$(1 - s)$	$\overline{W}$
Initial frequency	p^2	$2pq$	q^2	1
Frequency after selection	p^2	$2pq$	$(1 - s)q^2$	$1 - sq^2$

The frequency, p_1, of gene *A* in the next generation will be:

$$p_1 = (p^2 + pq)/(1 - sq^2) = p/(1 - sq^2)$$

The increment, Δp, of the frequency of gene *A* in one generation will be:

$$\Delta p = spq^2/(1 - sq^2)$$

Let, for example, genes *A* and *a* be equally frequent in the original population, so that $p = q = 0.5$. Also, let the fitness value of the dominants (*AA* and *Aa*) be unity, and suppose that the recessives (*aa*) have a fitness of 0 (a recessive lethal), 0.4 (a semilethal), 0.9, 0.99 (subvital), or 1.5 (supervital). The frequencies, p_1, of gene *A* in the next generation, and the increments of the gene frequency, will then be as follows:

Darwinian fitness	0	0.4	0.9	0.99	1.5
Selection coefficient (s)	1.0	0.6	0.1	0.01	−0.5
Frequency after one generation of selection (p_1)	0.67	0.59	0.5128	0.5012	0.444
Increment of gene frequency (Δp)	+0.17	+0.09	+0.0128	+0.0012	−0.056

For small selection coefficients, an approximate formula for the number of generations (n) necessary to change the frequency of a deleterious recessive gene from q_0 to q_n is derived from the equation:

$$ns = \frac{q_0 - q_n}{q_0 q_n} + \log_e\left(\frac{q_n}{1 - q_0} \cdot \frac{1 - q_n}{q_n}\right)$$

For the special case of complete selection, $s = 1.0$, against a recessive (i.e., selection against a recessive lethal), the formula becomes:

$$n = (q_0 - q_n)/q_0 q_n$$

The efficiency of selection depends on the gene frequency. Suppose that a recessive gene for some undesirable trait starts with a frequency of 0.5, and that the homozygous recessives are sterilized or otherwise eliminated from the population in every generation ($s = 1$). The frequency of this gene will then change as follows:

Generation	*Frequency*	*Generation*	*Frequency*	*Generation*	*Frequency*
0	0.500	8	0.100	50	0.020
1	0.333	10	0.083	100	0.010
2	0.250	20	0.045	200	0.005
3	0.200	30	0.031	1000	0.001
4	0.167	40	0.024		

The progress of selection is rapid at first, while the gene is frequent enough for an appreciable number of recessive homozygotes to be produced in the population, but it becomes slow as the gene frequency declines. It takes two generations to halve the frequency of a recessive lethal from 0.5 to 0.25, five generations from 0.20 to 0.10, fifty generations from 0.02 to 0.01, etc.

Very important in natural populations is normalizing selection against incompletely recessive genes or genes without dominance. Let the fitness of one of the homozygotes be $(1 - s)$, and of the heterozygote $(1 - hs)$, h being the coefficient of dominance (where the heterozygote is exactly intermediate in fitness, h is 0.5). We then have:

Genotype	A_1A_1	A_1A_2	A_2A_2
Darwinian fitness	1	$1 - hs$	$1 - s$
Initial frequency	p^2	$2pq$	q^2
Frequency after selection	p^2	$2pq\ (1 - hs)$	$(1 - s)q^2$

The frequency of A_1 will increase, and of A_2 decrease, in the next generation by:

$$\Delta p = hspq/(1 - hsq)$$

Selection acting against both the homozygous and the heterozygous carriers of a deleterious gene is decidedly more efficient than that acting against a completely recessive gene, especially if the frequency of the latter in the gene pool is low. Consider a selection of $s = 0.01$ (1 percent disadvantage) against a complete recessive, and $hs = 0.01$ ($s = 0.02$) against a gene without dominance ($h = 0.5$). The numbers of generations needed to effect a reduction in the frequencies of deleterious genes for the intervals specified are as follows:

From → To	*Recessive*	*No Dominance*
0.25 → 0.10	710	110
0.10 → 0.01	9,240	240
0.01 → 0.001	90,231	231
0.001 → 0.0001	900,230	230

For other forms of selection, such as selection in haploid organisms, gametic selection, and sex-linked genes, see Li (1955a), Lerner (1958), and Falconer (1960).

Experimental Models of Normalizing Selection in Drosophila

Natural selection in the laboratory is not a contradiction of terms. When man chooses individuals to be preserved and bred, and others to be eliminated or prevented from reproducing, he practices artificial selection. If an experimental population in the laboratory is placed in an environment in which some genotypes have an advantage over others, no discrimination by man toward producers and nonproducers is involved. Natural selection operates in the laboratory as well as in the wild.

Many investigators have contrived experimental populations of Drosophila in which selection could be studied. L'Héritier and Teissier (1934) constructed "population cages," in which containers of culture medium can be introduced or withdrawn without permitting the flies to escape. A mixture with known proportions of flies of desired kinds is introduced into the cage; samples of the population are taken at desired intervals, and the changing proportions of different phenotypes and genotypes are recorded. The experiments of Polivanov

(1964) with population cages containing the mutant Stubble and wild-type *Drosophila melanogaster* may serve as examples. They show very clearly the progress of the selection against Stubble, and at the same time display interesting deviations from the predicted course of the selection. Heterozygous Stubble flies (*Sb*/+) have short bristles, and wild-type flies (+/+) have long ones. Homozygous Stubble (*Sb*/*Sb*), is lethal. Polivanov set up four populations, the founders of which were *Sb*/+ heterozygotes. Thus the initial frequencies of *Sb* and + were equal, $p = q = 0.5$. The frequencies of *Sb* observed in different generations are shown in Fig. 4.1.

Assume for simplicity that the heterozygotes *Sb*/+ and the homozygotes +/+ are equal in fitness. Since homozygous *Sb*/*Sb* is lethal, one can predict the frequencies of *Sb* generation after generation with the aid of the formula given above: $n = (q_0 - q_n)/q_0q_n$. The decreasing frequencies predicted are indicated in Fig. 4.1 by black squares. The observed frequencies also show decreases of *Sb* in all four popu-

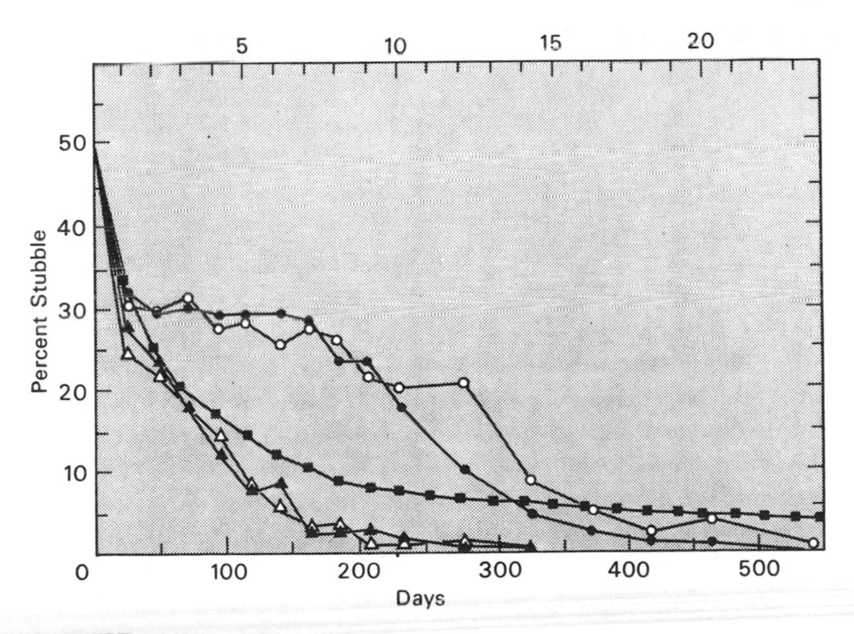

FIGURE 4.1

Selection against the mutant Stubble in experimental populations of Drosophila melanogaster. Circles—"monochromosomal" populations; and triangles—"polychromosomal" populations; squares—theoretically expected values. (From Polivanov)

lations. In two populations, however, symbolized in the Figure by triangles, *Sb* dwindled more rapidly than predicted. In the other two populations, symbolized by circles, *Sb* was for some generations more frequent than predicted.

These deviations are not accidental. The populations symbolized by triangles were polychromosomal, that is, they contained + chromosomes from several wild progenitors. The populations represented by circles were monochromosomal: their chromosomes with the + allele of *Sb* descended from a single progenitor. Why should this make any difference? We shall see that many "quasi-normal" chromosomes in natural populations are actually subvital, that is, cause slight but perceptible diminutions of fitness in double dose. In the monochromosomal populations the +/+ flies were actually less fit than the *Sb*/+ flies. Taking the fitness of *Sb*/+ to be unity, Polivanov estimated the fitness of +/+ flies during the first seven generations as 0.62. In polychromosomal populations +/+ flies were, on the contrary, more fit than their *Sb*/+ counterparts. This is not surprising, since *Sb* is a rather sharp mutant. Polivanov estimates the relative fitness of Sb/+ and +/+ in these populations as 1 : 1.25 to 1 : 1.44. Thus an ostensibly simple selective system reveals interesting complications.

Mutation Favored and Opposed by Selection

Biologically very important is the interaction of mutation with selection. Little needs to be said about a mutation that is increasing the fitness and is accordingly favored by selection. A favorable mutant allele will increase in frequency and eventually will displace and supplant the original allele. The initial increase will be more rapid for a dominant than for a recessive favorable allele, but the situation will be exactly reversed during the final stages of replacement.

Unfavorable mutants are discriminated against by normalizing selection. However, if a mutation arises at a rate u per generation, and is counteracted by a selection coefficient s, a state of genetic equilibrium will eventually be established, and both the original and the mutant alleles will continue to be present with certain frequencies. The equilibrium frequencies will be determined not only by the mutation and selection rates, but also by which, if either, allele is dominant or recessive. If the deleterious mutant allele is completely recessive—that is, if the fitness of the heterozygote, *Aa,* equals that of *AA,* the mutant allele will reach an equilibrium frequency at

$$q_a = \sqrt{u/s}$$

The mutant allele may be partially or completely dominant, that is, the fitness of the heterozygote may be between $(1 - hs)$ and 1 (where h is the coefficient of dominance). The equilibrium frequencies will then be reached at $p_A = u/hs$ (partial dominance) or $p_A = u/s$ (complete dominance).

What this means biologically is that a deleterious recessive mutant will reach a much higher equilibrium frequency in the gene pool than a dominant mutant. Let the mutation rate, u, be 10^{-5} (0.00001), and the selection coefficient, s, equal 0.1. A recessive mutant gene will then reach a frequency of 1 percent ($q = 0.01$), and a dominant mutant of only one-hundredth of 1 percent ($p = 0.0001$). These are, of course, the equilibrium frequencies of genes in the gene pool, and not of individuals in which these genes are manifested. Individuals heterozygous for a completely recessive mutant do not show its effects in their phenotype, whereas the opposite is true of heterozygotes for a dominant mutant.

The consequences of this are interesting. The smaller are the frequencies of recessive genes, the greater proportion will be concealed in heterozygotes ($2pq$), rather than manifested in homozygotes (q^2). For example, if a recessive and a dominant allele are equally frequent ($q = p = 0.5$), the heterozygotes are twice as common as either one of the homozygotes (0.50 and 0.25, respectively). Rare recessives, $q = 0.1$, 0.01, and 0.001, will be heterozygous with frequencies $2pq = 0.18$, 0.0198, and 0.001,998, and homozygous with frequencies of only $q^2 = 0.01$, 0.0001, and 0.000,001; hence, the heterozygotes will be 18 to almost 2000 times more frequent than the homozygotes.

On the other hand, if the mutation rates and the selection pressures opposing the mutants are equal, the dominants will be encountered and recognized more often than the recessives. Consider again what happens if u is 0.00001 and s is 0.1. Because the equilibrium frequency of a recessive mutant is $q = 0.01$, it will manifest itself in homozygous individuals with a frequency that will be $q^2 = 0.0001$, or 1 in 10,000. The equilibrium frequency of an equally deleterious dominant will be $p = 0.0001$, and the frequency of a heterozygous mutant individual, $2pq$, will be 0.0002, or 2 in 10,000, twice that of a recessive.

The matter can be approached from a different angle. Only dominant mutants whose Darwinian fitness is zero—that is, they are completely lethal or sterile—are extinguished by normalizing selection in the same generation in which they arise. In each generation a population will have only newly arisen completely dominant lethal or sterile mutants; no such mutants will be inherited from previous generations. With deleterious mutants whose fitness is greater than zero (semilethal or subvital), a population will contain some newly arisen and some inherited mutants. A variant allele arisen by mutation will persist in the population for a certain number of generations before its elimination. The persistence will be a function of fitness; it will be longer on the average for mildly deleterious mutants, and shorter for drastic ones. Fitness being equal, recessive mutants will persist longer than semidominant ones, and the latter longer than dominant ones, because the recessives, regardless of how drastic their effects may be when homozygous, are sheltered from normalizing selection when they are concealed in heterozygotes.

Populations of presumably all living species are said to carry genetic loads (Muller 1950). We shall see in the following chapters that the concept of genetic load, or genetic burden, is not easily definable in biologically meaningful and operationally useful ways. Not all mutants are deleterious, and the effects of some on the fitness of populations carrying them are ambivalent. Although a part of the genetic load may be said to exist because normalizing selection is not ideally efficient, other parts are maintained by various forms of balancing selection (see Chapter 5). Provisionally, we shall designate as genetic load simply the presence in populations of genetic variants that have deleterious effects on their carriers under some or under all conditions. Overt and concealed genetic loads will be distinguished; the former are detected by examination of individual members of a population, and the latter by study of their progenies.

Mutants in Natural Populations of Drosophila

The pioneer experimental work on what was later called genetic load or genetic burden in Drosophila was done by Tshetverikov (1926), H. Timofeeff-Ressovsky and N. W. Timofeeff-Ressovsky (1927), and Dubinin and his colleagues (Dubinin and fourteen collaborators 1934 and Dubinin, Romashov, et al. 1937) on populations of several species, mainly *Drosophila melanogaster*. Although Drosophila flies are, if anything, overtly less variable in nature than many other insect species, aberrant individuals do occur, and some of them are identical with mutants observed to arise in laboratories. No matter how carefully one examines the specimens collected in nature, however, recessive mutants that may be carried in the heterozygous condition will go undetected. Tshetverikov (1926) pointed out that a genetic analysis is needed to reveal such concealed variability. The simplest technique is to study the progenies of Drosophila females collected in nature. In their offspring one may discover the autosomal dominant mutants that these females and their mates carried, and also the sex-linked mutant genes. Inbreeding the F_1 individuals will permit detection in F_2 and F_3 generations of the recessive autosomal mutants. When practical, more refined methods, of the type described in Chapter 3 (Fig. 3.1), are used. They permit detection of concealed recessive mutants, such as lethals, semilethals, and all kinds of physiological as well as morphological variations.

The older work on mutants found in nature, or obtained by inbreeding Drosophilae collected in natural habitats, was ably summarized by Spencer (1947b). Spencer himself found a population of *Drosophila hydei* in which 6.5 percent of the males were vermilion-eyed. This sex-linked recessive mutant continued to occur in this population for at least 6 years. Some populations of this species contained several percent of the mutant bobbed. A variety of bobbed alleles was discovered; their frequencies varied from locality to locality.

Dubinin, Romashov, et al. (1937) examined some 130,000 wild *D. melanogaster* from several localities in southern Russia and found about 2800 "aberrant" individuals. A part of the latter had noninheritable

TABLE 4.1

Numbers of Drosophila melanogaster and D. subobscura females collected in their natural habitats that carried from 0 to 12 mutants (After Boesiger 1962 and Pentzos, Boesiger, and Kanellis 1967)

Species and Population	Number of Mutants											Average Number of Mutants per Fly
	0	1	2	3	4	5	6	7	8	9	10–12	
Melanogaster												
Banyuls 52A	7	24	59	38	20	4	6	2	...	...	...	2.54
52B	14	30	47	37	16	7	2	1	...	...	...	2.29
53A	5	7	20	30	55	41	27	21	9	2	1	4.45
53B	1	1	7	18	27	32	13	8	9	3	2	4.88
53C	...	5	13	12	8	10	7	4	3	1	...	4.00
Domme 57	...	6	15	17	18	13	9	9	3	1	4	4.36
Subobscura												
Litochoron	..	2	8	15	3	4	2	4	5	...	4	4.6
Samothraki	...	3	3	12	10	8	5	3	2	1	1	4.4
Thassos	...	5	30	59	62	25	4	2	1	1	...	3.5

abnormalities, but others were mutants, for the most part identical with well-known laboratory ones (extra bristles, ebony, sepia, yellow). The frequencies of the mutants varied from locality to locality and from year to year. Most mutants were recessives, although some semi-dominants were recorded; mutants that showed slight changes were more common than drastic and deleterious variants. At one collecting station—a pit with decomposing fruit—many individuals were homozygous for the gene divergent, which makes its carriers flightless.

Among the newer studies, those of Boesiger (1962) on *D. melanogaster* and of Pentzos, Boesiger, and Kanellis (1967) on *D. subobscura* are most carefully executed. In some populations of the former species in France, from 2.8 to 8.2 percent of individuals carry visibly detectable abnormalities; the corresponding percentages for the latter species in Greece are 6.3–9.5. After two generations of inbreeding, the females collected in their natural habitats were shown to have carried from none to as many as twelve mutants (Table 4.1).

Some comparable data exist for the housefly, *Musca domestica* (Milani 1967), and the mosquito *Aedes aegypti* (Craig and Hickey 1967). In the former, inbreeding reveals from zero to six mutants per progeny of a fertilized female; in the latter, the average number of mutants per female ranges in different samples from 0.72 to 2.96.

Overt Genetic Loads in Human Populations

Although a complete bibliography of this subject would contain hundreds or even thousands of entries, the genetic load in the human species is far from adequately known. Stevenson, Johnston, Stewart, and Golding (1966) and Kennedy (1967) have compiled data on the incidence of congenital malformations, a short summary of which is presented in Table 4.2. The incidence in different countries seems to be variable, but this may be largely spurious, depending on the criteria and the reliability of the procedures used. More intensive investigations show an average frequency more than five times as high as that derived from official records. What proportion of congenital defects is of genetic origin is also not well known. Neel (1958), who records 3.1 percent of "major congenital defect" in Japanese populations, considers that "a significant fraction of human

TABLE 4.2
Incidence of congenital defects in human populations (After Kennedy)

Source of Data	Number of Births Surveyed	Percentage of Malformations
I. Official records, birth certificates, retrospective questionnaires		
Belgium	740,956	0.67
Italy	2,660,990	0.15
United States and Canada	8,784,188	1.05
Total	12,186,134	0.83
II. Hospital and clinic records		
Britain and Ireland	640,413	1.18
Europe (excluding Germany)	2,560,937	1.37
Germany	2,154,964	0.97
United States and Canada	876,835	1.95
Other countries	652,462	0.99
Total	6,885,611	1.26
III. More intensive investigations		
Britain	170,224	2.88
Europe (excluding Germany)	78,610	2.96
Germany	8,516	2.20
United States	144,769	8.76
Other countries	121,264	2.85
Total	523,383	4.50

congenital defects are the segregants (phenodeviants) resulting from the existence and functioning of complex (multi-local) genetic homeostatic systems, of the type particularly discussed by Lerner (1954)."

McKusick (1966) lists 837 dominant, 531 recessive, and 119 sex-linked "phenotypes" in man, a great majority of which represent mild or serious genetic defects or diseases. Stevenson (1959, 1961) finds that, in Britain, 12–15 percent of all pregnancies that continue longer than 4 or 5 weeks end in abortions before the end of the twenty-seventh week, and more than 2 percent result in stillbirths. About 2.5 percent of all children are born with a "malformation detectable by the naked eye," and about the same proportion is recognized by 5 years of age. A considerable number of these are genetic. In Northern Ireland, about 26.5 percent of hospital beds are occupied by genetically handicapped persons, and 7.9 percent of consultations with medical specialists and 6.4 percent of those with general practitioners involve such persons. Genetic predisposition is involved in the etiology of many mental diseases, the incidence of which in human populations is considerable. The very careful study of Böök (1953) of an isolated population in northern Sweden showed the frequency of schizophrenia to be 2.63 ± 0.27, of oligophrenia 1.14 ± 0.13, and of epilepsy 0.35 ± 0.08 percent. The genetics of these nervous disorders is, however, far from clear. For reviews and bibliography see Pratt (1967), Petras and Curtis (1968), Erlenmeyer-Kimling and Paradowski (1966), and Shields (1968).

It was as recently as 1956 that Tjio and Levan established the correct chromosome number in man (46). Three years later, Lejeune, Turpin, and Gautier discovered the first chromosomal aberration. Individuals with the congenital malformation known under the misleading name of Mongolism (it has no relation to the Mongolian race) usually have 47 chromosomes. They are aneuploid mutants, trisomics, the extra chromosome being that designated No. 21 in the normal set. In little more than a decade since then, the study of chromosomal variations in man has burgeoned into a field with hundreds of publications and at least one special journal (reviews in Turpin and Lejeune 1965, Bartolos and Baramki 1967, Court-Brown 1967, and Reitalu 1968).

The fitness of aneuploid mutants in man is generally very low or even nonexistent. According to Stevenson, Johnston, et al. (1966), the frequency of trisomy-21 is at birth between 1 and 2 for 1000, ranging from between 0.3 and 0.5 per 1000 mothers 15–24 years old, to 6.34 for those aged 40–44 years, and as high as 16.65 for those older than 45. The frequency in the adult population is appreciably lower, owing to differential mortality, and since the affected persons seldom reproduce, their Darwinian fitness is close to zero.

Nondisjunction of the sex chromosomes gives rise to individuals with Turner's syndrome (45 chromosomes, a single X and no Y chromosome), Klinefelter's syndrome (47 chromosomes, two X's and a Y), and also XXX and XYY individuals (47 chromosomes). Court-Brown (1967) states that about 0.2 percent of the male baby population have a so-called Barr body in the nuclei of their buccal epithelial cells, and that a great majority of these are XXY Klinefelter individuals, while some are XXYY, XXXY, or even XXXXY in at least a fraction of their body cells. Persons with Turner's syndrome are probably always sterile, and those with Klinefelter's usually so. The relationship between abnormal sex chromosomes and mental disorder in man has been reviewed by Polani (1969).

Chromosomal aberrations in human individuals are more frequent than the above figures suggest. Failures of meiotic disjunction of the sex chromosomes and of chromosome 21 give rise to aneuploids at least some of which are viable, though usually more or less severely handicapped. The other chromosomes probably also undergo occasional nondisjunction, giving rise to aneuploid zygotes (trisomics and monosomics) that do not survive to birth. What proportion of spontaneous abortions results from this cause cannot be estimated with any degree of assurance. Persons heterozygous for translocations between chromosomes may enjoy normal health and have the normal number, 46, of chromosomes in their cells. A part of their progeny, however, will have some genes in excess (duplication) and some deficient, and will be malformed. This is the origin of the very severe abnormality called the *cri du chat* syndrome, which results from a deficiency for part of chromosome 5. Translocation heterozygosis is, thus, a part of the concealed genetic load (Lejeune 1969).

Concealed Genetic Loads in Human Populations

Incest, the mating of close relatives, is interdicted by law or custom almost universally in human societies. Only a minority of anthropologists believes the injunction to have arisen because the progenies of incestual unions were observed to be weak or malformed. Though this is indeed true in a statistical sense, to ascribe this discovery to primitive man would be to credit him with rather improbable acuity of discernment. Many individuals in human populations are heterozygous carriers of deleterious recessive variants. These variants have a greater probability of becoming homozygous in families in which the parents are relatives than in the progenies of mates not closely related. The homozygotes will only be a part of the progeny, however; in their siblings the recessives will be concealed in the heterozygous condition. To genetically untrained observers this is likely to make the causal relationship between inbreeding and the manifestation of deleterious genes far from evident. Only when the simple rules of Mendelian heredity are understood does it become comprehensible why rare recessive traits, such as albinism, occur particularly often in the progenies of those who have married cousins or other relatives.

Marriages of first cousins are common enough in at least some populations to make the collection of statistically meaningful data practicable. Aunt-nephew and uncle-niece marriages are generally infrequent, even where legally permitted. Marriages between second or third cousins are common, but the probability of homozygosis in the resulting children is much lower than in first-cousin marriages. In recent years, studies of progenies of marriages between relatives, as compared to those of unrelated parents, have been made in several countries. Despite heterogeneities in the results, due in large part to differences in the criteria used, the greater frequencies of stillbirths, infant deaths, and malformations among children of relatives are manifest. The review by Stevenson et al. (1966) records 12,779 stillbirths and infant deaths among 335,710 children of parents not known to be related, compared to 855 among 13,763 offspring of relatives. The frequencies are, consequently, 35.9 and 62.1 per 1000 births, respectively. The same review gives frequencies of 12.1 and 16.9 per 1000 births for congenital malformation among children of unrelated and of related parents, respectively.

Probably the most careful and detailed studies on the effects of inbreedings are those made by Schull and Neel (1962) in Japan. In addition to increased frequencies of malformations and neonatal deaths, they found "the average child of inbreeding" to differ slightly though significantly from the offspring of unrelated parents on a series of anthropometric traits, neuromuscular tests, and even school performance. In the populations of the cities of Hiroshima and Nagasaki slight but statistically significant differences were recorded in the following criteria, the "child of inbreeding" scoring less favorably than the control child:

Age when walked	Tapping rate
Weight	Color trail test score
Height	Maze tests score
Head girth	Verbal score
Chest girth	Performance score
	School performance
Calf girth	Language
Head length	Social studies
Head breadth	Mathematics
Head height	Science
Sitting height	Music
Knee height	Fine arts
Dynamometer grip	Musical education

Great care was taken to evaluate a possible inflation of differences between the inbred and the control children by socioeconomic factors. Yet, even when due allowances are made for the effects of such factors, the differences remain statistically meaningful. It is interesting that the study by the same authors (Schull and Neel 1966) of the semi-isolated population of the island of Kure, Japan, failed to disclose a significant effect of inbreeding (chiefly first-cousin marriages) on mortality during the first 15 years of life. A further discussion of attempts to measure the concealed genetic loads in human populations will be found in Chapter 5.

Genetic Loads in Maize and Other Plants

Many monoecious plants, though mostly cross-pollinated in nature, can produce seed also by spontaneous or artificial self-pollination. Selfing uncovers the concealed recessive mutants, because each mutant carried in heterozygous condition becomes homozygous in one-quarter of the progeny obtained by selfing. Crumpacker (1967) has published an excellent review and a summary of the extensive but scattered literature. Technically, the most easily detectable mutants are those that produce chlorophyll deficiencies, because they can be recorded in young seedlings from seeds obtained by selfing. In two species of rye grass (*Lolium perenne* and *L. multiflorum*), in timothy (*Phleum pratense*), orchard grass (*Dactylis glomerata*), *Festuca rubra*, clover (*Trifolium repens*), and cherry (*Prunus avium*) between 12 and 67 percent of the individuals tested were heterozygous for chlorophyll deficiencies or other seedling defects. Eiche (1955) examined 1,769,000 seedlings grown from seeds of 2031 Scotch pine trees (*Pinus sylvestris*) from 86 localities, mostly in Sweden. Some of these seeds came from cross-pollination, and others from selfing in nature. A total of 1368 mutant seedlings, or 0.077 percent, was recorded, but the incidence varied greatly among individual trees.

By far the most abundant data are available for Indian corn, *Zea mays*. Crumpacker (1967) summarizes 66 separate reports on the frequencies of recessive genes that give, when homozygous, defective and inviable seeds. Among 12,338 plants from field varieties tested, 11.0 percent were found heterozygous for such genes. A total of 106 reports describes tests of 18,697 plants for genes giving, in the homozygous condition, chlorophyll-deficient seedlings. The overall frequency of heterozygotes for such genes is 21.3 percent; in other words, about every fifth plant carries in its genotype such mutants, which are lethal or semilethal in double dose. In different samples, the frequencies varied from 0 to 64 percent. Chlorophyll deficiencies are of several different kinds and are produced by mutations of a number of different, nonallelic genes. About 180 such genes are known in maize. Crumpacker gives the following breakdown of the data (in percentages) according to the type of chlorophyll-deficient mutation:

White seedlings	5.5	Striped seedlings	3.1
Yellow seedlings	3.9	Pale green seedlings	3.3
Yellow and green seedlings	0.3	Virescent seedlings	2.6

The frequencies of the mutant alleles at any one locus are usually so low that in progenies of cross-pollinated plants chlorophyll-deficient seedlings are rarely found.

Assays of Viability Modifiers Concealed in Natural Populations

A technique for quantitative studies of recessive genetic variants concealed in apparently normal individuals in Drosophila populations has been described in Chapter 3 and illustrated in Fig. 3.1. Variations of this technique are available and have been used for the analysis of natural populations, in five species of Drosophila (*Drosophila melanogaster, pseudoobscura, persimilis, prosaltans,* and *willistoni*). All the techniques are similar in principle: by means of a series of crosses to laboratory stocks with suitable mutant gene markers in certain chromosomes, a class of flies is obtained carrying in duplicate (i.e., in the homozygous condition) a chromosome that was carried singly in an individual collected in a natural habitat of the species. This class of flies develops in cultures together with siblings that carry the chromosome under test in single dose (heterozygote) or do not carry it at all. The frequencies of these classes are theoretically predictable from the nature of the crosses made.

Thus, with the procedure diagramed in Fig. 3.1, one-quarter of the zygotes in the F_3 generation will carry the markers *Cy, L,* and *Pm*, two quarters either *Cy* and *L* or *Pm*, and one-quarter will be flies without mutant markers, homozygous for the "wild" chromosomes tested. These frequencies are expected, of course, only if the homozygous class is equal in viability to the classes carrying mutant markers. Suppose, however, that the wild chromosome carries a recessive lethal; a culture will then contain no wild-type flies. If the wild chromosome is semilethal (viability less than one-half normal), the wild-type class will appear with a frequency between 0 and 12.5 percent; a subvital chromosome will result in a frequency above 12.5 but below 25.0; and, finally, a supervital chromosome will raise the frequency above 25.0 percent.

For many purposes, it is more convenient to express the results of experiments in percentages, not of wild-type flies in the cultures, but of the average (or "normal") viability of flies heterozygous for pairs of chromosomes taken at random from the gene pool of the population. Such a "normal" viability standard is arrived at by making intercrosses of flies that carry the *Cy L* marker chromosomes and also chromosomes derived from different individuals sampled in the population. Because the mutant markers generally reduce, at least slightly, the viabilities of their carriers, the average normal viability corresponds to somewhat more than 25.0 percent of wild-type flies in the cultures (Fig. 3.1).

All species of Drosophila examined with the aid of the technique just described have yielded qualitatively, though not quantitatively, similar results. Figure 4.2 shows a sharply bimodal distribution of the

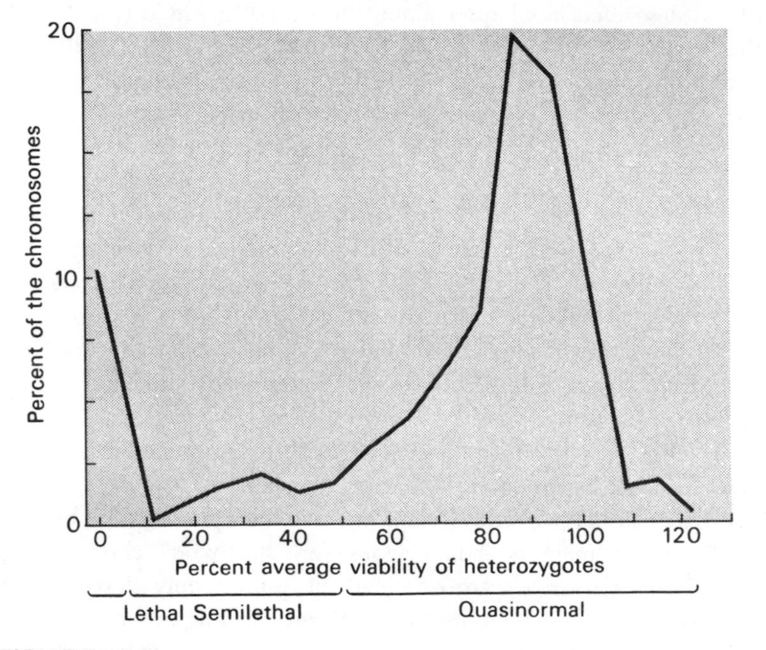

FIGURE 4.2

Viabilities in double dose of second chromosomes of Drosophila pseudoobscura from natural populations. (After Dobzhansky and Spassky)

viabilities in 284 second chromosomes of *Drosophila pseudoobscura* from Texas and California. Many chromosomes (17 percent in the experiment diagramed in Fig. 4.2) are lethal in double dose and produce only an occasional homozygous survivor, or none at all. Quasinormal chromosomes, with viabilities between 50 and 125 percent of the normal average, are most frequent. The concavity between the two peaks corresponds to the semilethal range—between 0 and 50 percent of normal viability; only about 7 percent of the chromosomes are found in this range.

A summary of the data on the frequencies of lethal and semilethal second chromosomes in populations of *Drosophila melanogaster* is given in Table 4.3. The frequencies vary from 11 to 62 percent—almost a sixfold difference. Some of these variations are probably due to differences in experimental techniques; others reflect the ecological situ-

TABLE 4.3

Percentages of second chromosomes in natural populations of Drosophila melanogaster that are lethal or semilethal in double dose

Population	Chromosomes Studied	Percentage Lethal or Semilethal	Authority
Caucasus, U.S.S.R.	2971	15.6	Dubinin et al. (1932-1936)
Caucasus, U.S.S.R.	1040	25.4	Berg (1939)
Uman, U.S.S.R.	2700	24.3	Olenov et al. (1937, 1938)
Crimea, U.S.S.R.	1630	24.8	Dubinin (1938, 1939)
Amherst, U.S.A.	2352	35.8	Ives and Band (1945-1959)
Maine, U.S.A.	226	42.9	Ives (1945)
Wisconsin, U.S.A.	231	34.2	Greenberg and Crow (1960)
Ohio, U.S.A.	177	49.7	Ives (1945)
Florida, U.S.A.	468	61.3	Ives (1945, 1951)
New Mexico, U.S.A.	203	62.1	Ives (1945)
Korea	611	11.2	Paik (1960)
Hiroshima, Japan	1901	11.6	Minamori and Saito (1964)
Kofu and Katsunuma, Japan	2457	16.6	Oshima (1967)
Kofu and Katsunuma, Japan	2773	21.5	Watanabe (1969b)
Egypt	301	29.4	Dawood (1961)
Lipari, Italy	215	34.1	Karlik and Sperlich (1962)
Israel	1222	34.7	Goldschmidt et al. (1955)

ations in which the populations live, which cause some of these populations to dwindle periodically to very few individuals and permit others to maintain relatively high densities. The incidence of lethals and semilethals in the third chromosomes of the same species is less well known; Wallace, Zouros, and Krimbas (1966) found in a South American population about equal frequencies (55 percent) of second and of third chromosomes being lethal or semilethal in double dose, and Band and Ives (1963) also found the frequencies in the two chromosomes to be about equal in the population of Amherst, Massachusetts. Even the tiny fourth chromosome has a 3 percent frequency of lethals (Hochman 1961). This means that in many, probably most, natural populations almost every individual fly carries one or more recessive lethal or semilethal variant in its genotype.

As seen in Fig. 4.2, chromosomes with quasi-normal viability in double dose are most frequent. The percentage of wild-type flies observed in test cultures for a given chromosome is, of course, subject to sampling errors; therefore, part of the variation in the viability scores among quasi-normal chromosomes comes from sampling errors, part from unavoidable environmental variations in different cultures,

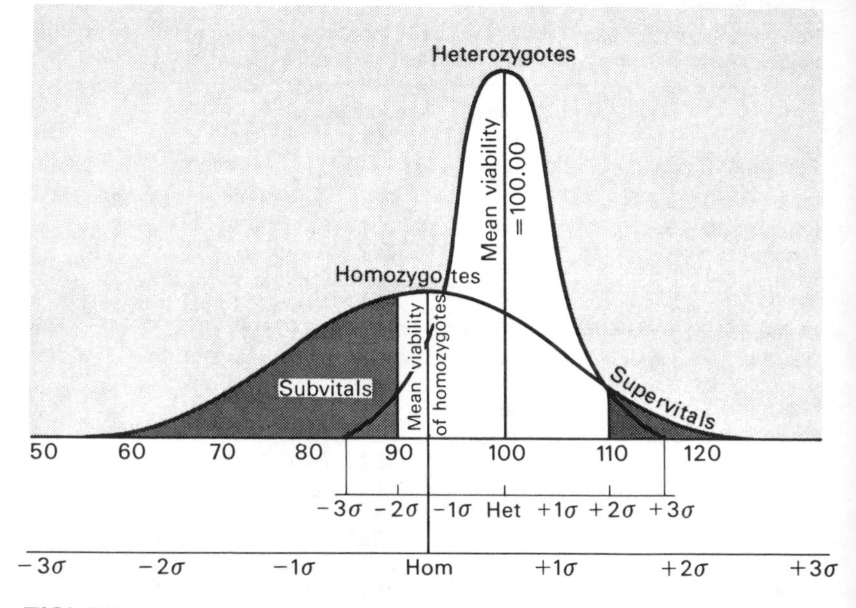

FIGURE 4.3

Technique of estimation of the frequencies of chromosomes which are subvital or supervital in double dose. Further explanation in text. (After Wallace and Madden)

and part from genetic differences between the chromosomes. Wallace and Madden (1953) developed a simple but ingenious method for analysis of the data (see Fig. 4.3). As stated above, the "normal" viability (or normal fecundity, or development rate, etc.) is defined as the average observed in an array of individuals with random combinations of the chromosomes from a given population in a specific environment. This is not an arbitrary but a biologically and operationally meaningful definition. In nature, "normal" Drosophila flies or butterflies or mice or men are not imaginary ideal types with perfect viability, optimum fertility, and fastest development, but arrays of genotypically as well as phenotypically diverse individuals, which arise by drawing sets of genes from the population's gene pool. Such arrays are obtained in laboratory experiments, as indicated above, by intercrossing females and males with a marker chromosomes and different wild chromosomes. The experiments are made by raising flies with each chromosome combination, and with each homozygous chromosome, in several replicate cultures. The numbers of the flies and the percentage variations observed in different cultures provide estimates of the variances due to sampling and to environmental fluctuations. By subtracting these from the total observed variance, we obtain an estimate of the genetic or "real" variance.

At this point, we must decide what ranges of viabilities we shall consider subvital, normal, and supervital. Since the frequencies of quasi-normal chromosomes with different viabilities form an approximately normal bell-shaped curve (Fig. 4.3), the choice of boundary points is rather arbitrary. Wallace and Madden (1953 and Wallace 1968) have proposed the following ingenious method. We have defined the mean "normal" viability as the average viability of random heterozygous combinations of chromosomes from a given population. Wallace also includes in the "normal" range all variants within two standard deviations (calculated from the genetic variance) above and below the average viability of the heterozygotes. Thus, about 95 percent of the heterozygous combinations are included among "normals." Now it is easy to calculate what proportion of the quasi-normal homozygotes fall below this "normal" range; these are considered subvital. Similarly, the ones above the "normal" range are supervital.

Table 4.4 shows the frequencies of chromosomes with different viabilities in double dose in populations of two species of Drosophila in a certain locality in California (estimates of Dobzhansky and Spassky 1953, corrected by Sankaranarayanan 1965). The most frequent class consists of the subvitals, followed by lethals, semilethals, and normals; supervitals are least frequent. Very few individuals in nature carry

TABLE 4.4

Percentages of chromosomes in natural populations of Drosophila pseudoobscura and D. persimilis with different effects in double dose

Species and Effect	Chromosome Second	Third	Fourth	Species and Effect	Chromosome Second	Third	Fourth
Pseudoobscura				Persimilis			
Lethal and semilethal	33.0	25.0	22.7	Lethal and semilethal	25.5	22.7	28.1
Subvital	62.6	58.7	51.8	Subvital	49.8	61.7	70.7
Normal	4.3	16.3	22.3	Normal	24.5	13.5	0.9
Supervital	<0.1	<0.1	<0.1	Supervital	0.2	2.1	0.3
Female sterile	10.6	13.6	4.3	Female sterile	18.3	14.3	18.3
Male sterile	8.3	10.5	11.8	Male sterile	13.2	15.7	8.4

only "normal" and supervital chromosomes. The properties of these rare chromosomes will be discussed further in the next chapter.

Table 4.4 does not mention the "first," that is, the X chromosome. Because Drosophila males carry only a single X, and most of the genes in this chromosome are not present in the Y chromosome, X chromosomes represent a special situation as far as accumulation of deleterious genetic variants is concerned. A recessive lethal mutant may be harmless in the heterozygous condition in females, but it will kill one-half of their sons. In hymenopteran insects, females are diploid whereas males are haploid; a deleterious recessive in any chromosome may be harmless in heterozygous females but will manifest itself in males. Normalizing natural selection will accordingly operate differently on the two-thirds of the gene pool carried in females and the one-third carried in males. If the mutation rate is u and the selection coefficient against a mutant that is recessive in females is s, the equilibrium frequency of the deleterious allele will be approximately $3u/s$. Deleterious sex-linked recessives are expected to be found in nature only three times as frequently as equally deleterious dominants. And yet, Laidlaw, Gomes, and Kerr (1956) discovered that honey-bee females are frequently heterozygous for genes that are lethal in double dose. These genes have sex-limited manifestations, that is, they do not affect the two sexes in the same way. For example, Kerr and Kerr (1952) and Drescher (1964) found in natural populations of *Drosophila melanogaster* some X chromosomes that are decidedly subvital or even semilethal in homozygous females but do not appreciably harm the males.

Assays of Fertility Modifiers in Natural Populations

The techniques used for assays of concealed recessive viability modifiers are applicable to the detection of other genetic variants as well. As shown in Fig. 3.1, one-quarter of the zygotes obtained in the F_3 generation carry in double dose a chromosome present singly in an individual in nature. Provided that these zygotes are not lethal, one can examine their morphological and physiological characters for possible genetic differences. The simplest tests distinguish completely sterile genotypes from fertile ones. Females or males from the test cultures are placed together with individuals of the opposite sex known to be fertile, and the numbers of cultures that do and do not produce progenies are noted. Table 4.4 shows, for natural populations of *Drosophila pseudoobscura* and *D. persimilis,* the frequencies of chromosomes that, though giving quasi-normal viability in homozygotes, make these homozygotes completely sterile as females or as males, but rarely have this effect in both sexes.

Temin (1966) in *Drosophila melanogaster* and Marinkovic (1967) in *D. pseudoobscura* made studies of the fecundity and fertility modifiers carried in natural populations. In the latter species, the average egg laying capacity of heterozygous females, between the sixth and the twelfth day after emergence from the pupae, is 207.5 ± 5.9 eggs, and that of females homozygous for second chromosomes permitting quasi-normal viability is 161.5 ± 5.7 eggs. Furthermore, only 44.9 ± 2.2 percent of the eggs of homozygous females develop to adult stage, compared to 69.9 ± 1.8 percent in the control; 5.1 percent of the homozygous females studied were completely sterile, owing to the combination of drastically lowered fecundity and low viability of the few eggs deposited.

One of the frequent effects of homozygosis for chromosomes extracted from natural populations of Drosophila is deceleration of development. In extreme cases the time required for egg-to-adult development is more than twice as long in homozygotes as in an average heterozygote. So great a retardation would probably be fatal in nature. Accelerations of development are relatively rare, though Marien (1958) was more successful in selecting for faster than for slower development. Among the less common variations are homozyogtes of giant and of dwarf body size. One homozygote in *Drosophila pseudoobscura* was so sensitive to ether that it died before its heterozygous siblings were anesthetized.

The Classical Model of the Genetic Population Structure

It has been pointed out above that normalizing natural selection is a conservative force. It perpetuates what has successfully passed the scrutiny of the environment and eliminates changes that impair Darwinian fitness. Although mutants presumably arise in all living species, most of them reduce the fitness of their carriers. However, a number of generations may intervene between the origin of a mutant and its elimination. This number is, on the average, smaller for more deleterious than for less deleterious mutants, and smaller for dominants than for recessives. This process of genetic elimination was termed "genetic death" by Muller (1950). This designation is acceptable if one keeps in mind that a "genetic death" does not always produce a cadaver. This should be remembered, for example, in estimating the "genetic deaths" expected in human populations over many generations as a result of radiation exposures. Weak or abnormal sexual drive, delay in sexual maturity, shorter reproductive period, lowered fecundity or complete sterility, and similar factors, cause genetic elimination just as effectively as differential mortality.

Muller (1950) and Haldane before him (1933, 1937) emphasized an important, and at first sight paradoxical, fact. Completely lethal or sterile mutants and mutants only mildly subvital cause the same amounts of genetic elimination, if the mutation rates that give rise to these variants are the same. The reason is that the inflow of a given kind of mutant into the gene pool must sooner or later match the outflow of the same mutant. We have seen above that the equilibrium frequency of a deleterious mutant in a population depends both on its fitness and on the mutation rate. The average number of generations that intervene between the appearance and the elimination of a mutant is a function of fitness. The number of eliminations (genetic deaths) is equal to the mutation rate for recessives and is double the mutation rate for dominants. Many mutants are incompletely recessive or—what amounts to the same thing—incompletely dominant. The rate of elimination will accordingly be between u and $2u$, where u is the mutation rate.

There are many gene loci that undergo mutation, and the aggregate mutation rate is quite considerable, $n\bar{u}$, where n is the number of loci and $\bar{u}$ the average mutation rate. Hence, mutation imposes a heavy genetic burden. Muller and Kaplan (1966) have given a most explicit statement of the classical model:

> Since even such tiny steps, so slightly and to the ordinary view imperceptibly altering the effectiveness of an individual gene, can affect fitness enough to become established, it follows that in the great majority of cases it is after all valid to speak of a "normal gene" and a "normal type". . . . The greatest part of the continuing genetic variation observed within a genetically united

population must usually have been caused by multiple genes each of which, mutating separately, has given rise to its own small mutational load before the mutants are eliminated by selection. . . . Of course, this variation is in evolution a very necessary "evil," since it allows natural selection a grasp by which in time of changed needs or opportunities the constitution of the population may be altered adaptively.

CHAPTER FIVE

BALANCING SELECTION AND CHROMOSOMAL POLYMORPHISMS

The Maintenance of Genetic Variability

Imagine a universe in which the environment is absolutely constant and uniform. In such a universe, evolution might culminate in an ideal adaptedness. Populations of every species would consist of genetically identical individuals, homozygous for all normal or wild-type genes, and free of what the classical model of genetic population structure regarded as "a very necessary evil" of genetic variability. Suppression of all mutability would be the climax of such evolution. It would have realized the *eidos,* the ideal types of human and other species, postulated more than two thousand years ago in Plato's philosophy. At least implicitly, the classical model conjures such perfect types as genetic ideals for each living species. Yet, as Mayr (1963) has rightly said, "The replacement of typological by population thinking is perhaps the greatest conceptual revolution that has taken place in biology." At any rate, there is hope that it is taking place.

In the world as it really is, the environment is neither constant nor uniform, and no genotype is a paragon of adaptedness in all environments. At any one time level, diverse genotypes are needed to exploit the environments varying in space. They are also needed to maintain the adaptedness to environments varying in time. Some genetic changes arising by mutation seem to be unconditionally harmful, that is, deleterious in all existing external and genetic environments. Such variability is an unavoidable evil. Other variability is necessary but not evil. As we have seen, normalizing selection keeps the level of "evil" variability as low as possible. Balancing selection is a complex of several selective processes that maintain, enhance, or regulate genetic variability, most of which is adaptively beneficial.

There are, then, two major kinds of genetic variability in natural populations: first, that maintained by mutation pressure and kept within bounds by normalizing selection; and, second, that maintained by balancing selection, though arising ultimately also by mutation. The latter variability, when sharply marked and discontinuous, is termed genetic polymorphism, defined as "the occurrence together in the same locality of two or more discontinuous forms of a species in such proportions that the rarest of them cannot be maintained merely by recurrent mutation" (Ford 1964, 1965).

There are several kinds of balancing selection. Of the two most important types, the relatively better known and more often discussed is heterotic balance, which is due to heterozygotes for certain gene alleles or gene complexes having fitness superior to that of the respective homozygotes. A less familiar form of balancing selection that is possibly even more important in nature than the heterotic type is diversifying (also called disruptive) selection. Most living species face a variety of environments, feed on different foods, grow in different soils, avoid or resist different enemies and parasites, etc. Diversifying selection favors different genotypes in different subenvironments or ecological niches; it occurs in experiments "when we maintain a single population by choosing more than one class of individuals to provide parents of each generation" (Thoday 1953, 1959).

Some diversifying selection is frequency-dependent. A genotype may be favored when rare and discriminated against when it becomes common or prevalent. The result will be a stable polymorphism. A frequency-dependent selection favoring the common and disfavoring the rare genotypes would, of course, lead to fixation of the favored genes and not to polymorphism. Diversifying and frequency-dependent natural selection will be discussed in Chapter 6.

Heterotic Balance

The life cycle of sexually reproducing organisms consists of alternating phases. One phase (diplophase, zygote) has twice as many chromosomes as the others (haplophase, gamete, sex cell). Many mutants are more or less deleterious both in the homozygous and in the heterozygous condition, that is, when present in double dose as well as in single dose, in the diplophase. Yet some, though only a minority, of genetic variants give rise to hybrid vigor, heterosis, high Darwinian fitness in the heterozygous condition, although the same variants reduce fitness when homozygous. The behavior of such variants in cross-fertilizing populations is very different from that of variants neutral or deleterious in heterozygotes and hence subject to normalizing selection.

Let the fitness of the heterozygote A_1A_2 be unity and of the two homozygotes less than unity. The operation of the heterotic balancing selection will then be as follows:

				Total
Genotype	A_1A_1	A_1A_2	A_2A_2	Population
Darwinian fitness	$1-s$	1	$1-t$	W
Initial frequency	p^2	$2pq$	q^2	1
Frequency after selection	$p^2 - sp^2$	$2pq$	$q^2 - tq^2$	$1 - sp^2 - tq^2$

The outcome of the selection will be, not fixation or elimination of either A_1 or A_2, but a genetic equilibrium at which both will occur in the population with predictable frequencies. Indeed, the rate of the gene frequency change per generation will be:

$$\Delta q = (q - tq^2)/(1 - sp^2 - tq^2) - q = pq(sp - tq)/(1 - sp^2 + tq^2)$$

At equilibrium no changes occur, the frequencies are constant, and Δp and Δq are zero. From the equation one obtains, then, $ps = qt$, and $p = t/(s + t)$ and $q = s/(s + t)$. When these frequencies of A_1 and A_2 are reached, a stable polymorphism is established. It should be noted that this form of natural selection will have some consequences that appear, at first sight, peculiar. It will cause retention in the population of all kinds of harmful genes, including those for crippling and even lethal hereditary diseases, provided only that the heterozygotes are at least slightly superior in fitness to the homozygotes. For example, suppose that A_2A_2 is lethal, and A_1A_1 is 95 percent as fit as the heterozygote A_1A_2. We have, then, $t = 1$ and $s = 0.05$. Equilibrium is reached when the lethal A_2 has the frequency $q = 0.05/(1 + 0.05) = 0.0476$. At equilibrium, the population will produce q^2, or 0.0023, that is, 23 per 10,000 lethal homozygotes.

The equilibrium frequencies of p and q depend only on the relative fitness of the two homozygotes, and not on the degree of heterosis. Thus, if both homozygotes are equally fit, the polymorphism is established at $p = q = 0.5$, regardless of whether the heterozygote is only slightly or is greatly superior to the homozygotes. The degree of heterotic superiority influences, not the equilibrium frequencies, but the number of generations needed to attain a stable equilibrium after the first appearance of a heterotic mutant or some environmental disturbance. It should also be noted that to possess a high Darwinian fitness the heterozygote need not be superior in all components of the adaptive value. Suppose, for example, that a heterozygote A_1A_2 is less viable but more fecund than A_1A_1, and more viable but less fecund than A_2A_2. If the product of relative viability and relative fecundity is higher in A_1A_2 than in A_1A_1 and A_2A_2, heterotic balanced polymorphism will be established and maintained.

Chromosomal Inversion Polymorphism in Drosophila

In natural populations of outbreeding sexual species, polymorphisms abound in all kinds of externally visible morphological, physiological, and biochemical traits. We choose as a paradigm a rather recondite trait, chromosomal inversions in Drosophila flies, because these polymorphisms have been extensively studied both in nature and in experiment.

Very early in the work of the T. H. Morgan school on *Drosophila melanogaster*, inversions, or C-factors as they were then called, were detected as suppressors of crossing over. Sturtevant (1926) first showed that one of the C-factors had the gene arrangement in a part of the genetic linkage map of a certain chromosome inverted. With the introduction in the early nineteen thirties of the technique of examination of the giant chromosomes in the larval salivary glands, detection and description of inversions and other chromosomal aberrations in Drosophila became easy and accurate.

Suppose that an individual has the gene arrangements *ABCDEFGHI* and *AEDCBFGHI* in two homologous chromosomes; when these chromosomes pair, they must form a loop, shown in the upper right corner of Fig. 5.1. A second inversion may occur in the same chromosome. The location of the second inversion may be outside the limits of the first: *AEDCBFGHI* → *AEDCBFHGI*. Such inversions are called independent. An individual heterozygous for *ABCDEFGHI* and *AEDCBFHGI* will have a double loop, shown second from the top in Fig. 5.1. The

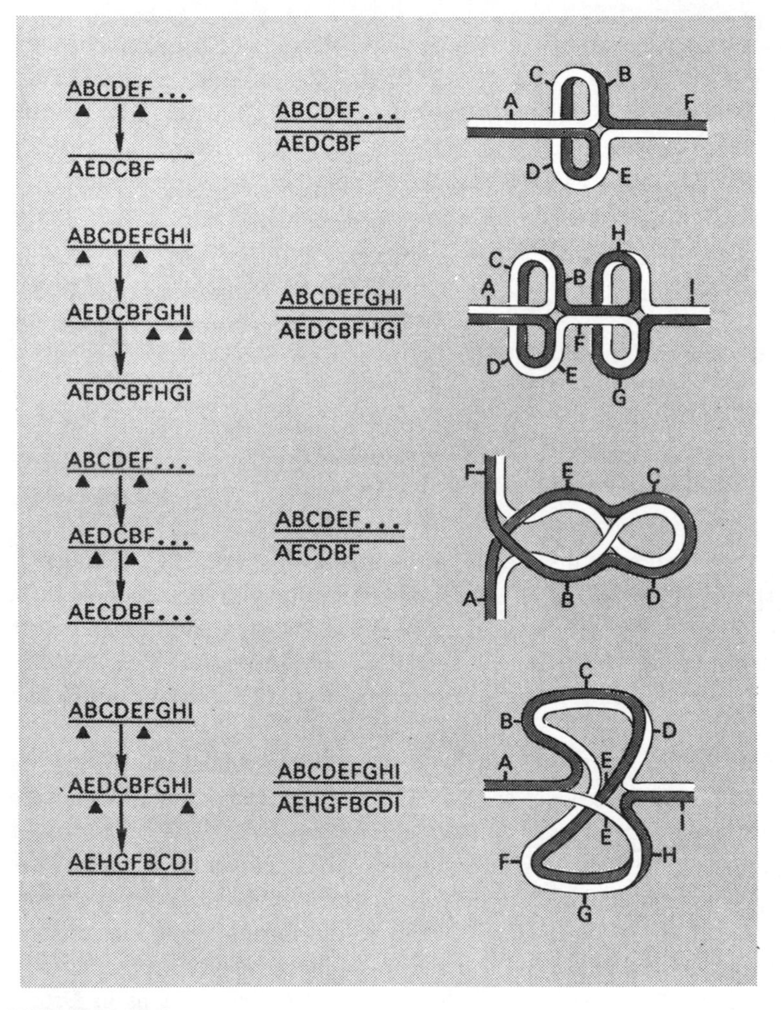

FIGURE 5.1

Chromosome pairing in the salivary gland cells of individuals heterozygous for inversions. Upper row—a single inversion; second from the top—two independent inversions; third from the top—two included inversions; lower row—overlapping inversions.

second inversion may occur inside the first, forming included inversions: *ABCDEFGHI* → *AEDCBFGHI* → *AECDBFGHI* (second from the bottom in Fig. 5.1). Finally, the second inversion may have one end inside and the other end outside the limits of the first; such inversions are overlapping: *ABCDEFGHI* $\rightarrow$ *AEDCBFGHI* $\rightarrow$ *AEHGFBCDI* (the lower right corner in Fig. 5.1).

Overlapping inversions have interesting properties. Suppose we observe in different strains the arrangements *ABCDEFGHI*, *AEDCBFGHI*, and *AEHGFBCDI*. The first can arise from the second or give rise to the second through a single inversion. The same is true for the second and the third. But the third can arise from the first, or vice versa, only through the second arrangement as the probable intermediate step in the line of descent. If only the first and the third gene arrangements are found in some species, it is probable that the second remains to be discovered or that it existed in the past. If all three are found, the phylogenetic relationships are $1 \rightarrow 2 \rightarrow 3$ or $3 \rightarrow 2 \rightarrow 1$ or $1 \leftarrow 2 \rightarrow 3$, but not $1 \rightleftharpoons 3$. The existence of previously unknown gene arrangements was predicted in two different species, *Drosophila pseudoobscura* and *D. azteca*, and the predictions were verified by subsequent findings in nature (Dobzhansky and Sturtevant 1938 and Dobzhansky 1941).

Figure 5.2 shows a phylogenetic chart of gene arrangements found in one of the chromosomes (the third) in natural populations of two closely related species, *D. pseudoobscura* and *D. persimilis*. Only one of the gene arrangements, the Standard, occurs in both species, and only one, the Hypothetical, has been predicted as a necessary "missing link" but not yet actually found. Any two arrangements connected in Fig. 5.2 by a single arrow give a single inversion loop in the giant chromosomes of the salivary glands of heterozygous larvae. The four gene arrangements in the central portion of the diagram, the Standard, the Hypothetical, the Tree Line, and the Santa Cruz, may plausibly be supposed to be the ancestral ones, and the remainder to be derived from them as shown by the arrows.

None of the gene arrangements occurs over the entire distribution area of its species; and in no natural population is the complete collection of arrangements found. The geographic distribution of the gene arrangements will be discussed in Chapter 9. In some localities as many as eight arrangements occur together, and structural inversion homozygotes (flies having two chromosomes with the same gene arrangements) and inversion heterozygotes (flies with two chromosomes of a pair having different gene arrangements) are encountered in nature. The chromosomal inversions thus give rise to a remarkable polymorphism in fly populations (Dobzhansky and Epling 1944).

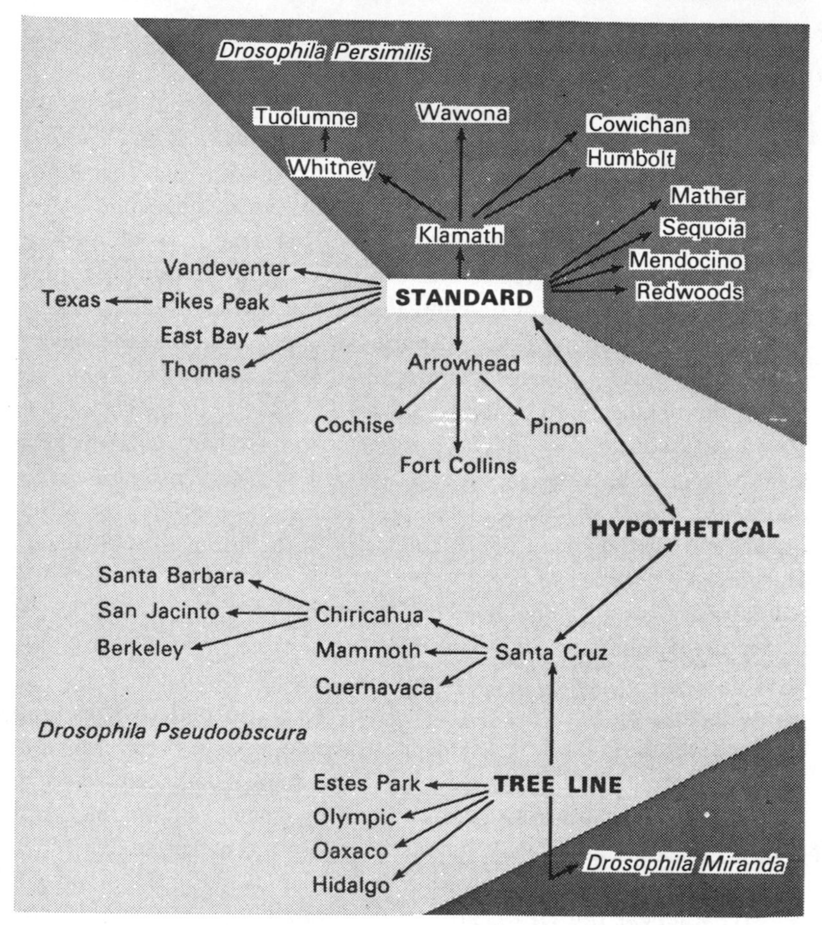

FIGURE 5.2

Phylogenetic relationships of the gene arrangements in the third chromosomes of Drosophila pseudoobscura, D. persimilis, and D. miranda.

Chromosomal inversion polymorphisms have been observed in natural populations of many species of Drosophila. Carson (personal communications) finds that, of the 68 species endemic to the Hawaiian archipelago and studied by him, about one-third have at least one polymorphic inversion. The proportion is higher among continental species, especially in the subgenus Sophophora, in which chromosomally monomorphic species are exceptional. The fairly extensive older literature has been reviewed by da Cunha (1955), Dobzhansky (1961), and Carson (1965). Among the most significant recent additions are the studies of Brncic and Koref-Santibañez (1964, 1965) on species of the *mesophragmatica* group; of Brncic (1966) on *Drosophila flavopilosa,* which is specialized to live in the flowers of a solanaceous plant, *Cestrum parqui*; of Prevosti (1964), Krimbas (1964), and Sperlich and Feuerbach (1966) on *D. subobscura*; of Stalker (1960, 1964a,b, 1965) on species of the *melanica* group; of Miller and Voelker (1968) on species of the *affinis* group; and of Carson, Clayton, and Stalker (1967) on species endemic to the Hawaiian Islands.

The polymorphism assumes quite different characters in different species of Drosophila; these variations are as yet not adequately understood, and many intriguing problems await investigation. Some species of Drosophila have become widespread geographically, owing to their association with man, and particularly with the fruit and vegetable trades. In most of such "domestic" species (*D. melanogaster, ananassae, immigrans, hydei, busckii*) the chromsomal polymorphism is little differentiated geographically, so that populations from remote and climatically unlike countries carry the same chromosomal variants, often with similar frequencies. In three domestic species (*D. simulans, virilis, repleta*) chromosomal polymorphisms are absent, and the populations everywhere are chromosomally monomorphic.

What the causal connections are between the uniformity or lack of chromosomal polymorphism and the ability of a species to adapt to man-made environments is a matter of speculation. Perhaps these chromosomally monomorphic species have reached some kind of "general-purpose genotypes" (Baker 1965), which confer upon them wide ranges of environmental tolerance. "Wild" species of Drosophila, which are not associated with man, show again a variety of conditions. Many, though not all of them, are chromosomally highly polymorphic, and the polymorphism is notably differentiated geographically (Chapter 9). In *Drosophila willistoni, paulistorum, robusta, subobscura,* and some others, inverted sections occur in every chromosome of the set, longer chromosomes having more inversions, approximately in proportion to their length, than shorter chromosomes. In other species, both wild and domestic, of which *D. pseudoobscura, persimilis, nebulosa,* and *busckii* are examples, the inversions are concentrated mostly in a single chromosome, which may or may not be the longest in the chromosomal set. The possibility that the variable chromosomes are especially prone to undergo breakage was ruled out for *D. pseudoobscura* long ago by Helfer (1941). Other conjectures have also been advanced, but none of them is supported by compelling evidence.

Inversion Polymorphism as an Adaptive Trait

Individuals of chromosomally polymorphic species of Drosophila look outwardly identical, regardless of the gene arrangements in their chromosomes. Inversion heterozygosis can be detected in two ways. First, it causes a suppression, partial or complete, of the detectable gene recombination in the chromosomes involved. The second—and the easier—way of detection is to examine the gene arrangement cytologically by looking at the chromosomes under the microscope. The absence of visible changes in the external appearance of a fly with this trait led originally to the belief that inversions are adaptively neutral, a misinterpretation found in the second edition of the ancestor of this book (Dobzhansky 1941). The following facts have resulted in abandonment of this view.

If samples of a population of a species are taken repeatedly in the same locality, the relative frequencies of some chromosomes with different gene arrangements can be seen to undergo cyclic seasonal changes. Figure 5.3 summarizes the observations on populations of *Drosophila pseudoobscura* in a locality on Mount San Jacinto in California. These studies were conducted from 1939 to 1946, but later observations in the same and in other localities disclosed similar situations (Dobzhansky, Anderson, and Pavlovsky 1966 and references therein). The three commonest gene arrangements in this population are Standard (ST), Arrowhead (AR), and Chiricahua (CH). The frequency of ST decreases, and that of CH increases, from March to June; the opposite change takes place during the hot season, from June to August. Dubinin and Tiniakov (1945) and Borisov (1969) observed seasonal changes in the incidence of inversions also in populations of *D. funebris* near Moscow, Russia. We are witnessing here changes in the gene pool of populations; they are evolutionary changes by definition.

The seasonal genetic changes seemed startling at the time of their discovery, because their rapidity necessitated the assumption of very high

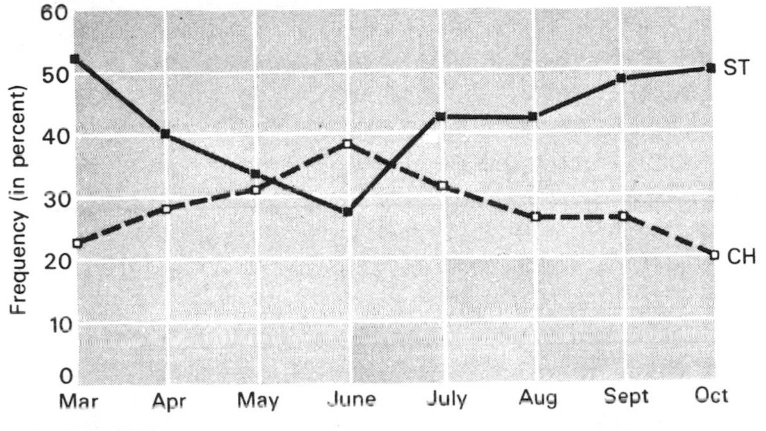

FIGURE 5.3

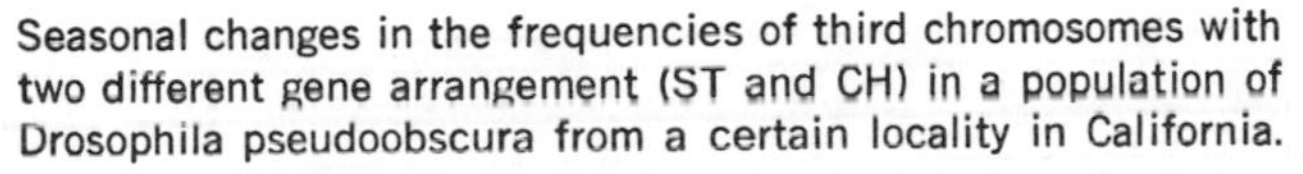
Seasonal changes in the frequencies of third chromosomes with two different gene arrangement (ST and CH) in a population of Drosophila pseudoobscura from a certain locality in California.

selection pressures. As Ford (1964) rightly remarked, "Their magnitude would indeed be far beyond that envisaged twenty or twenty-five years ago, but quite in keeping with that now being recognized as usual in wild populations." Except for normalizing selection, which counteracts the spread of hereditary malformations, natural selection was believed to act too slowly to be noticeable within a human lifetime. Evidence of strong selection was obtained, however, in experimental populations, maintained for several to many generations in population cages (see Chapter 4). As described previously, a mixture of flies with known proportions of chromosomal or genic variants is introduced into the cage. The population reaches a maximum size consistent with the amount of food given, usually within a single generation, and samples of the population are taken from time to time for study.

An example of the results obtained in experimental populations of *Drosophila pseudoobscura* appears in Fig. 5.4. Four populations were started with about 10 percent Standard (ST) and 90 percent Chiricahua (CH) third chromosomes derived from the population of a certain locality in California. Within four months, a period that corresponds under the conditions of the experiment to about four generations, the frequency of ST approximately trebled. Thereafter it rose more slowly, and reached an apparently stable equilibrium at close to 70 percent ST and 30 percent CH. Making certain simplifying assumptions, the most

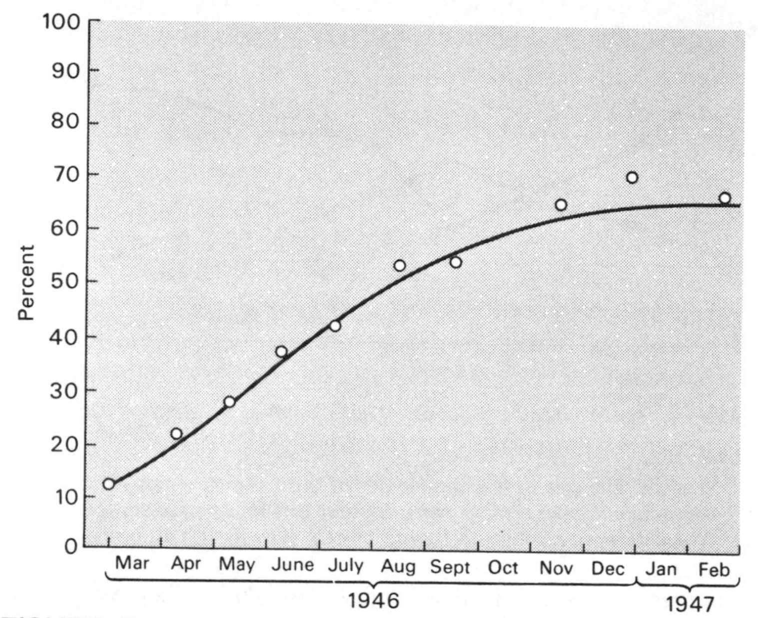

FIGURE 5.4

Frequency changes of third chromosomes with different gene arrangements in an experimental laboratory population of Drosophila pseudoobscura. The curve shows the frequencies of chromosomes with ST, competing with those with CH gene arrangements.

questionable of which is that the fitness of a genotype is independent of its frequency in the population, we can estimate from the slopes of the selection curves in Fig. 5.4 the fitnesses of the inversion homo- and heterozygotes (Wright and Dobzhansky 1946 and Dobzhansky and Pavlovsky 1953). The estimates turn out, averaging the results in the four experimental populations, as follows:

Chromosome	*Fitness (W)*	*Selection Coefficient (s)*
ST/CH	1	0
ST/ST	0.89	0.11
CH/CH	0.41	0.59

The differences between the selective values of the naturally occurring chromosomal variants are often quite striking. For example, Anderson, Oshima, et al. (1968) experimented with populations of *D. pseudoobscura* that contained third chromosomes with five different gene arrangements: ST, AR, CH, TL, and PP. The estimates of fitness obtained in one of the populations turned out as follows:

Heterokaryotypes				*Homokaryotypes*	
ST/AR	1.00	AR/PP	0.33	AR/AR	0.58
ST/CH	0.76	TL/PP	0.27	PP/PP	0.45
ST/PP	0.61	AR/CH	0.26	ST/ST	0.35
ST/TL	0.51	CH/PP	0.26	TL/TL	0.35
AR/TL	0.45	CH/TL	0.19	CH/CH	0.26

Although these estimates are subject to large experimental and sampling errors, it is notable that the three fittest combinations are heterozygotes. All but one homozygote, and six out of ten heterozygotes, are less than half as fit as the fittest heterozygote (ST/AR). The selective forces acting on these populations are powerful indeed.

The Superior Fitness of Heterokaryotypes

Although the evidence is unequivocal that at least some heterokaryotypes are superior in fitness to homokaryotypes, comparatively little is known about the physiological and ecological sources of this superiority. Several possibilities come to mind: differential survival between the egg and the adult stage, higher fecundity, greater sexual activity, superior longevity, or any combination of these. For the ST and CH gene arrangements in *Drosophila pseudoobscura*, experiments in population cages indicate the fitness sequence, ST/CH > ST/ST > CH/CH. Moos (1955) found the preadult viability to be ST/CH = ST/ST > CH/CH; the development rate ST/CH > ST/ST = CH/CH; the fecundity at 25°C ST/CH = ST/ST > CH/CH, and at 16°C ST/CH > ST/ST > CH/CH; and the longevity ST/CH > ST/ST = CH/CH. The overall fitness, computed from Moos's observations, taking ST/CH to be unity, is ST/ST = 1.08 and CH/CH = 0.44. Estimates from the changes observed in the frequencies of the chromosomes in the population cages are ST/CH = 1, ST/ST = 0.90, CH/CH = 0.41. The agreement between the two sets of estimates is probably about as close as could be expected. Brncic, Koref Santibañez, et al. (1969) found that the rate of development of inversion heterozygotes in *D. pavani* is faster than that of homozygotes.

Spiess and his colleagues (Spiess and Langer 1964a,b, Spiess, Langer, and Spiess 1966, and E. B. Spiess and L. D. Spiess 1967) and Kaul and Parsons (1965) have made careful studies of the mating propensity of the carriers of different karyotypes in *Drosophila pseudoobscura* and *D. persimilis*. The picture revealed is quite complex: relative mating speeds are different for males and for females, they are not the same at different temperatures, they depend on the age of the flies, etc. Certain facts, however, stand out. Male heterozygotes are consistently superior in mating speed when compared to male homozygotes, whereas in females a corresponding superiority is less consistent or is absent. Males seem to play more active and females more passive roles in courtship, although it is up to the female to accept or reject a courting male. As an illustration, consider the differences in the mating propensity index for karyotypes of *D. pseudoobscura*:

Karyotypes	*Males*	*Females*
AR/ST – AR/AR	+17.3	+ 3.7
AR/TL – AR/AR	+17.6	− 7.2
AR/CH – AR/AR	+ 1.3	+11.7
AP/PP – AR/AR	+ 2.9	−14.8
ST/TL – ST/ST	+24.2	− 3.8
ST/CH – ST/ST	+ 5.1	+ 6.7
ST/PP – ST/ST	+10.6	+ 5.2
Average	+11.3	+ 0.2

Differential mortality between the egg and the adult stage may change the proportion of karyotypes among the eggs compared to that among adult flies. To test this possibility, samples of eggs deposited by the flies in the population cages are removed and raised under optimal conditions, to enable all or most of them to develop into grown-up larvae and adults. Examination of the chromosomes reveals the karyotypes to be present in proportions demanded by the binomial square rule, p^2 ST/ST : $2pq$ ST/CH : q^2 CH/CH. The flies mate at random with respect to the chromosomal type. Now, samples are taken of the adult flies developed in the population cages, where there is stringent competition of larvae for a limited food supply, and the chromosomal constitution of these flies is determined. Among the adult flies, the heterozygotes are more, and the homozygotes are less, numerous than they should be according to the binomial square rule (Dobzhansky 1947).

In nature, as in experimental populations, differential mortality favoring the heterozygotes could maintain the polymorphism. Dobzhansky and Levene (1948) collected females of *Drosophila pseudoobscura* inseminated by males in their natural habitats and allowed them to produce offspring in the laboratory, under optimal conditions. In these offspring, the inversion homozygotes and heterozygotes were present with frequencies conforming to the binomial square rule. A small but significant excess of heterozygotes, and a corresponding deficiency of homozygotes, were found, however, among the adult male flies captured outdoors. This is not an invariable rule, and other polymorphic populations show no evidence of differential mortality.

The color polymorphism in the marine copepod *Tisbe reticulata*, though not known to be connected with chromosomal rearrangements, may be considered here. The studies of Battaglia (1958, 1964; Battaglia, Lazzaretto, and Malesani-Tajoli 1966, and references therein) have disclosed a situation remarkably paralleling that in Drosophila. Four color forms, determined by three alleles of a gene, occur in the lagoon of Venice. They are the recessive *trifasciata* (vv), the dominants *violacea* (V^VV^V or V^Vv) and *maculata* (V^MV^M or V^Mv), and the double dominant *violacea-maculata* (V^VV^M). The heterozygotes are superior in viability to the homozygotes. This is particularly clear in the progenies of *violacea-maculata*, which should theoretically produce 1 : 2 : 1 ratios of the homo- and heterozygotes. They actually produce, depending on the degree of crowding in the cultures, ratios from which the following viability estimates are derived:

Crowding	V^VV^V	V^VV^M	V^MV^M
Low	0.89	1	0.90
Intermediate	0.67	1	0.76
High	0.66	1	0.62

If the viability of *violacea-maculata* is taken to be unity, the viabilities of the heterozygotes and the homozygotes for the recessive allele are:

$$V^Vv = 0.94, \qquad V^Mv = 0.92, \qquad vv = 0.85$$

The relative frequencies of the color forms differ in different parts of the lagoon of Venice, and Battaglia adduces evidence that these variations are responses to different salinities of the water. In experimental cultures only *violacea* survives at low salinity of 0.15 percent; at 0.25 percent the viability sequence is *maculata* > *violacea* > *trifasciata*, whereas at 0.45 percent it is *trifasciata* > *maculata* > *violacea*.

Adaptedness of Polymorphic and Monomorphic Populations

Consider again a population of *Drosophila pseudoobscura* polymorphic for ST and CH third chromosomes. The estimates of the Darwinian fitness of the three karyotypes obtained above are ST/CH = 1, ST/ST = 0.89, and CH/CH = 0.41. Note that these estimates were obtained in populations kept at 25°C, whereas at 16°C the three karyotypes were not detectably different in fitness. The average fitness of a population, $\overline{W}$, would be greatest, 1.00, if it consisted of heterokaryotypes only. With sexual reproduction this situation cannot endure for more than a single generation, since homokaryotypes will inevitably appear. (Populations consisting only of heterozygotes are, however, possible with asexual reproduction or parthenogenesis; see Chapter 12.) It is easy to show that $\overline{W}$ is maximized when ST and CH reach equilibrium values of 0.84 and 0.16, respectively, by calculating the frequencies of the three karyotypes for different values of p and q, multiplying them by fitness, and computing the average:

Chromosome Frequency ST = p, CH = q	*Karyotype Frequency* ST/CH	ST/ST	CH/CH	*Mean Fitness* ($\overline{W}$)
$p = 0.99, q = 0.01$	0.020	0.980	>0.000	0.892
$p = 0.10, q = 0.90$	0.180	0.010	0.810	0.521
$p = 0.84, q = 0.16$	0.269	0.706	0.026	0.908

The question that logically arises is, in what way is a population with a higher $\overline{W}$ better off than one with a lower $\overline{W}$ value? Apparently flourishing populations homozygous (monomorphic) for CH can easily be maintained in population cages. Beardmore, Dobzhansky, and Pavlovsky (1960) compared the productivity of monomorphic AR and CH and of polymorphic AR + CH populations kept in laboratory population cages. These cages have fifteen cups with culture medium, a fresh cup being inserted on alternate days or at other desired intervals. A polymorphic AR + CH population produced, on the average, 327 flies per cup, monomorphic AR 230 flies, and monomorphic CH only 202 flies. In terms of weight (biomass) the productivities were 261 milligrams for the polymorphic and 190 and 164 milligrams, respectively, for the two comparable monomorphic populations. An individual fly was, if anything, slightly smaller and lighter in the polymorphic than in the monomorphic populations. Nevertheless, the conclusion is inevitable that polymorphic populations can generate more flies and a greater biomass than monomorphic ones for the same amount of food.

The experiments just described were made at a temperature of 25°C. Battaglia and Smith (1961) obtained the same results at 16°C. Far from being a confirmation of the earlier work, their results were rather unexpected; as stated previously, if one judges by the lack of changes in the relative frequencies of the chromosomal types at 16°, the Darwinian fitnesses of the karyotypes at that temperature are at least not very different from each other.

A statistic, r_m, the innate capacity for increase or the intrinsic rate of natural increase (Andrewartha and Birch 1954), has been used by Dobzhansky, Lewontin, and Pavlovsky (1964) and by Ohba (1967) to estimate the adaptedness of polymorphic and monomorphic populations of *Drosophila pseudoobscura*. What this statistic measures (to oversimplify the matter considerably) is the rate of growth of a population when the quantities of food, space, and other resources are not limiting, and predators and parasites are excluded. Realistically, such a situation may occur in Drosophila when a small number of overwintering individuals encounter amounts of food that become limiting only after the population increases manifold. Other examples are readily imaginable. The value r_m is obtained from equation:

$$\int_0^\infty e^{-r_m x} l_x m_x \, \sigma x = 1$$

where e is the base of natural logarithms, l_x the probability at birth of being alive at age x, and m_x the number of female offspring produced in unit time by a female aged x. With an animal like Drosophila, one finds out the proportion of eggs that survive to become adults, the time the development takes, the adult longevity, and the numbers of eggs deposited by a female per day at different ages. In the experiments of Dobzhansky, Lewontin, and Pavlovsky, the tests were made deliberately in a rather stringent environment, since the adults tested developed in crowded population cages. The parameters r_m, at 25°C, were found for two polymorphic and four monomorphic populations:

AR + CH	Polymorphic	0.220
AR + PP	Polymorphic	0.214
AR	Monomorphic	0.207
AR	Monomorphic	0.205
CH	Monomorphic	0.192
PP	Monomorphic	0.170

In Ohba's experiments (1967) the environment was made as close as possible to the optimal state. The r_m values were higher, but the polymorphic populations were no longer superior:

AR + CH	Polymorphic	0.269 and 0.266
AR	Monomorphic	0.275
CH	Monomorphic	0.263

Darwinian fitness is a measure of the reproductive success of the carriers of a given genotype in relation to that of the carriers of other genotypes. It is a comparative measure; the fitness of a population is meaningful if this population competes with others. The innate capacity for increase, on the other hand, is an absolute measure of one aspect of the adaptedness of a population under a given set of environmental conditions. It is not surprising that Darwinian fitness and adaptedness tend to be correlated, and yet it should be realized that they measure different qualities or properties of life (Dobzhansky 1968a,b).

A knowledge of the adaptedness of populations monomorphic for certain genotypes or karyotypes may or may not predict the Darwinian fitness of the genotypes in a polymorphic population, or vice versa. Ayala (1969a) has estimated the abilities of populations of *D. pseudoobscura* monomorphic and polymorphic for AR and CH gene arrangements in the third chromosome to compete with another species, *D. serrata.* A serial transfer technique has been used. Known numbers of adults of both species are placed in a culture bottle, allowed to deposit eggs, and transferred to fresh cultures at weekly intervals. When young adults emerge in the cultures where the eggs were deposited, the flies are etherized, counted, and added to the bottle with the adult and ovipositing individuals. The numbers and proportions of the two species are determined weekly. *Drosophila serrata* is a stronger competitor, and its proportions gradually increase. The fitness of *D. serrata* relative to monomorphic and polymorphic populations of *D. pseudoobscura* was found to be as follows:

Monomorphic AR	1.74
Polymorphic AR + CH	1.62
Monomorphic CH	1.95

Polymorphic Inversions as Supergenes

Despite their outward uniformity, representatives of the same species of Drosophila with different gene arrangements in their chromosomes may be perceptibly, even strikingly, different in their adaptive properties. The nature and the origin of these differences pose challenging questions. As pointed out above, the most prominent genetic effect of heterozygosis for inversions is suppression of gene recombination in the chromosomes that carry them. If the inverted and the noninverted chromosomes differed in a single gene, the suppression of recombination would be neither useful nor harmful. The suppression is beneficial, however, if the chromosomes carry supergenes, that is, complexes of linked genes that are favorable in some particular combinations but not in others. Suppose that the chromosomes $A_1B_1C_1D_1$ and $A_2B_2C_2D_2$ yield a highly fit heterozygote, but the recombination products, such as $A_1B_1C_2D_2$ and others, are inferior in fitness. Such mutually concordant gene complexes are *coadapted* to each other. Inversions binding together coadapted gene complexes will be favored by natural selection.

Experimental validation of the coadapted supergene hypothesis was achieved through studies of chromosomes of *Drosophila pseudoobscura* of different geographic origins. It will be shown in Chapter 7 that some of the gene arrangements are geographically widely distributed. Experimental populations may, then, be planned in two ways: the competing chromosomes may be derived from flies collected either in the same locality or in more or less remote localities. With chromosomes of geographically uniform origin, experiments like the one depicted in Fig. 8.4A give repeatable results; provided that the environment is adequately controlled, the slope of the selection curve, the equilibrium level, and the fitness estimates are nearly the same.

With chromosomes of geographically unlike derivations, however, the outcome becomes strangely unpredictable. Figure 8.4B shows the results for four experimental populations, all started simultaneously with the four the results of which are depicted in Fig. 8.4A. The initial frequencies were again 20 percent ST chromosomes from California, and 80 percent CH chromosomes, but this time from Mexican localities. Only one of the populations, that symbolized by triangles, achieved equilibrium. The three others followed erratic courses, tending toward fixation of ST and elimination of CH chromosomes. Calculations of the fitness of the three karyotypes give irregular results, the heterozygote (heterokaryotype) ST/CH no longer being the fittest. The CH chromosomes from California and from Mexico look identical under the microscope, and yet the experiment shows that they must carry different complexes of genes.

Wallace (1954) made the observation that "triads" of overlapping inversions rarely occur with high frequencies in the same population of *Drosophila pseudoobscura*. The following gene arrangements would constitute a triad: $A_1B_1C_1D_1E_1F_1G_1H_1$, $A_2E_2D_2C_2B_2F_2G_2H_2$, and $A_3E_3D_3G_3F_3B_3C_3H_3$. Suppose now that the heterokaryotype $B_1C_1D_1E_1/E_2D_2C_2B_2$ has a high fitness, and so does $C_2B_2F_2G_2/G_3F_3B_3C_3$. The part of the chromosome $F_1G_1H_1/F_2G_2H_2$ is, however, free to combine in the first heterozygote, and the part $A_2E_2D_2/A_3E_3D_3$ can undergo free crossing over in the second, thus disrupting the coadaptation. No such disruption will occur in the heterokaryotype with a double inversion (the first and the third of the gene arrangements above). Although Levitan, Carson, and Stalker (1954) found the rule of avoidance of triads inapplicable to *D. robusta,* it seems to fit the situation in several other species.

A more direct demonstration of coadapted supergenes carried in inversions was made in *Drosophila willistoni* and *D. paulistorum* (Dobzhansky and Pavlovsky 1958). In some South American populations of these species, more than half of the individuals found in nature are heterozygous for certain inversions. High frequencies of the inversion heterozygotes are maintained also in experimental populations set up in the laboratory with strains from any natural locality. Quite different is the behavior of populations started with hybrids between strains derived from localities some hundreds of kilometers apart. In such populations the inversion heterozygotes become much less frequent than in the parental populations. The interpretation of these findings is simple. In each locality, the chromosomes with different gene arrangements are coadapted to yield highly fit heterozygotes. The chromosomes of the inhabitants of geographically remote localities need not be coadapted, since they do not form heterozygotes, except in contrived laboratory populations. In these artificial populations, chromosomes with the same gene arrangements but with different complexes of genes can undergo recombination freely, thus disrupting the coadapted supergenes.

There is no way to tell in how many loci the coadapted supergenes, such as those locked in the ST and CH chromosomes of *Drosophila pseudoobscura*, may differ. An inversion, to be retained in a population in which it first arises and then to spread to other populations of a species, should presumably possess some adaptive advantage. One possible source of such an advantage is position effect (Sperlich 1958, 1966a,b). The action of a gene in the development may depend not only on its own structure but also on its neighbors in the linear series of genes in a chromosome (see Lewis 1967 for a review). It has been known for a long time that inversions and translocations that arise by mutation in laboratory experiments are often accompanied by lethal or visible effects resembling gene mutations. Hence it is conceivable that the inversions found in nature are selectively favorable because of the changes induced by position effects. The difficulty is that this hypothesis demands a coincidence of at least two rare events—mutational origin of an inversion and of adaptively valuable position effects.

A more plausible hypothesis is that a newly arisen inversion may be favored by natural selection if it causes an adaptively valuable linkage disequilibrium, that is, if it binds together two or more genes in the same chromosome that give fit heterozygotes with other chromosomes present in the same breeding population. Variant genes present in natural populations do produce positive or negative fitness effects, depending on the other genes with which they are associated in the same chromosome—this has been demonstrated, for example, by the experiments in synthetic lethals and the release of genetic variability by recombination (Chapter 3). Crossing over and recombination of linked genes are undesirable when they break up favorable gene associations.

Prakash and Lewontin (1968) have given most direct evidence of coadaptation and linkage disequilibrium in third-chromosome inversions in *Drosophila pseudoobscura*. It can be seen in Fig 5.2 that the gene arrangements in this chromosome belong to two groups or "phylads," those descended from Standard and those derived from Santa Cruz. Two genes, one responsible for the larval protein Pt-10 and the other for the enzyme alpha-amylase, are each represented by different alleles in the two phylads. The geological age of the phylads is not known, but Prakash and Lewontin are on safe ground in concluding that "different inversions are genetically differentiated and the differentiation has been maintained over several million generations by natural selection."

Vann (1966) and Sperlich (1966a) submitted the coadaptation hypothesis to different tests. Homozygous lines of *Drosophila melanogaster* and *D. pseudoobscura* were obtained by inbreeding or by special crosses with marker genes. Inversions and translocations were then induced by X-ray treatment. Most of these chromosomal aberrations were lost within a small number of generations in experimental populations. They were retained for longer periods, or established balanced polymorphisms, when the homozygous lines carrying them were outcrossed to various more or less unrelated lines of the same species. The heterozygotes for an inverted chromosome and one with the original gene arrangement were then also heterozygotes for some, presumably fairly numerous, genes. Such heterozygotes had, at least in some instances, heterotic properties.

The problems of the adaptive consequences of linkage disequilibrium, and particularly of the origin of inversion polymorphisms, have also been studied by mathematical analysis and by computer simulation. Papers of Lewontin (1967b), Fraser and Burnell (1967), and Bodmer and Felsenstein (1967) give critical summaries and further references to the pertinent literature.

Inversion Polymorphism in Organisms Other than Drosophila

Inversion polymorphisms are particularly frequent in natural populations of Drosophilidae and related families of the order of flies (Diptera). This is probably not accidental. Male meiosis in these insects does not involve the formation of chiasmata between paired homologous chromosomes, there is no crossing over, and blocks of genes in the same chromosome are completely linked. Chiasmata and recombination of linked genes do occur at meiosis in females. The consequences of chiasma formation are different for paracentric inversions (inversions of chromosome sections not including the centromere) and for pericentric inversions (sections including the centromere).

The situation with a paracentric inversion is shown schematically in Fig. 5.5. Of the four chromosomes resulting from the meiotic divisions, one has two centromeres, one has none, and two are normal non-cross-over chromosomes. It happens that one of the normal chromosomes is always included in the egg nucleus, while the other three are eliminated into the polar bodies. The fertility of females, as well as of males heterozygous for paracentric inversions, is consequently undiminished in comparison to that of individuals free of inversions. Inversion heterozygosis being per se a neutral trait, a small amount of hybrid vigor suffices to maintain the polymorphism.

The situation is different with pericentric inversions. Here chiasmata inside these inversions give rise to chromosomes lacking some genes and carrying other genes in excess, and consequently to a partial sterility of heterozygous females (Fig. 5.6).

Inversion polymorphism is, nevertheless, not restricted to forms lacking chiasmata in male meiosis. Giant chromosomes of larval salivary gland cells were discovered originally, not in Drosophila, but in midges of the family Chironomidae. Many species of this family are highly polymorphic for inversions (Keyl 1962, Martin 1965, Blaylock 1966, and references therein). Yet chiasmata are formed in male meiosis in Chironomus. Beerman (1956) found the solution to this puzzle—the fertility of male inversion heterozygotes in Chironomus is not reduced, presumably because the sperms containing abnormal chromatids fail to function. Inversion polymorphisms have also been observed in natural populations of mosquitoes (Kitzmiller 1967, and Kitzmiller, Frizzi, and Baker 1967), of simuliids or black flies (Rothfels 1956, Pasternak 1964, and Grinchuk 1967), of cecidomyids (Kraczkiewicz 1950), of Liriomyza (Mainx, Fiala, and Kogerer 1956), of *Phryne cincta* (Wolf 1968), and of other flies.

In organisms that lack giant chromosomes in the salivary glands, the detection of inversions is difficult and may be impossible. Crossing over within a paracentric inversion gives, as shown in Fig. 5.5, a dicentric and an acentric chromatid, which appear at the anaphase of the first meiotic division as a "bridge" and as an acentric fragment lagging between the disjoining groups of chromosomes. Although chromosome bridges and fragments may arise also because of abnormal "stickiness" of the chromosomes (Newman 1966), their formation is usually a reliable indicator of inversion heterozygosity. Such heterozygosity has been detected in populations of many species of plants; classical examples are *Paris quadrifolia* (Geitler 1938) and species of

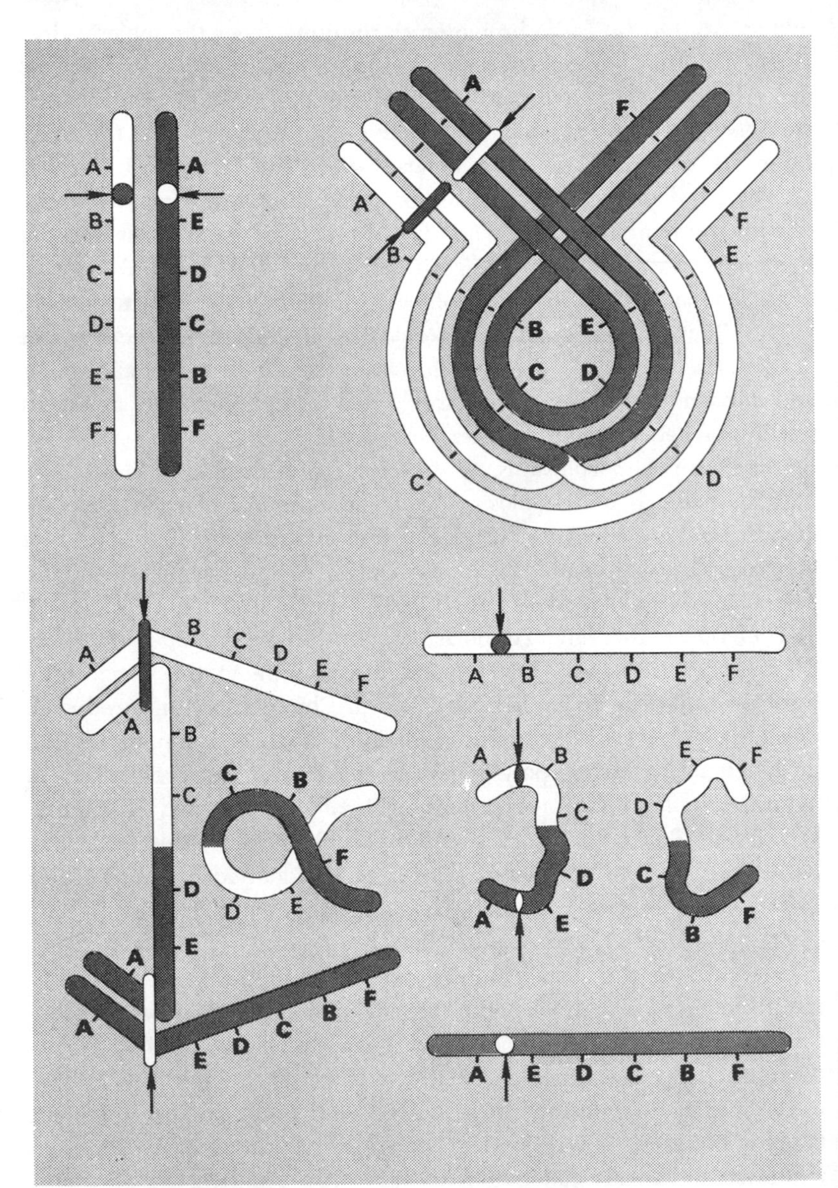

FIGURE 5.5

Crossing over in a heterozygote for a paracentric inversion. The centromeres are marked by arrows.

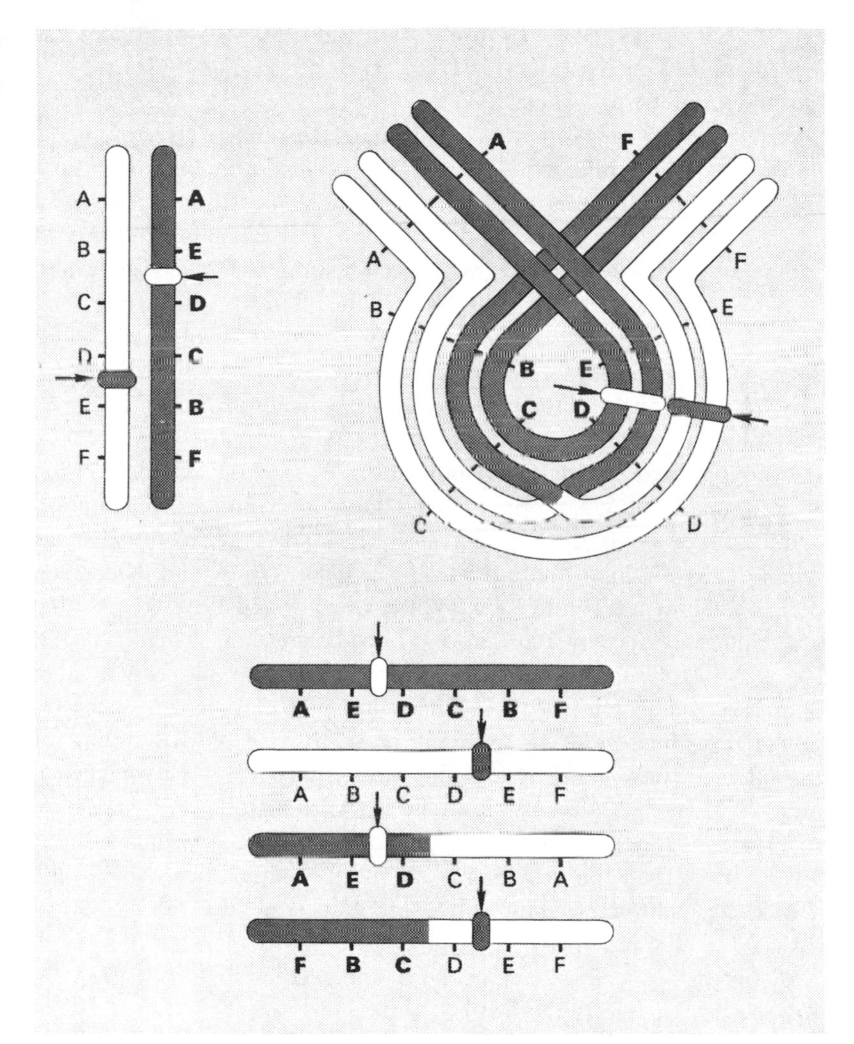

FIGURE 5.6

Crossing over in a heterozygote for a pericentric inversion. The centromeres are marked by arrows.

Paeonia (Stebbins 1939, Walters 1942, and Marquardt 1952), of Polygonatum (Suomalainen 1947b), and of Trillium (Haga and Kurabayashi 1954, and Haga 1956).

On the other hand, the absence of anaphase bridges and fragments at meiosis is not valid evidence of lack of inversion heterozygosity. This is true simply because chiasmata in the chromosome parts outside the inverted segments do not result in chromosome bridges and fragments; inversion heterozygosity, which suppresses chiasma formation within the inversions, will be undetectable by observations on meiotic division figures. Hence, although no instances of bridges and fragments in human meiosis have been securely established as yet, it is by no means certain that inversion polymorphism is absent or even rare in human populations.

Although chromosomal polymorphisms in natural populations of several species of grasshoppers have long been known, our understanding of them is due to the masterly studies of White and his collaborators (White 1949, 1954, 1957, 1968; White and Nickerson 1951; Lewontin and White 1960; White, Cheney, and Key 1963; White, Lewontin, and Andrew 1963; and White, Carson, and Cheney 1964). The polymorphisms manifest themselves mainly in the presence at meiosis of unequal bivalents, which consist of chromosome partners having the centromeres in different positions, one of them more or less median (metacentric) and the other subterminal (acrocentric). Such centromere shifts are best interpreted as due to pericentric inversions, in which the two breaks in the chromosome occur at different distances from the centromere. A possible alternative explanation is that the centromeres themselves are transposed to different locations in the chromosome.

At any rate, no chiasmata occur in the sections between the centromeres, and therefore the fertility of the heterozygotes is not reduced. In some instances, evidence has been obtained that inversion heterozygotes are in fact more viable than the corresponding homozygotes. Furthermore, in the Australian grasshopper species *Moraba scurra* the inversions in different chromosomes interact to produce a series of viability values (Lewontin and White 1960). With the two variants of one chromosome denoted as A1 and A2, and those of another chromosome as B1 and B2, the population of a certain locality showed the following relative viabilities (the most viable type being taken as unity):

	A1A1	A1A2	A2A2
B1B1	0.84	1.00	0.81
B1B2	0.64	1.00	0.97
B2B2	0.39	0.68	0.92

Unequal bivalents, interpretable most plausibly as due to pericentric inversions, have been found in natural populations of the beetle Pissodes (Manna and Smith 1959), of mice Leggada and Mastomys (Matthey 1964a, 1966a, b), and of the black rat, *Rattus rattus* (Yosida, Nakamura, and Fukaya 1965). They have also been found in laboratory colonies of Norway rats, *Rattus norvegicus* (Yosida and Amano 1965, and Bianchi and Molina 1966), and of *Peromyscus maniculatus* (Ohno, Weiler, et al. 1966). Whether these inversions originated in nature or in captivity is not wholly clear.

Translocations and Other Chromosomal Polymorphisms

Let two nonhomologous chromosomes be symbolized as AB and CD. A translocation, that is, an interchange of blocks of genes, transforms them into AD and BC. The translocation heterozygote will form at meiosis a ring of chromosomes:

AB–BC–CD–DA

Six kinds of gametes may be formed, depending on whether alternate or adjacent chromosomes go to the same pole at meiotic divisions:

1. AB, CD	3. AB, DA	5. CD, BC
2. BC, DA	4. AB, BC	6. DA, CD

Gametes 1 and 2 are regular or euploid, since they contain all the chromosome parts once and only once; the remainder (3–6) are exceptional (or aneuploid), carrying some chromosome parts in duplicate and being deficient for other parts. In plants, pollen grains and ovules with duplications and deficiencies are usually aborted, unless the duplicated and deficient parts are very short or the plant is a polyploid. Animal gametes may function regardless of whether they carry normal or abnormal chromosome sets, but the zygote formed by the union of an euploid and an aneuploid gamete will die or develop into a malformed or weak individual. Translocation heterozygotes are accordingly semisterile, or a part of their progeny is of low fitness. Gametes, especially male ones, are produced in most organisms in so great an abundance that the loss of a fraction of them is more easily tolerable than the elimination of many zygotes.

By and large, a greater amount of heterosis is needed for balancing selection to maintain translocation than inversion polymorphisms. And yet some of the genetic systems in the living world are versatile enough for this handicap of translocation polymorphisms to be overcome. Indeed, the disjunction of chromosomes at meiosis is under genetic control, and translocation heterozygotes can be selected to produce a majority of the euploid (1 and 2 in the above scheme) and few or no aneuploid gametes (3–6). An elegant demonstration of this has been obtained in some plants. In rye (Thompson 1956, Hrishi and Müntzing 1960, and Rees 1961) and also in Chrysanthemum (Rana 1965, and Rana and Jain 1965), translocation heterozygotes occur in populations, and the proportions of regular and exceptional gametes vary from strain to strain. By artificial selection Sun and Rees (1967) have achieved in rye, which is normally a cross-fertilizing plant, striking reductions of the frequencies of aneuploid gametes.

In most animal populations translocation heterozygotes occur only as rare mutants, presumably to be eliminated by normalizing selection. Among many thousands of chromosome complements of several species of Drosophila examined in salivary gland cells by the present writer, only two translocation heterozygotes were seen: one in *Drosophila ananassae* from Brazil and the other in *D. pseudoobscura* from Arizona. Translocation heterozygotes in isolated individuals were found in the grasshoppers *Gesonula punctifrons* (Sarkar 1955) and *Chorthippus brunneus* (Lewis and John 1964). In man, healthy members of some families are translocation heterozygotes; grossly abnormal births occur, however, in such families because of the formation of duplication-deficiency gametes (Hauschteck, Mürset, et al. 1966, Court-Brown 1967, and Aya, Kuroti, et al. 1967). John and Lewis (1958, 1959) found translocation heterozygotes quite abundant in some colonies of the cockroach *Periplaneta americana* and in *Blaberus doscoidalis,* and adduced good evidence that these heterozygotes are superior in fitness to the homozygotes.

In Chapter 12 we shall discuss briefly the bizzare genetic systems evolved in some evening primroses (Oenothera), which are founded on translocation heterozygosis. Translocation polymorphisms have been found in natural populations of many plants. They were known for a long time in Jimson weed, *Datura stramonium* (Blakeslee, Bergner, and Avery 1937) and *D. meteloides* (Snow and Dunford 1961), in Paeonia (Walters 1942), in Trillium (Haga and Kurabayashi 1954), and in Clarkia (Håkansson 1942 and Mooring 1958, 1961). In at least some of these translocations the distribution of the chromosomes at meiosis is so regular that few pollen grains and ovules are aborted because of the duplications and deficiencies in their gene complements; other translocation heterozygotes do cause such abortion, but the fitness of the heterozygotes appears to be high enough to more than compensate for this loss.

An interesting kind of translocation involves the union of two rod-shaped chromosomes into a single V-shaped body, or vice versa. If each of two chromosomes has a subterminal centromere, breaks in the vicinity of the centromeres, followed by reunions of fragments, may result in one chromosome with a median centromere, and another centromere with only a little chromosomal material, which is subsequently lost. Some cytologists accept the existence of chromosomes with terminal centromeres; unions of two such chromosomes may occur because of centric fusions, while chromosomes with median centromeres may give rise by centric fission to two chromosomes with terminal centromeres. Finally, some groups of animals have diffuse instead of localized centromeres, removing most of the constraints on the fragmentations and unions of the chromosomes. In any event, increases and reductions of the chromosome numbers in related species are fairly ubiquitious in the animal as well as in the plant kingdom.

Consider, however, what happens to new chromosomes formed by centric fusion or separation. Let a chromosome with a median centromere be AB, and the two semihomologous chromosomes with terminal centromeres be A and B, respectively. Heterozygous individuals will be formed, in which the AB chromosome will pair at meiosis with A and with B. If this trivalent association disjoins so that some gametes receive AB and others A + B, the fertility of the heterozygote will not be affected. On the other hand, formation of gametes AB +A, AB + B, A, and B alone will lead to inviable or abnormal zygotes.

It appears that such trivalents give viable disjunction products often enough for them to occur in natural populations of some species. Staiger (1954, 1955) showed that about 1 percent of individuals of the snail *Purpura lapillus* near Roscoff, France, are translocation heterozygotes. In the isopode *Jaera albifrons* the chromosome numbers vary from 19 to 29 (Lécher 1967). In African mice of the subgenus Leggada, Matthey (1963, 1964a, 1966a) found individuals with 31, 32, 33, and 34 chromosomes in the populations of one species, and with 20, 21, and 22 chromosomes in populations of another species. White, Carson, and Cheney (1964) made a detailed study of the chromosomal races of the grasshopper *Moraba viatica* having 17 and 19 chromosomes, respectively; the distribution areas of the two overlap in a narrow belt, and there individuals with trivalent associations are found. The fertility of the heterozygotes is reduced, a fact that, according to the authors, explains the narrowness of the overlap. Manna and Smith (1959) found numerical chromosomal polymorphisms in bark weevils, Pissodes, and Smith (1962, 1966) in the ladybird beetle *Chilocorus stigma* and closely related species. The populations of the latter are monomorphic (26 chromosomes) in the eastern United States, but in Canada there is a gradient from east to west, with chromosome numbers dwindling to 22, 20, and 14.

Duplication and deficiencies for chromosome sections are rarely involved in polymorphisms in natural populations, and there is no proven case of such polymorphism being maintained by balancing selection. Five kinds of Y chromosomes have long been known in *Drosophila pseudoobscura,* and three in *D. persimilis* (Dobzhansky 1937). The differences in length between these Y chromosomes are due to greater or lesser amounts of heterochromatic materials. Variations in the size of the Y chromosomes have also been recorded in healthy individuals in human populations (Gripenberg 1964; Makino, Sasaki, et al. 1963; Makino and Takagi 1965, and Cohen, Shaw, and MacCluer 1966). If the average length of chromosomes 17-20 in the normal set is taken as unity, the Y chromosome common in Japanese populations has a length of 0.87, but some individuals have Y's measuring more than 1.0 on such a scale.

From Intraspecific Polymorphism to Species Differential

No two species have the same genes, and present indications are that the numbers of genes that differentiate even closely related species are, in general, fairly large (Chapter 11). By contrast, species may have either the same or different numbers of chromosomes, and the gene arrangements in these chromosomes may be either similar or strikingly different. Speciation is usually, but not necessarily, accompanied by chromosomal differentiation. Carson, Clayton, and Stalker (1967) have called species with identical gene arrangements homosequential. These authors examined the disc patterns in the salivary gland chromosomes in many species of Drosophila endemic in the Hawaiian Islands. Four groups of homosequential species, with 3, 2, 2, and 2 species, respectively, were found. Kastritsis (1969) found the neotropical species *Drosophila guaramunu* and *D. griseolineata* to be also homosequential.

Among some 17,138 plant species examined cytologically up to 1955, the haploid numbers range from 2 (in *Haplopappus gracilis* and some molds) and 3 (in several species of Crepis), to more than 250 in a species of fern. The most frequent numbers are 9, 8, 11, 12, 7, 14, and 13, in that order (Grant 1963). Among 3317 animal species studied up to 1951, the numbers range from 2 pairs in some flatworms and scale insects (Icerya) to 127 in the crab Eupagurus (Makino 1951 and White 1954). It is frequently stated that the nematode worm *Ascaris megalocephala* has 2 or even a single chromosome in its sex cells; however, these chromosomes are compound and fall apart into more numerous elements in the somatic cells.

More important than changes in chromosome numbers are changes in the internal organization of the chromosomes, that is, in the gene arrangement. Such changes are brought about mainly by inversions and translocations of chromosome segments. We have seen above that the former are more likely to establish balanced polymorphisms in outbreeding populations than are the latter. A higher degree of heterozygous advantage is needed to compensate for the loss of fertility in translocation than in inversion heterozygotes. At least in Drosophila, species differences are due far more often to inversion than to translocation of blocks of genes. Nevertheless, translocations have played a not insignificant role in chromosome repatterning. If polyploidy is temporarily omitted from consideration, it can be said that changes in chromosome numbers are brought about by various kinds of translocations. Multiplications and deletions of whole chromosomes are even less likely than translocations to be retained in populations, except in polyploids which reproduce asexually or by parthenogenesis (see Chapter 12).

Gene Arrangements in Drosophila Species Hybrids

If species of Drosophila can be hybridized, the giant chromosomes in the salivary gland cells permit more precise comparison of the chromosome structures in the species crossed than is attainable by any other method. All degrees of differentiation are found. As stated in the preceding section, some pairs of species are homosequential, that is, show no recognizable differences in the gene arrangement. Sturtevant (1929) found the genetic maps of the chromosomes of *Drosophila melanogaster* and *D. simulans* to be similar, except for a long inversion in the third chromosome. Pätau (1935), Kerkis (1936), and Horton (1939) studied the salivary gland chromosomes in the hybrids. Aside from the inversion in the third chromosome, these authors found 24 short sections, each involving a few stainable discs, that differ in the two species. Among these, 6 sections seem to be minute inversions. The remainder are undefined changes that may be minute inversions, translocations, or qualitative changes in the chromosomal materials.

Drosophila pseudoobscura and *D. persimilis* differ in two moderately long inversions in the X and in the second chromosome. Usually there is also a second inversion in the X, and an inversion in the third chromosome, that is, a total of four inversions. The variations in the numbers of the inversions that distinguish the species are due to differences between the strains of the parental species. No small undefined differences, like those in the hybrids between *D. melanogaster* and *D. simulans,* are present (Tan 1935, and Dobzhansky and Epling 1944).

The morphological resemblance between *D. pseudoobscura* and *D. persimilis,* on one hand, and *D. miranda,* on the other, is at least as close as that between *D. melanogaster* and *D. simulans.* Their metaphase chromosomes are identical, except that one of the autosomes of *D. pseudoobscura* is present only once in the chromosome group of the *D. miranda* male; *D. pseudoobscura* is XX and XY in the female and the male, respectively, whereas the *D. miranda* female is $X^1X^1X^2X^2$ and the male X^1X^2Y. The Y chromosome of *D. miranda* harbors some material homologous to that in the X^2 of the same species and in the corresponding autosomal pair of *D. pseudoobscura.* The origin of *D. miranda* must have involved a translocation of that autosome onto the Y chromosome, with subsequent rearrangement of the autosomal material by repeated inversions.

Dobzhansky and Tan (1936) have compared the gene arrangements in the *D. pseudoobscura* and *D. miranda* chromosomes in the salivary glands of hybrid larvae. The differences are so profound that the chromosomes either fail to pair entirely or else form extremely complex pairing configurations. Genes that in one species lie adjacent may, in the other species, be far apart in the same chromosome. Some small blocks of genes that are located in the same chromosome in one species have apparently been translocated to different chromosomes in the other. Finally, homologues of certain sections in *D. pseudoobscura* have not been detected in *D. miranda*, and vice versa. It seems, then, that some chromosome sections have been so thoroughly rebuilt by repeated inversions or translocations that their disc patterns in the salivary gland chromosomes no longer resemble each other, and pairing of the homologous genes does not take place.

A comparison of the chromosomes of *D. pseudoobscura* and *D. miranda* is shown in Fig. 5.7. If the chromosome sections that seem to be present in one species only are disregarded, the differences between the two species are due chiefly to repeated inversions and to a lesser extent to translocation of blocks of genes. To derive the gene arrangement in the chromosomes of one species from that in the other, a minimum of 49 breakage points (and perhaps twice as many) is necessary.

Patterson, Stone, and their colleagues have done excellent analysis of the evolution of the chromosomes in 12 species and subspecies of the *virilis* group of Drosophila (reviews in Patterson and Stone 1952, and Stone, Guest, and Wilson 1960). At least 92, and possibly as many as 120, inversions have occurred in the phylogeny of this group. If the gene arrangements in the chromosomes of the Oriental species *Drosophilia virilis* are used as the standard, the probable phylogeny of the group can be traced as shown in Fig. 5.8. Three of the 15 chromosome complements in this figure are hypothetical—Primitive I, II, and III. They are reconstructed with the aid of the overlapping inversions analysis described above and illustrated in Figs. 5.1 and 5.2. Primitive I differs from the existing *D. virilis* in a single inversion in the second chromosome. This is the ancestral gene arrangement of the whole group. Two more inversions, A and B in the X chromosome, transform Primitive I into Primitive II, which is the ancestor of the closely knit triplet of North American species *D. texana, D. novamexicana,* and *D. americana.* These species differ in, respectively, 6, 7, and 9 inversions from the Primitive II ancestors. Primitive III differs from Primitive I by 6 inversions in the autosomes, plus several rearrangements in the X chromosome that defied analysis because of their complexity. Primitive III is the ancestor of the Japanese *D.*

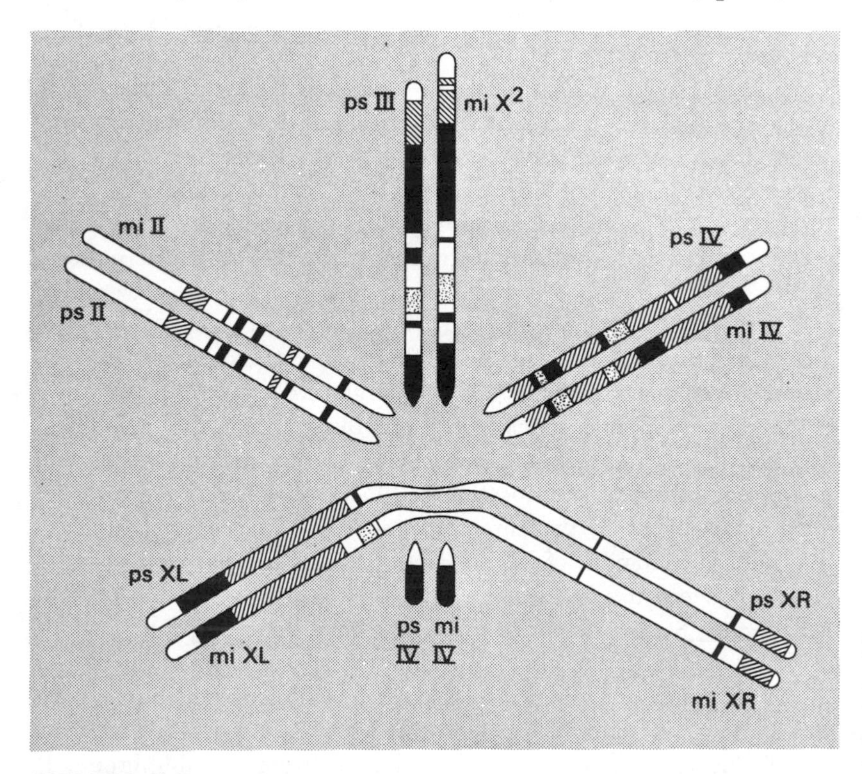

FIGURE 5.7

The gene arrangements in the homologous chromosomes of Drosophila pseudoobscura and D. miranda. Sections with similar gene arrangements are shown white, inverted sections cross-hatched, translocations stippled, and those of obscure homologies black.

On facing page: Phylogenetic relationships of the gene arrangements in species of the virilis group of the genus Drosophila. The chromosomes of Drosophila virilis are taken as the standard; the inverted sections in various lines of descent are marked by letters. (After Stone, Guest, and Wilson)

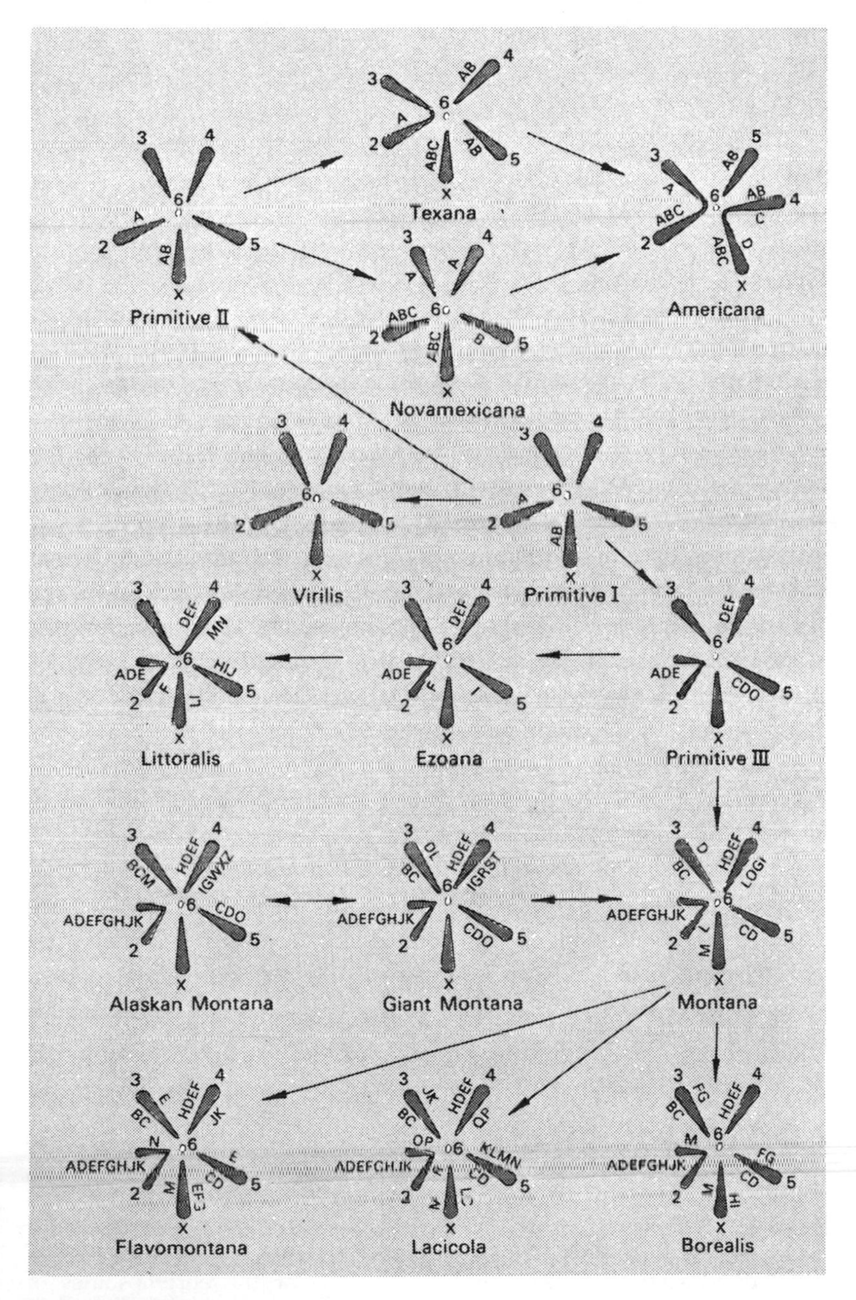
Texana
Americana
Primitive II
Novamexicana
Virilis
Primitive I
Littoralis
Ezoana
Primitive III
Alaskan Montana
Giant Montana
Montana
Flavomontana
Lacicola
Borealis

FIGURE 5.8

ezoana and the North American *D. montana.* The latter is in turn the presumed ancestor of 5 more species, which have accumulated more and more inversions as their evolution progressed. Some descent relationships that are uncertain are indicated in Fig. 5.8 by double arrows.

A great majority of the inversions in the phylogeny of the *virilis* group are paracentric, not including the centromere. This is as it should be, because crossing over within pericentric inversions, which include the centromere, results in partial sterility of the heterozygotes for such inversions (see above). And yet, the transition from Primitive I to Primitive III involved a pericentric inversion, which transformed the ancestral rodlike chromosome with a subterminal centromere into a V-shaped chromosome with a submedian centromere. This V-shaped chromosome was then passed to all the descendants of Primitive III, as shown in Fig. 5.8. For the same reason—partial sterility of heterozygotes—translocations are infrequent. Nevertheless, a translocation transformed the rod-shaped chromosomes 2 and 3 of Primitive II into a V-shaped chromosome in *D. texana.* Another translocation formed the X chromosome and chromosome 4 of *D. texana* into a V-shaped complex in *D. americana.* Still another translocation effected the union of chromosomes 3 and 4 of *D. ezoana* into a single chromosome in *D. littoralis* (Fig. 5.8).

A series of studies by Wasserman (1960, 1963, and references therein) on gene arrangements in the chromosomes of 46 species of the *repleta* group of Drosophila has revealed a slightly different situation. Although as many as 144 paracentric inversions have participated in the evolution of the chromosome complements, the gene arrangements in related species are generally less differentiated than those in the species of the *virilis* group, and therefore in turn have diverged less than those in *D. pseudoobscura* and *D. miranda.*

Gene Arrangements in Drosophila Species That Cannot be Crossed

In the preceding section reference was made to the pioneering work of Sturtevant (1929) on linkage maps of the chromosomes of *Drosophila melanogaster* and *D. simulans.* Though very laborious, this method of comparison is applicable to species that cannot be crossed. Mutants that arise in different species frequently produce similar phenotypic changes. The inference is that the genes that produce similar mutations are homologous. The reliability of this method is limited, because phenotypically similar ("mimic") mutants are often produced at different gene loci in the same species.

Nevertheless, Sturtevant and Novitski (1941) were able to homologize the linkage maps of several species of Drosophila. They concluded that a chromosome complement of five pairs of rod-shaped (acrocentric) and one pair of dotlike chromosomes is the ancestral condition in the genus Drosophila (Primitive I in Fig. 5.8). This karyotype recurs in many unrelated species of different subgenera and species groups. Other karyotypes in Drosophila arose from this primitive one through translocations that combined in various ways the six chromosome limbs or "elements." Although the "elements" are retained quite tenaciously, the genes within them are found in rather different

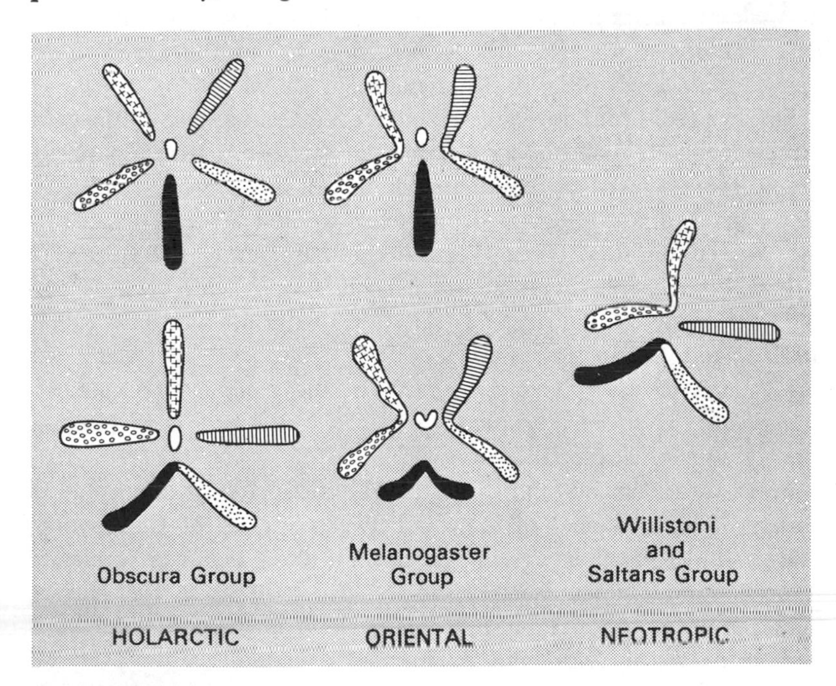

FIGURE 5.9
Homologies of the chromosome sets in the subgroups of the subgenus Sophophora of the genus Drosophila.

linear orders in different species. This is a consequence of the prevalence of paracentric inversions and the relative rarity of pericentric ones and of translocations (see above).

Figure 5.9 shows chromosome complements (represented as haploid) in species of the subgenus Sophophora of the genus Drosophila. One of the species of the Holarctic *obscura* group, namely, the European *Drosophila subobscura,* has the ancestral karyotype of five rods and a dot. Other species in this group, including the American *D. pseudoobscura, D. persimilis,* and *D. miranda,* have a new X chromosome, which corresponds to the X and one of the autosomes of *D. subobscura.* (The X chromosome of the ancestral set is shown in Fig. 5.9 in black.) In the *melanogaster* species group, native apparently in the Oriental region, the four pairs of acrocentric autosomes are fused into two pairs of V-shaped metacentric ones. In some species, such as *D. ananassae,* a pericentric inversion has transformed the acrocentric X chromosome into a small metacentric. Finally, in the *willistoni* and *saltans* species groups, native in tropical America, only three pairs of chromosomes are left, the dotlike "element" of the ancestral set being translocated to the X chromosome.

Stalker (1965) introduced a method of comparative study of the gene arrangements, by painstaking matching of the disc patterns in photographs of sections of the giant chromosomes in the salivary gland cells. In the hands of a cytological virtuoso, this method gives results hardly less dependable than comparing the chromosomes in the cells of a hybrid. By such photographic comparison, supplemented where possible by study of hybrids, Stalker (1965, 1966), Kastritsis (1966, 1969a), and Carson and Stalker (1968) succeeded in tracing the phylogenies of the gene arrangements in the *melanica* (6 species), *tripunctata* (11 species), *guarani* (7 species), and *grimshawi* (29 species) groups of Drosophila.

Beerman (1955a,b) and Keyl (1961, 1962, 1965) discovered a variety of chromosomal differences between species of Chironomus, midges, and Kitzmiller, Frizzi, and Baker (1967) between species of Anopheles mosquitoes. Comparisons of disc patterns in the salivary glands of pure species and also, where possible, of interspecific hybrids were made. As in Drosophila, paracentric inversions are the principal agents of change in the gene arrangements. Thus, *Chironomus tentans* and *Ch. pallidivittatus* have similar gene arrangements in the right limbs of chromosome 1 and differ in one inversion each in the left limbs of chromosomes 1 and 3, and in two inversions each in the right limbs of chromosomes 2 and 3; the gene arrangements in the left limb of chromosome 2 and in chromosome 4 are too different to make analysis possible. The X and Y chromosomes have homologous banding but differ in several inversions; in *Ch. tentans* it is sometimes chromosome 1 and sometimes chromosome 2 that serves as the sex-determining pair. Among 20 species of Chironomus, Keyl (1962) found 4 translocations and numerous inversion differences, as well as differences in the appearance of homologous bands in salivary gland chromosomes, apparently resembling the differences found between *Drosophila melanogaster* and *D. simulans* (see above).

Changes in Chromosome Structure in Speciation

There are several mechanisms whereby evolutionary changes in the chromosome numbers can be effected. Polyploidy makes the numbers in related species multiples of some basic haploid number, *n*. Among classical examples of polyploid series of species are wheats and related grasses with $2n = 14$, 28, and 42, and Chrysanthemum with $2n = 18$, 36, 54, 70, and 90. Polyploidy is frequent among plants and relatively rare in animals (Chapter 11); it is a one-way street, moving toward higher but rarely if ever toward lower numbers.

In most organisms every chromosome has one and only one centromere, and this organelle cannot arise *de novo*. Yet some groups, both among animals (lepidopterans, trichopterans, homopterans, heteropterans and some insects, some scorpions, and, as mentioned above, nematodes) and among plants (the family Juncaceae, some algae), have chromosomes with multiple or diffuse centromeres (a review in White 1954, Suomalainen 1966). This makes possible large differences in chromosome numbers among related species. For example, the butterfly genus Erebia has species with haploid numbers 11, 14, 17, 19, 21, 22, 28, 29, and 40, and the saturnid moths with 13, 14, 19, 29, 30, 31, 33, and 49 (Makino 1951). Such differences can be mistaken for polyploidy, but, as far as is known, they involve little reconstruction of the chromosomal apparatus and no increase or decrease in the amount of genic materials.

When the chromosomes have a single centromere each, the freedom of changes in the chromosome number and in gene alignment is more limited. The requirement is that each chromosome has one and only one centromere. In the elegant study of Gerassimova (1939) two translocations were obtained in the progeny of irradiated *Crepis tectorum*. This species has four pairs of chromosomes, denoted A, B, C, and D. One of the translocations exchanged sections of chromosomes A and D, and the other of B and C. Translocation heterozygotes are semisterile, because of the formation of aneuploid gametes (see above), but translocation homozygotes are fertile. By intercrossing the two translocation strains, a double-translocation heterozygote, involving all four chromosomes, was obtained. The fertility of this heterozygote was quite low. Among its progeny, however, there were individuals homozygous for both translocations. They were not only viable and fertile with each other, but also formed semisterile hybrids when crossed with the parental form. Gerassimova rightly considered her double-translocation homozygote deserving of a new species name, *Crepis nova*. A similar synthetic species was obtained by Stebbins (1957b) in the progeny of highly sterile hybrids between the grasses *Elymus glaucus* and *Sitanion jubatum*. Even earlier (1936) Kozhevnikov combined two translocations in *Drosophila melanogaster* to produce a "*Drosophila artificialis*." The latter, when crossed to the chromosomally normal *D. melanogaster*, produces no viable offspring at all. Unfortunately, only one-quarter of the eggs produced by intercrossing *D. artificialis* females and males are viable. As Grant (1963) has said, "The chromosome structural arrangement is the umbilical cord of the species."

CHAPTER SIX

BALANCING SELECTION AND GENETIC LOAD

Heterogeneity of Environment

Mathematical models usually assume, and experimental studies endeavor to provide, uniform and constant environments in which the process of selection can take place. This convention is convenient, because the Darwinian fitness of a genotype in a constant environment is also constant. Unfortunately, it is also an oversimplification. Not only a species or a population but even a single individual faces a variety of environments in its lifetime.

We have seen in Chapter 2 that the genotype determines, not a single phenotype, but a norm of reaction, a repertoire of phenotypes that can arise in different environments or successions of environments. Within a certain range of environments these phenotypes are generally adaptive (adaptive modifications), whereas outside this range they are adaptively ambiguous or maladapted (morphoses in the terminology of Schmalhausen 1949). The range of environments may be greater or smaller than that to which the genotypes available in a population react by adaptive modifications. Environments often occur in "patches" succeeding each other in space or in time. If the patches are large or prolonged enough so that an average individual member of the population is likely to spend its lifetime in a single patch, the environment is coarse-grained; if an individual encounters many patches, it is fine-grained. In his perceptive theoretical study, Levins (1968 and references therein) has pointed out that coarse-grained and fine-grained environments call for different evolutionary strategies for adaptation.

The coarseness or fineness depends on the mobility of the organism, and on the topography and many other features of the environment itself. If the range of environments is smaller than the tolerance of individuals with a given phenotype, then the optimal strategy is a single phenotype, monomorphism. This phenotype probably has its highest adaptedness near the middle of its environmental range, and does at least moderately well in other environments. The range of environments may, however, be so great that a phenotype will do well in some of them and poorly in others. In a fine-grained environment the optimum strategy may still be a single phenotype, specialized for some of the subenvironments, especially those most frequently encountered. In a coarse-grained environment the optimum is sometimes a "mixed" strategy, that is, a multiplicity of phenotypes. Several phenotypes and genotypes may be called for, with frequencies dependent on those of the environments. This will usually, though not always, mean genetic polymorphism. Even in coarse-grained environments, however, there are also conditions in which a single phenotype is optimal.

Adaptive Norm and Stabilizing Selection

An evolutionary strategy that favors phenotypic monomorphism need not rest on genetic uniformity. Classical genticists assumed that "normal" or "wild-type" representatives of a species are similar in appearance because they are also similar genetically. Discovery of the concealed genetic variability in natural populations of Drosophila has shown how far off the mark this assumption was. Experiments in which different "normal" chromosomes from the same population were allowed to undergo gene recombination gave products some of which were subvital, semilethal, and even lethal. These chromsomes evidently carried different genes. Ford (1964) made an elegant study on phenotypically indistinguishable populations of the moth *Triphaena comes* from certain islands off the coast of Scotland. Since they proved to be quite distinct genetically, their phenotypic similarity evidently is reached by different genetic means.

Phenotypically monomorphic species or populations have single adaptive norms. An adaptive norm is, however, not a single genotype but usually a great array of genotypes. The common property of this array is that the constituent genotypes and the phenotypes engendered by them enable their carriers to survive and to reproduce in the environments that the species inhabits. Normalizing selection (Chapter 4) protects the adaptive norm by trimming off ill-adapted variants, the expressed genetic load of the population. As stated in Chapter 4, Schmalhausen (1949), Mather (1955), and Waddington (1957) distinguish also stabilizing (or canalizing) and diversifying (or disruptive) selections. Consider a character, such as stature or growth rate or fertility, the variation of which follows a bell-shaped probability curve. Stabilizing selection favors the modal phenotype and discriminates against both extremes. Diversifying selection, on the other hand, favors the extremes and discriminates against the intermediates. (I prefer the name diversifying to "disruptive" because, far from being disruptive, this form of selection is a constructive factor in adaptive evolution!)

The phenotypic repertoires of genotypes are of at least three kinds. First, homeostatic buffering or canalization (cf. Chapter 2) may give the same phenotype within the entire range of the environments in which the genotype is viable. Thus, almost all human genotypes ensure the development of a four-chambered heart, a suckling instinct in the infant, a capacity to think in symbols and to learn a language, etc. Second, a continuous range of phenotypes may develop, depending on the state of the environment. Weight, muscular development, educational achievement in man, and body size in many animals and plants are examples. Third, there may be an environmental threshold below which one phenotype and above which another phenotype develop, or a "stochastic switch" when an environment determines only the probability of the development of certain phenotypes (Levins 1968). Thus, the castes (workers, soldiers, queens) of some social insects are genetically similar and are determined by the quality or quantity of the food given to the larvae. Flowering in many plants and diapause in many insects (Andrewartha 1952) are induced by the length of daylight and darkness (photoperiod). In the marine worm Bonellia sex is determined by the environment—larvae growing on the mother's proboscis become males, and free-living larvae females.

Diversifying Selection: Theoretical

A theoretical model of polymorphism due to heterosis, that is, superior fitness of heterozygotes over the corresponding homozygotes, was discussed in Chapter 5. A stable equilibrium can be maintained by balancing selection even in uniform environments. If the fitness of the three genotypes A_1A_1, A_1A_2, and A_2A_2 is $1 - s$, 1, and $1 - t$, respectively, selection will drive the alleles A_1 and A_2 to equilibrium frequencies of $t/(s + t)$ and $s/(s + t)$, respectively. The mean fitness of the population, $\overline{W}$, is maximized when the equilibrium frequencies are reached. It remains, nevertheless, below that of the fittest genotype, A_1A_2.

Although the diversifying form of balancing natural selection may be widespread and important in nature, it is much less well understood and documented than the heterotic form. Ludwig (1950), Mather (1955), and Ford (1964) have discussed the possible role of diversifying selection in evolution. Levene (1953), Li (1955a,b), Prout (1968), and Levins and MacArthur (1966) constructed mathematical models of some special cases; a general treatment is yet to be made. Levene assumes a panmictic, random mating population that has available to it two or more environments. An insect whose larvae feed on different plants growing in the same locality presents such a situation. Now suppose that genotypes A_1A_1, A_1A_2, and A_2A_2 survive at different rates in the different environments. Levene has shown that a sufficient, though not necessary, condition for a stable polymorphism of A_1 and A_2 is that the harmonic mean of the viabilities of the heterozygotes be greater than that of the homozygotes. He gives an example in which this condition is not satisfied, but an equilibrium is nevertheless achieved. Let the fitness, W, of the genotypes in two environments be:

Environment	A_1A_1	A_1A_2	A_2A_2
I	2	1	1.1
II	0.5	1	1.1

In neither environment is the heterozygote superior. If the two environments are equally frequent, and the initial frequency of A_2 is less than 0.65 (65 percent), a stable equilibrium will be established at frequencies of A_2 and A_1 equal to 0.4 and 0.6. The frequency 0.65 of A_2 is the point of unstable equilibrium; above this point no balanced polymorphism is established, A_2 reaches fixation (frequency 1.0), and A_1 is eliminated. Here, then, is a situation in which the outcome of the selection depends not only on the Darwinian fitness of the competing genotypes and the environment but also on the gene frequencies at which the selection starts operating.

Li (1955a) considered an instance of what Wallace (1968a) has later called marginal overdominance. A heterozygote may have an average advantage over a range of environments, even though it is not superior to the homozygotes in any one of them. Assume a panmictic population, in which 20, 30, and 50 percent of individuals live in environments I, II, and III, in which the viabilities of the three genotypes are as follows:

Environment	A_1A_1	A_1A_2	A_2A_2
I	1.2	1	0.9
II	0.6	1	0.8
III	0.7	1	1.1

The heterozygote is superior in only one niche, and this is not the most frequent one. Li finds, however, that in the population as a whole a stable polymorphism is established at the frequencies 0.158 of A_1 and 0.842 of A_2.

An important constraint in Levene's and Li's models is that, though the selection takes place in different environments, the survivors meet and mate at random regardless of the environment in which they developed. In coarse-grained environments this constraint is relaxed. Most individuals that develop in a given patch (e.g., on a certain food plant) mate and produce progeny in the same patch. Or the carriers of different genotypes may preferentially seek the environments most favorable to themselves. This makes stable polymorphism more likely to be achieved than in Levene's model. Levins and MacArthur (1966) and Levins (1968) have made some headway in analyzing such situations.

Dempster (1955) and Haldane and Jayakar (1963a,b) found that stable polymorphism is possible if a regular cyclic variation in the selection favors one or the other allele in successive generations, or if the arithmetic mean of fitnesses of recessive homozygotes in different generations is higher, while the geometric mean is lower, than unity. As an illustration of this situation, Haldane and Jayakar state, "An occasional severe epidemic of *falciparum* malaria might suffice to keep the gene for hemoglobin S from disappearing, even if for generations on end heterozygotes for sickling were at a disadvantage compared with persons homozygous for hemoglobin A." Merrell and Rodell (1968) have described a "seasonal selection" in the frog *Rana pipiens*. Its color variant *burnsi* is due to a dominant gene. No evidence of heterozygous advantage is found, but *burnsi* survives winter better than the prevalent, wild-type *pipiens*. The frequency of *burnsi* is greater in spring, and decreases toward fall.

Experiments and Observations on Diversifying Selection

A series of brilliant studies on diversifying selection (which the authors called disruptive) has been carried out on *Drosophila melanogaster* by Thoday and his colleagues (Thoday and Boam 1959, Millicent and Thoday 1961, and Gibson and Thoday 1962, 1963). The trait used in their experiments was the number of sternopleural bristles on the sides of the thorax. The selection procedure adopted in one of their experiments was as follows. In two populations, 1(H) and 2(H), males with the highest bristle numbers were mated in each generation to females that also had the highest bristle numbers but were derived from two other populations, 3(L) and 4(L). At the same time, males with the lowest numbers of bristles in 3(L) and 4(L) were mated to females with the lowest bristle numbers for 1(H) and 2(H). The populations were all started from a common source; 1(H) and 2(H) were selected for more bristles, and 3(L) and 4(L) for fewer bristles. The critical circumstance was that in each generation there was a

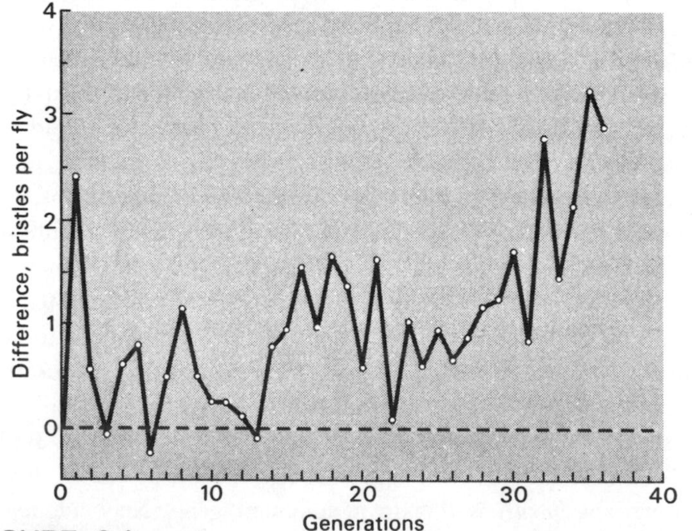

FIGURE 6.1

Diversifying selection for bristle numbers in Drosophila melanogaster. The diagram shows the differences between the mean bristle numbers per fly in the "high" and "low" lines. (After Thoday)

50 percent gene exchange between the H and the L populations. In effect, the selection was carried in a single population, in which individuals with extreme bristle numbers were selected, and those with modal numbers were discarded.

Figure 6.1 summarizes the results of one of the experiments. The difference between the mean numbers of bristles in the H and the L lines gradually increased. Further analysis showed that the variation in the bristle numbers was due mainly to only two genes on the second chromosome. A combination of alleles of both genes decreasing the bristle number acted as a recessive synthetic lethal, while two plus alleles of both genes on the same chromosome functioned as a dominant synthetic lethal (see Chapter 4).

Prout (1962), Scharloo (1964), Druger (1967), and others showed that both diversifying and stabilizing selection can modify very appreciably the manifestation of certain mutants in Drosophila, increasing or decreasing the variability. The most convincing evidence of the operation of diversifying selection in nature comes, however, from studies on mimicry in the swallowtail butterflies *Papilio dardanus* and *P. memnon* (Clarke and Sheppard 1960, 1962, Clarke, Sheppard, and Thornton 1968 and Sheppard 1969; see also a masterful summary in Ford 1964). The females of these butterflies mimic in shape and coloration butterflies of different genera and families, which are protected by being distasteful to predators. Furthermore, in different parts of Africa (*P. dardanus*) and of tropical Asia (*P. memnon*) they mimic different models. To be successful, a mimetic resemblance must be as close as possible to the model; accordingly the females have very different color patterns in different parts of the species distribution area. Males are nonmimetic, geographically relatively uniform, and quite different from the females, except in territories (Madagascar, Ethiopia) where the females are also nonmimetic and resemble the males in color patterns. In addition, in many places the females are polymorphic, mimicking different models.

Genetic analysis showed that the variety of female forms is determined by what at first sight appears to be a series of multiple alleles of a single gene locus, the manifestation of which is sex-controlled, being different in females but not in males. More detailed study disclosed a more complex and interesting situation. In the first place, instead of there being a single locus, the variety of mimetic forms is due to supergenes, closely linked alleles of several loci. Moreover, although the forms that are found in the population show clear-cut dominance and recessiveness, intercrosses of different populations give incomplete dominance. The products of such interpopulational hybridization have their mimicry disrupted by the breakdown of the gene patterns characteristic of separate populations. One may wonder why the mimicry is restricted to the females and the males are left unprotected; at any rate, natural selection has acted to diversify the females and to make them resemble butterflies of quite different species, rather than to be like the males of their own species.

Frequency-Dependent Selection

An interesting suggestion of a possible mechanism maintaining the polymorphism of shell colors and patterns in the land snail Cepaea has been made by Clarke (1962). Bird predators learn to search for prey of a certain appearance, and therefore kill a disproportionately high percentage of the common varieties of the snail and a low percentage of the rare ones. Fitness is therefore decreasing as a color type becomes frequent. This is, then, a frequency-dependent selection.

Many plants (e.g., some Nicotiana, Petunia, Oenothera, and Trifolium) and also some animals (the ascidian Ciona) have evolved self-incompatibility mechanisms that enforce outcrossing and prevent close inbreeding. The key feature of these mechanisms is a series of multiple alleles of a gene, which make the pollen carrying a given allele fail to germinate and to grow on the stigma of a plant with the same allele. Since the stigma tissue is diploid and pollen grains are haploid, a minimum of three alleles is needed to make the mechanism operative. Even if the three alleles are equally frequent in a population, however, many of the pollinations will fail to effect fertilization. A fourth, a fifth, etc., allele appearing by mutation will therefore have selective advantages; furthermore, the advantages will be greatest when a new allele is rare because pollen carrying this allele will function on almost every plant in the population. Conversely, as an allele grows in frequency, its advantage will diminish and eventually disappear. This process was analyzed by Wright (1964b), using the observations of Emerson (1939) on *Oenothera organensis*. The total number of individuals of this curious species is only of the order of 500, growing in some canyons of the Organ Mountains in New Mexico. Yet

Emerson found 47 self-incompatibility alleles in the subpopulations of this species.

Petit (1958) and Ehrman (in Ehrman, Spassky, et al. 1965) independently discovered, in *Drosophila melanogaster* and *D. pseudoobscura* respectively, that when two kinds of males are present in a confined space the rare type has an advantage in mating. This Petit-Ehrman effect has since been found by these and other authors, in several species of Drosophila and under a variety of circumstances (Ehrman 1967, Petit and Ehrman 1969, Spiess 1968, Spiess and Spiess 1969, and references therein). An example of the results obtained is presented in Table 6.1. Equal numbers of females and males of two chromosomal types, CH and AR, of *D. pseudoobscura* are introduced into observation chambers, and the matings occurring during 3–4 hours after introduction are recorded. During this time no female copulates more than once, but a male can do so repeatedly. The chi squares (χ^2) in the table serve to evaluate the probability of the observed deviations from random mating being due to errors of sampling; a chi square of 3.84 corresponds to a 5 percent, and of 6.64 to a 1 percent, probability level. It can be seen that CH and AR females mate very nearly in proportion to their numbers, regardless of their relative frequencies. This is not true of the males. Each type is very successful in mating when it is rare; when the two kinds of males are nearly equally numerous, neither kind has an advantage.

In laboratory experiments, rare males have mating advantages when

TABLE 6.1

Numbers of matings recorded in observation chambers with twenty females and twenty males of Drosophila pseudoobscura, among which individuals of two strains, CH and AR, were present in different proportions (After Ehrman)

Proportions	Males Mated			Females Mated		
CH:AR	CH	AR	χ^2	CH	AR	χ^2
18 : 2	78	27	28.81	93	12	0.24
16 : 4	75	43	19.93	94	24	0.01
14 : 6	63	46	7.73	72	37	0.81
12 : 8	42	75	28.32	69	48	0.05
10 : 10	49	66	2.51	56	59	0.08
8 : 12	57	63	2.81	58	62	3.47
6 : 14	37	71	0.93	42	66	4.06
4 : 16	46	64	32.73	32	78	5.68
2 : 18	33	80	46.30	12	101	0.05

the two competing kinds differ in chromosomal inversions (see Chapter 5), in geographic origins, or in mutant genes, and even when they belong to the same strain but were raised in different environments. The nature of the stimuli responsible for this behavior is as yet obscure. Ehrman experimented with two superimposed observation chambers, of which the upper contained many males of one kind and a minority of a second kind; the lower chamber had many males of the type that was rare in the upper chamber. The matings were scored in the upper chamber. When the two chambers were separated only by a layer of cheesecloth, the "rare" males in the upper chamber no longer had the mating advantage. In such an arrangement, when a current of air is passed from one chamber to the other, the "rare" males lose their advantage if the air comes from the chamber that has their kind in excess, but not in the opposite case.

Nothing is known about possible mating advantages of rare genotypes in natural environments. If they exist in the natural habitats of the flies, the resulting frequency-dependent selection may be a potent instrumentality for maintaining the polymorphic equilibria of gene alleles without heterosis. Even mildly deleterious alleles could be maintained in natural populations by these means. Rare alleles will grow in frequencies until the mating advantages of their carriers decrease and disappear. More research in this field is evidently needed.

Frequency-dependent selection operates, at least in laboratory populations, for the beetle *Tribolium castaneum* (Sokal and Huber 1963), *Drosophila ananassae* polymorphic for certain inversions (Tobari 1966 and Tobari and Kojima 1967), and perhaps *D. melanogaster* polymorphic for variants of the esterase-6 enzyme (Yarbrough and Kojima 1967). Indications of frequency dependence are found also for inversions in *D. persimilis* (Spiess 1957) and *D. pseudoobscura* (Pavlovsky and Dobzhansky 1966). By and large, the fitness is higher when a genotype (or a karyotype) is rare than when it is common. When equilibrium is reached, the fitnesses of the polymorphs become uniform. Sokal and Huber (1963) found, however, more complex relations in Tribolium populations containing the mutant sooty (s) and its wild-type allele. When the frequency of sooty is $q = 0.25$, the viabilities of the three genotypes follow the sequence $+/+ > s/s > +/s$; at $q = 0.5$ they become $+/+ > +/s > s/s$; and at $q = 0.75$ the sequence is $+/s > +/+ > s/s$.

Little is known about the ecological and physiological factors that make the fitnesses of some genotypes frequency dependent. These factors may well be different in different cases. Carriers of one genotype may injure those of another in various ways. Dramatic examples of such hostile interactions, resulting in "xenocide" and "suicide" of one or both competitors, are described by Sokoloff, Lerner, and Ho (1965) and by Sokoloff and Lerner (1967) in mixed cultures of two species of flour beetles, *Tribolium castaneum* and *T. confusum*. These beetles engage in cannibalism, the egg and pupal stages being most liable to be devoured. When fed on corn flour, *T. confusum* is superior to *T. castaneum*, apparently because the latter requires some nutrient that is in short supply in corn. Nevertheless, *T. castaneum* can maintain itself on corn medium by cannibalizing *T. confusum* and thus gaining the otherwise deficient nutrient. When a mixture of the two species is placed on corn, *T. castaneum* eliminates *T. confusum* (xenocide). Yet, having done so, the winner is deprived of necessary nutrition, and its population approaches extinction (suicide). Some plants excrete into the soil substances that are toxic to other species, and even to individuals of the same species (Bonner 1950).

Mutual facilitation appears to be at least as common as mutual damage, and is of particular interest to us as a mechanism that can maintain genetic variability. If two or several genotypes are at their most efficient when exploring somewhat different environmental resources, a genetically mixed population may be better off than a genetically uniform one. Although the physiological reasons are difficult to pin-point, mixed plantings of varieties of barley (Gustafsson 1953), rice (Roy 1960), and oats (Jensen 1965) often give appreciably higher yields than plantings of any one of the component varieties.

Lewontin (1955) in *Drosophila melanogaster* and Lewontin and Matsuo (1963) in *D. busckii* studied the proportions of larvae surviving in limited amounts of culture medium. The same numbers of larvae were introduced into all cultures, but in some cultures all the larvae were of the same strain and in others they were mixtures of two different strains. For several genetic mixtures, the average survival was higher than that in pure cultures. This is, of course, not an invariable rule. Dawood and Strickberger (1964) and Weisbrot (1966) studied the influence of the presence in the culture medium of metabolic waste products, or of killed larvae, of some strains of *D. melanogaster* on the survival of larvae of the same and of different strains. In some combinations the survival of larvae exposed to the waste products of a foreign strain was very significantly reduced. Levene, Pavlovsky, and Dobzhansky (1958) compared the adaptive values of the homo- and heterokaryotypes for ST, AR, and CH inversions in the third chromosome of *D. pseudoobscura.* The values for populations that have pairs of these chromosome types (i.e., ST + AR, ST + CH, and AR + CH) are appreciably different from those for populations having all three. In other words, the fitness of a karyotype depends on what other karyotypes are present in the same medium, and in what frequencies they occur.

Overdominance and Heterosis

We must now return to consideration of the genetic variability maintained in populations by the heterotic form of balancing selection. The paradigm of this kind of genetic variability discussed in Chapter 5 was rather complex—inversions and translocations found in natural populations of many animals and plants. Such chromosomal variants, though inherited as if they were alleles of one gene, actually represent supergenes. They are maintained in populations because the heterozygotes (heterokaryotypes) for the supergenes are superior in fitness to the homozygotes (homokaryotypes).

Superior fitness of heterozygotes may arise also through interaction of alleles at a single locus. A heterozygote can be similar in phenotype to one of the homozygotes (complete dominance), intermediate between the homozygotes (incomplete or lack of dominance), or outside the range of the homozygotes (overdominance). If the phenotype trait is viability or fitness, overdominance may mean that the heterozygote is heterotic and superior to both homozygotes. The fact that "overdominance" is properly applied only to phenotypes produced by the interaction of alleles of a single gene makes the term operationally of limited value. What we ordinarily observe are hybrids, not between gene alleles, but between strains, varieties, or populations that differ in many genes. The phenotypic effects of each gene are embedded in the matrix of the genotype as a whole. It is at least difficult, and usually impossible, to distinguish between a heterotic gene allele and a complex of linked genes, a supergene, in a chromosome that causes the heterozygote to be superior in fitness. In the following discussion we shall use the term heterosis in preference to "overdominance."

Heterozygous Effects of Newly Arisen Mutants

A simple but essential distinction must be made in appraising the effects of a genetic variant on fitness. The array of variants newly arisen by mutation will inevitably be different from the variants that have persisted in a population for many generations. The former have not passed the winnowing process of normalizing natural selection, whereas the latter are the survivors of this force. The carriers of new variants are therefore likely to be, on the average, less fit than the carriers of long-persisting variants. The evidence is overwhelming that most mutations arising in any living species are deleterious (cf. Chapters 3 and 4). The problem here at issue has a narrower focus: Do the recessive genes that are deleterious or lethal when homozygous cause a reduction of fitness also in the heterozygous condition? As we saw in Chapter 4, normalizing natural selection is far more effective in lowering the incidence in the gene pool of variants that incapacitate the heterozygotes than in reducing the frequencies of those that harm only the homozygotes. In contrast, superior fitness in heterozygotes maintains a heterotic allele in the population even if the homozygote is lethal.

Stern, Carson, et al. (1952) examined the viability effects in the heterozygous condition of 75 sex-linked lethals in *Drosophila melanogaster*. Males with these lethals in their single X chromosomes do not survive. Tests were made to compare the viability in crowded cultures of two classes of females: one class carried a lethal X chromosome and an X with a visible gene marker, and the other class had a nonlethal X and an X with the same marker. On the average, the survival rate of the females heterozygous for lethals was 2.5 percent lower than that of females free of lethals. Stern and his collaborators pointed out, however, that not all lethals in this sample were deleterious in heterozygotes. Wallace (1968a) analyzed their data further and showed that about 29 percent of the lethals may have been not only neutral but also slightly heterotic. Wallace's own observations (1965, 1968) on lethals in the second chromosome of *D. melanogaster* disclosed a similar situation; of 32 lethals tested, 2 were semidominant and markedly deleterious in the heterozygous conditions, whereas the others resulted in an average lowering of the viability.

That some newly arisen mutants, including those that are lethal when homozygous, may increase fitness in the heterozygous condition is nevertheless certain. Gustafsson and his colleagues (Gustafsson 1946, 1951, 1963a,b, Gustafsson, Nybom, and Wettstein 1950, and references therein) have discovered several examples of this kind. Thus, some chlorophyll-deficient mutants in barley, though lethal when homozygous, are superior in the heterozygous condition to noncarriers in respect to productivity and to competitive ability in dense stands.

Mukai and Burdick (1959, 1961) and Tano and Burdick (1965) established, by most careful and extensive tests, the heterotic effects of one particular lethal in the second chromosome of *Drosophila melanogaster*. In the same species, Torroja (1966) found three sex-linked lethals that appeared to be heterotic. In all these cases, the evidence really shows the lethals to be heterotic in a particular genetic system and in a particular environment. An unconditionally heterotic gene is probably a will-o'-the-wisp.

Newly Arisen Mutants in Relation to the Genetic Background

Most individuals in populations of sexually reproducing, diploid or polyploid, and outbreeding species are heterozygous for alleles at many gene loci. The proportion of heterozygous genes can be reduced by inbreeding or by means of more sophisticated genetic techniques described in Chapters 3 and 4. An entirely homozygous individual is, however, a freak, if found at all in natural populations. As will be shown in this section, heterozygosis exists even in self-fertilizing species, that is, under most rigorous inbreeding. High heterozygosity is the state of the genetic systems in most organisms, established during the eons of their evolutionary development. It is reasonable that for newly arisen mutants the effects of heterozygosis on fitness may well be different, depending on the degree of heterozygosity prevailing in the genotype into which they are introduced.

The trail-blazing work on this problem was done by Wallace (1958, 1965). Strains of *Drosophila melanogaster* homozygous for second chromosomes were obtained by means of a technique similar to that diagramed in Fig. 3.1. The chromosomes chosen for the experiments had homozygotes of quasi-normal viability. Males homozygous for these chromosomes were given a dose of 500 r of X-rays and crossed to unirradiated females with marker genes *Cy* and *L* (Curly and Lobe) in one second chromosome and *Pm* (Plum) in the other. Unirradiated control males with the same chromosomes were similarly crossed. By means of a series of crosses, again similar in principle to those in Fig. 3.1, cultures were obtained segregating four classes of flies, theoretically in equal numbers: Curly-Lobe Plum, Curly-Lobe, Plum, and wild type. In the experimental series, however, the Curly-Lobe and wild-type classes carried one irradiated wild-type chromosome, whereas in the control series both chromosomes were unirradiated. Denoting a given wild-type chromosome as $+_1$ when unirradiated and as $+_1{}^*$ when irradiated, we have:

Experimental:	$Cy\ L/Pm$	$Cy\ L/+_1{}^*$	$Pm/+_1$	$+_1/+_1{}^*$
Control:	$Cy\ L/Pm$	$Cy\ L/+_1$	$Pm/+_1$	$+_1/+_1$

Since the mutant gene markers depress slightly the survival rates of their carriers, the meaningful way to compare the irradiated and control cultures is to measure the viabilities of the different classes relative to that of the *Cy L/Pm* flies, taken as 1.000. The choice of the *Cy L/Pm* class as the viability standard is valid because it has exactly the same second chromosomes in the experimentals and in the controls. In one experiment the average ratios found in 764 experimental and 766 control cultures were as follows:

Experimental		*Control*	
$Cy\ L/+^*$	1.115	$Cy\ L/+$	1.094
$Pm/+$	1.137	$Pm/+$	1.146
$+/+^*$	1.033	$+/+$	1.008

The viabilities of the wild-type and the Curly-Lobe classes carrying one irradiated second chromosome exceed the viabilities of the corresponding classes in the unirradiated controls. The difference may seem slight, but, because of the large numbers of flies counted, it is significant at the 4 percent probability level for the Curly-Lobe class, and at the 2 percent level for the wild-type class. It follows that the average effect of newly induced mutations in the heterozygous condition is a viability increase. With the total count of flies in seven experiments exceeding 3.25 million, the probability of the observed result being due to chance has been reduced for the wild-type class to 0.2 percent (Wallace 1968a). On the other hand, when irradiated second chromosomes were introduced in a heterozygous genetic background (hybrids of different strains of the same species), no heterotic effects were observed (Wallace 1963a).

Two working hypotheses suggest themselves. First, a degree of heterozygosity above a certain minimum level may be requisite for high fitness in normally outbreeding species. This idea was advanced by Lerner (1954) before Wallace's work was published, and supported by Lerner's analysis of selection for high productivity in poultry and other animals. Second, since only a part of the newly arising mutations may be heterotic, there is a better chance of such mutations improving highly inbred or homozygous strains than strains already having a heterozygosity that has accumulated under the control of balancing selection.

The findings of Wallace met with some disbelief and even hostility. Falk (1961), who used a quite different experimental design, interpreted his results at first as contradicting Wallace's findings but later (Falk and Ben-Zeev 1966) as confirming them. Crenshaw (1965b) obtained concordant evidence in *Tribolium confusum*. The best supporting evidence came, however, from the ingenious experiments, on spontaneous rather than radiation-induced mutants, carried out by Mukai and his colleagues (Mukai and Yamazaki 1964, 1968, Mukai, Chigusa, and Yoshikawa 1964, 1965, and Mukai, Yoshikawa, and Sano 1966).

The experiments of Mukai (1964) on the accumulation of mutations that modify the viability of *Drosophila melanogaster* were described in Chapter 3. He used a variant of the *Cy L/Pm* technique, shown in Fig. 3.1 for the second chromosome, and concluded that as many as 15 percent of these chromosomes acquire a new mutation in every generation. In agreement with Wallace, Mukai found that the mutants are, on the average, deleterious in the homozygous condition. The more generations over which the mutations are allowed to accumulate, the lower becomes the average viability of the homozygotes.

The central question is, then, what effects these mutants have in heterozygotes. Mukai and his colleagues had 104 separate lines that initially contained the same second chromosome; this chromosome was "normal" in the sense that its homozygous carriers had good viability. Generation after generation, single males from each of the 104 lines, carrying the descendants of the "normal" chromosome and chromosomes with the mutant *Pm*, were outcrossed to *Cy L/Pm* females. From time to time the viability of the homozygotes for each of the 104 chromosomes was tested as shown in Fig. 3.1. The viability in the homozygous condition of some of the chromosomes declined more or less conspicuously, whereas others remained "normal." Mukai believes that the latter escaped mutations affecting their viability.

The next step in the experiment was to test the viability of heterozygotes for a collection of chromosomes on (1) homozygous genetic backgrounds, (2) heterozygous genetic backgrounds from the same population, and (3) heterozygous genetic backgrounds from different populations. Heterozygotes (1) carried a supposedly "normal" chromosome and chromosomes changed by mutations; series (2) and (3) carried different chromosomes from the same population or from different populations. Figure 6.2 gives a concise summary of the results. The curve for the homozygous genetic backgrounds lies above that for the intrapopulational heterozygotes, and above its control. Not only are the newly arisen mutations heterotic when placed together with "normal" chromosomes, but also the lower the viability in the homo-

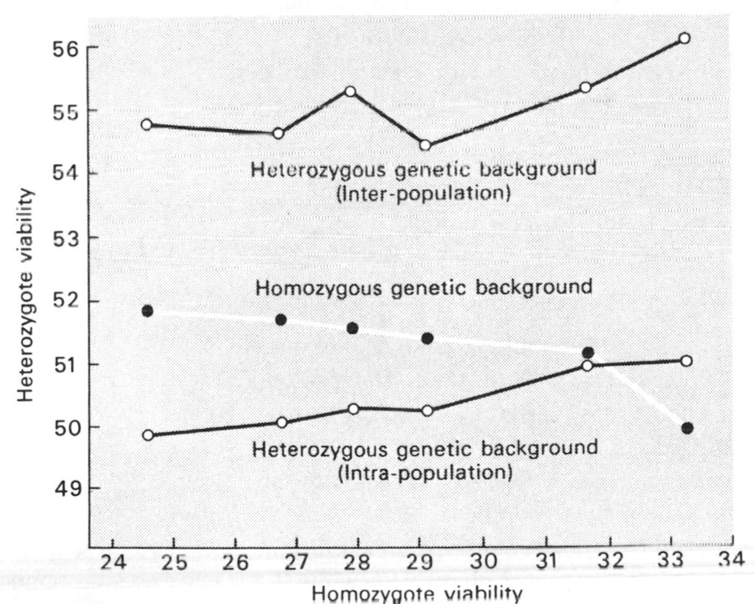

FIGURE 6.2

Relationships between the viabilities of homozygous and heterozygous carriers of certain chromosomes of Drosophila melanogaster, depending on the genetic background. (After Mukai et al.)

zygous condition the better are the heterozygotes. The opposite occurs in the interpopulational series—here the viability of the homozygous carriers of the respective chromosomes decreases. The viability of the interpopulational heterozygotes (3) is superior to the viabilities in series (1) and (2); this is the expected heterosis.

The earlier work of Wallace (1958) indicated that the $Cy\ L/+^*$ class was slightly more viable than the $Cy\ L/+$. This finding was unexpected, since the flies of both classes were heterozygous for many genes. A follow-up test by Wallace (1963b), however, showed no viability difference.

The work of Mukai and his colleagues (see especially Mukai 1969b) leaves an important question unanswered. Why should series (1) in their experiments show heterosis, and series (2) deleterious effects of the mutant genes (Mukai and his colleagues call these *cis*- and *trans*-heterozygotes, respectively)? It is certainly a misconception to regard the "normal" chromosome in Mukai's experiments as somehow mutation-free; any genotype has a history of origin from mutational elements. Why, therefore, should there be heterosis in combinations of this chromosome with its mutant derivatives but not in combinations of these mutant derivatives with each other? There is an obvious need for more work in this field.

Deleterious and Heterotic Mutants in Experimental Populations

We saw in Chapter 4 that normalizing selection is far more efficient in eliminating mutants with deleterious effects in heterozygotes than completely recessive mutants. Heterotic mutants, unless they are lost by chance soon after their origin, are multiplied by heterotic balancing selection. Even if heterotic mutants are a small minority in the mutation spectrum, they are expected to increase in frequencies as selection does its work. Wallace (1962) found that newly arisen lethals in the second chromosome of *Drosophila melanogaster* had an average coefficient of dominance of about 0.07, while in his experimental populations maintained for several years the coefficient became reduced to about 0.02. Some of the lethals in such old populations must have been newly arisen, whereas others originated several, and still others arose many, generations before the tests; one may surmise that the lethals that persisted longest were least deleterious, neutral, or heterotic. Most convincing evidence in this connection has been obtained by Wills (1968) in populations of the yeast *Saccharomyces cerevisiae*. Mutants induced by frameshift mutagens (cf. Chapter 2) tend to be completely

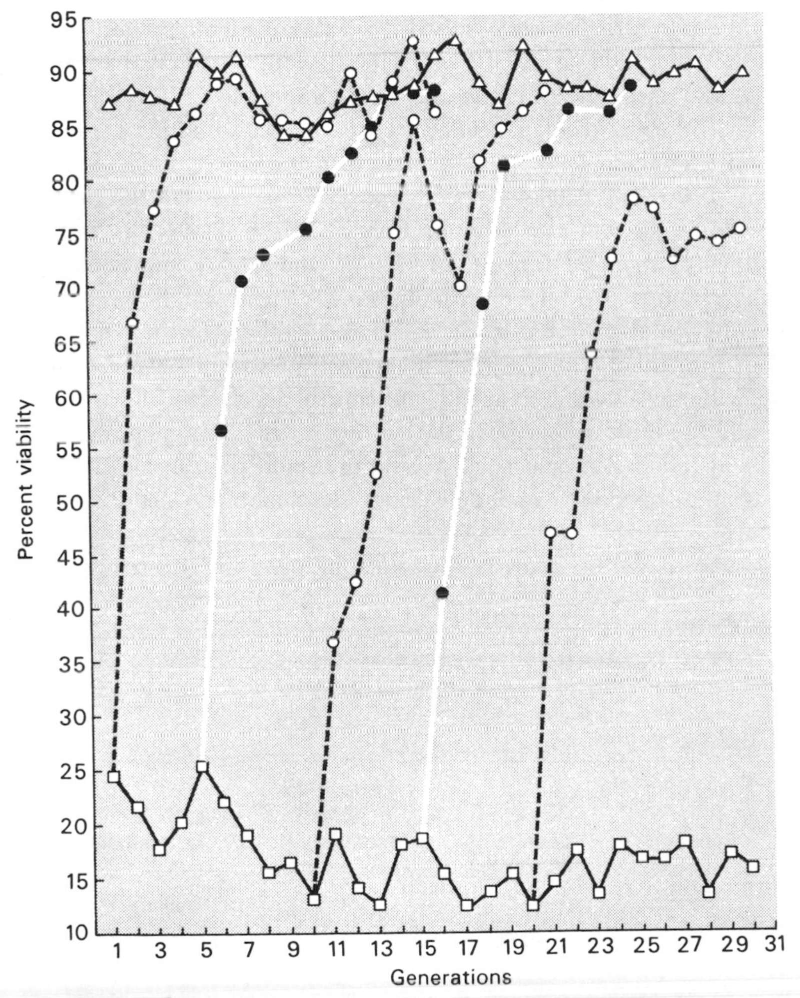

FIGURE 6.3

Egg-to-adult survival in the control (triangles) and irradiated (squares) populations of Drosophila melanogaster. Circles—populations with radiation histories that were gradually recovering normal viabilities after the irradiation was discontinued. (After Sankaranarayanan)

recessive in diploid heterozygotes. Base-substitution mutagens give mutants that are deleterious under some environmental conditions and heterotic under others.

Sankaranarayanan (1964, 1965, 1966) and Salceda (1967) have carried out an impressive series of studies on irradiated populations of *Drosophila melanogaster*. In order to minimize competition among larvae for food and space, the populations were maintained by introducing into each culture bottle only enough eggs to yield 50–70 adult flies; the numbers of eggs needed to produce these adults varied for different populations. In three populations the adult males received 2000 r, 4000 r, and 6000 r, respectively, of X-rays per generation. The egg-to-adult survival rate dropped from about 90 percent in the control down to about 50 percent, 28 percent, and 16 percent, depending on the amounts of radiation administered (Fig. 6.3). From time to time subpopulations were made which received no further radiation; as shown in the figure, the viability recovered within several generations, without, however, reaching the control level. The sharp reduction, and the rapid recovery, of the survival rates are accounted for by dominant lethals, chromosomal aberrations such as translocations resulting in semisterility, and deleterious dominant and semidominant mutants.

Consider now the recessive mutants induced by the irradiation.

TABLE 6.2

Percentages of lethal and semilethal second chromosomes of Drosophila melanogaster in populations having received 120,000 r of X-rays after the irradiation was stopped (After Sankaranaryanan 1966)

Population	X-rays per Generation (r units)	Generations after Irradiation Was Discontinued	Lethals and Semilethals
Control	0	0	11.6 ± 1.5
ER-20	6000	0	87.9 ± 3.1
		19	75.9 ± 3.6
		45	40.1 ± 4.1
ER-30	4000	0	90.3 ± 2.9
		19	59.8 ± 4.6
		45	46.0 ± 4.3
ER-60	2000	0	57.0 ± 4.8
		19	40.9 ± 4.7

Table 6.2 shows the proportions of second chromosomes that were lethal or semilethal in the generation in which the radiation was discontinued and 19 and 45 generations thereafter. Except for the control, all populations received an aggregate dose of 120,000 r of X-rays, but at different rates per generation. At the time of relaxation (generation 0), as many as 90 percent of the chromosomes were lethal or semilethal. Thereafter the percentage fell sharply, but even 45 generations later it was still almost 4 times greater than in the control. At the same time, the frequency of allelic lethals increased instead of decreasing. What does this mean? The most reasonable inference is that, out of a variety of lethal genes induced by the irradiation, many were eliminated but some were not only preserved but even increased in frequencies. This inference has been upheld by Salceda (1967); out of 50 lethals from each population tested, some lethals were found as many as 8, 9, 10, and even 15 times in the same population. Since Salceda's material came from populations of thousands of individuals kept in population cages (see Chapter 5), these high frequencies can hardly be explained by random genetic drift (see Chapter 8).

Similar conclusions were reached by Mourad (1964) and Torroja (1964), who studied irradiated populations of *Drosophila pseudoobscura*. Wallace and Madden (1965) obtained evidence suggesting that some genes which in the homozygous condition cause sterility have positive selective values when heterozygous.

Deleterious and Heterotic Mutants in Natural Populations

In Chapter 4 we saw that natural populations of Drosophila and other outbreeding species carry great accumulations of lethals and other deleterious variants concealed in the heterozygous state. What effects may these variants have on the viability and fitness of the heterozygotes? For about two decades this has been a bone of contention in population genetics.

Muller (1950) insisted that complete recessivity is very rare, and that the heterozygous carriers of deleterious "recessives" are really handicapped compared to noncarriers. He did not completely reject the existence of overdominant heterotic heterozygotes. However, since heterotic balanced polymorphisms make some individuals in the population less fit than the fittest heterozygote, Muller regarded polymorphism as a temporary miscarriage of natural selection, bound to be corrected when a mutant allele appears that yields a fitness at least as high as that of the formerly heterotic heterozygote. This conception was the keystone of Muller's evolutionary views, and particularly of his eugenic ideals. The evidence he used in his arguments pertained almost exclusively to newly arisen mutants, rather than those recovered from natural populations.

Cordeiro (1952) in *Drosophila willistoni,* as well as Greenberg and Crow (1960), Hiraizumi and Crow (1960), and Morton, Chung, and Friedman (1968) in *D. melanogaster,* found an average reduction of viability, and Temin (1966) a reduction of fertility, in heterozygotes for lethal, semilethal, subvital, and sterility factors. Furthermore, these authors argued that the degree of dominance (the h coefficient; see Chapter 4) is greater for subvital than for completely lethal chromosomes. Morton and his colleagues (Morton 1964, Morton and Chung 1959, and Morton, Krieger, and Mi 1966) found no evidence of heterotic balancing selection maintaining polymorphisms in human populations, such as blood group polymorphisms (see, however, Morton and Chung 1959).

In contrast to the above, Wallace and Dobzhansky (1962) and Dobzhansky and Spassky (1963) found in *Drosophila pseudoobscura* no correlation between the viabilities of the homozygotes for a sample of chromosomes extracted from natural populations and of the heterozygotes for these same chromosomes. Marinkovic (1967) obtained similar results for fecundity and fertility. Band and Ives (1963, 1968), Band (1963, 1964), and Tobari (1966) found in *D. melanogaster* that the same chromosome may be deleterious in some but heterotic in other natural or experimental environments. Oshima and Hiroyoshi (1967) and Watanabe (1969a,b) discovered in some localities in Japan that certain lethals persist in the populations year after year. Magalhães, de Toledo, and Cunha 1965 and Magalhães, J. S. de Toledo, et al. 1965 observed what appeared to be a semidominance of lethals in *D. willistoni* at certain seasons; later they found the situation to be more complex, because a surprisingly high proportion of the lethal chromosomes contain synthetic lethals (see Chapter 3). Among eight lethals in the fly *Coelopa frigida* tested by Burnet (1962), one was deleterious, at least one and probably two heterotic, and the remainder apparently neutral when heterozygous. Müntzing (1963) studied in detail a chromosome in rye that was lethal in double dose but heterotic in single dose. For reviews of other pertinent literature, see Dobzhansky (1964a,b) and Wallace (1968a).

Crow and Temin (1964) used a different approach to detect the average deleterious effects of lethal chromosomes in the heterozygous condition in Drosophila populations. Most of the lethal chromosomes extracted from a natural population are not allelic, their lethal effects being due to mutations in different genes. Flies with two lethal chromosomes are viable if the lethals are not allelic, but they die if they have two allelic lethals. If one knew the incidence in a population of chromosomes lethal in double dose (Q), and the proportion of such chromosomes with allelic lethals (A), then the rate of elimination of the lethals because of homozygosis, QA, should be equal to the frequency of origin of lethals per chromosome, U. Now, the data available for populations of several species agree in showing that the mutation rate, U, is larger than QA. This is expected if most lethals have at least mildly deleterious effects in heterozygotes; since the heterozygotes for rare genes far outnumber the homozygotes, a slight depression of the fitness of heterozygotes makes elimination more effective in the heterozygous condition than in the homozygotes. Crow and Temin calculate the average deleterious effect of lethals in the heterozygous condition (h) for different species to be between 0.015 and 0.018.

Unfortunately, this approach is not free of serious pitfalls. Not only does it ignore the existence of synthetic lethals, but also it assumes that all the lethals in a sample from the same "locality" came from one panmictic population. Yet Dobzhansky and Wright (1943) and Wallace (1966a, 1968a) have shown that the mobility of the flies in their natural habitats is so limited that local inbreeding doubtless takes place. The effects ascribed to semidominance of the lethals may well be due to such inbreeding.

A weakness of most experimental studies on the manifestation of near recessives in the heterozygous condition is that the experimental environments deviate, often grossly, from natural ones. To make matters worse, in many studies the chromosomes to be tested are placed on the genetic background of quite unrelated laboratory strains. This ignores the fact that natural selection operates, not with genes in a vacuum, but with genes and chromosomes in the external environments in which a population lives, and in genetic environments that are, in their turn, products of natural selection.

Dobzhansky and Spassky (1968) have made an attempt to overcome at least the second of these disabilities. Using a technique analogous to that diagramed in Fig. 3.1, they extracted 20 second chromosomes that were lethal and 25 that were quasi-normal when homozygous from a population of *Drosophila pseudoobscura* from a certain locality in California. Similarly, 25 lethal and 25 quasi-normal chromosomes were obtained from a population in Arizona. By means of a series of crosses, six kinds of strains were obtained with (1) California chromosomes on the genetic background of the California population, (2) California chromosomes on the Arizona background, (3) California chromosomes on the background of a mixture of strains of Mexico and Guatemala, (4) Arizona chromosomes on the Arizona background, (5) Arizona chromosomes on the California background, and (6) Arizona chromosomes on the Mexico-Guatemala background.

Six pairs of test crosses were then made. One cross in each pair involved a lethal and the other a quasi-normal second chromosome of the same geographic origin and genetic background. In all crosses the males had one wild-type second chromosome and one second chromosome with a mutant marker. The females were from populations of California, Arizona, and of mixed Mexico + Guatemala origin. The progeny of each cross should, theoretically, segregate in the ratio 1 : 1 of flies with and without the mutant marker. The marker-free flies have all their chromosomes of known geographic origin.

More than half a million flies were classified and counted. The results can best be summarized in the following tabulation, where asterisks indicate the differences significant at the 5 percent probability level or better:

Second Chromosomes	*Genetic Background*	*Viability in Heterozygotes*
California	California	Lethals > Quasi-normals*
	Arizona	Quasi-normals > Lethals
	Mexico	Quasi-normals > Lethals*
Arizona	Arizona	Quasi-normals = Lethals
	California	Quasi-normals > Lethals*
	Mexico	Quasi-normals > Lethals

Lethal chromosomes are, on the average, neutral or slightly heterotic on the genetic background of the population in which they were found; they are, on the average, slightly deleterious on foreign geographic backgrounds.

Results similar in principle have been obtained by Golubovsky (1969) and Watanabe (1969b). Watanabe tested 2773 second chromosomes of *Drosophila melanogaster* from a population in Japan on their native genetic background, and 1638 chromosomes from the same population on a foreign genetic background. The frequencies (in percentages) of lethal and semilethal chromosomes found were as follows:

	Native Background	*Foreign Background*
Lethals	14.9	17.4
Semilethals	6.6	18.6

The deleterious effects of these chromosomes are, on the average, stronger on the foreign than on the native background.

The most securely established and, according to some, the only example of heterotic polymorphism in man is that of hemoglobin S-hemoglobin A (see Chapter 2). The gene allele responsible for hemoglobin S is quite frequent in populations that live in, or are descended from, the tropical lowlands of Africa. Homozygosis for this allele causes a severe sickle-cell anemia, and the homozygotes rarely survive adolescence. The heterozygotes have both hemoglobins, A and S, and not only enjoy nearly normal health but also are to some extent protected against *falciparum* malaria. Their adaptive value is accordingly superior to normal (A) homozygotes in lands where *falciparum* malaria is pandemic (Allison 1955, 1964, and Livingstone 1964, 1967).

Similarly, Mediterranean anemia (thalassemia) is due to homozygosis of a gene allele that makes hemoglobin C. This gene has a high incidence in populations of the Mediterranean lands and of some parts of southern Asia. Although the homozygote is lethal, the heterozygote enjoys protection against malaria (Montalenti 1965 and references therein).

The etiology of schizophrenia is a subject of unceasing polemics. Although its precise mode of inheritance is still obscure, the involvement of genetic factors is established beyond reasonable doubt. The incidence of this grave psychiatric disability in human populations is so high that the possibility must be seriously considered that at least some of the genetic factors enjoy heterozygous advantage (Erlenmeyer-Kimling and Paradowski 1966, Gottesman 1968, and references therein). The same hypothesis is applicable to a very common metabolic disorder, *diabetes mellitus* (Steinberg 1959). Also, the data of Myrianthopulos and Aronson (1966) strongly suggest that Tay-Sachs disease, a grave neurological disorder, is maintained by a reproductive advantage of the heterozygous carriers despite lethality in the homozygous condition.

Shifting Concepts of Genetic Loads

Haldane pointed out in an insightful article (1937) that deleterious mutations impair the well-being of a population, not in proportion to the reduction of the viability or fitness of their individual carriers, but rather in proportion to the frequency of their origin. This paradox is easily explained. We saw in Chapter 4 that deleterious mutations opposed by normalizing selection eventually reach equilibrium states. At equilibrium the numbers of mutations that arise per generation are, on the average, equal to the numbers eliminated by selection. Consider, for example, a deleterious recessive that arises at a rate u and is opposed by a selection s. Its equilibrium frequency in the population will be $q = \sqrt{u/s}$, and the frequency of homozygotes will be q^2. If the fitness of a homozygote is impaired to the extent s, the population suffers impairment $sq^2 = su/s$ or u, the mutation rate.

In 1950, Muller went further. "Our load of mutations," he said, consists of accumulated deleterious genetic variants arising by mutation; all such variants are eliminated eventually by "genetic death." The aggregate genetic death that "our" (i.e., the human) population suffers approximately equals the total mutation rate for all deleterious recessives, and is approximately double the mutation rate for the dominants because the latter are eliminated mainly in heterozygotes.

Since mutation rates are increasing (because, e.g., of radiation exposures), so are the genetic loads and genetic deaths. Calculations have predicted that mankind will suffer ghastly numbers of genetic deaths from its increasing genetic load. It has not always been made clear in such calculations that, as pointed out in Chapter 4, a genetic death does not necessarily produce a cadaver; if the carrier of a certain genotype gives birth to one child fewer than the carrier of another genotype, we are witnessing a genetic death.

A formal definition given by Crow (1958) states, "The genetic load of a population is . . . the proportion by which the population fitness (or whatever trait is being considered) is decreased in comparison with an optimum genotype." Later Crow and Kimura (1963) amended this as follows: "the proportion by which the fitness of the average genotype in the population is reduced in comparison with the best genotype." An obvious difficulty is that the operational usefulness of the "optimum" or "best" genotype as a yardstick is doubtful at best. At least in sexual outbreeding species, any two individuals, identical twins excepted, have different genotypes. How does one find, then, the individual who possesses the optimum genotype? Even supposing that this individual could be found, in the next generation there will be another single individual with another optimal genotype. It is also questionable whether this best genotype will conserve its matchless qualities in all environments in which the species lives.

The genetic load, defined as a departure from the optimal genotype, is part and parcel of the typological thinking that is the basis of the classical model of genetic population structure (cf. Chapter 5). As Wright (1960) pointed out in his penetrating critique of this model,

> If we assume that there is one best genotype and that this is homozygous for all type genes, it follows that all mutational changes from this are injurious and selected against. For each mutation there will be on the average one elimination ("genetic death") to restore the status quo. . . . If we define damage in terms of number of genetic deaths, it follows that all mutations produce equal damage in the long run and it merely becomes necessary to estimate the number of mutations produced by a given amount of radiation to appraise the damage.

The inadequacy of this model is never more apparent than when it is applied to man. Here the genetic load in a technical sense is too easily equated with its social repercussions. The "genetic death" of a child or an adult who succumbs to an agonizing hereditary disease is not at all equivalent in human terms to an early abortion or to a semisterility or sterility that interferes with childbearing. Genetic variation in man can be evaluated only in terms of the balance between the contribution that a carrier of a genotype makes to society and his social cost.

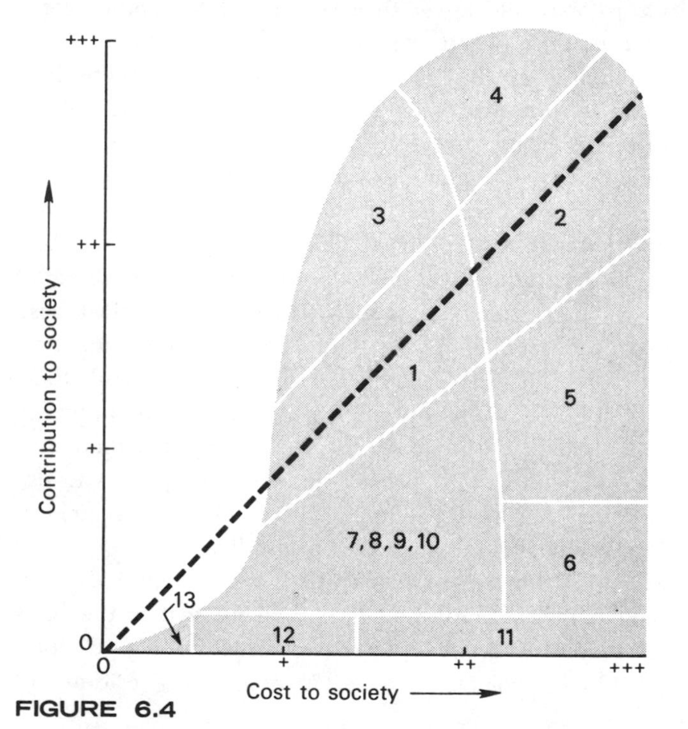

FIGURE 6.4

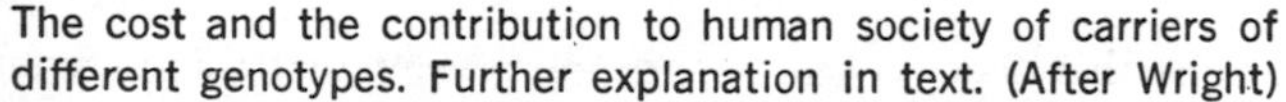

The cost and the contribution to human society of carriers of different genotypes. Further explanation in text. (After Wright)

Wright (1960) illustrates this relationship with the scheme in Fig. 6.4, where the dashed line symbolizes approximate equality of contribution and cost. The meanings of the subdivisions are as follows:

1-2. The bulk of the population (the adaptive norm): approximate balance of contribution and cost at modest levels.

3-4. The genetic elite: high to extraordinary contribution at moderate to high cost.

5. Contribution less than cost (e.g., unearned wealth).

6. Contribution much less than cost (criminals, charlatans, etc.)

7-10. Genetic load: subnormal health, low intelligence, early physical breakdown, major psychoses.

11-13. Genetic load: complete incapacity throughout lifetime, death before maturity, and death before birth.

Fraser (1962) rightly points out that the concept of genetic load "might be more realistic if the average rather than the optimum fitness (of a population) were considered." If one wishes to adhere to Crow's definition of a genetic load as given above, another term, such as "genetic burden" (Dobzhansky 1964a,b), might be more apt to evaluate the biological consequences of genetic variation on the overall adaptedness of a population or a species.

The mean adaptedness of a population is a meaningful and operationally accessible yardstick. One may distinguish further the adaptive norm, genetic burden or load, and genetic elite. The fact that the boundaries between these categories can be established only by convention is an advantage rather than a disadvantage. Consider, for example, classes 5 and 6 in Wright's classification. For some purposes these classes may be considered a part of the social and, by extension, the genetic burden, whereas from the more technical standpoint of public health they belong to the adaptive norm.

Perhaps the most serious drawback of the genetic load and, for that matter, genetic burden concepts is their failure to consider genetic variability in relation to the ecological factors that determine the death rates and population sizes in nature. As pointed out by Wallace (1968a,b).

> Novel sources of premature death need not be met by a corresponding increase in the number of progeny produced. Population size, not progeny size, takes up the immediate shock of both genetic and nongenetic deaths. . . . The important point that emerges is that both genetic and nongenetic loads are absorbed not only by progeny size but by population size as well. Furthermore, with changing population size, the distinction between genetic and environmental loads is to a large extent removed; one type of load can be exchanged for another.

We shall return to a consideration of the ecological aspects of selection in Chapter 7.

Lethal Equivalents and B : A Ratios

Morton, Crow, and Muller (1956) and Morton (1960) proposed to measure the genetic loads of different populations and even of species by expressing them in terms of "lethal equivalents." At the same time, they suggested a method to estimate the relative weights of the two components of the genetic loads—that kept up by mutation pressure and that maintained by heterotic balancing selection. These works can be esteemed as seminal, because of the great amount of experimentation, observation, and theoretical thought that they have evoked. On the other hand, doubts about the validity of the measures and the methods proposed may make it difficult to justify such an accolade.

A lethal equivalent is defined as "a group of mutant genes of such number that, if dispersed in different individuals, they would cause on the average one death, e.g., one lethal mutant, or two mutants each with 50 percent probability of causing death, etc. . . . The total mutational damage per gamete is the average number of lethal equivalents in the zygote that would result from doubling the chromosomes of this gamete." The probability of survival of a zygote, considering a single gene with a "normal" and a deleterious allele, is:

$$1 - qFs - q^2(1 - F)s - 2q(1 - q)(1 - F)sh$$

where q is the frequency of the deleterious allele, s the selection coefficient, and h the coefficient of dominance. F is the coefficient of inbreeding, which equals 0.25 for children of siblings, 0.0625 for children of first cousins, 0.0156 for children of second cousins, etc. Also, qFs is the probability of death due to homozygosis from inbreeding (consanguinity), $q^2(1 - F)s$ the probability of death due to homozygosis under random mating, and $2q(1 - q)(1 - F)sh$ the probability of death due to deleterious effects in the heterozygotes.

A population carries deleterious mutant alleles of many genes. Furthermore, some deaths are due to environmental accidents (x). Morton, Crow, and Muller have assumed that all the genetic and environmental causes of death act independently of one another. The probability (S) of survival of a zygote in a population is then:

$$S = e^{-\Sigma x - F\Sigma qs - (1-F)\Sigma q^2 s - 2(1-F)\Sigma q(1-q)sh}$$

This expression can be conveniently written as:

$$-\ln S = A + BF$$

where A is the mortality from genetic causes and environmental accidents in the population under random mating ($F = O$), and B is the concealed genetic load that would be expressed under complete homozygosis ($F = 1$). The values of A and B separately are:

$$A = \Sigma x + \Sigma q^2 s + 2\Sigma q(1-q)sh$$
$$B = \Sigma qs - \Sigma q^2 s - 2\Sigma q(1-q)sh$$

The total genetic load, expressed under this interpretation in lethal equivalents, is B plus the genetic components of A (i.e., excluding the environmental accidents, Σx). The relative magnitudes of A and B, that is, of the mortality or loss of fitness under random mating and with inbreeding, may give us an insight into the nature of the genetic load.

Consider as before (Chapter 5) a polymorphism for two alleles of a gene maintained by heterotic balancing selection, with selection coefficients s and t acting against the homozygotes. At equilibrium, the loss of viability or fitness will be $A = st/(s + t)$. If the population were to become completely homozygous (i.e., if the heterotic heterozygotes disappeared), the loss of viability or fitness would be $B = 2st/(s + t)$. The $B : A$ ratio is, then, 2 (with three alleles giving equally heterotic heterozygotes it would be 3, and with n alleles it would be n). By contrast, with deleterious genes maintained by mutation pressure and opposed by normalizing selection, the $B : A$ ratio may be much larger. We have seen that, at equilibrium, the heterozygous carriers of deleterious recessives far outnumber the homozygotes. Even if recessivity is incomplete and the heterozygotes are opposed by a selection sh, the $B : A$ ratio will be approximately $1/2h$. If the coefficients of dominance, h, are assumed to lie between 1 and 5 percent, the $B : A$ ratios will be between 10 and 50.

Accurate measurements of inbreeding effects in normally outbreeding species, such as man, were scarce before the work of Morton, Crow, and Muller gave impetus for the collection of such data. These authors utilized whatever observations were available on mortality and morbidity in the progenies of parents not known to be relatives, and among children of relatives, mostly cousin marriages. The average genetic load that a person carries was estimated as between 3 and 5 lethal equivalents (or between 1.5 and 2.5 per gamete). The $B : A$ ratios in different samples ranged from 7.94 to 24.41. This was taken as evidence that genetic loads in human populations are due chiefly to recurrent mutations; polymorphisms maintained by heterotic balancing selection were deemed rare and unimportant.

These conclusions were soon challenged on several grounds. A detailed and critical review of data on human populations has been given by Schull and Neel (1965). Data most carefully collected and analyzed by these investigators themselves for two Japanese cities have yielded the following estimates of lethal equivalents per gamete:

City	Stillbirths	Neonatal Deaths	Infantile and Juvenile Deaths	Total
Hiroshima	0.14	0.29	0.63	1.06
Nagasaki	– 0.02	0.20	0.15	0.33

Not only does the genetic load of the Hiroshima population seem to be three times heavier than that of Nagasaki, but also the $B:A$ ratios for these populations turn out to be 6.08 and 1.08, respectively. Taken at face value, these figures indicate a predominance of mutational load in Hiroshima and of balanced load in Nagasaki. Neel and Schull (1962) pointed out, however, that the same $B:A$ ratio (10 in their theoretical example) may arise under various conditions, with predominance of either the mutational or the balanced component of the genetic load.

By definition (see above), the lethal equivalents measure the genetic load that would be expressed at complete homozygosis ($F = 1$). In reality, the estimates can be obtained only by extrapolation from observations on much lower degrees of inbreeding. In man, marriages of first cousins ($F = 0.0625$) furnish the greater part of the data. Epistatic interactions between the components of the genetic loads could make the extrapolations seriously faulty. Suppose, for example, that the survival rate from disease A is 60 percent, and from disease B is 40 percent. If the survival and mortality from the two diseases are independent, 24 percent of those contracting both diseases simultaneously can be expected to survive. However, it may be that a double insult will leave few or no survivors (positive synergism); conversely, one disease may somehow counteract the other (negative synergism).

This problem is more easily approached in experimental animals or plants than in human materials. Dobzhansky, Spassky, and Tidwell (1963), Stone, Wilson, and Gerstenberg (1963), Torroja (1964), Mettler, Moyer, and Kojima (1966), and Spassky, Dobzhansky, and Anderson (1965) worked with *Drosophila pseudoobscura*; Nicoletti and Giardina (1964), Crow (1968), and Temin, Meyer, et al. (1969) with *D. melanogaster*; Malogolowkin-Cohen, Levene, et al. (1964) with *D. willistoni;* and Levene, Lerner, et al. (1965) with *Tribolium castaneum* and *T. confusum.* In almost all cases evidence of appreciable interactions was found.

The estimates of the numbers of lethal equivalents (B) derived from the publications just cited vary from 0.490 ± 0.094 to 0.876 ± 0.247 for *Drosophila pseudoobscura* and from 0.828 ± 0.129 to 1.435 ± 0.160 for *D. willistoni.* The $B : A$ ratios range from 3.30 ± 1.59 to 7.39 ± 2.67 in *D. pseudoobscura* and from 4.84 ± 0.86 to 14.97 ± 6.67 in *D. willistoni.* The experiments on Tribolium were made on two species (*T. castaneum* and *T. confusum*), in two environments that differ in humidity, and in a variety of outbred and inbred strains. The mortalities were recorded separately at each stage of the life cycle. The mean estimate of B is 0.75 for outbred strains of *T. castaneum* and 0.44 for *T. confusum.* The mean $B : A$ ratios are 3.7 for outbred *T. castaneum* and 1.7 for *T. confusum.*

An overview of the available data leads, then, to a conclusion that could not be (or at least has not been) anticipated on theoretical grounds. The genetic loads in human populations, judged by the criteria used, are of the same order of magnitude as those in the lowly flies and beetles, Drosophila and Tribolium. On the other hand, Sorensen (1969) estimates the load in the Douglas fir, *Pseudotsuga menziesii,* to be between 3 and 27 lethal equivalents, with a median value of about 10. Actually, the genetic loads have not yet been measured with any precision in any of these species. Furthermore, the hope that the $B : A$ ratio may permit one to discriminate between the mutational and the balanced components of the loads has not been realized (see Levene 1963 and Levene, Lerner, et al. 1965).

Balance and Classical Models Contrasted

In Chapter 4 we discussed the classical model of the genetic population structure, which assumes that there exists a "normal type" of each species or population, carrying the "normal genes." Evidence delineated in Chapters 5 and 6 places the validity of this model in doubt. Many genes in natural populations are not represented by homozygous "normal" alleles. Heterozygosis is common, because polymorphisms maintained by various forms of balancing selection are widespread or even ubiquitous. The balance model of the genetic population structure acknowledges genetic diversity as a fundamental phenomenon of nature. The gene pool of a population is envisaged as an array of alleles at many, perhaps at most, gene loci. None of these alleles may be the universally "normal" ones; the fitness conferred by many of these gene alleles on their carriers depends on what other alleles at the same and at other loci are present in the genotype and, of course, on the environment in which the carriers develop and live. There is no "normal type," only an adaptive norm composed of an array of genotypes, the common property of which is that they yield a satisfactory fitness in most of the environments which the population frequently encounters. The adaptive norm is not sharply delimited from the genetic load on one side and the genetic elite on the other. Many genotypes pass from one of these arbitrary classes to the others when the environment is altered.

Following Wallace (1968a), it is convenient to contrast the classical and the balance models by considering the extreme but perhaps not entirely unrealistic situations depicted in Fig. 6.5. In classical Dro-

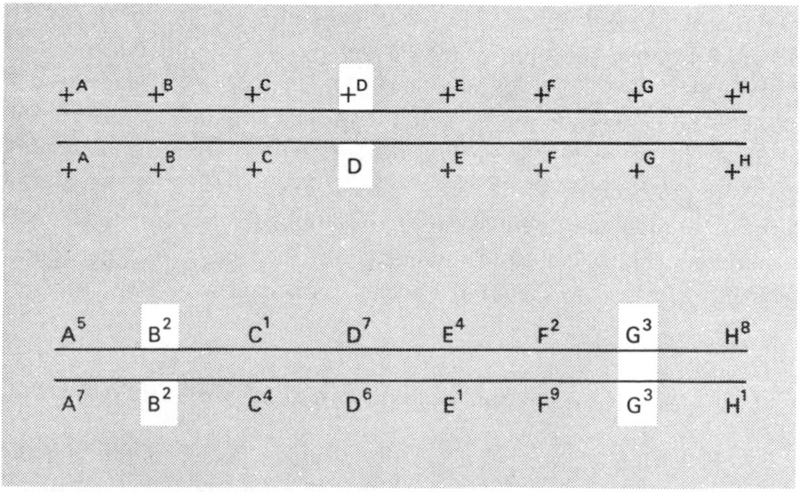

FIGURE 6.5

The classical (above) and the balance (below) models of the genetic population structure. In the former only a single gene locus is heterozygous ($D/+^D$); in the latter only two loci are shown to be homozygous. (B^2/B^2 and G^3/G^3, after Wallace)

sophila genetics it has been customary to symbolize the normal, or wild-type, allele of each gene by a + (plus). The ideal normal type of the classical model would be then homozygous for all + alleles. This ideal is not always realized, however, because of mutation pressure. Many, perhaps all, individuals will have here and there a mutant allele at some locus or loci (Fig. 6.5, upper part). The balance model envisages a greater or lesser proportion of the gene loci being in the heterozygous state, frequently though not invariably for pairs of interacting alleles that give heterosis. How great this proportion is likely to be will be discussed in the next chapter. Some inbreeding occurs in all populations; in the lower part of Fig. 6.5 three-quarters of the loci are represented as heterozygous, and none of the alleles is symbolized by a +.

CHAPTER SEVEN

DIRECTIONAL SELECTION

Sieve and Beanbag Analogies

The action of natural selection has often been compared to that of a sieve. Selection allegedly creates nothing new; it merely allows relatively ill-adapted genetic variants to be lost, and retains the better adapted ones. A related analogy is dubbed "beanbag genetics." According to Mayr (1963):

> The procedure of the classical Mendelian genetics, of studying each gene locus separately and independently, was a simplification necessary to permit the determination of the laws of inheritance and to obtain basic information on the physiology of the gene. . . . The Mendelian was apt to compare the genetic contents of a population to a bag full of colored beans. Mutation was the exchange of one kind of bean for another.

Haldane (1964) promptly made a spirited defense of "beanbag genetics;" it has furnished many useful models of genetic processes, expressed in rigorous mathematical language.

The disagreement is illusory. The sieve analogy does apply in some circumstances. When a bacterial culture of many millions of cells is exposed to a sufficient concentration of an antibiotic, there is perhaps a single mutant cell, a differently colored "bean" in a very large bag, that survives and gives a culture of a resistant variety. The same thing may well occur in a population of an insect pest treated with DDT or another insecticide. On the other hand, Mayr rightly stresses that genes not only act but also interact, a fact of which Haldane was fully aware. Darwinian fitness is a property not of a gene but of a genotype and of the phenotypes conditioned by this genotype. Selection favors, or discriminates against, genotypes, that is, gene patterns. The coadaptation of supergenes, discussed in Chapter 5, and frequency-dependent selection, dealt with in Chapter 6, go beyond individual genotypes and operate in gene pools of populations. To sustain the sieve analogy, one would have to imagine an extraordinary "sieve," in which the loss or retention of a particle would depend on the other particles it contains.

Consider the emergence of man from his Australopithecus-like ancestors. This was hardly a matter of the superposition of a few lucky mutations; major evolutionary advances probably always involve reconstructions of the genetic system. That selection can work only with raw materials arisen ultimately by mutation is manifestly true. But it is also true that populations, particularly those of diploid outbreeding species, have stored in them a profusion of genetic variability. A temporary suppression of the mutation process, even if it could be brought about, would have no immediate effect on evolutionary plasticity. Rapidly evolving groups need not have high mutation rates, nor should evolutionary stasis be taken as evidence of insufficient mutability. Polymorphisms maintained by various forms of balancing selection provide strikingly sensitive genetic systems that can react to changes in the environment without waiting for new mutations to appear. In Drosophila even seasonal changes in the environment evoke genetic alterations (Chapter 5). These alterations are, of course, cyclic and reversible. The gene pool is in constant motion; if a simile is desired, a stormy sea is more appropriate than a beanbag.

The Efficacy of Artificial Selection

Darwin used artificial selection as a model of the natural process; a mathematical theory of selection must almost necessarily be derived from experiments on artificial selection. Perhaps the longest (50 years) selection experiment reported in detail is that of Woodworth, Leng, and Jugenheimer (1952). Selection for high and low protein and oil contents in corn gave results summarized in Table 7.1. Not only was the selection highly effective, but also no "plateau" seems to have been reached even after 50 generations.

An equally spectacular selection experiment, increasing the egg production in a White Leghorn flock between the years 1933 and 1965, has been reported by Lerner (1958, 1968). Spring-hatched pullets of the foundation stock laid on the average 125.6 eggs a year, of which only 24.1 eggs were laid before January 1. In 1965 the average annual production was 249.6 eggs, and the number laid by January 1 almost quintupled compared to the 1933 record.

TABLE 7.1

Progress of selection for protein and oil contents in corn (in percentages) (After Woodworth, Leng, and Jugenheimer 1952)

	High		Low	
Generation	Protein	Oil	Protein	Oil
Initial	10.9	4.7	10.9	4.7
5	13.8	6.2	9.6	3.4
10	14.3	7.4	8.6	2.7
15	13.8	7.5	7.9	2.1
20	15.7	8.5	8.7	2.1
25	16.7	9.9	9.1	1.7
30	18.2	10.2	6.5	1.4
35	20.1	11.8	2.1	1.2
40	22.9	10.2	8.0	1.2
45	17.8	13.7	5.8	1.0
50	19.4	15.4	4.9	1.0

A summary of the history of improvement of the mold *Penicillium chrysogenum*, a producer of penicillin, is given by Dubinin (1961). The productivity of the original strain is taken as 100; selection of a spontaneous mutant increased it to 250. A mutant induced by X-ray almost doubled the productivity to 500, and an ultraviolet-induced mutant further raised it to 900. Another mutant, also induced by ultraviolet, decreased the productivity to 675, but freed the strain from an undesirable yellow pigment. A sequence of further mutations induced by ultraviolet and nitrogen mustard raised the productivity to 2000, 2500, 3000, and finally to 5000 units.

Insect strains can be selected for resistance to DDT and other insecticides (see below) or to poisons such as copper sulfate (Yanagishima 1961). One may wonder whence the genetic variance selectable for such resistances comes. Has nature equipped insect pests with genes for resistance in anticipation of chemists inventing ever new insecticides? Of course, this is not so; resistance to insecticides is achieved by modifications of physiological processes that originally served quite different functions. For example, houseflies resistant to DDT contain dehydrochlorinase, an enzyme that is also present, though in much lower quantities, in nonresistant strains (review in Bender and Gaensslen 1967).

How successful a selection program will prove to be is not easily predictable. Indeed, "no geneticist has succeeded, despite several efforts, in producing rat-sized mice, while dog fanciers, not in the least versed in genetics, have produced a fifty-fold range of variability, stretching from the four-pound Chihuahua to the St. Bernard weighing 200 pounds" (Lerner and Donald 1966). Waddington (1960) selected a strain of *Drosophila melanogaster* that survived feeding on a medium with up to 7 percent sodium chloride; by contrast, *D. pseudoobscura*, after 30 generations of selection for salt resistance, rarely survived on a medium with 3 percent sodium chloride (Dobzhansky and Spassky 1967b). Yet, it is remarkable that almost any trait, at least in sexually reproducing organisms, is changeable by artificial selection if applied systematically and for many generations.

The Heritability and Prediction of Selection Gains

Consider a continuously varying trait, such as size of the body or of its parts, rate of growth, productivity, or Darwinian fitness. The selection of such traits has been practiced for several millennia and studied scientifically for more than a century. Modern genetic theory derives from the works of Wright (1921), Fisher (1930), and Mather and Harrison (1949).

Usually, though not always, continuously varying traits have a polygenic basis. The genes concerned have individually small phenotypic effects, and as a rule give intermediate heterozygotes instead of dominance or recessiveness. Again as a rule, these genes cannot be isolated and studied one by one, although Thoday and his colleagues (Spickett and Thoday 1966, Thoday 1967, and references therein) have achieved just such a feat.

The observed phenotypic variance (σ_P^2, the average squared deviation from the mean) usually has a component due to environmental modifications (σ_E^2) and also a genotypic component (σ_G^2). The latter is, in turn, a composite. The additive or genetic variance (σ_A^2) is "traceable to the differences between the average values of the different alleles in all genetic combinations in which they appear" (Lerner 1958). The remainder of the genotypic variance is due to dominance, recessiveness, or overdominance (σ_D^2), and to the epistatic interaction of genes at different loci (σ_I^2). Fisher's (1930) so-called fundamental theorem of natural selection states that "rate of increase in fitness of any organism at any time is equal to its additive genetic variance in fitness at that time."

An important statistic is the heritability, h^2, which is the ratio of the additive to the phenotypic variance, σ_A^2/σ_P^2. Several methods for the estimation of heritability have been devised. They are alike in principle; the expression of the trait concerned is compared in the population submitted to the selection, in the individuals selected, and in the progeny of these individuals. Heritability of 1.00, or 100 percent, means that the progeny is exactly like the selected parents; zero heritability means absence of correlation between the parents and their offspring. In reality, the heritabilities usually fall between these limits. Examples of heritabilities compiled from Falconer (1960a) and Brewbaker (1964) are as follows:

Trait	Heritability
Amount of white spotting in Friesian cattle	0.95
Slaughter weight in cattle	0.85
Plant height in corn	0.70
Root length in radishes	0.65
Egg weight in poultry	0.60
Thickness of back fat in pigs	0.55
Weight of fleece in sheep	0.40
Ovary response to gonadotrophic horomone in rats	0.35
Milk production in cattle	0.30
Yield in corn	0.25
Egg production in poultry	0.20
Egg production in Drosophila	0.20
Ear length in corn	0.17
Litter size in mice	0.15
Conception rate in cattle	0.05

Heritability is not an immutable property of a trait. For environmentally labile traits heritability increases as the environment in which the population lives becomes more uniform, and decreases in heterogeneous environments. Inbreeding may lower heritability, and hybridization and mutation may increase it. Furthermore, the heritability measured in a single generation may not be the same as the heritability realized over several consecutive generations of selection. A great deal of experimental and theoretical work has been carried out, especially by the Edinburgh school of geneticists, to test the predictive value of the genetic theory of selection (F. W. Robertson 1955, 1957; Clayton, Morris, and A. Robertson 1957; A. Robertson 1960; Falconer 1960a, also Lerner 1958, and references therein). Can selection gains for a given trait be predicted if the heritability of the trait is known? The answer generally is that such predictions are borne out fairly well for several initial generations of selection. As the selection progresses, however, the gains are less and less in accord with expectation, being generally below those anticipated. Eventually the selection brings no response, and a "plateau" or "ceiling" is reached.

The simplest (and often the valid) explanation of selection plateaus is depletion of the store of genetic variation. The limiting case would occur if the selection favored a single genotype among many present in the original population, and this genotype were established to the exclusion of all others. One would then expect that in a population reaching a plateau for a given trait the heritability of this trait would be zero. One would also expect that relaxation (suspension) of the selection would leave the selection gains intact in the generations that followed. These expectations are in fact realized in some experiments; but in others heritability continues to be present, and when the selection is discontinued the selected trait gradually relapses toward the original population mean. Lerner (1954) defined "the property of the population to equilibrate its genetic composition and to resist sudden changes" as genetic homeostasis.

Mutability and Selection Advances

It has been mentioned above that outbred sexual diploid populations generally have sufficient stores of genetic variability to respond to challenges of selection. As pointed out particularly by Mather (1953 and earlier works), the entire contents of these stores are not immediately available for selection to work with. If a trait is influenced by many genes, several genes modifying this trait in plus and minus directions may be linked in the same chromosomes. This variability can be "released" gradually, as crossing over generates chromosomes with only plus or only minus modifiers. In a population that has reached a selection plateau, the selection may suddenly become effective again, and then a new plateau is achieved. Such resumption of the activity of selection may be due to either or both of two causes: recombination of tightly linked genes, and the origin of new mutations. It is often difficult to discriminate between these possibilities

The induction of useful mutants by X-ray or other mutagens was described in Chapter 3. Attempts to augment the supply of genetic variance by similar means have repeatedly been made. The most successful of these was reported by Scossirolli (1954, 1965). His starting material was a population of *Drosophila melanogaster,* which was previously selected for increased numbers of sternopleural bristles and had reached a plateau. Irradiation of this population ushered in a spectacular spurt of selection effectiveness. On the other hand, the attempt of Abplanalp, Lowry, et al. (1964) to achieve a similar enhancement of selection effectiveness for egg production in poultry gave a negative result.

The experiments of Ayala (1966, 1967, and references therein) are particularly interesting because the traits selected in them are closely affiliated to the adaptedness of the populations for living in stressful environments—to be sure, artificial. Populations of two related species, *Drosophila serrata* and *D. birchii,* were maintained by a serial transfer technique. Adult flies were introduced into regular culture bottles and transferred three times a week to fresh bottles; when the eggs deposited developed into new adults, these were counted, weighed, and added to the surviving older adults. Once in 2 weeks the adult populations were etherized, weighed, and counted. The productivity and the longevity, as well as the total population size, were thus recorded. The culture bottles were crowded, and the competition was intense. The resulting natural selection effected a steady improvement in the above parameters of adaptedness. Some populations received 2000 r of X-rays, while others served as unirradiated controls. In Table 7.2 it can be seen that the improvements were noticeably greater in the irradiated than in the control populations. It is hardly necessary to add that this result should be viewed in the perspective of the severity

TABLE 7.2

Productivity achieved by natural selection in unirradiated control and irradiated Drosophila populations (After Ayala 1966, 1967)

	Number of Individuals		Biomass, mg	
Population	Total Population	Produces per Week	Total Population	Produces per Week
D. serrata				
Control	1294 ± 50	595 ± 21	868 ± 32	316 ± 27
Irradiated I	1955 ± 65	941 ± 30	1309 ± 48	527 ± 22
Irradiated II	2558 ± 98	1133 ± 27	1676 ± 60	614 ± 20
D. birchii				
Control	992 ± 67	533 ± 22	651 ± 39	275 ± 15
Irradiated I	1800 ± 167	933 ± 32	1223 ± 81	497 ± 36
Irradiated II	1756 ± 103	833 ± 31	1149 ± 70	457 ± 31

of the selection to which these populations were exposed; the relatively few favorable mutants induced were doubtless far outnumbered by the unfavorable ones that were selected out.

Genetic Homeostasis

Selection of *Drosophila pseudoobscura* for positive and for negative response to light (phototaxis) has given results represented in Fig. 7.1. In every generation, batches of 300 females and 300 males are forced to run through a maze, in which a fly must make 15 choices of light or dark passages. A fly that always chooses light is given a score of 16, and one choosing darkness a score of 1. The population was originally neutral to light, with an average score of almost exactly 8.5. To select for negative phototaxis 25 pairs of flies with lowest scores, and for positive phototaxis 25 pairs with highest scores, are chosen to be parents of the next generation. The realized heritability of behavior with respect to light is quite low, 0.09. Neverthcless, as the figure shows, genetically photopositive and photonegative populations have

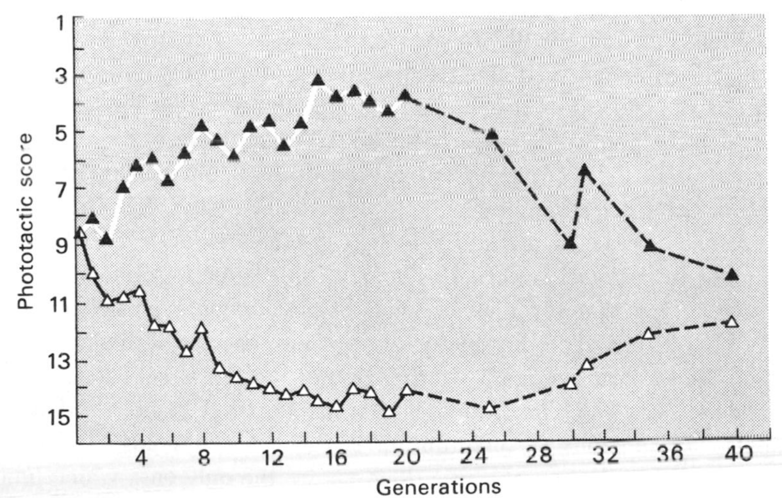

FIGURE 7.1

Selection for positive (light triangles) and negative (dark triangles) phototaxis Drosophila pseudoobscura. After the twentieth generation the selection was relaxed (dashed lines). (After Dobzhansky and Spassky)

been obtained (Dobzhansky and Spassky 1967a, 1969). After 20 generations of selection, very few flies were making equal numbers of light and dark choices.

Starting in the twentieth generation, the selection was relaxed. The parents of the following generations were taken regardless of their phototactic response. In about every fifth generation thereafter the populations were tested for their responses to light. It can be seen (dashed lines in Fig. 7.1) that the divergence in the phototactic response achieved by selection was almost obliterated by the fortieth generation. In point of fact, the population selected for negative phototaxis ended by being slightly photopositive. Very similar results have been obtained with selection for reaction to gravity (geotaxis).

Animal and plant breeders have found, sometimes to their displeasure, that relaxation of selection is quite often, though not invariably, followed by a gradual loss of what the selection had previously accomplished. Moreover, the deceleration of selection gains and the reaching of a "plateau" may occur without depletion of genetic variance. Lerner (1954) has analyzed this situation in a brilliant book. Genetic homostasis may be brought about by several mechanisms that are not mutually exclusive alternatives. The mean values of some characters in a population may be established and preserved by natural selection. For example, the mean size or weight of an animal, the rate of its development, and the mean number of progeny produced are those that yield the best achievable adaptedness in the environments in which the population lives. Artificial selection imposed on a population displaces these characters from their optima and is at loggerheads with natural selection; unless complete homozygosis for the genes concerned is reached, natural selection operates to undo the damage wrought by artificial selection as soon as the latter is relaxed.

A genotype that gives an optimal expression of a particular trait may be a heterozygote for one or more genes, and it will beget less fit homozygotes in the progeny. Lerner quotes the statement of a poultry breeder: "What one bird does, flocks can be bred to do," and comments, "Presumably the assumption is that the physiological factors limiting the production of any individual are the only ones which limit the production of a group. However, the laws governing the physiology of populations transcend those governing the physiology of individuals."

It is not necessary to suppose that the trait artificially selected for, in itself, conflicts with natural selection. Even if it is important in nature for a Drosophila population to be on the average photo- and geotactically neutral, it is not evident that such neutrality is also imperative for populations in laboratory bottles or cages. However, selection for one trait often produces correlated effects on other traits (see the next section). We have used above, as a paradigm of selection processes, the selection in bacterial cultures of mutants resistant to antibiotics or to the attacks of bacteriophages. In populations of diploid and outbreeding forms selection may be a more complex operation. Selection for one trait may unintentionally induce alterations elsewhere in the genetic system. Such alterations may be expected often to conflict with the exigencies of general adaptedness to the environment. Lerner (1954) has stressed particularly that high adaptedness of a population may depend on a certain "level of obligate heterozygosity," that is, on most of the component individuals being heterozygous for many gene alleles maintained, as we would now say, by the heterotic form of balancing natural selection. The level "for each species and probably for each less inclusive Mendelian population is determined by its evolutionary history." This obligate level may be interfered with by rapid changes imposed by artificial or natural selection.

Correlated Responses to Selection

Darwin stressed that "the whole organism is so tied together during its growth and development, that when slight variations in any one part occur, and are accumulated through natural selection, other parts become modified. This is a very important subject, most imperfectly understood." Our present, still imperfect, understanding of such correlations derives mainly from works of Mather and Harrison (1949), Clayton, Knight, et al. (1957), Lerner (1958), and Falconer (1960a,b and references therein). The simplest cause is the pleiotropism of many genetic variants (see Chapter 2). For example, mutants in Drosophila that arrest the formation of the brown pigment in the eyes have pale or colorless testicular envelopes. Tight linkage of genes carried in the same chromosomes is a less simple but quite frequent cause.

The character selected for by Mather and Harrison (1949) was the number of abdominal bristles in *Drosophila melanogaster*; correlated responses included sterility and the numbers of spermathecae. Berg

(1960) studied "pleiades" of correlated characters in several plant species; Belyaev and Evsikov (1967) found the pelage colors in mink to be correlated with fertility and other physiological traits of these animals. Selection in Tribolium for modifications of the development rate also changes the body size, productivity, and viability of the beetles (Dawson 1966). Dobzhansky and Spassky (1967b) found in *D. pseudoobscura* that selection for the ability of larvae to develop on food with increased concentrations of sodium chloride interferes with the progress of simultaneous selection for geotaxis.

The literature dealing with correlated responses to selection for desirable traits in domesticated animals and agricultural plants is extensive and scattered; there are some useful references, in addition to those cited above, in Schwanitz (1967). Correlated responses due to pleiotropisms (physiological correlations) may make it impossible to endow a breed of domestic animals or plants with combinations of characteristics that would be desirable to man but are physiologically incompatible (see Lerner and Donald 1966). Correlations due to linkage can be overcome more easily, if desired crossovers are obtained; nevertheless, they may seriously slow down a selection program (Shultz 1953). Under natural selection, correlated responses may help to explain the combinations, found in many living species, of high adaptedness in some respects with other apparently maladapted traits.

Transference of the gains achieved by selection in one environment to other environments is a kindred problem. Falconer (1960a) selected mice, and Park, Hansen, et al. (1966) bred rats, kept on abundant and on meager diets, for large and for small body sizes. The selection was effective in both directions in both environments. Mice selected for large size on meager diet were bigger than those similarly selected on abundant diet when raised on a low plane of nutrition. Mice selected on either type of diet for small size were equally small when given a meager diet; those selected for small size on low diet were larger when grown on abundant diet.

In analogous experiments of Druger (1967), *Drosophila pseudoobscura* was selected at two temperatures, 25°C and 16°C, for long and for short wings (and large and small body size correlated with wing length). Control flies are larger when grown at low than at high temperature. Selection responses were obtained in both directions at both temperatures. When the selection was made in the same direction as the environmental modification, however, the carry-over of the selection gains to the other environment was incomplete. Results similar in principle were obtained also by Frahm and Kojima (1966) and by Tantawy and El-Helw (1966).

Canalizing selection (Chapter 4) tends to stabilize the development pattern of a species, to make certain traits develop similarly in most environments that this species encounters in its habitats. The result is eventually that some traits show little or no variation, being uniform in all representatives of the species. Are such traits genetically frozen, or can they be altered by selection? To consider a specific case, the number of secondary vibrissae (whiskers) in the house mouse is fixed at 19. However, a mutant gene Tabby reduces this number and makes it variable. Selection on the genetic background of Tabby permits genetic changes in positive as well as negative directions. Eventually, normal (i.e., non-Tabby) mice obtained from the selected strains have numbers of vibrissae deviating from 19 (Dun and Fraser 1959, and Kindred 1967). The constancy of the vibrissae number is evidently a result of developmental buffering, and not of lack of genotypic variance.

Waddington's (1953) so-called "genetic assimilation of an acquired character" is a similar tour de force, but achieved by manipulation of the external rather than of the genetic environment. Certain strains of *Drosophila melangogaster* exposed to 40°C for some hours during the pupal stage show a break in a particular wing vein, giving them a resemblance to the mutant crossveinless. Selection for and against the induction of this phenocopy (Chapter 2) was successful. Eventually the strain giving a strongly expressed phenocopy after the heat treatment showed some breaks in the crossvein at normal temperature as well. The analogy with alleged Lamarckian inheritance is superficial, at best, as Waddington clearly realizes. In a series of most careful studies Milkman (1962, 1965, 1966) demonstrated that polygenic variation modifying the crossveins in Drosophila is widespread in natural populations.

Industrial Melanism

Darwin argued that natural selection must be genetically altering the living species. He did not claim to have observed the alterations; he supposed that the changes are too slow to be noticeable in a human lifetime. In general this remains true; yet some instances are on record in which changes wrought by natural selection have been witnessed. Changes in the chromosomal constitution of Drosophila populations were discussed in Chapter 5. Industrial melanism in moths, however, provides the most striking example, outside the realm of microorganisms.

A melanic, black variant of the moth *Biston betularia* was recorded in Manchester, England, as far back as 1848. The original form of this insect, light gray peppered with black dots, was until then the only one known. By 1895, however, about 98 percent of these moths in Manchester were black, and the melanic variants spread and became common or predominant in most of the industrial districts of England. The appearance and the spread in industrial districts of melanic forms are by now recorded for roughly 100 species of moths, mostly in Britain but also on the European continent and in North America. So spectacular a phenomenon could not fail to bring forth speculations, some of which were, it is now evident, on the wrong track. It was surmised, for example, that melanic mutations were produced by salts of heavy metals (lead and manganese) on the foliage of the vegetation polluted with industrial fumes. This would require mutation rates some orders of magnitude greater than are known to be induced by strong mutagens.

The modern understanding is derived chiefly from the splendid works of Kettlewell (1955, 1961, 1965), Clarke and Sheppard (1966), and Ford (1964, 1965, and references therein). In brief, the melanic variants are colored protectively where they rest on vegetation blackened by industrial pollutants, whereas the light-colored ancestral forms are better camouflaged on light backgrounds. Natural selection favors the variants best protected from predators by being inconspicuous because their colors match the most frequent backgrounds. The chief predators are birds, which hunt by sight and which were observed to pick moths resting on tree trunks.

Kettlewell has given an experimental demonstration that the color camouflage is really effective. He released equal numbers of melanic and light variants of *Biston betularia* in two woods; in one wood the bark of the trees was free of soot and overgrown with light lichens, and in the other the trees were blackened by soot and lichens were absent. In the first wood 12.5 percent of the light and 6.3 percent of the melanic specimens were recaptured alive; in the second wood there were 13.1 percent of lights and 27.5 percent of melanics among those recaptured. Among 190 moths observed to be seized by birds in the unpolluted wood, 164 were melanics and only 26 were light-colored. In the polluted area, 43 light and 15 of the equally numerous melanics fell victims to birds.

With few exceptions, the melanic variants were shown by breeding experiments to differ from the original light forms in single genes, and the dark colorations proved to be dominant over the light ones. Before the nineteenth century mutations from light to melanic forms were doubtless appearing spontaneously in *Biston betularia* and in other species, and were constantly eliminated by natural selection because their carriers were not camouflaged and were exposed to predators. The Industrial Revolution, however, caused widespread pollution of the countryside. The direction of the selection became reversed: the melanics are now protectively colored, whereas the formerly "normal" light forms are at a disadvantage in industrial districts. Here, then, is natural selection modifying a wild species toward better adaptedness in man-made environments.

Substitution of the melanic for the light alleles is, however, not the whole story. Since the alleles for melanism are dominant, the melanics in nature are mostly heterozygotes, at least until they become more frequent than the original form. It appears that during the half century between 1900 and 1950 the heterozygotes evolved heterosis, which they did not possess formerly. Records have been preserved of crosses of melanic to light *Biston betularia* made at the beginning of the century. The progeny showed a 1 : 1 segregation of the dark and the light. Similar crosses made by Kettlewell in 1953, however, gave a significant excess of melanics. Furthermore, the heterozygotes preserved in old collections are appreciably less darkly pigmented than those found in nature now. The advantage of melanism in polluted areas is evidently so great that natural selection promotes the spread of more extreme alleles as well as of some darkening modifying genes, in addition to the major gene for melanism. Hence the genetic difference between the melanic and the nonmelanic forms, as they occur in Britain, involves at present not a single gene but a gene complex. This has been proved by outcrossing the British melanics to a race or a closely related species of Biston from North America. The result was a breakdown of the dominance; instead of clear-cut melanic and nonmelanic segregation, a continuous series of individuals ranging from pale to black was obtained.

Resistance to Insecticides

Fairly numerous species of animals and plants are pests. Some of them injure man directly, as parasites or vectors of disease; many more damage crops and other food sources or cause harm in various additional ways. Pest control is a stern necessity; it aims to restrict the abundance of such species as much as possible or even to exterminate them entirely. Most control techniques utilize various chemicals, insecticides and fungicides, some of them of remarkable potency. Pest populations have responded to these ingenious human inventions, however, by evolving strains resistant to the particular chemical substances that achieved satisfactory measures of control in previous generations. The sequel is what appears to be an endless race between novel means of control and the evolutionary versatility of the pests.

The red scale (*Aonidiella aurantiae*) yielded the first well-established instance of insecticide resistance. Fumigation of citrus trees attacked by these insects with hydrocyanic gas was an effective control technique until 1914, when a resistant population was observed in an orchard near Corona, California. The resistance then gradually spread to other citrus-producing areas of the state, although in some localities fumigation continued to give satisfactory results. Quayle (1938) and Dickson (1940) isolated resistant and nonresistant strains, and showed that F_1 hybrids between them are intermediate in resistance. A segregation is observed in the F_2 generation; the resistance appears to be caused by a single gene without dominance. Cyanide-resistant strains appeared also in other species of scale insects.

The invention of DDT and other highly potent insecticides was followed by their widespread use and misuse. One of the foremost problems in pest control became the emergence of strains resistant to these insecticides. The literature accumulated on this subject is immense. The early work was reviewed by Crow (1957) and Olenov (1958). Brown (1967) has published an excellent review of genetic studies on the resistance of insect vectors of disease. Strains of houseflies resistant to DDT appeared apparently independently in many countries, as have strains resistant to dieldrin, to organophosphorus compounds such as malathion, to carbamates, and to nicotine sulfate. Strains resistant to DDT and dieldrin have been found and studied genetically in several species of anopheline mosquitoes, in *Culex fatigans* and *Aedes aegypti*, in the cockroach *Blatella germanica*, in the cattle tick *Boophilus microplus*, in bedbugs, and in body lice. A variety of genetic situations was found in these organisms, as well as in species of Drosophila used for pilot experiments. The resistance may be due to a single gene, or it may be polygenic. Oshima and Hiroyoshi (1956) succeeded in locating the genes for DDT and for nicotine resistance on different chromosomes of *Drosophila melanogaster* and *D. virilis*. Usually the alleles for resistance show no dominance, although examples of dominant and of recessive alleles are also known.

Some authors looked for evidence as to whether mutations that confer resistance are present in pest populations before the insecticide is applied, or arise after the application. This question is usually insoluble, and it is of no particular importance anyway. The numbers of individuals in the populations of some pests are so great that even rare mutants are likely to be present in many localities. More important, from both the practical and the theoretical standpoint, is whether natural selection completely replaces the alleles that make the insect population sensitive with those that make it resistant to a given insecticide. If it is true, as almost certainly is the case, that the resistant variants are at least slightly lower in fitness than the nonresistant ones in the absence of insecticide treatments, then some of the latter may well be preserved in populations. When the application of a given insecticide is discontinued because it is no longer effective, the selection pressure that favored the alleles for resistance is relaxed, and in fact may be reversed in favor of the original alleles for sensitivity. Hence, after some generations of reversed selection the population may become sensitive again. Pertinent evidence in this connection is reviewed by Keiding (1967). It appears that in at least some cases the relaxation is indeed followed by a weakening of the resistance, but full sensitivity is not restored.

Host-Parasite Coevolution

A biotic community includes usually several, and sometimes many, species, which stand to each other in relations of more or less close mutual dependence. Host-parasite, prey-predator, carrier-symbiont, flowering plant-pollinating insect are obvious examples. Ecologically related or interdependent species may undergo coevolution (Ehrlich and Raven 1964), making their interactions mutually advantageous, or at least minimizing the damage that one or both associates may suffer. The spread and transformation of the myxomatosis disease in rabbits in Australia constitutes a fascinating instance of coevolution, the main stages of which have been observed and recorded (Fenner 1959, 1965).

The rabbit (*Oryctolagus cuniculus*) population in Australia is descended almost exclusively from some two dozen individuals brought from England in 1859. By 1950, the population was estimated in hundreds of millions. Myxomatosis, a virus disease, was introduced as a control measure in 1950. It occurs naturally in the Brazilian rabbit, *Sylvilagus brasiliensis,* in which it causes a relatively benign infection. Transferred to European or Australian rabbits, it gives a lethal disease with gross lesions. The virus spread in Australia and within a few years reduced the rabbit population to about 1 percent of its former strength. However, in 1955 it was noticed that the disease was becoming less virulent. By 1958, only 54 percent of artificially inoculated rabbits showed "severe" symptoms, the rest having "moderate" or "mild" ones (in 1953, 95 percent had "severe" symptoms), and the rabbit population began increasing again. The rabbits have developed genetic resistance, while the virus is becoming less virulent.

Selective processes are taking place both in the host and in the parasite. The virus, which is transferred from rabbit to rabbit by mosquito bites, does not multiply in the mosquitoes; the insects serve merely as "flying needles." The selection favoring more resistant genetic variants among the rabbits is easily understood. The selection for reduced virulence in the virus is no less real. The virus can be transferred only from a living to another living rabbit; death of the host results in death of the virus. Hence the less virulent virus strains that do not kill their hosts, or at least leave them alive longer, have an advantage in transmission. The low virulence of the virus in the Brazilian rabbit species is probably an outcome of such coevolution for mutual accommodation.

Abundant data are available on populations of the stem rust of wheat, *Puccinia graminis tritici.* The wheat rust has what phytopathologists call "races," of which at least 189 are known. The races are identified by their ability to infect, and by the lesions produced on certain wheat varieties; each wheat variety is immune to some but susceptible to other races of the rust (Stakman 1947). A census of the frequencies of the races has been taken in the United States and other countries for some decades (Stakman, Loegering, et al. 1943), and during this period the rust populations have changed (Table 7.3). Breeders select and introduce new varieties of wheat, which are more or less resistant to rust "races" prevalent at the time and in the geographic regions where the selective breeding is practiced. Thus, rust race 56 was first found in the United States in 1928, soon after the susceptible variety Ceres was planted on a large scale. The rapid increase of No. 56 caused a drastic reduction first of the yield, and then of the plantings, of the Ceres variety. Races 36 and 49 were prevalent until 1933, but later became uncommon. No. 56 was rare until 1934 but became widespread after that; No. 34 was common between 1933 and 1935, etc. Here, then, is a competition between artificial selection by man of the wheat plant, and natural selection in the rust fungus.

Attempts to find blight-resistant American chestnuts were reported by Nienstaedt and Graves (1955).

TABLE 7.3

Percentage frequencies of "physiologic races" of the rust Puccinia graminis tritici in the United States in different years (After Stakman, Loegering, et al. 1943)

	Race								
Year	11	17	19	21	34	36	38	49	56
1930	4.0	0.3	0.6	6.7	0.6	36	30	20	0.2
1931	22	0.6	1.3	4.0	2.1	28	15	25	1.0
1932	4.9	1.4	4.9	1.6	0.9	9.6	46	27	2.1
1933	1.7	1.4	1.4	4.5	7.1	3.7	33	37	3.7
1934	0.6	0.6	0.3	1.8	22	21	2.8	1.3	33
1935	19	1.5	1.3	7.4	18	6.1	4.6	1.4	44
1936	12	4.4	1.2	0.8	4.2	3.0	22	1.2	47
1937	8.4	6.1	3.1	0.6	1.1	6.0	8.7	7.4	56
1938	2.0	3.0	6.4	1.0	0.8	1.2	16	0.9	66
1939	3.2	10	3.3	0.4	0.6	0.8	24	0.6	56
1940	1.2	34	2.2	0	0.5	1.8	10	1.2	44
1941	1.3	51	3.8	0	0	2.5	6.0	2.4	32
1942	0.3	27	6.2	0	0.2	2.3	27.3	3.9	31
1943	0.1	23	1.9	0	0.1	0.4	24.4	0.3	49
1944	0	21	6.6	0	0	0.2	26.1	0.2	43

Natural Selection in Plant Populations

Human activities have altered, sometimes radically, the environments of many living species. These species had either to respond by adaptive genetic changes or to die out. Bradshaw and his colleagues have made a series of studies on races of some common species of British plants that became adapted to live in harsh man-made environments (Bradshaw, McNeilly, and Gregory 1965, Jain and Bradshaw 1966, McNeilly and Bradshaw 1968, and McNeilly 1968). The soils near lead, copper, and zinc mines contain the salts of these metals in amounts that are highly toxic to plants. Nevertheless, some species of plants grow successfully on these soils. In particular, the grass *Agrostis tenuis* growing on soils contaminated with lead or copper, and *Anthoxanthum odoratum* (sweet vernal) on zinc-contaminated soil, have been investigated. The heavy metal tolerance of the plants on contaminated soils is genetic rather than due to acclimatization. It is absent in populations of the same species growing only a short distance away from the mines.

What is remarkable is the precise localization of the metal-tolerant genotypes on the contaminated soils. Copper-tolerant *Agrostis tenuis* grows on copper mine workings over an area about 400 × 100 meters in size. The surrounding grassland has little copper in the soil, and the Agrostis there is intolerant. The transition from full tolerance to intolerance may occur over a distance of only a few meters. The sharpness of the transition depends, however, on the gene exchange between the tolerant and the intolerant colonies. Where the prevailing winds carry the pollen of the tolerant plants toward the normally intolerant populations, the transition is more gradual than where the prevailing wind direction is the opposite. The selection against the intolerant genotypes on the contaminated soils is very severe; the tolerant genotypes are selected against on uncontaminated soils rather less rigidly.

A parallel situation, but one involving selection by natural rather than man-made environments, is described by Aston and Bradshaw (1966). *Agrostis stolonifera* grows both in locations exposed to winds from the sea and in nearby locations protected from these winds. The plants from wind-exposed locations have much shorter stolons than those growing in protected places, and this difference is also genetic. The sharpness of the transition is again dependent on the wind direction and the consequent pollen transport and gene flow. The selection evidently favors plants with short or long stolons, respectively, in the habitats exposed and protected from winds.

Haldane's Dilemma and Load "Space"

Haldane (1957) pointed out in a seminal paper that evolution entails a "cost." If certain assumptions are granted, the substitution of one gene allele for another or of one chromosomal variant for another can be said to result in some genetic deaths. Moreover, this cost may be so great as either to slow down the evolutionary process or to seriously depress the vigor of the population on which the selection acts. If the substitution of alleles at many gene loci is involved, the change may require too many generations and too long a time to be realized. Kimura (1960, 1961) reaffirmed Haldane's inferences and introduced the concept of substitutional genetic load. This is a measure of the lowering of the Darwinian fitness in a population undergoing selection, compared to the supposedly optimal fitness reached when the substitution is completed.

Haldane considers as a model the substitution of the dominant allele for melanism in *Biston betularia* for the recessive light allele (see above). In his words, "It is convenient to think of natural selection provisionally in terms of juvenile deaths. If it acts in this way, by killing the less fit genotypes, we shall calculate how many must be killed while a new gene is spreading through a population." The total number of deaths, D, needed for the substitution is

$$D = -\ln p_0$$

where p_0 is the initial frequency of the favored dominant. For a gene without dominance $D = -2 \ln p_0$, and for a recessive gene $D = p_0^{-1} - \ln p_0 + O(k)$, where k is the selection coefficient in favor of the optimal genotype, O. Since genes that are favored by selection but were previously unfavorable are initially rare, and so are favorable genes first arising by mutations, the frequency p_0 is usually small. Taking it to be of the order of 5×10^{-5}, Haldane calculates that the values of D will lie between 10 and 100, and takes 30 as representative. He notes that, unless the selection is very strong, D is nearly independent of the selection coefficients.

In biological terms, the value $D = 30$ means that 30 times the average number of individuals comprising the population in a generation must die to achieve the substitution. If the deaths due to the selection remove, on the average, 10 percent of the individuals, about 300 generations will be needed to accomplish the substitution at a single gene locus. In reality, evolutionary changes involve the substitution of many genes. The ravages of death necessary to substitute favored alleles at many independently acting genes will be so enormous that either the population must become extinct, or else unwieldy lengths of time and numbers of generations will be required for it to become improved by the selection. If species differ in hundreds or even thousands of genes, it is hard to understand how new species can ever be formed. Haldane was too good a biologist not to see that the result he reached creates a dilemma. He concluded his paper thus: "I am quite aware that my conclusions will probably need drastic revision. But I am convinced that quantitative arguments of the kind here put forward should play a part in all future discussions of evolution."

A cognate predicament arises if we consider the "costs," in terms of deaths (or reduction of fertility), of maintaining heterotic balanced polymorphisms (Chapter 6). Suppose that the three genotypes A_1A_1, A_1A_2, and A_2A_2 have fitness 0.9, 1, and 0.9. The mean fitness of the population at equilibrium is, then, 0.95. A population that consists of the "optimum genotype" A_1A_2, that is, of heterozygotes alone, would have a fitness of 1.00. Such a population cannot exist for more than a single generation in a sexual outbreeding species. Sacrificing 5 percent of the zygotes to maintain heterotic balance is not a prohibitively large cost under most circumstances.

We have seen, however, that there are valid reasons to suppose that many natural populations maintain many unfixed gene alleles balanced by heterosis. Suppose that there are 100 polymorphic genes like the above; if the reductions of fitness they produce are independent and multiplicative, the average fitness of a population containing them will be only 0.95^{100}, or about 0.006. In other words, only 6 zygotes out of every thousand produced will survive, and 994 will suffer genetic death. The calculation does not take into account deaths due to environmental accidents, which befall presumably almost every population. If environmental and genetic deaths are also independent, very few if any organisms are fecund enough to withstand such devastations. Higher vertebrates and man certainly do not possess enough "load space" to maintain more than a very few balanced polymorphs.

What Proportions of Genes Are Polymorphic and Heterozygous?

Evidence reviewed in the foregoing chapters shows, conclusively in my opinion, that much of the genetic variability in natural populations of outbreeding sexual species is maintained by several forms of balancing selection. This evidence upholds the balance model, rather than the classical model, of the genetic population structure. The "normal" species genotype is a will-o'-the-wisp, like Plato's *eidos*. With many gene loci, no single allele can be regarded as "normal"; the adaptive norm of a species or a population is an array of heterozygotes at many loci. A representative genotype in natural population is more like the one in Fig. 6.5B than that in 6.5A.

One is now naturally led to inquire just what proportion of the gene loci are polymorphic, and what is the approximate number of genes for which an average individual of a species is heterozygous. A path toward such explicit quantification of our models of the genetic population structure has been opened by the brilliant work of Hubby and Lewontin (1966 and Lewontin and Hubby 1966). They strove to find a technique that satisfies the following criteria: "(1) Phenotypic differences caused by allelic substitution at single loci must be detectable in single individuals. (2) Allelic substitutions at one locus must be distinguishable from substitutions at other loci. (3) A substantial portion of (ideally all) allelic substitutions must be distinguishable from each other. (4) Loci studied must be an unbiased sample of the genome with respect to physiological effects and degree of variation."

Studies of the electrophoretic mobility of enzymes and other proteins offer the best available approach to fulfilling these requirements. We have seen in Chapter 2 that gene mutations come from the substitution, loss, or addition of a single or several nucleotides in a DNA chain in a chromosome. In turn, this results in the substitution of one or more amino acids in the protein coded by the altered part of the DNA chain. A certain proportion, which Lewontin and Hubby estimate as approximately one-half, of the amino acid substitutions change the electric charge of an altered protein, and make it move more rapidly or more slowly in the electric field. It is evidently important to detect genes represented in a population by more than a single allele, as well as fixed, invariant genes. Yet Mendelian genetics is concerned with gene differences; the operation employed to discover a gene is hybridization: parents differing in some trait are crossed, and the distribution of the trait in the hybrid progeny is observed. The technique of electrophoresis also detects proteins that are uniform in all individuals studied; it allows, "as a first order of approximation, to equate a protein without any detectable variation to a gene without detectable variation" (Hubby and Lewontin 1966). To avoid bias, an enzyme or other protein is chosen for the study when chemical techniques for its detection are available, without prior knowledge of its variability or fixity.

Prakash, Lewontin, and Hubby (1969) have examined 24 gene-loci-controlling enzymes or larval proteins. Samples of natural populations of *Drosophila pseudoobscura* were taken in three localities in the western United States and in one locality in Colombia, South America. Among these genes, approximately 40 percent are polymorphic, that is, are represented by more than a single distinguishable allele in at least one population (Table 7.4). In the populations of separate localities, only that of Colombia has a lower polymorphism; this population is isolated from the main body of the species by a distribution gap of some 2400 kilometers. From the detected frequencies of homozygotes and heterozygotes at different loci, including those exhibiting no polymorphisms, the percentages of genes heterozygous in an average individual can easily be computed. As shown in Table 7.4, this average, considering all the populations, is 12.3 percent.

It must be emphasized that the above values are patently underestimates; if only one-half of the variant proteins are discriminated by the technique of electrophoresis, the percentages must be approximately doubled. Evidently the crucial problem is whether the genes studied are indeed a fair sample of at least the structural genes composing the hereditary endowment of the species. Since care was taken to avoid preferential choice of variable genes, there seems to be no valid reason to suspect a bias. A problem no less urgent is whether these spectacularly high proportions of polymorphic and heterozygous genes occur in organisms other than *Drosophila pseudoobscura*. Johnson, Kanapi, et al. (1968) in *D. ananassae*, Stone, Johnson, et al. (1968) in *D. nasuta*, O'Brien and MacIntyre (1969) in *D. melanogaster* and *D. simulans*, and my colleagues Ayala, Richmond, Perez, and Mourão (unpublished) in four species of *D. willistoni* groups have obtained results very similar to those in *D. pseudoobscura*.

What about man? Harris (1966) studied ten enzymes in the blood

TABLE 7.4

Percentages of polymorphic genes, and estimated percentages of genes heterozygous in an average individual, in populations of Drosophila pseudoobscura (After Prakash, Lewontin, and Hubby 1969)

Population	Number of Genes	Percent Polymorphic	Percent Heterozygous
California	11	46	14.0
Colorado	10	42	11.0
Texas	9	38	12.0
Colombia	6	25	4.4
Mean	9	38	12.3

TABLE 7.5

Enzyme polymorphisms in a human population (After Harris 1967)

Enzyme	Number of Alleles	Frequency of Commonest Phenotype
Red-cell acid phosphatase	3	0.43
Phosphoglucomatase		
PGM_1	2	0.58
PGM_3	2	0.53
Placental alkaline phosphatase	3	0.41
Acetyl transferase	2	0.50
Adenylate kinase	2	0.90
Serum cholinesterase		
E_1	2	0.96
E_2	2	0.96
6-Phosphogluconate dehydrogenase	2	0.90

serum and placentae in the population of England. Four of the ten (red-cell acid phosphatase, two phosphoglucomutases, adenylate kinase) proved polymorphic. The average proportion of genes for which an individual is expected to be heterozygous may also be estimated from blood group polymorphisms and is about 16 percent (Lewontin 1967a). Table 7.5 lists nine gene loci responsible for enzyme polymorphisms, each with two or more alleles, the rarest of which has a gametic frequency greater than 0.01 (Harris 1967). Very rare alleles may be mutants controlled by normalizing selection, and 1 percent frequency is arbitrarily taken as a minimum for polymorphism maintained presumably by balancing selection. In five of the nine the frequency of the commonest phenotype (either a homozygote for the commonest allele or a heterozygote for two alleles) is between 40 and 60 percent. If all the genes in Table 7.5 are considered, the probability of two individuals taken at random having the same enzyme phenotype is only 7 per 1000. The only mammal other than man for which comparable data are available is the house mouse. Selander, Hunt, and Yang (1969) have analyzed 41 genes in Danish populations, and found 41 per cent of them polymorphic. The estimated proportion of the genes heterozygous in an average individual is 8.5 percent.

Suppose that a Drosophila zygote has 10,000 gene pairs, and a human zygote 20,000. Both figures are underestimates (see Chapter 3). The figures for the estimated proportions of the genes that are polymorphic in populations or heterozygous per individual are also, in all probability, underestimates. It then turns out that natural populations of Drosophila are polymorphic for at least 3000, and human populations for 6000, genes; an individual Drosophila is heterozygous on the average for some 1150 genes, and an individual human being for 3200.

The experimental findings in Drosophila and in man are clearly in accord with the balance rather than the classical model of the genetic population structure. The maintenance of the abundant polymorphism and heterozygosity in populations demands, however, an explanation. Several possibilities may be considered. If the heterozygotes for pairs of alleles that yield variant enzymes are superior in fitness to the homozygotes, the polymorphisms can, theoretically, be maintained by the heterotic form of balancing selection. The difficulty is, as we have seen, that if the effects on fitness of the polymorphisms of different genes are assumed to be independent and multiplicative, the genetic load becomes too heavy to carry. Suppose that each of the 3000 polymorphisms in Drosophila, and each of the 6000 in man, reduce the fitness of the population by as little as 1 percent. The aggregate reduction would be 0.99^{3000} and 0.99^{6000}, or 10^{-13} and 10^{-26}, respectively, of the optimal fitness. No organism, let alone the human species, could withstand such an eclipse of its reproductive potential.

Some forms of frequency-dependent balancing selection impose no genetic load when equilibrium frequencies are reached. Suppose, for example, that the carriers of genes for some variant enzymes have an advantage in mating when they are rare in the population, but a disadvantage when they are too common (see Chapter 6). Selection will tend toward an equilibrium when the mating success of the different genotypes is uniform. Although Yarbrough and Kojima (1967) obtained an indication of such selection for the esterase-6 gene in *Drosophila melanogaster*, it seems far-fetched to assume a similar selection for all of the thousands of polymorphic loci. Diversifying selection can also maintain polymorphisms if the various genotypes have superior fitness in different ecological niches or subenvironments. This might impose no genetic load if each genotype could unfailingly choose the environment to which it is best adapted, but more and more load results as environmental misplacement becomes frequent.

The easiest way to cut the Gordian knot is, of course, to assume that a great majority of the polymorphisms observed involve gene variants that are selectively neutral, that is, have no appreciable effects on the fitness of their carriers. This option has in fact been taken by Kimura and Crow (Kimura 1968, Kimura and Crow 1969, and Crow 1969). The problem of adaptively neutral genetic variants will be taken up in Chapter 8.

Rigid versus Flexible Selection

In a scientific theory simplicity is evidently a desirable trait. In a sense, all science is an attempt to simplify the world in order to make it comprehensible. Yet simplicity is no warrant of validity; some brilliant scientists have been deluded into imagining that nature must have adopted the simplest solution that their minds were able to devise for her problems. Our strategy must be to try out simple hypotheses and theories first, but be ready for more complex ones if evidence indicates the need for them. Haldane's dilemma is a product of a theoretical study; the load-space dilemma arises from a series of experimental investigations climaxing in the research of Lewontin, Hubby, Harris, and others. These dilemmas are, hopefully, blessings in disguise, since they are forcing geneticists and evolutionists to look for paradigms of the natural selection process other than the classical one of "genetic death." Although no generally compelling new paradigms have yet emerged, the search for them is an inspiring intellectual adventure.

Defining the genetic load as the decrease of fitness in comparison with that of the optimum genotype leads to predicaments some of which have been discussed in Chapter 6. Wright, Wallace, the present writer, and others attempted to escape these predicaments by using the mean fitness, rather than the elusive optimal fitness, as an operationally meaningful standard to distinguish the adaptive norm from the genetic burden and the genetic elite. Li (1963) has exposed another paradox: "If we calculate the genetic load in terms of the highest fitness value, the population will never have any gain no matter how beneficial the mutation is. In fact, the more beneficial the mutation, the greater the genetic load, implying that the population is suffering from a greater amount of genetic elimination and is worse off from an 'optimum' genotype." Brues (1964) presented arguments essentially similar to those of Li, and contrasted the "cost of evolution with the cost of not evolving," the latter being much the greater.

In science, as in everyday life, it is possible to have before one's eyes all the components of a pattern and still fail to see the pattern. Feller (1967) the mathematician and Wallace (1968a,b) the biologist pointed out that this is in effect what has been happening to evolutionary geneticists, who have thought in terms of relative fitnesses and of relative frequencies of genes, and neglected the absolute population sizes and the factors controlling them in nature. Darwin, in deriving his theory of natural selection, used the quite general observation that only a fraction, and often a small fraction, of the progeny brought into being survive and become the parents of the next generation. For a bisexual population to be numerically stable, each of the breeding females must leave, on the average, one daughter, who will survive and reproduce. If the average is more than one daughter per mother, the population will expand; if less than one, it will contract. Most of the progeny die out or fail to reproduce because of environmental and genetic causes. The key question is, Are these causes always independent? Actually, both independence and interdependence occur in nature.

Consider a population that produces 1,000,000 fertilized eggs or seeds. Because of a limited food supply, restricted space, and other density-dependent factors, in a certain generation only 10,000 individuals reach maturity. In another generation the environment becomes more permissive, and 100,000 individuals mature. Suppose further that this population carries a burden of recessive lethal genes, such that 5 percent of the fertilized eggs are homozygous for these lethals. These 5 percent will die, regardless of whether only 10,000 or 100,000 individuals are able to reach maturity. Wallace calls this hard (I prefer to call it rigid) selection. By contrast, suppose that 5 percent of the zygotes have a genotype that enables them to survive under conditions such that a total of 100,000 individuals mature, but that makes them die where only 10,000 reach maturity. This is soft (according to Wallace) or flexible selection.

In what sense can a genotype subject to flexible selection be considered a part of the genetic load or burden of the population? The carriers of this genotype are among the zygotes eliminated when only 10,000 individuals survive, but when 100,000 individuals mature this genotype is included among the survivors. The genotype is not responsible for "genetic deaths," and it does not affect the size of the population. It will endanger the population only if it becomes very frequent in a succession of generations in a benign environment, after which the population is exposed to more stringent environmental conditions.

Unambiguous examples of flexible selection operating in nature are not easy to come by. This means, not that such selection is rare or unimportant, but only that interactions of genetic selection and ecological variables are little known. Consider again the relative fitnesses of the heterokaryotypes and homokaryotypes in *Drosophila pseudoobscura* described in Chapter 5. If the fitness of the best karyotype is taken to be 1.00, some other karyotypes are found to have much lower fitnesses. The homozygotes for the CH gene arrangement have a fitness of only 0.26, far down in the semilethal range. Does this mean that populations monomorphic for CH chromosomes cannot survive? Quite the contrary—such populations have been maintained in laboratory population cages, and at first sight they seem to be as flourishing as polymorphic populations, in which a majority of the individuals have fitness close to unity. To be sure, more detailed studies have disclosed that polymorphic populations usually have higher intrinsic rates of increase and produce more individuals from the same amount of food than monomorphic ones (see Chapter 5). All these observations are mutually consistent. Here the selection is flexible. Monomorphic CH populations have, under the conditions of the laboratory population cages, an adaptedness quite sufficient to survive and even flourish; yet when the populations are made chromosomally polymorphic, the comparatively low Darwinian fitness of the carriers of CH chromosomes depresses their relative frequency or even eliminates them entirely.

Selection Thresholds

Sved, Reed, and Bodmer (1967), King (1967), Sved (1968), Maynard-Smith (1968a), and others have devised mathematical models of the operation of natural selection. Although these models are not identical, we may discuss them conjointly, since they postulate various forms of flexibility of selection. Consider a population with a fairly large number, such as 1000, of polymorphic genes maintained by a heterotic balancing selection. Suppose that the fitness of the heterozygote for each gene is 1 percent higher than that of the homozygotes (selection coefficients $s = t = 0.01$). With random mating, the numbers of genes for which individuals in this population will be heterozygous will form a probability distribution with a mean of 500. Individuals heterozygous and homozygous for most or for all of these genes will seldom or never be produced. Assume now that all genotypes with a number of heterozygous genes above a certain minimum have similar fitness, or at any rate that the fitness does not grow or decrease in a direct proportion to the number of heterozygous or homozygous loci.

Such a selection threshold has in fact been inferred by Lerner (1954), chiefly from experiments on the genetics of poultry, and called by him the obligate level of heterozygosity. The selection does not act on each gene independently from the others; it discriminates against individuals that carry more than a certain number of unfavorable genes. Each "genetic death," if one wishes to continue using this expression, removes from the population not a single such gene but an array of them. King (1967) concludes, "One thousand or more polymorphisms can be maintained through an average heterozygote superiority of about 1 percent per locus, with a very small total effect on the variance of fitness in the population." Sved, Reed, and Bodmer (1967) say cautiously that this number "could not be much greater than 1000." The reason for this caveat is that the theory leads one to expect losses of fitness on inbreeding that are perhaps greater than those observed in experiments.

In discussing Haldane's dilemma, Sved (1968) makes the following point:

The only condition required for a gene to be selectively advantageous is that individuals possessing the gene be on the average fitter than those not possessing it. If the population size is controlled in some density-dependent manner, then the increase in the numbers of individuals possessing the gene could be purely at the expense of those not possessing it. This could, for example, be due to inherited differences in ability to compete for some limiting resource. The mean fitness of the population, as measured by the immediate change in the population size, might be raised, conceivably lowered, or very likely left undamaged as a result of the gene substitution.

Sved then proceeds to show that, with selection working as he has stated, there is no obvious upper limit to the number of loci or the rate at which the allelic substitution may take place.

The arguments of Maynard-Smith (1968a) are essentially similar to those of Sved, but he discusses briefly the possibility that gene substitutions not controlled by density-dependent factors are free of any "cost" of genetic deaths. Suppose that the distribution area of a species or a population is limited by a boundary beyond which it is unable to stand the climatic conditions, such as cold or dryness. Mutations in one or more genes that confer such an ability will make possible a spread into previously uninhabitable territories. In such a case the total species population expands and does not suffer losses on account of a substitutional load.

In summary, the problems of the maintenance of genetic variability in natural populations, and of the ways in which natural selection acts, are as yet far from solved. These are basic problems of any causal theory of evolution. Both theoretical and experimental studies in this field, however, have resulted in gratifying or even spectacular advances in recent years. Moreover, it is probable that the state of the knowledge as outlined above will be surpassed in the near future.

CHAPTER EIGHT

RANDOM DRIFT AND FOUNDER PRINCIPLE

Directed and Random Processes

The elementary components of evolutionary changes are alterations of the frequencies of gene alleles or chromosomal variants in the gene pool of a population. Consider the simplest case of a single gene; the change can be symbolized as a movement of a point on a frequency scale in either direction from 0 (absence of an allele) to 1 (its fixation). If two, three, or n genes are considered, the point must be envisaged as moving in a two-, three-, or n-dimensional space. Gene frequency changes may be brought about by deterministic (directed) or stochastic (random) causes. With the former, the gene frequency in any generation is predictable, provided that the pressures impelling the change are known; with the latter, the frequency is indeterminate, and only the variance of the possible frequencies can be anticipated. Wright's (1955) classification of the "modes of change in gene frequency" is as follows:

1. *Directed processes*
   a. Recurrent mutation
   b. Recurrent migration and crossbreeding
   c. Selection
2. *Random processes*
   a. Fluctuations in mutation rate
   b. Fluctuations in migration
   c. Fluctuations in selection
   d. Accidents of sampling
3. *Unique events*
   a. Novel favorable mutation
   b. Unique hybridization
   c. Swamping by mass immigration
   d. Unique selective incident
   e. Unique reduction in numbers.

For didactic reasons, the operation of each of the above "modes" can be studied in isolation from the others. Only rarely, however, are evolutionary changes in nature due exclusively to one mode; far more often, the interaction of several modes is involved. This simple consideration has been emphasized repeatedly by Wright, because its disregard has led to some unenlightening and at times acrimonious polemics. An evolutionary change need not be due either to directed or to random processes; quite probably it is the result of a combination of both types. The theoretically desirable and rarely achieved aim of investigation is to quantify the respective contributions of the different factors of gene frequency change, as well as their interactions.

Random Genetic Drift

The idea of genetic drift was adumbrated by Brooks (1899) and by A. L. Hagedoorn and A. C. Hagedoorn (1921). It was developed independently by Fisher (1928, 1930), Dubinin and Romaschov (1932 and Dubinin 1931), and especially Wright (1921, 1931, 1932) and Malécot (1948, 1959). Sometimes referred to as the "Sewall Wright principle," it has been misused in a way Wright himself never intended, namely, as a spurious "explanation" of evolutionary changes that seem to be devoid of adaptive significance, and therefore hard to explain by natural selection. In the absence of mutation, selection, and migration, the frequencies of genetic variants in a population remain, in accord with the Hardy-Weinberg principle (Chapter 4), constant generation after generation. This is strictly true, however, only in ideal, infinitely large populations. In reality, no population is infinite and many are small.

Consider sexual diploid populations of 500,000, 5000, and 50 individuals, respectively. If the population sizes remain constant, the next generation will come from samples of 1,000,000, 10,000, and 100 gametes from the gene pools of the previous generation. Suppose that two alleles, A and a, are in some generation equally frequent in these populations ($p = q = 0.5$). The sampling process introduces a variance of pq/N and a standard deviation $\sqrt{pq/N}$, where N is the number of gametes sampled. The frequencies of alleles A and a in the next generation will, therefore, be 0.5000 ± 0.0005, 0.500 ± 0.005, and 0.50 ± 0.05, respectively, in the three populations. This means that in about 95 of 100 samples the gene frequencies will be between 0.4990 and 0.5010 in the large, between 0.490 and 0.510 in the intermediate, and between

0.40 and 0.60 in the small population. The variation in gene frequency may be considered negligible in the large, small in the intermediate, but appreciable in the small population.

Evidently the gamete sampling process occurs in every generation, and the variance pq/N grows in proportion to the number of generations elapsed. Imagine many isolated colonies of a species, all of which start with equal frequencies, $p = q = 0.5$, of two alleles of some gene. The average frequency of the alleles, even after many generations, will still be 0.5. If the populations of the colonies are large, the spread of the frequencies among individual colonies will be small, following a high-peaked bell-shaped curve with a mode at 0.5. However, in small colonies one of the alleles may be lost or fixed ($p = 0$ and $q = 1$, or vice versa). A lost allele can be reintroduced only by mutation or migration. On the assumption that neither occurs, the frequencies of the colonies with different proportions of A and a will reach a U-shaped distribution, as shown in Figs. 8.1A and 8.1B. In other words, more and more colonies will contain only allele A or only allele a, and intermediate allele frequencies will be progressively infrequent. Given enough time, the colonies will be of two kinds, some monomorphic for A and others monomorphic for a. This result will be achieved sooner if the colonies are small than if they are large.

Interactions of Random Drift, Selection, Mutation, and Migration

These interactions have been examined especially by Wright (1931, 1940, 1948, 1966). His mathematics are too abstruse to be presented here, but his premises and conclusions are admirably simple and clear. The smaller the population size, the greater are random variations in gene frequencies, and the less effective become weak selection pressures. In small populations, alleles favored by selection may be lost and less favored ones may reach fixation. In large populations even very small selective advantages and disadvantages will eventually be effective, but a more rigorous selection is needed to overcome the random drift in small populations.

The relations between population size and selection intensity are illustrated in the diagrams in Fig. 8.1. The abscissae indicate the gene frequencies from 0 (loss) to 1 (fixation). The ordinates may be inter-

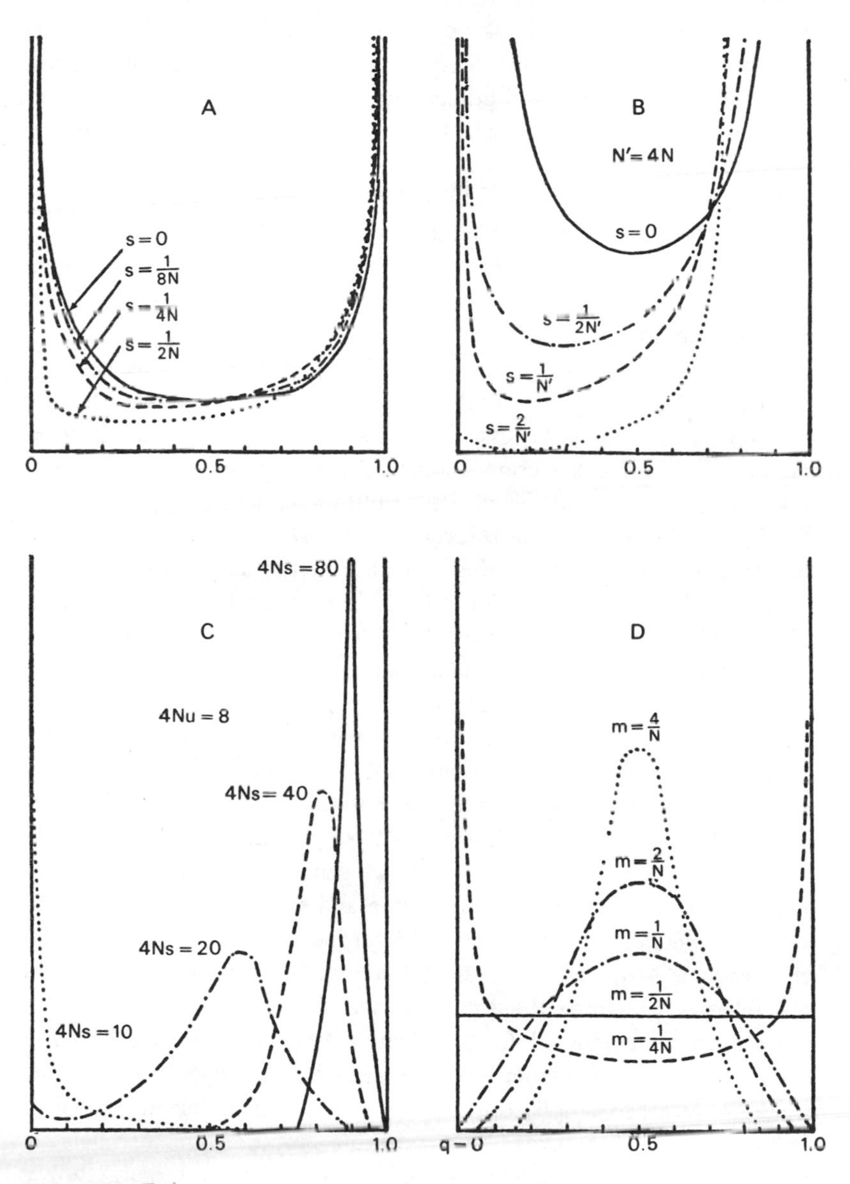

FIGURE 8.1

Distribution of gene frequencies in populations of different size under different selection, mutation, and migration pressures. (From Wright)

preted in any one of three ways. First, one may consider the fate of the different genes in a single population, for natural populations vary with respect to many genes. Some of these genes may reach fixation, others may be lost, and still others remain unfixed, represented by two or more alleles with different frequencies ($0<q<1$). The ordinates indicate, then, the frequencies of the different genes in a population. Second, one may follow the fate of the same gene in different populations, in the colonies into which a species is broken up. The ordinates refer then to the frequencies of the subgroups in which a given gene frequency is reached. Third, the ordinates may show how often, in the long run, any one gene comes to possess a given frequency.

In small populations (Fig. 8.1A) the gene frequency curves are U-shaped. A majority of variable genes are either fixed or lost most of the time. The effectiveness of weak selection is low in small populations. With small selection coefficients, of the order $s = 1/8N$ to $s = 1/2N$, the shape of the curves is little modified. Genes are lost or fixed at random, with little reference to the selection pressure. Figure 8.1B represents the action of a selection of the same intensity as in Fig. 8.1A, but in a population that is four times larger ($N' = 4N$). Here a selection of the order $s = 2/N'$ is rather effective; the curve no longer is U-shaped but has a maximum at the right, indicating that the gene alleles favored by selection largely supplant the less favored ones.

The interaction of mutation pressure with selection and the population size factor is even more complex. If the selection pressure is of the same order as the mutation rate, and both are small, the random variations of the gene frequencies in small populations become important. The curve is U-shaped, indicating that gene alleles reach fixation or loss largely irrespective of the mutation and selection pressures. Mutation and selection rates that may be regarded as small are such as make the products $4Nu$ and $4Ns$ less than unity. The greater the mutation and selection pressures, the greater is their effectiveness.

An example borrowed from Wright (1931) is reproduced in Fig. 8.1C. It refers to a population of intermediate size with a moderate mutation rate ($4Nu = 8$) opposed by a selection of varying intensity. With selection of the order $4Ns = 10$, the mutation pressure gets the upper hand, and the allele favored by selection is largely lost. A doubling of the selection intensity ($4Ns = 20$) changes the situation. The gene frequencies fluctuate over a great range of values. In a species segregated into numerous isolated colonies this means a differentiation into local races, some of which will possess characteristics favored by selection and others relatively unfavorable ones. With selection becoming more stringent ($4Ns = 40$ and $4Ns = 80$), the amplitude of variation is gradually restricted, the gene frequencies being kept within rather narrow limits centered on the equilibrium values.

In the three models just described, it was assumed that the colonies into which a species is subdivided are completely isolated from each other. In reality this rarely if ever happens. Even populations of oceanic islands receive occasional migrants from the mainland or from other islands. Wright's Fig. 8.1D shows the effects of migration. If the migration coefficient is $m = 1/4N$ (one migrant individual on the average in four generations), the isolation is effective and the curve of gene frequencies becomes U-shaped. With $m = 1/2N$ (one migrant in alternate generations) the fixation of alleles is slowed down. With more frequent migration the distribution of the gene frequencies shows less and less variance. This may seem to make random drift inconsequential as an evolutionary agent. It should be noted, however, that Wright's m coefficient really means that the migrants to the various colonies come at random from all the colonies into which a species is divided. In reality, this is not so; the exchange of migrants occurs mainly between adjacent colonies, which are likely to have rather more similar gene frequencies than colonies remote in space.

Small population sizes are not the only factors making random drift possible. Fluctuations of selection and mutation rates are other random processes that may lead to interesting results (Kimura 1954). Assume that there are genes each with two alleles, A_1 and A_2, B_1 and B_2, etc., which are adaptively neutral on the average, that is, over a long series of generations. The mean selection coefficients are then, zero: $s = 0$. However, from time to time one or the other of the alternatives alleles becomes slightly advantageous or disadvantageous, so that there is a variance of selection coefficients, V_s.

Kimura gives an example of a population numerically so large that sampling variations can be neglected, in which the alternative alleles are initially equally frequent ($p = q = 0.5$), and in which the variance of the selection coefficients is $V_s = 0.0483$. For about 27 generations the distribution curve of the gene frequencies remains unimodal and thereafter becomes U-shaped. After 100 generations most genes reach "quasi-fixation," the most probable frequencies of the alternative alleles being 0.0007 and 0.9993.

In another example, one allele is, on the average, favored by a mean selection coefficient $s = 0.1$, but the selection has a variance $V_s = 0.0025$. One may then assume either many genes in the same population, with the favored alleles being initially rare, or many populations in which the favored allele of a certain gene is initially rare. The frequencies of the favored allele or alleles will, of course, gradually increase. However, whereas with constant selection all frequencies will be predictable and alike in a given generation, with variable selection the frequency distributions will become more and more changeable.

The Biological Meaning of the Value N

The genetically effective population size, N, is not identical with the census size, that is, the total number of individuals in a colony. The simplest model is that of a colony isolated from the rest of the species by secular barriers, within which the population is panmictic and remains stationary in numbers from generation to generation. For such a colony, N is equal to the number of individuals of the generation now living that will be the actual progenitors of the following generation, or to the number of the actual parents of the generation now living. In reality, the surviving progeny of some parents is large, that of others is small, and that of still others has been destroyed entirely. If N_0 is the number of parents, and K that of their gametes giving rise to the surviving offspring ($K = 2$ on the average), the effective population can be expressed as

$$N = \frac{4N_0 - 2}{2 + \sigma_K^2}$$

(Wright 1940). Populations of most species vary in numbers, often within an enormous range, from generation to generation. Such variations are connected either with seasonal cycles (as in insects that produce several generations per year), with climate, or with fluctuations in the abundance of parasites, enemies, or prey. Wright has shown that the effective N is much closer to the number of individuals attained at the maximum contraction of the population than to that attained at the maximum expansion. For example, a population may increase tenfold in each of the succeeding generations and then return to its initial size, its minimum being N_0 and its maximum $N_0 \times 10^8$. The effective size is then $N = 6.3\ N_0$ (Wright 1940).

Crow and Morton (1955) and Kimura and Crow (1963) have

defined the effective N in two ways, which are usually though not always equivalent. First, in a finite population, two alleles of the same gene inherited by an individual have a certain probability of being identical by virtue of descent from the same ancestor. Second, random drift owing to the sampling variance in the process of gene transmission from generation to generation leads eventually to random loss or fixation of some gene alleles. The effective number may then be defined "as the size of an idealized population that would have the same amount of inbreeding or of random gene frequency drift as the population under consideration." Kimura and Crow give formulae for effective N's in populations with different kinds of reproductive biology—monoecious and with separate sexes, constant and changing census numbers, etc. The variables in these formulae are census numbers (numbers of individuals in different generations), mean numbers of progeny per parent, variance of these mean numbers, and a measure of departure from the Hardy-Weinberg proportions.

The relationships between the census and the genetically effective population numbers are exceedingly complex. They are determined by ecological variables and by the reproductive biology of the species. The customary way of representing the distribution area of a species on a geographic map as a continuous territory is misleading. Rarely if ever is the population density uniform throughout the gross distribution area. Almost always one finds a mosaic of more or less discrete colonies separated by tracts where the species is rare or absent. The genetically effective sizes may vary greatly in different colonies. In this case the determining variables are not only the numbers of individuals per colony but also the rates of dispersal and migration, and the consequent gene flow between colonies. The dispersal is, in turn, a function of the vagility (mobility) of the species in different development stages.

Although the literature on territorial differentiation and vagility is extensive, most of the data have not been collected and, to my knowledge, not reviewed from the evolutionary point of view (see, however, the discussions in Andrewartha and Birch 1954, Mayr 1963, and Grant 1963). The most important fact to keep in mind is that animals capable of moving hundreds and even thousands of miles may nevertheless breed in close proximity to the places where they were born. It is well known that many migratory birds return to their old nesting places for breeding year after year. Some fishes, such as salmon, are born in streams in the headwaters of rivers; migrate to the ocean, where they grow to maturity; and then return to the exact stream in which they were born (Foerster 1968 and references therein). Twitty and his colleagues (Twitty 1959 and Twitty, Grant, and Anderson 1967) have described a remarkable homing ability in the newt, *Taricha rivularis*. It should be kept in mind, however, that the degree of precision of this homing migration differs from species to species. Mayr (1942) contrasts the behavior of geese (species of Anser and Branta) with that of ducks (Anas, Spatula). Mated pairs of geese migrate together to their winter quarters and return together to their nesting places; among the ducks the pair formation takes place in the winter quarters, and the mates are often natives of places remote from each other. The result is formation of many distinct races in geese and little race differentiation among ducks.

Tinkle (1965) and Merrell (1968) compared the absolute and the effective population numbers in the lizard *Uta stansburiana* and in the frog *Rana pipiens*, respectively. In the lizard the numbers of breeding adults were generally similar to the total numbers in the study areas. By contrast, in the frog the numbers of the egg masses counted were only fractions of the numbers of available females, so that the effective size was much smaller than the census size.

Colwell (1951) examined the dispersal of pine pollen by wind; most of the pollen is transported only 10–30 feet downwind from the source, although some pollen grains were found at distances of 150 and more feet. The seeds of a pine tree usually come from a male parent in the immediate neighborhood. The distance over which the pollen of Phlox is transported by insects is, on the average, only 1.1 meters; the maximum is 3.6 meters (Levin and Kerster 1968). Kerster and Levin (1968) estimate the genetically effective size of "neighborhood" colonies of *Lithospermum carolinense* as between 3.9 and 4.6.

Random Drift in Experimental Populations

Interactions of random drift and natural selection were studied by Kerr and Wright (1954) in small laboratory populations of *Drosophila melanogaster*. The populations were propagated by 4 females and 4 males, taken to serve as progenitors of the next generation. Although 4 pairs of flies can produce hundreds of offspring, only 4 new pairs were taken at random in each progeny. The sex-linked recessive mutant forked bristles (*f*), the dominant Bar eye (*B*), and the autosomal recessives spineless (*ss*) and aristapedia (ss^a) were used as markers.

In the first experiment, 96 populations were started, each with 1 *f/f*, 2 *f*/+, and 1 +/+ females and 2 *f* and 2 + males. The initial frequencies of *f* and of its wild type allele were, hence, 0.5. Because of the random choice of the parents in the following generations, however, the frequencies of *f* rose in some populations and of + in others. After 16 generations, only *f* was present in 29 populations and only + in 41 populations, while 26 populations had both + and *f*. This result is expected if the effective population size was about 83 percent of the actual number, that is, 6.64 rather than the actual 8 individuals. Since the number of the populations in which *f* was fixed was not much smaller than that of the populations in which the + allele achieved fixation, the gene *f* was not greatly discriminated against by natural selection.

In contrast, the second experiment showed that the Bar eye condition was distinctly disadvantageous. Here 108 experimental populations were started, each with 4 females heterozygous for Bar (*B*/+), 2 *B* males and 2 + males. The initial frequencies of *B* and of + were, then, equal at 0.5. After 10 generations, *B* was lost in 95 populations and normal eyes in only 3 populations, while 10 populations continued to have flies both with Bar and with normal eyes. The observed rates of loss and fixation were as expected if the genetically effective population size was about 72 percent of the actual one.

In the third experiment, 113 populations were started, each with 4 females and 4 males heterozygous for spineless and for aristapedia (ss/ss^a). The heterozygotes for these two alleles of the same locus are heterotic, that is, more viable than either homozygous *ss/ss* or ss^a/ss^a. As a consequence, the *ss* allele reached fixation in only 8 populations and the ss^a allele in no population after 10 generations. In 105 populations the *ss* and ss^a alleles continued to occur, the mean frequency of ss^a being 38.8 percent, and the highest frequency 87.5 percent. The ratio of fitnesses of the three genotypes was approximately 0.40 *ss/ss* : 1 ss/ss^a : 0.14 ss^a/ss^a. The three experiments show the interactions of drift and selection with diagrammatic clarity.

The experiments of Buri (1956) were similar in principle. He used two alleles of brown (eye color) in *Drosophila melanogaster*, and the populations were perpetuated by groups of 16 individuals in each generation. From an initial frequency of 0.5 for each allele, the gene frequency distribution spread over the range of values from 0 to 1, and the frequencies in the neighborhood of 0.5 occurred no more often than the higher and lower ones after about 10 generations.

Prout (1954) approached the problem in a different way. Three populations of *Drosophila melanogaster* were kept by Bruce Wallace for many generations in laboratory population cages; two of them were "large," containing about 10,000 flies, and one "small," with about 1000 adult flies. The small and one of the large populations were irradiated with gamma rays of radium. A considerable proportion of the second chromosomes in these populations were lethal in double dose, and Prout studied the frequencies of allelic lethals among the lethal second chromosomes. Random drift in small populations may decrease the number of different lethal genes present, and therefore increase the frequencies of allelic lethals among the lethals that remain (see below). This is what was actually observed. The data were consistent with an estimate that the effective population size was about 256 in the population containing approximately 1000 flies. In the populations with 10,000 flies, the effective size could not be distinguished from infinite.

Isolation by Distance

The model of the population structure considered above was that of a species subdivided into partially or completely isolated colonies. Wright (1943a, b, and Dobzhansky and Wright 1943, 1947) and Malécot (1948, 1959) envisaged also a population distributed uniformly over a large territory. This territory may be two-dimensional (continuous forest, steppe, surface waters in large lakes or seas) or unidimensional (a river or a shoreline for water-dwelling forms, river banks or coasts for terrestrial ones). There are, then, no impediments to migration, except for the isolation by distance; this is due to the limited locomotor abilities (vagility) of the organism concerned, which may be several orders of magnitude smaller than the total distribution area of the species.

Kimura and Weiss (1964) proposed a "stepping-stone" model, which is a compromise between the isolated-colonies and the isolation-by-distance types. This third model assumes that a species is distributed discontinuously, forming numerous colonies in a two-dimensional space or in one dimension, and that in any one generation the individuals born in any colony can migrate at most one step, to the adjacent colonies.

The papers cited above must be consulted by those who wish to follow in detail the mathematical treatment of the various models. The situations encountered are quite complex, especially if the interactions of natural selection with isolation by distance and other factors are to be analyzed. In point of fact, more mathematical work in this field is needed to clarify the situation.

Here again, however, the basic postulates and at least the main conclusions are simple enough. Wright considers that, in a population distributed uniformly over a two-dimensional territory, the parents of an individual are drawn from a certain average area. It may be assumed that this average area is a circle with a diameter D and a population of N breeding individuals. The grandparents come from a larger territory with a greater population, $\sqrt{2}\,D$ and $2N$, respectively. For n generations the territory becomes $\sqrt{n}\,D$, and the population nN. The amount of local differentiation of the populations due to genetic drift is much less in a continuously inhabited territory than is expected in isolated colonies. With fewer than 100 breeding individuals within a circle of diameter D, ($N<100$), considerable differentiation will occur, provided that mutation and selection do not overpower the random drift. With N greater than 1000 there will be little differentiation, and with N greater than 10,000 substantially no differentiation. Genetic differentiation of the colonies may occur also under the stepping-stone model, more easily with a unidimensional distribution than with two-dimensional ones.

Drift and Migration in Local Populations of Drosophila

Isolation-by-distance, stepping stone, and isolated-colonies models describe, more or less realistically, the breeding structures of natural populations of different organisms. All these models predict that local populations (sometimes called demes) may become genetically differentiated by random drift. This will happen provided that the vagility (migratory power) of the organism is so limited in relation to its population density that the genetically effective population size, N, of the colonies, demes, or neighborhoods is more or less small. With the same population densities, the effective N's will evidently be smaller in animals that move little, or in plants whose seeds or spores are not widely dispersed, than in those that traverse on the average greater distances between their birthplaces and the places where they leave their offspring. With constant vagility, the effective N's will be greater for species building dense populations per unit area than for species continuously or at least periodically rare.

Wright, Dobzhansky, Hovanitz (1942), Robertson (1962), Kimura, Maruyama, and Crow (1963), and Wallace (1968a) all pointed out that small effective population sizes may lead to experimentally detectable genetic phenomena. We know (Chapter 4) that natural populations of Drosophila carry many recessive or quasi-recessive autosomal lethals. In populations of effectively infinite size, the equilibrium value for a recessive autosomal lethal equals the square root of the mutation rate giving rise to the lethal. In populations of small effective sizes the gene frequencies vary within a range that is inversely related to the population size (N). In a small colony some of the lethals will be altogether absent, others may be as common as their mutation rates would permit them to be in large populations, and still others may occur even more frequently. Wright showed that in a species segregated into small colonies each lethal will at any given time be present in a certain proportion of the colonies, but absent in others. In other words, each colony will contain only some of the lethals that exist in the species. Moreover, the average equilibrium frequencies of all lethals in a species as a whole will be smaller if the species is subdivided into small colonies than if it represents a very large undivided population. For a lethal that arises by mutation once in 100,000 gametes the situation will be as follows:

Population Size, N	*Equilibrium Frequency, q*	*Percentage of Colonies Free of the Lethal*
1,000,000 or more	0.0032	0
100,000	0.0030	0
10,000	0.0020	15
1,000	0.0008	87
100	0.00026	99
10	0.00008	99.9
Self-fertilization	0.00002	99.996

If a species is a large panmictic population, or if it is subdivided into sizable breeding units, there should be no differentiation with respect either to kind or to frequencies of recessive lethals. The frequencies of the lethals should everywhere equal the square roots of their mutation rates. This situation is approached in the tropical species, *Drosophila willistoni*. Pavan, Cordeiro, et al. (1951) examined the incidence of lethals, semilethals, and sterility genes in populations of this species from diverse regions of Brazil. Although some variations were encountered, no systematic differences between the populations were brought to light. *Drosophila willistoni* is very common and widespread throughout Brazil, and its breeding populations are large. Pavan and Knapp (1954) found that among lethal second chromosomes extracted from populations of remote localities the frequency of allelic lethals was 0.00157. Lethals from remote localities are independent in origin by mutations. Among lethals from the same populations, some of which might have been identical by descent, the rate of allelism (0.00169) was not significantly higher. Interesting and meaningful exceptions were the populations of small isolated islands of Angra dos Reis, where a higher frequency (0.00514) of allelic lethals was found. Here a significant proportion of the lethals were doubtless identical by descent.

Spencer (1947a) found very high frequencies of heterozygotes for the recessive mutant stubble bristles in the population of *Drosophila immigrans* in a certain locality in Pennsylvania, but not in other populations of the same species. Such an accumulation of mutant heterozygotes in one population may well have been due to smallness of the genetically effective size of the population.

Dobzhansky and Wright 1941, 1943 and Wright, Dobzhansky, and Hovanitz 1942 found different sets of recessive lethal third chromosomes in *Drosophila pseudoobscura* populations from different localities. Population samples were taken at intervals of about one month during the breeding season, at nine stations on Mount San Jacinto, California. A "station" was a territory at most a hundred yards square, in which traps were exposed always in the same positions. The nine stations were in three groups or "localities." The distances between the localities were from 10 to 15 miles, and the distances between the stations within a locality varied from ¼ mile to 2 miles. There were no barriers to migration from station to station or from locality to locality. Some lethal-carrying third chromosomes were detected in every sample. The strains containing these lethals were then intercrossed to determine which of the lethals were allelic (mutations of the same gene loci) and which were not allelic (mutants at different loci). The rates of allelism, expressed in percentages of the intercrosses with allelic lethals, were as follows:

Within a station	Collected simultaneously	2.53
	1–11 months apart	1.97
Within a locality	Collected simultaneously	1.30
	1–11 months apart	0.69
Between localities		0.57
Between regions		0.44

Lethals found in populations within territories less than 2 miles apart are alleles more frequently than those found 10-15 miles apart. In fact, no significant difference in the chances of allelism was observed among lethals collected in different localities on San Jacinto and those collected on San Jacinto and in the Death Valley regions (a distance of 200 miles or more apart).

The most refined studies on the rates of allelism of recessive lethals in relation to distance have been made in *Drosophila melanogaster* by Wallace, Zouros, and Krimbas (1966) and Wallace 1966b. The rates of allelism, in percentages were as follows:

Within a collecting site	4.61
Sites 30 meters apart	3.65
Sites 60 meters apart	3.24
Sites 90 meters apart	2.75

On the assumption that allelic lethals in geographically remote populations arose independently by similar mutations, the rates of their allelism permit estimation of the minimum numbers of genes that produce lethals in a given chromosome of a certain species. Such a minimum estimate for the third chromosome of *Drosophila pseudoobscura* is 289 (Dobzhansky and Wright 1941). From laboratory experiments we know that the gross mutation rate producing recessive lethals and semilethals in the third chromosome of *D. pseudoobscura* (i.e., the sum of the mutation rates at all loci) is 0.307 ± 0.036 percent, or about 3 new lethals per 1000 gametes per generation. With 289 loci producing the lethals, the mutation rate per locus is $u = 0.000,0106$

(i.e., 0.003,07 ÷ 289). In an infinitely large population the equilibrium frequency of a lethal is $q = \sqrt{u} = 0.003{,}209$. Yet only about 15 percent of the chromosomes in the California populations carry lethals. The concentration per locus is, then,

$$q = 0.15 \div 298 = 0.000{,}519$$

Taken at face value, the discrepancy between the observed and the computed equilibrium frequencies of the lethals indicates low effective sizes of the populations. The assumption here, however, is that the lethals are completely recessive, and we saw in Chapter 6 that this is not always the case. Furthermore, a discrepancy could arise also as the result of very local inbreeding, such as matings of siblings soon after their emergence from the pupae. These so far undefined variables do not explain, however, the well-established fact that lethals found within a population of a small territory are alleles more frequently than those from distant localities. This finding strongly indicates that the genetically effective size of the populations sampled was limited. Using the isolation-by-distance model, Wright (in Dobzhansky and Wright 1943) has computed from these data on the frequencies of allelism of the lethals that the parents of an individual of *D. pseudoobscura* in the localities studied are drawn from a population of some 500–1000 individuals.

The genetically effective sizes of at least some natural populations of Drosophila are small enough to make possible some differentiation by random drift. This conclusion does not rest on inferences from genetic data alone; it is supported also by ecological evidence. Timofeeff-Ressovsky (1939), Dobzhansky and Wright (1943, 1947), Dubinin and Tiniakov (1946), Burla, da Cunha, et al. (1950), Burla and Greuter (1959), and Wallace (1966a) b, 1968a) released in natural habitats of their species known numbers of Drosophila flies marked with easily visible mutant genes that do not incapacitate their carriers. At intervals after the release, baits that attract the flies were exposed at various distances from the point of the release; the numbers of marked and wild flies that came to the baits were recorded.

The rates of dispersal are greater in *Drosophila pseudoobscura* and *D. funebris* than in *D. willistoni or D. melanogaster* and are more rapid in each species at higher than at lower temperatures. One year after the release of mutant flies of *D. pseudoobscura* in a locality in the Sierra Nevada of California, about half of the progeny of these flies were found within a circle with a radius of approximately 0.86 kilometer from the point of release. About 95 percent of the progeny were estimated to be located within a circle having a radius of 1.76 kilometers, and 99 percent within a circle of 2.2 kilometers. *Drosophila melanogaster* has considerably lower vagility than *D. pseudoobscura* (Wallace 1966a, b). In both species, the population densities are low in many places, especially at unfavorable seasons. The slow migration, owing to random wanderings of the flies in search of food and oviposition sites is, however, not the only means of dispersal. Some individuals are occasionally transported involuntarily by winds over much greater distances (a review in Carlquist 1966). How frequent such long-distance transport is, and how much gene flow results from it, are not known. It may be more important historically, enabling the species to spread to new territories, than in determining the composition of the gene pools of local colonies. Genetic differentiation of partially isolated colonies by random drift is probably a widespread phenomenon.

Drift and Migration in Human Populations

This is a controversial subject, owing, at least in part, to sheer misapprehension. Gene frequency differences observed between human populations need not be produced by selection alone or by drift alone; as indicated previously, they may be due to the interaction of both forces. Random drift is not limited to traits that are adaptively neutral; as shown above, a genetic trait that fluctuates from being advantageous in some generations to disadvantageous in others may also be subject to drift. To discover that some genetic variants differ in fitness under some circumstances does not rule out the possibility of drift. Moreover, although limitation of effective population size favors drift, the latter can also occur in large populations.

Glass, et al. (1952) and Glass (1954) made an elegant study of the Dunkers, a religious sect the members of which only rarely marry outside the community. The Dunkers, who now number about 3000 people in Franklin County, Pennsylvania, are descended from some 50 families that migrated from Western Germany to the United States between 1719 and 1729. The incidence of some blood group alleles among the Dunkers is significantly different both from that in the part of Germany from which their ancestors came, and from that of the surrounding American population. For the "classical" O A B blood system, the following allele frequencies have been found:

	i	I^A	I^B
Dunkers	60	38	2
Western Germany	64	29	7
United States	70	26	4

The frequencies for these alleles among the Dunkers are not like those among Germans or among Americans, nor are they intermediate between the two. The MN blood types have frequencies of about 30 percent MM, 50 percent MN, and 20 percent NN, nearly equal among Germans and among Americans. Yet among the Dunkers 44.5 percent MM and 13.5 percent NN have been found.

Steinberg, Bleibtreu, et al. (1967) observed even greater variations in the M and N frequencies among colonies of another religious isolate—the Hutterites, living in the northern United States (Montana and South Dakota) and Canada (Alberta, Saskatchewan, and Manitoba). In point of fact, the frequencies of the M allele varied from 0.49 to 0.90, which is not far from the total frequency range of this allele in mankind as a whole.

Birdsell (1950), studying aboriginal tribes in Australia, and Giles, Walsh, and Bradley (1966) in New Guinea, found significant gene frequency differences between neighboring populations. In the Pitjandjara tribe in south-central Australia the frequency of the I^A allele reaches 49 percent, whereas the adjacent tribe of Ngadadjara has only 28 percent. The Nangatadjara and the Ngadadjara have 96 percent of the M^N allele, whereas the Aranda, living not far away, have only 62 percent.

It is well known that the recessive gene for albinism reaches high frequencies in some isolated human populations, such as the San Blas Indians of Panama, where about 7 persons per 1000 are albinos. In Europe the frequencies of albinos are of the order of 1 : 10,000 to 1 : 30,000. Woolf and Dukepoo (1969 and Woolf 1965) found 26 albinos in a population of about 5000 in a group of villages of Hopi Indians in Arizona, a frequency of 1 : 192. The authors suggest "cultural selection" as a possible (though in my opinion not very likely) explanation of this high frequency. Although many albinos in Hopi communities never marry, male albinos stay in the villages while normally pigmented individuals work in the fields, and thus have opportunities to beget children.

Cavalli-Sforza and Edwards (1964 and Cavalli-Sforza 1969) put forward as an "admitted oversimplification" the working hypothesis "that genetic diversity among present-day populations has been largely caused by random processes (both random genetic drift and the random variation of selective values)." Analysis of blood group polymorphisms in 37 mountain villages in the valley of Parma, Italy, revealed significant heterogeneities for 8 out of the 9 gene alleles studied (the OAB, MN, and Rh blood group systems). Most of the villages existed as early as the eleventh century, and the rate of immigration from the outside to the Parma valley was very low. The populations of the different villages in the valley do not intermarry freely; examination of marriage licenses shows that the probability of marriage is a function of the distance between the birthplaces of the persons concerned. In Fig. 8.2 the logarithm of the probability of

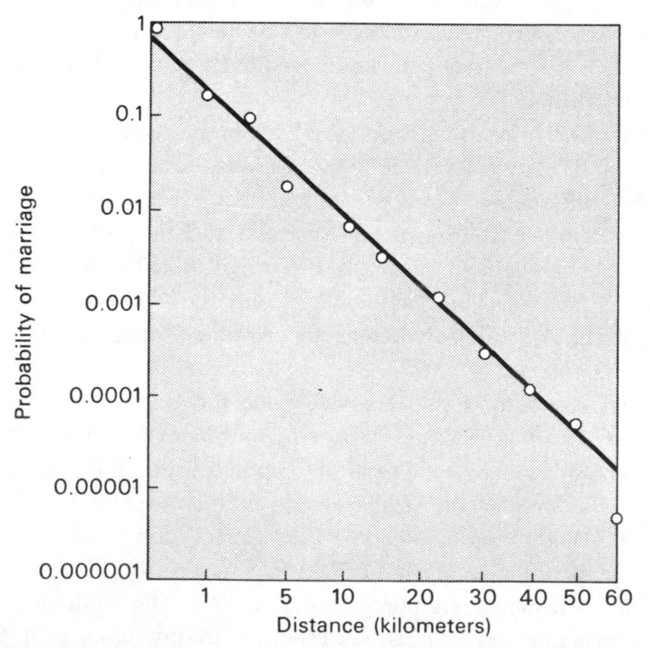

FIGURE 8.2

Probability of marriage and distance between the birthplaces of marriage partners. (After Cavalli-Sforza from Wallace)

marriage is plotted against the square root of the distance between the birthplaces of the marriage partners (prepared by Wallace 1968a, from Cavalli-Sforza's data). The observed points fit a straight line. Using Wright's isolation-by-distance model (see above) Cavalli Sforza, Barrai, and Edwards (1964) estimated the effective population sizes in the Parma valley as between 214 and 266. Alström and Lindelius (1966 and references therein) have made similar studies in Sweden.

A more ambitious extension of the work by Cavalli-Sforza et al. is the construction of evolutionary trees of the populations of the world. The points of branching and the lengths of the branches reflect the genetic "distances" between the populations, measured in terms of the gene frequency differences or differences in anthropometric characters. A sample of 15 populations, 3 from each continent, were chosen because of the availability from these populations of frequency data regarding 18 alleles (OAB, MN, Rh, Diego, and Duffy blood groups). The "tree" obtained (Fig. 8.3) is, as the authors say, "probably acceptable to many anthropologists."

The Founder Principle

The term founder principle was proposed by Mayr (see 1963) for "the establishment of a new population by a few original founders (in an extreme case, by a single fertilized female) which carry only a small fraction of the total genetic variation of the parental population." This may be regarded as a special case of Wright's random drift (2d) or a unique event (3e in Wright's classification; see the first section of this chapter). Populations of many species on oceanic islands, though they may now number in millions, are descendants of single migrants or small groups of migrants, introduced long ago by accidental long-distance dispersal. The same may be true of inhabitants of isolated bodies of water and of various kinds of ecological "islands," such as forests isolated on the tops of mountain ranges or surrounded by treeless plains. Isolated colonies are also formed often on the peripheries of species distribution ranges, where a species reaches the limits of its ecological tolerance. Such colonies are exposed to the risk of extinction, because the environment may occasionally become vigorous beyond their tolerance limits; the colonies are repop-

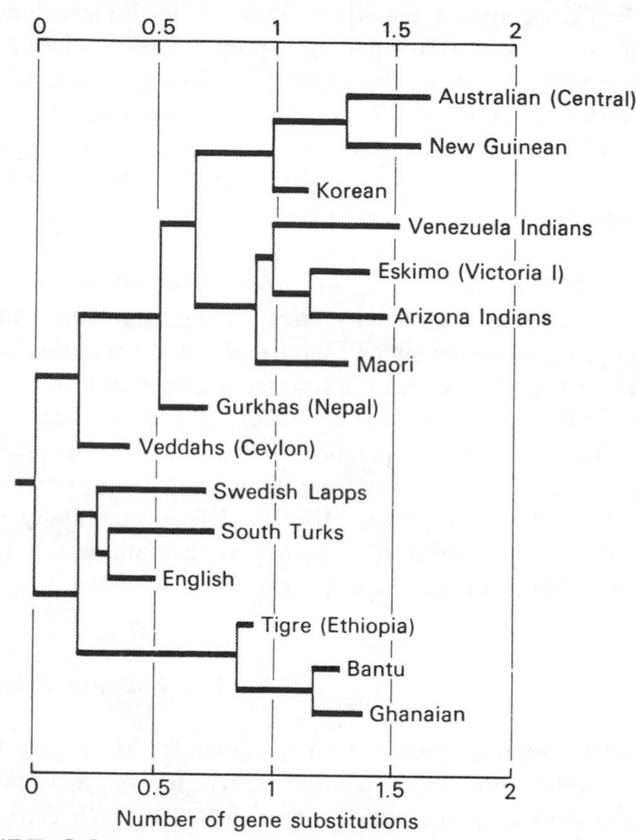

FIGURE 8.3

A phylogenetic tree of human population inferred from comparisons of the frequencies of certain genes in these populations. (After Cavalli-Sforza and Edwards)

ulated by new migrant founders from other parts of the distribution area. In primitive mankind, whole tribes were probably destroyed by various calamities and re-established from a few survivors or migrants from other tribes.

The immediate and obvious result of a population being reduced to very small numbers of individuals is, in normally outbred species, close inbreeding and loss of vigor. The ultimate result may be extinction (Voipio 1950, Suomalainen 1958, Andrewartha and Birch 1954, and references therein). If a colony escapes extinction, natural selection will work to restore its adaptedness. The selection must operate, however, barring new mutations, with the genetic variability that happens to be introduced or that is preserved in the founders. In the new, depauperate gene pool many gene alleles will find themselves in new genetic environments. Mayr (1954) dubbed this situation "genetic revolution." Both Mayr (1963 and references therein) and Carson (1959, 1965) surmised that such revolutionary reorganizations may initiate the formation of new species. This aspect will be considered in Chapter 11; here we shall examine some experimental studies of the operation of the founder principle.

We saw in Chapter 5 that chromosomal polymorphisms in the natural populations of some species of Drosophila are maintained by heterotic balancing selection. Heterokaryotypes, which have two chromosomes of a pair with different gene arrangements, are, as a rule, superior in fitness to homokaryotypes, which have the same gene arrangements in both members of a chromosome pair. This difference can be demonstrated also in experimental populations, maintained in laboratory population cages. Natural selection in such artificial populations establishes balanced equilibria, at which the different kinds of chromosomes achieve stable frequencies, determined by the relative fitnesses of the homo- and the heterokaryotypes (see Chapter 5 for more detail).

The rule of superior fitness of heterokaryotypes holds, however, chiefly in populations in which all the chromosomes are descended from ancestors collected together in the same natural locality. Such populations are said to be of uniform geographic origin.

Experimental populations can also be made with chromosomes of diverse geographic origins, derived from different populations. The outcome of selection in populations of mixed origin is remarkably erratic (Dobzhansky and Pavlovsky 1957 and references therein). The contrasting behaviors of populations of uniform and of mixed origins can be seen in Fig. 8.4. Eight experimental populations of *Drosophila pseudoobscura* have been made polymorphic for third chromosomes, some of which had the gene arrangement called Standard (ST) and others the Chiricahua (CH). All populations were started with 20 percent ST and 80 percent CH chromosomes. The four populations in Fig. 8.4A were of uniform geographic origin; their ancestors came from a certain locality in California. The four populations in Fig. 8.4B were of mixed origin—their ST chromosomes were

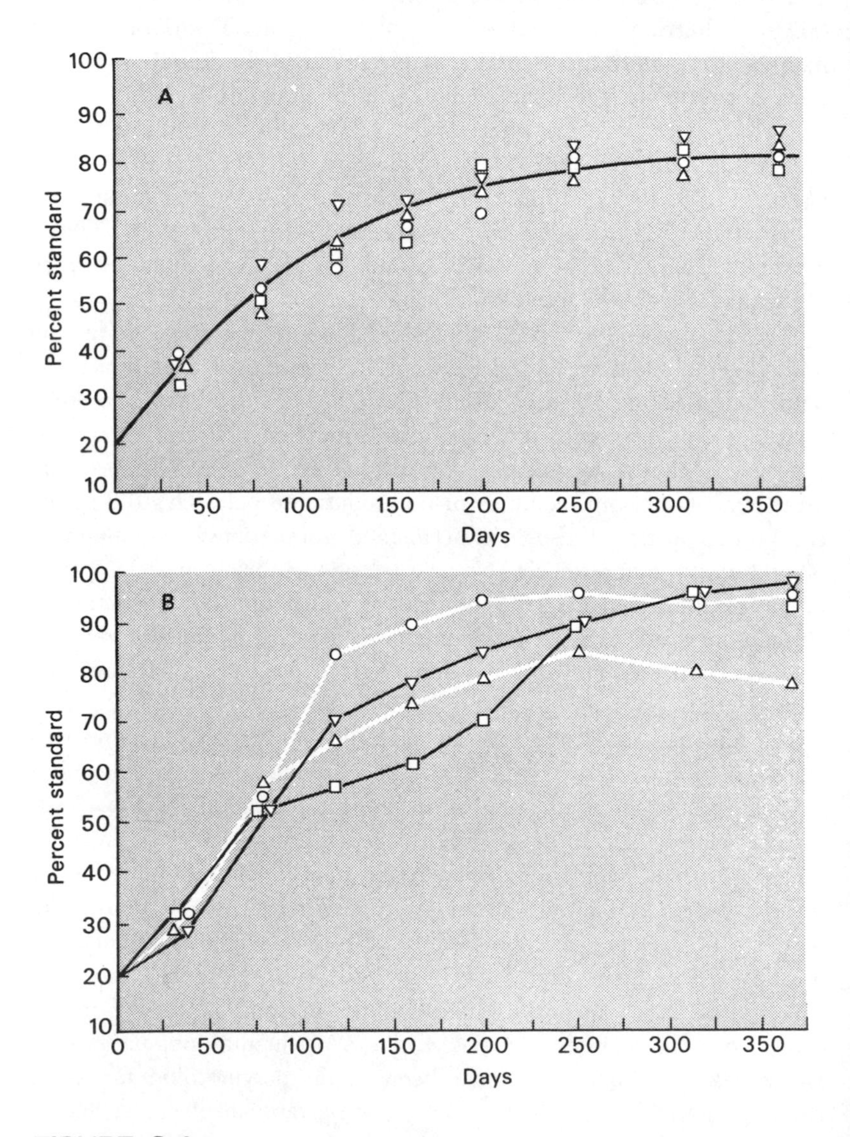

FIGURE 8.4

Natural selection in experimental populations of Drosophila pseudoobscura polymorphic for ST and CH gene arrangements in third chromosomes. Chromosomes of uniform geographic origin (A) and of different geographic origins (B). (After Dobzhansky and Pavlovsky)

derived from a population in California, and their CH chromosomes from a population in Mexico.

The changes observed in the geographically uniform populations were similar, within the limits of sampling errors. The fitnesses of the karyotypes, computed from the data in Fig. 8.4A, are as follows:

ST/ST	ST/CH	CH/CH
0.895	1	0.413

The populations of different geographic origins (Fig. 8.4B) behaved quite similarly at first, but after the 100th day a divergence was observed. In no two populations were the changes quite similar. Only in one population was an apparently stable, balanced equilibrium eventually reached. In three others the CH chromosomes were on the way to elimination.

Why should there be so striking a difference between the behaviors of populations of uniform and mixed origins? In the former, the changes are repeatable and predictable, and in the latter indeterminate. As a working hypothesis, we can assume that the fitness of the homo- and heterokaryotypes for third chromosomes with ST and CH gene arrangements depends not only on these chromosomes themselves but also on the rest of the genetic system, that is, on the genes in all other chromosomes. Now, the gene contents of the ST and CH chromosomes in any natural population where they occur together are coadapted, fitted to each other and to the genes in other chromosomes by long-continued natural selection to yield highly fit heterokaryotypes. The populations of California and Mexico need not have their chromosomes coadapted, because hybrids between these populations are produced only in laboratory experiments. In how many genes the populations of California differ from those of Mexico we do not know, but it is reasonable to suppose that the number is fairly high, to make these populations adapted to their respective climatic and other environments. Experimental populations obtained by hybridization of natural populations of diverse geographic origins may thus undergo Mayr's "genetic revolutions" (see above).

Natural selection works in experimental populations, of uniform as well as of mixed origins. In the former it simply establishes the most favorable frequencies of the chromosomes on a relatively uniform genetic background. Gene recombination in mixed populations generates a great variety of genetic backgrounds. If we suppose that the parental populations differ in 100 genes, 3^{100} genotypes are potentially possible. The number of flies produced per generation in the experimental populations is between 1000 and 4000—infinitesimally small compared to the potentially possible numbers of genotypes. It is a matter of chance which ones of the possible genotypes appear in which population, or do not arise at all. Natural selection must, however, work with what is available, and it finds different genetic materials on hand in different populations. The result is the apparent indeterminacy of the outcomes in different populations.

It is possible, fortunately, to test the validity of this hypothesis experimentally. The variability of the outcome of selection in populations of mixed origin should be greater if they are started with few founders, and less if the founders are numerous. It is impossible to arrange populations with 3^{100} founders, which would give entirely uniform results. Dobzhansky and Pavlovsky (1957) made 10 populations with 4000 founders, and 10 with only 20 founders. The founders in all populations were F_2 hybrids of *Drosophila pseudoobscura* from California and from Texas, and all populations contained originally 50 percent of their third chromosomes with the Arrowhead (AR) gene arrangement from California, and 50 percent with Pikes Peak (PP) chromosomes from Texas. The results are shown in Fig. 8.5. "Large" and "small" refer exclusively to the numbers of founders; because of the high fertility of Drosophila, all population cages will come to contain about the same number of flies. It can be seen in the figure that the diversity of selection outcomes was greater among the small than among the large populations. This is statistically significant.

The same hypothesis was verified by Dobzhansky and Spassky (1962) in a still different way. In their experiments, 10 populations were started, each with 20 founders. The founders were hybrids of California strains with AR, and Texas strains with PP, chromosomes. In 5 populations, however, the founders were "multichromosomal," that is, hybrids of flies from 10 different California and 10 Texas strains. In the other 5 populations the founders were only "bichromosomal," descended from a single California and a single Texas individual. All the populations were put through several cycles of expansion in numbers, and new starts from 20 founders each. Multichromosomal populations should have considerably more genetic variability in their gene pools than bichromosomal ones and, therefore, are

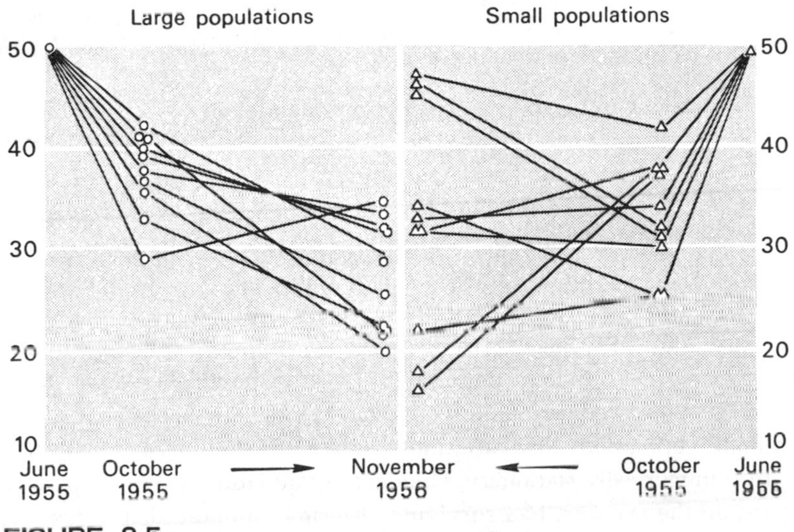

FIGURE 8.5

Experimental demonstration of the operation of the founder effect. Further explanation in text. (After Dobzhansky and Pavlovsky)

expected to show a greater diversity of selection outcomes. The evidence of the experiments was, indeed, in accord with this expectation.

Solima-Simmons (1966) added a still different corroboration. She outcrossed the two populations that diverged most among the multichromosomal ones described above, and used the F_2 hybrids to make 5 new populations, starting with 20 founders each; 5 other populations, which served as controls, were also started with 20 founders each, taken, however, from only one of the populations, without intercrossing. The populations were then put through 9 cycles of alternating expansions and reductions to 20 founders. The hybrid populations developed significant genetic divergence, whereas the nonhybrid ones showed no divergence.

The experiments of Dawson and Lerner (1966) were made with the beetles *Tribolium castaneum* and *T. confusum* competing in the same cultures. Under the experimental conditions used, *T. castaneum* outbreeds and eventually eliminates *T. confusum*. The authors made 10 populations with 10 pairs of each species as founders, and another 10 populations with 2 pairs as founders. The outcomes were much more variable in the populations started with fewer founders.

The Decay of Genetic Variability

We saw in Chapter 4 that the pre-Mendelian concept of blending, or "blood heredity," created a quandary for the theory of evolution by natural selection, or for that matter any evolution theory. Indeed, the blending of parental heredities in the offspring would cause a decay of the hereditary variability in a sexual population far more rapidly than any known process could restore it. The demonstration by Hardy and Weinberg that the mechanism of Mendelian segregation averts the blending of parental heredities eliminated this quandary. Only in an ideal, infinite population, however, could hereditary variation be preserved indefinitely by the Mendelian mechanism alone. In finite populations this variation is subject to slow decay.

In a numerically stationary Mendelian population a pair of parents gives, on the average, two surviving offspring. Suppose that a gene A_1 mutates to A_2, and gives a single heterozygous individual A_1A_2. The individual must mate with a nonmutant, $A_1A_1 \times A_1A_2$. The offspring is expected to consist of equal numbers of A_1A_1 and A_1A_2. Because of chance, however, the number of survivors may be 0, 1, 2, 3, or more individuals, forming a Poisson series. If there are no survivors, the mutant allele is lost; if 1 survives, the probability of loss is 0.5; with 2 individuals surviving the probability of loss is 0.25, and with r survivors it is 2^{-r}. The aggregate probability is 0.3679 that a single mutant will be lost in a stationary population in every generation, 0.3679 that it will be represented in the next generation also by a single individual, 0.1839 that it will be represented by 2 individuals, 0.0613 by 3, 0.0153 by 4, etc. If the mutant is not lost in the first generation, it is exposed to the risk of extinction in the next one and in the following generations. To be sure, some mutants will, owing to chance, be represented by 2, 3, or more individuals. But since the loss of a gene is irreversible, most mutants are not preserved in the populations. As shown by Fisher (1930), this is true not only for neutral and harmful but even for advantageous mutants (see Table 8.1).

The decay of the variability may be visualized in another way. Imagine a population in which every gene is represented only once, that is to say, every individual carries 2 different alleles neither of which is present in any other individual. If the population is numerically stationary, 36.79 percent of the alleles will be lost in the next

TABLE 8.1
Probability of extinction and of survival of a mutation appearing in a single individual (After Fisher)

Genera-tion	Probability of Extinction No Advantage	Probability of Extinction 1% Advantage	Probability of Survival No Advantage	Probability of Survival 1% Advantage
1	0.3679	0.3642	0.6321	0.6358
3	0.6259	0.6197	0.3741	0.3803
7	0.7905	0.7825	0.2095	0.2175
15	0.8873	0.8783	0.1127	0.1217
31	0.9411	0.9313	0.0589	0.0687
63	0.9698	0.9591	0.0302	0.0409
127	0.9847	0.9729	0.0153	0.0271
Limit	1.0000	0.9803	0.0000	0.0197

generation, some will be represented once, and others twice, thrice, and more times. The loss of some and the increase in frequency of other alleles will continue in subsequent generations as well. If no mutation occurs, the population will ultimately become homozygous for a single allele.

Wright (1931) and later Kojima and Kelleher (1962) studied the process of decay of variability in populations that are stable and those that are fluctuating in numbers. In a stable population of N breeding individuals, $1/2N$ genes either reach fixation or are lost in every generation. Suppose that in a population each of many genes is represented by 2 alleles that are equivalent with respect to selection, and that the frequency of each allele is 50 percent ($q = 0.5$). With no mutations, the frequencies of the alleles will fluctuate up and down from 50 percent, some becoming more and others less frequent. Eventually, gene frequencies from 0 to 100 percent will become equally numerous, and $1/4N$ of the genes will reach fixation and $1/4N$ will be lost in every generation. Wright's formula describing this process is

$$L_T = L_0 e^{-T/2N}$$

where L_0 and L_T are the numbers of the unfixed genes in the initial and in the T generation of breeding, respectively, N is the effective population number, and e is the base of natural logarithms.

If the population size fluctuates from generation to generation, the chance that a mutant allele will be retained is greater if the mutation occurs when the population is expanding than if it occurs during the contraction phase.

The Opposition of Drift and Mutation

Random drift and mutation may now be envisaged as opposing forces, tending toward loss and replenishment, respectively, of the genetic variability. Kimura and Crow (1964) and Wright (1966) made a theoretical study of the variety of alleles that can be maintained in populations of various effective sizes. A gene can give rise to many different alleles by substitutions of single nucleotides in the DNA chain. The accumulation of variant alleles in the gene pool of the population will be counteracted, however, by their loss because of drift, until a steady state is reached at which equal numbers of alleles are added and lost (see Li 1955a). Suppose now that the alleles are all neutral, that is, do not modify the Darwinian fitness of their carriers. The effective number of the alleles maintained is then simply $4Nu + 1$, where N is the effective population size and u the mutation rate. Table 8.2 shows that this number depends greatly on the second parameter, as well as on the population size. It should be noted that the "effective" number is less than the "actual" or total number, the latter being swelled by very rare alleles that have not reached the steady state. Even so, with gene loci numbering in tens or hundreds of thousands, the genetic variety will be impressively large, except in very small populations.

If the mutant alleles are deleterious, their variety will of course be reduced. More interesting is the situation in which alleles produce

TABLE 8.2

Numbers of alleles per gene maintained in populations of various effective sizes by different mutation rates, when the alleles are neutral ($s = 0$) or heterotic ($s = 0.1$) (After Wright 1966)

Population Size (N)	Mutation Rate (u)	Neutral	Heterotic	Population Size (N)	Mutation Rate (u)	Neutral	Heterotic
10^3	10^{-7}	1.0	4.6	10^5	10^{-7}	1.0	51
	10^{-6}	1.0	5.3		10^{-6}	1.4	62
	10^{-5}	1.0	6.3		10^{-5}	5.0	81
	10^{-4}	1.4	11.2		10^{-4}	41	137
10^4	10^{-7}	1.0	15.3	10^6	10^{-7}	1.4	176
	10^{-6}	1.0	17.7		10^{-6}	5	219
	10^{-5}	1.4	22.0		10^{-5}	41	323
	10^{-4}	5.0	31.9		10^{-4}	401	689

heterozygotes of superior fitness and thus can be assisted by heterotic balancing selection. Wright (1966) makes the simplifying assumptions that all combinations of alleles induce a similar degree of heterosis, all homozygotes having a fitness of 0.99 and all heterozygotes of 1.00, and that the mutation rates at all loci are the same. The numbers of alleles that can be maintained even in small populations and with low mutation rates are then impressively large.

Evolution by Random Walk

In Chapter 7 we reviewed some of the rapidly accumulating evidence that high proportions of genes are polymorphic in populations of Drosophila, man, and presumably other sexually reproducing species. We have also seen that these findings raise difficult theoretical problems. How are the multitudinous polymorphisms maintained? How is it possible for an average individual to be heterozygous for thousands of genes? The simplest explanation is that polymorphisms are maintained by balancing selection. This explanation contravenes the classical theory of genetic population structure, which would force us to conclude that numerous polymorphisms impose on the population a genetic burden too heavy to carry. King (1967), Sved, Reed, and Bodmer (1967); Wallace (1968a,b), and others have pointed out that the difficulty really stems from an oversimplified view of natural selection adopted by the classical theory. It ignores the ecologic factors regulating population size, the different modes of operation of rigid and flexible selection, and the importance of selection thresholds (see Chapter 7).

A quite different way to escape difficulty is advocated by King and Jukes (1969), Kimura and Crow (1969 and Crow 1969), and Arnheim and Taylor (1969). King and Jukes call their theory "non-Darwinian" evolution. This term is equivocal; if one restricts "Darwinian" to evolution by natural selection, there are many and diverse "non-Darwinian" theories (Lamarckism, autogenesis, nomogenesis, etc.). What King and Jukes actually postulate is evolution by random walk: given many genetic variants that are selectively neutral or very nearly so, their frequencies in populations will drift at random until some of them are lost and others reach fixation.

Consider an idealized diploid population of genetically effective size N. Suppose that a gene in this population has in some generation $2N$

alleles, no two of them identical, and yet all exactly equivalent in their effects on fitness. In the next and the following generations more and more alleles will be lost by chance (Table 8.1), but other alleles, also by chance, will be multiplied and represented by more and more copies. If there are enough generations of such random walk, only a single allele will remain, all $2N$ copies descended from one original gene. The chances of any one of the $2N$ originally different but selectively equivalent alleles being ultimately victorious and reaching fixation are equal, or $1/2N$. The average number of generations until fixation occurs in a population is $4N$ (Kimura and Ohta 1969). In a small isolated population, with effective N of the order of tens or hundreds, this is not unrealistic. To have an allele fixed in an entire species, such as man, however, the time required may be too long, even on a geological scale.

Suppose now that mutations occur in the population, but that all mutant alleles are adaptively equivalent to the original one and to each other. The chances of any one of the mutant alleles being eventually fixed are still $1/2N$. Let the mutation rate yielding new but adaptively neutral alleles be u. The number of mutant alleles appearing in every generation will then be $2Nu$, and among them $2Nu/2N = u$ will eventually reach fixation. In other words, the number of alleles reaching fixation in every generation will, on the average, be equal, in an idealized equilibrium population, to that of new alleles arising by mutation.

Kimura (1969) considers, not mutant gene alleles, but nucleotide sites, some of which undergo substitution and produce mutations, all of them assumed to be adaptively neutral. A haploid gene complement in a mammal is estimated to contain 3–4 billion nucleotide pairs, enough to code for some 2 million polypeptide chains of 500 amino acids on the average. Assuming an effective population size of 10,000, and a rate of nucleotide substitution equal to 2 per gamete per generation, Kimura arrives at an impressively large figure of some 80 thousand heterozygous nucleotide sites per individual.

In another paper (1968) Kimura gives estimates of the average numbers of gene alleles maintained in populations of different sizes (N). The number of gene alleles depends on the product of the mutation rate (u) and the genetically effective population size (N_e), always assuming that the alleles are neutral with respect to fitness. Some of the estimates are shown in Table 8.3.

TABLE 8.3

Average numbers of gene alleles expected in populations of different actual and genetically effective (N_e) sizes and different mutation rates (u) (After Kimura 1968)

$N_e u$	Population Size			
	10^3	10^4	10^5	10^6
0.01	1.30	1.39	1.49	1.58
0.10	3.83	4.75	5.68	6.60
1.00	23.1	32.3	41.5	50.7
10.00	134.7	226.1	318.1	410.2

On the assumption that many or most mutational changes that arise are adaptively neutral, evolution by random walk must be regarded as an important, or even as the prevalent, evolutionary mode. The ubiquity of polymorphisms and the high heterozygosity found in natural populations are then no longer difficult to explain. It is the validity of the basic assumption that is very much in doubt.

The Problem of Adaptively Neutral Traits

Anyone who compares the species of some large genus of animal or plants finds that many, and indeed most, of the traits in which the species differ seem to be trivial and to play no role in the survival or reproduction of the creatures involved. Why, for example, are the anterior scutellar bristles in many species of Drosophila always convergent and in other species invariably divergent? Or what adaptive significance can be ascribed to minor differences in the shape of the acorns in related species of oaks, Quercus? Or, for that matter, what is the selective value of facial differences among persons, provided, at least, that they are equivalent from the standpoint of sexual and social acceptability?

To concede that the above and many other traits are selectively neutral is to surrender before even starting on a quest to discover their significance. A visible "trait" may, indeed, be of no importance in itself, but it may be an outward manifestation, a part of the pleiotropic effects, of a variant gene contributing significantly to the organism's welfare (for examples, see Chapter 2). A "trait" is really an abstraction, a semantic device employed by a biologist to facilitate the description of a kind of organism that he studies; the importance of a trait depends on the developmental nexus of which it is a part (Dobzhansky 1956a). What appears to be the adaptive neutrality of a trait may, then, be simply a measure of our ignorance. On the other hand, to assert that no trait and no genetic change can be selectively neutral is to create a dogma, not to say a superstition. The adaptive role of a character must be demonstrated; it cannot be assumed a priori.

The theory of evolution by natural selection was held in low esteem, even by some pioneers of genetics, during, roughly, the first quarter of the current century. Many evolutionists of that time assumed that evolutionary changes stem from some unknown internal "law" (e.g., Berg 1969). Many of these changes neither help nor hinder the organism's survival—they are adaptively irrelevant. In the nineteen thirties, Wright's random drift idea appeared to offer a ready explanation of adaptively neutral changes, although, as pointed out at the beginning of this chapter, Wright himself stressed the interactions of deterministic and stochastic processes, and not drift alone. For about a quarter of a century thereafter, a hyperselectionism became fashionable. All differences between populations and species were assumed to be products of natural selection. Even so judicious an author as Mayr (1963) wrote, "Selective neutrality can be excluded almost automatically wherever polymorphism or character clines (gradients) are found in natural populations," and "For these reasons, it appears probable that random fixation is of negligible evolutionary importance."

Theories of non-Darwinian evolution by random walk have started a new swing of the pendulum in the opposite direction (see, however, a critique by Richmond 1970). King and Jukes (1969), perhaps to some extent playing the role of devil's advocates, state, "Evolutionary change is not imposed upon DNA from without; it arises from within. Natural selection is the editor, rather than the composer, of the genetic message. One thing the editor does not do is to remove changes which it is unable to perceive." But can one really believe that at least the general sense and the form of the genetic message have not been composed during the billions of years of evolution controlled by natural selection?

Perhaps the strongest theoretical argument for the neutrality of some mutants is based on the degeneracy of the genetic code. As we have seen, six nucleotide triplets code for each leucine, serine, and arginine (Table 1.2). Mutations that change a triplet to its synonym, for example, CUU to CUC or CUA or CUG, should not be "perceived" by natural selection, because the protein formed will have the same amino acid in the same position. Almost one-fourth of the 549 possible nucleotide substitutions will yield synonymous codons. Even this argument, however, is not quite convincing. Because the synonymous codons often require different transfer RNA's for their translation, "synonymous" mutants need not be selectively equivalent. Transfer RNA's are not equally abundant in different tissues and different species (see, e.g., Holland, Taylor, and Buck 1967 and Caskey, Beaudet, and Nirenberg 1968).

The different amino acids are also not equally abundant in proteins. King and Jukes (1969) reasoned that, if many of the mutational substitutions are selectively neutral, the abundance of the amino acids may be correlated with the number of triplets coding for them. Indeed, they find such a correlation in 53 vertebrate proteins for which the amino acid composition is known. The commonest amino acids are serine and leucine, each coded by six synonymous triplets. Methionine and tryptophan, each with a single triplet, are least frequent. To be sure, the correlation is not perfect; for example, arginine is comparatively rare even though it is coded by six triplets. The argument is, however, ambiguous. King and Jukes themselves point out that the causal relationships may be the reverse of that assumed above: some amino acids may be common and others rare, not because they have more or fewer triplets, but because the genetic code may have evolved, in remote ancestors of the organisms now living, to match the abundance of the amino acids favored by natural selection.

The problem of neutral mutational changes will eventually be solved by relevant experimental evidence rather than by theoretical arguments. One source of such evidence is comparison of amino acid sequences in homologous proteins in different organisms. Cytochrome c is an enzyme performing similar functions in cellular respiration in most diverse organisms. Margoliash and his colleagues (Margoliash 1963, Margoliash and Smith 1965, Fitch and Margoliash 1967a,b, and references therein) compared the amino acid sequences in the cytochromes c in a variety of organisms—man, horse, pig, rabbit, chicken, tuna fish, a moth (Samia), a fungus (Neurospora), and a yeast. The cytochromes c of the vertebrates are chains of 104 amino acids, with a prosthetic heme group attached at positions 14 and 17. The yeast cytochrome c is a chain of 108 amino acids. In all these cytochromes 55, or 53 percent, of the positions are invariant, that is, occupied by the same amino acids. So great a similarity has a negligible probability of being due to chance; all the cytochromes are homologous (i.e., evolved from the cytochrome in a remote common ancestor).

Amino acids at other positions are variable. Organisms relatively close in the biological system (e.g., different mammals) have more similar cytochrome c's than do remote forms, such as mammals and yeasts. The invariant portions of the cytochrome c molecules are assumed to be essential for the function that the enzyme performs. The variable parts are, in Margoliash's words, "undoubtedly compatible with a variety of sequences." The hypothesis that the amino acid substitutions in the variable parts are due to chance was submitted by King and Jukes (1969) to the following test. By using the Poisson distribution formula, the numbers of substitutions expected by chance can be calculated and compared with the numbers actually observed. The two sets of values are in remarkably good agreement:

Changes per site	0	1	2	3	4	5	6	7	8	9
Expected	6	16	20	18	12	6	3	1	0.3	0.1
Observed	6	17	18	19	10	6	3	1	1	0

In vitro experiments appear to show that the cytochrome c's of various organisms are fully interchangeable. Does it follow that the amino acid replacements in the evolutionary process were merely the "noise" of adaptively neutral mutations? It is by no means certain that man would enjoy normal health if his cytochrome c were replaced by horse or yeast cytochrome, or vice versa.

Comparative studies on hemoglobins, particularly in vertebrates, reveal an interesting, but equally ambiguous, situation. Reference was made in Chapter 2 to variant hemoglobins found in some human families or individuals. A majority of the approximately 100 hemoglobins discovered differ from the ubiquitous hemoglobin in single amino acid replacements. Some of the variants are known to cause ill health, whereas for others the evidence is inconclusive. The hemoglobins of other animals differ in more numerous substitutions, as shown in Table 8.4. The alpha chain of human hemoglobin is identical with that in the chimpanzee, and differs by a single amino acid substitution from that in the gorilla (glutamic in man, and aspartic acid in gorilla at position 23), by 17–25 substitutions from that of the other

TABLE 8.4

Numbers of amino acid differences between the hemoglobin alpha chains of some animals. (After Dayhoff and Eck 1968)

	Gorilla	Horse	Donkey	Cattle	Sheep	Llama	Pig	Rabbit	Mouse	Carp
Human	1	18	20	17	20	20	18	25	17	71
Gorilla	0	19	21	18	20	19	19	26	18	70
Horse		0	2	18	18	16	17	25	23	70
Donkey			0	20	19	18	18	27	25	70
Cattle				0	13	22	17	26	20	68
Sheep					0	23	18	29	24	70
Llama						0	20	23	18	60
Pig							0	26	24	70
Rabbit								0	28	75
Mouse									0	71

mammals in the table, and by 71 substitutions from that of the fish (carp). According to Perutz and Lehman (1968), if one compares the known sequences in the hemoglobins and myoglobins of various mammals, only 7 out of more than 140 amino acid sites have remained invariant. These invariant sites include the attachments of the heme groups and the points of contact between the different chains. Can man get along with gorilla hemoglobin and vice versa? This is not an unfair question, since, as pointed out above, many mutants with amino acid substitutions have been observed in man. Perhaps a gorilla-like human variant will some day be discovered.

Prakash, Lewontin, and Hubby (1969) argue most persuasively that the allozyme polymorphisms in natural populations of Drosophila are maintained by some forms of balancing natural selection, rather than by random drift. Let us recall that it was the discovery of the hitherto unperceived abundance of polymorphisms and heterozygosity in populations (see Chapter 7) that stimulated the revival of random drift or random walk theories. Yet Prakash et al. find identical frequencies of allozyme polymorphs maintained in remote parts of the distribution area of *Drosophila pseudoobscura*. Genetic drift should have diversified these frequencies unless the flies migrate freely over large distances. The migration rates are known, however, to be small (Dobzhansky and Wright 1943, 1947). Prakash et al. conclude, "The hypothesis of widespread balancing selection at most of the polymorphic loci fits most easily all of the observations on the central, marginal and isolated populations, although we cannot exclude, for some loci, a model of selectively neutral isoalleles." The problem of evolution by random walk invites further studies.

CHAPTER NINE

POPULATIONS, RACES, AND SUBSPECIES

Individual and Group Variability

Immanuel Kant, who was a naturalist before he became the prince of philosophers, wrote in 1775 the following remarkably perceptive lines:

> Negroes and whites are not different species of humans (they belong presumably to one stock), but they are different races, for each perpetuates itself in every area, and they generate between them children that are necessarily hybrid, or blending (mulattoes). On the other hand, blonds or brunettes are not different races of whites, for a blond man can also get from a brunette woman altogether blond children, even though each of these deviations maintains itself throughout protracted generations under any and all transplantations.

It appears that Kant had a clearer idea about the distinction between individual variability and the variability of populations than many authors writing today.

In outbreeding sexual species—man, of course, included—no two individuals (identical twins excepted) have the same genotype. Parents differ genetically from each other and from their children, as do the latter among themselves. In asexual and some parthenogenetic species, on the contrary, there may be clones numbering thousands or even millions of individuals with identical genotypes. Clones are "pure races," two or more of which may occur sympatrically, that is, in the same territory. In sexual species, arrays of genetically identical individuals that could be considered pure races do not exist and never existed (although some anthropologists and biologists fancied their existence, usually in the past since they could not find them at present). Races are neither individuals nor particular genotypes; they are genetically distinct Mendelian populations. As a general rule, races are allopatric, living in different territories. Genetic differences between geographic races are maintained, at least in part, by their geographic separation.

Mankind is one of the exceptions to this rule. Before the advent of civilization, human races were allopatric, separated in space, like races of most other sexually reproducing animals and plants. Civilization created a variety of social forces that make possible, at least for a time, the sympatric coexistence of human races. Races or breeds of domesticated animals and plants are also often sympatric, their separation being maintained by the more or less deliberate effort of husbandmen.

A race is a Mendelian population, not a single genotype; it consists of individuals who differ genetically among themselves. It is important to realize that similar genetic elements are involved in individual and in race differences. For example, blonds and brunettes occur as individual variants, polymorphs, in many human populations, and races differ in the incidence of these variants. Some populations are polymorphic for blue and brown eye colors, whereas in others only brown eyes are found. Blue-eyed individuals are not a race distinct from brown-eyed ones, yet eye color is one of the traits distinguishing races. Blood groups are classic examples of intrapopulational polymorphisms and also of interpopulation variability. Individuals with O blood type are not racially distinct from those with A or B blood; however, races do differ in the incidence of these blood groups.

Some race differences, though by no means all, are qualitative: the frequency of a gene allele may be zero in one race and unity in another. If one examines phenotypic traits rather than genes, qualitative differences are more common. Natives of central Africa all have darker skins than do natives of Europe. Skin color, however, is a trait determined polygenically. Dark and light pigmentation alleles occur as polymorphs in most populations, and some populations are intermediate in phenotype, and presumably also in genotype, between the "blacks" and the "whites."

The recognition of the similarity of the genetic basis of individual and of group variability is one of the achievements of genetics. The classical race concept was typological. Thus, Hooton in 1926 (quoted after Count 1950) defined race as "a great division of mankind, the members of which, though individually varying, are characterized as a group by a certain combination of morphological and metrical features, principally non-adaptive, which have been derived from their common descent," and said, "One must conceive of race not as the combination of features which gives to each person his individual appearance, but rather a vague physical background, usually more or less obscured or overlaid by individual variations in single subjects, and realized best in a composite picture."

Some anthropologists claimed an almost preternatural ability to see Celtic, Nordic, Dinaric, Armenoid, Lapponoid, Cromanoid, Oriental, Berberic, etc., "types," all living together and interbreeding in populations of Ireland, Poland, and other countries. Criminals were a "race" different from law-abiding "races" living next door or even in the same families. Typology is at the bottom of the vulgar notion that any so-called Negro in the United States (though actually a majority of his genes may be derived from his white ancestors) has a basic and unremediable negroid nature, just as any Jew partakes of some Jewishness, etc. There are no Platonic types of Negroidness or Jewishness or of every race of squirrels or butterflies. Individuals are not mere reflections of their racial types; individual differences are fundamental biological realities, from which race differences are derived.

Curiously enough, whereas typological thinking leads some people to believe in indelible racial nature, it leads others to deny that races exist. A flurry of polemics among systematic zoologists was started by Wilson and Brown's (1953) critique of the practice of describing and naming subspecies of animals. (A subspecies is a race to which a Latin name has been given.) Some anthropologists rejected the existence of races in mankind (several critical articles are reprinted in the volume edited by Montagu 1964). Now, two issues are involved here: first, under what conditions races should be named; and, second, whether a biological phenomenon called race really exists. Races and species are given names in order that those who study them can tell each other what they are talking and writing about. But it is a typological fallacy to think that all individuals given the same group name are identical. As stated above, races are Mendelian populations in which no two individuals are identical. Furthermore, these populations are by no means always separated by geographic or other boundaries, and they are frequently connected by gradual transitions. The obvious fact is, however, that members of the same species who inhabit different parts of the world are often visibly and genetically different. This, in the simplest terms possible, is what race is as a biological phenomenon.

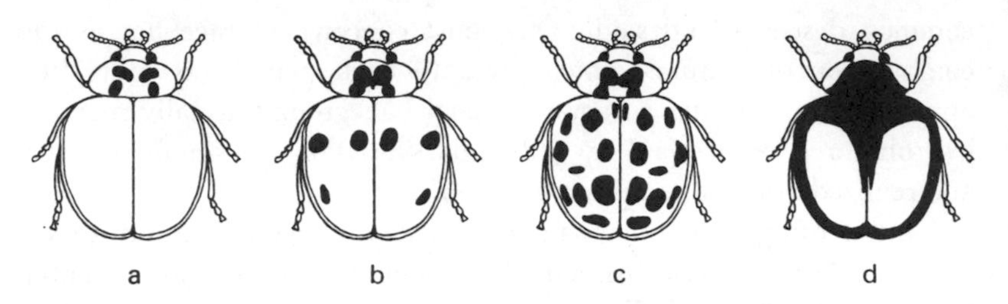

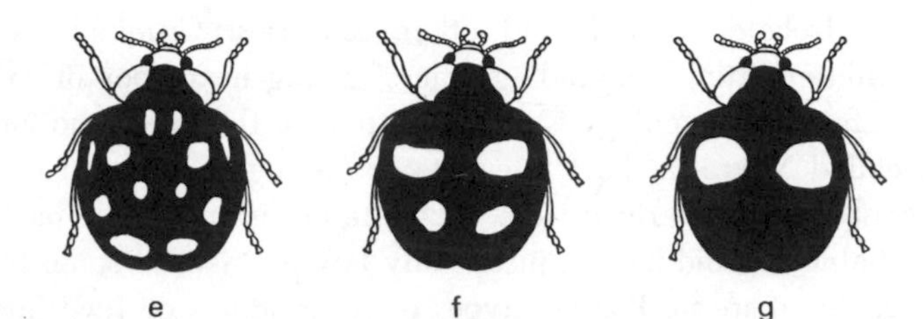

FIGURE 9.1
Color patterns in Harmonia axyridis. The form succinea a–c, aulica d, axyridis e, spectabilis f, and conspicua g.

Gene Frequency Gradients in Harmonia

We consider first an example of racial variation due to frequency differences of alleles of a single gene. This is a paradigm of race variation in general. More complex situations will be considered later.

Variant color patterns in the Asiatic beetle *Harmonia axyridis* (Fig. 9.1) are so striking that some of them were originally described as separate species and even genera. Hosino (1940) and Tan (1946) showed, however, that the variation is caused by five alleles of a single gene; one of the alleles gives the yellow and the yellow black-spotted forms (*succinea, firgida, 19-signata*), another the black yellow-spotted *axyridis,* a third the yellow black-rimmed *aulica,* and the other two the black orange-spotted *spectabilis* and *conspicua* (Fig. 9.1). The alleles exhibit "mosaic dominance"—a heterozygote for any two alleles has black pigment in all parts of the wing covers (elytra) which are black in the carriers of either allele.

Table 9.1, compiled from data of Dobzhansky (1933) and of Komai, Chino, and Hosino (1950), shows the geographic distribution of the color variants. In Siberia west of Lake Baikal the populations consist almost entirely of the black yellow-spotted *axyridis*. From Baikal eastward the population is mostly yellow or yellow black-spotted. On the Pacific Coast (Khabarovsk and Vladivostok) there appear also the black *spectabilis* and *conspicua*. In Japan (the localities in Table 9.1 are arranged from north to south) the frequencies of *succinea* decrease and those of *conspicua* increase as one goes southward. The black-rimmed *aulica* is found in eastern Siberia, China, and Japan, but always in low frequencies.

The variant color patterns are not only sharply distinct but also discrete; there are no intermediates between them. Yet it would not do to call them races, since they interbreed freely and the progeny of a pair may segregate into two or more variants. By contrast, the populations of different territories are racially distinct, because their gene pools contains different frequencies of the alleles responsible for the color forms (Table 9.1). If we are presented with a single specimen of *Harmonia axyridis*, we shall only occasionally be able to tell from which population it came. For example, any form other than the black yellow-spotted *axyridis* is unlikely to have been found in the

TABLE 9.1

Geographic variation of the frequencies of color pattern (in percentages) in Harmonia axyridis

Region	succinea	axyridis	spectabilis	conspicua	Number Examined
Altai Mountains	0.05	99.95	...	...	4,013
Yeniseihk Province	0.9	99.1	...	...	116
Irkutsk Province	15.1	84.9	...	...	73
West Transbaikalia	50.8	49.2	...	...	61
Amur Province	100.0	...	...	...	41
Khabarovsk	74.5	0.2	13.4	10.7	597
Vladivostok	85.6	0.8	6.0	6.8	765
Korea	81.5	...	6.2	12.5	64
Mukden, China	90.7	...	4.5	4.6	1,865
Peking, China	83.3	...	8.9	7.3	9,635
Szechwan, China	42.6	0.01	28.8	25.1	1,074
Soochow, China	66.6	...	16.5	16.1	6,231
Sapporo, Japan	42.9	1.0	21.6	34.3	1,184
Akita, Japan	60.0	2.2	9.6	28.2	135
Takasino, Japan	36.9	3.5	11.8	47.7	5,758
Sanage, Japan	23.1	14.5	25.9	36.2	24,443
Kyoto, Japan	15.3	5.1	15.8	63.7	2,494
Matuyama, Japan	10.7	5.8	19.1	64.1	534
Fukuoka, Japan	2.3	2.2	11.1	83.6	995

Altai Mountains, while a specimen of *axyridis* probably does not come from Peking, China. One would prefer to have a sample of as many individuals as possible, taken at random in the same neighborhood. The frequencies of the color patterns in such a sample will define its geographic origin, and generally the more precisely the larger is the sample.

What causes brought about the racial variation in *Harmonia* is unknown. Komai and Hosino (1951) and Komai (1954) presented evidence that natural selection must be at work. Population samples from Suwa, Japan, collected from 1912 to 1954, show a gradual decline in the frequency of *succinea* from above 40 to below 30 percent, and an increase of *conspicua* from 42 to 56 percent. In 1948–1950 samples were collected on different trees and crops; *axyridis* was more frequent on pine trees than on wheat, whereas *conspicua* showed the opposite preference.

Chromosomal Races in Drosophila Pseudoobscura

The example of racial variation that we are about to consider is analogous to that of the color patterns in Harmonia. However, instead of alleles of a single gene it is concerned with a series of supergenes—coadapted combinations of several or many genes locked in inverted sections of chromosomes and therefore inherited as single units. In Chapters 5 and 8 the inversion polymorphism in the third chromosome of *Drosophila pseudoobscura* was described. The distribution area of this species extends from British Columbia to Guatemala, and from the Pacific to Texas (an isolated colony in the Andes of Colombia may or may not have been introduced by man.) Not counting very rare or local ones, eight gene arrangements are found over more or less extensive territories (Fig. 5.2), but not one of these is species-wide in distribution.

A typical situation is illustrated in Fig. 9.2; one observes gradients (clines) of frequencies of the various gene arrangements in traveling across the species distribution area. On the Pacific Coast of California the commonest gene arrangement is Standard (ST), followed by Arrowhead (AR) and Chiricahua (CH). Eastward from California, in Arizona and New Mexico, ST dwindles and AR increases in fre-

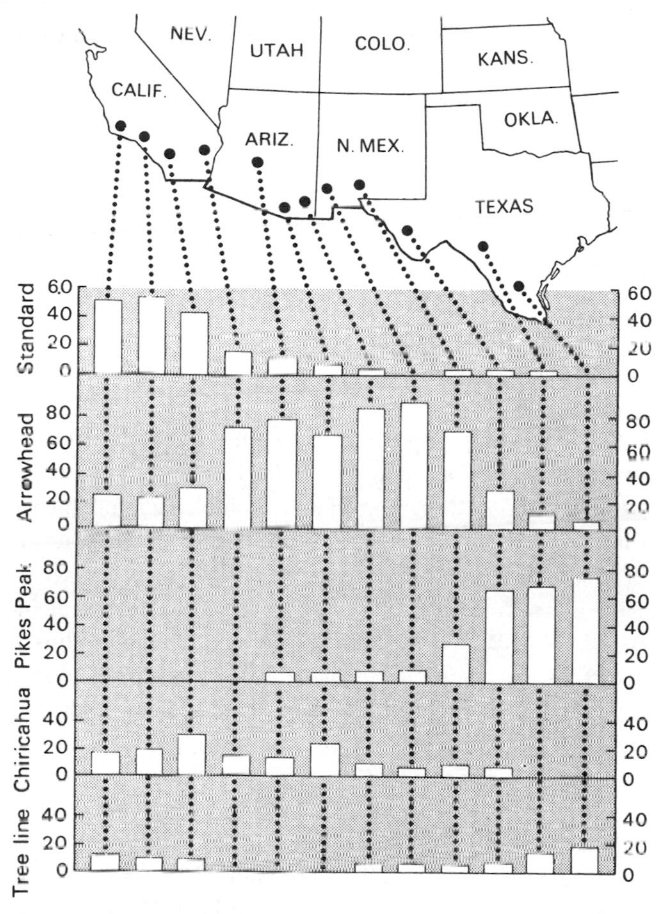

FIGURE 9.2

Frequencies of different gene arrangements in third chromosomes of Drosophila pseudoobscura in southwestern United States.

quencies. Pikes Peak (PP) chromosomes also become more frequent eastward, and are predominant along the eastern face of the Rocky Mountains and in Texas. In North Mexico CH is the most frequent chromosome, while in Central Mexico there appear chromosomes not found at all in the United States (Dobzhansky and Epling 1944).

Where climatic and ecological conditions change rapidly, chromosome frequency gradients can be observed over distances much shorter than those in Fig. 9.2. For example, as one ascends the slope of the

Sierra Nevada range in California, ST chromosomes change from about 45 percent at an elevation of 850 feet to only 10 percent at 10,000 feet, while AR chromosomes increase along the same transect from about 25 to 50 percent of the total (Dobzhansky 1948). This strongly suggests that the frequency gradient reflects in this case a response of the populations to natural selection. The same conclusion is compelled by the evidence of the frequency changes observed in some localities, mentioned briefly in Chapter 5. These changes are cyclic, following the seasons. Thus, in some localities on Mount San Jacinto, in California, ST chromosomes wane and CH wax in frequencies during spring months, whereas the reverse changes occur during the hot summer months.

The changes taking place in summer are easily reproduced in experiments in laboratory population cages (see Chapter 5) kept at 25°C. At this temperature, the ST/ST homokaryotype has a higher fitness than the CH/CH homokaryotype. It proved to be more difficult to reproduce the changes that occur in nature in spring, when CH/CH must have a fitness greater than that of ST/ST. This was achieved, however, by Birch (1955), who used experimental populations in which larval crowding was minimized. This condition makes sense—food is abundant in the habitats of the San Jacinto populations in spring, when these populations expand in number; they contract during the hot season (Dobzhansky 1956b).

Thus far, a satisfactory explanation has not emerged for the changes of longer duration, observed in the populations of *Drosophila pseudoobscura* in the western United States between 1940 and the present (Dobzhansky, Anderson, and Pavlovsky 1966, and references therein). In 1940, PP chromosomes were exceedingly rare along the Pacific Coast, from California to British Columbia (only 4 such chromosomes were found among more than 20,000 studied). By 1957–1963 they were present in all localities where adequate samples were made, their overall incidence in California having been between 6 and 8 percent. They have apparently slightly declined since then, although they remain the dominant chromosomal type in the Rocky Mountains and Texas. As shown in Fig. 9.3, the frequencies of CH have declined almost everywhere in the Pacific Coast states, as did the incidence of another gene arrangement, called Santa Cruz. The frequencies of ST and AR also underwent considerable changes, the former generally increasing and the latter decreasing in frequency.

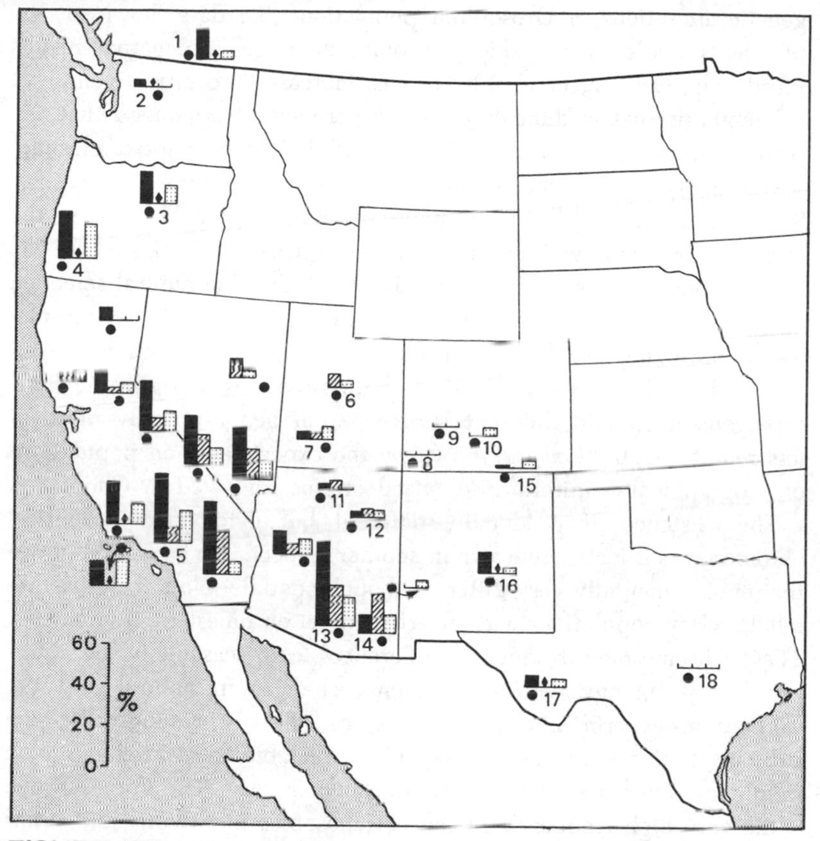

FIGURE 9.3

Changes in the frequencies of the CH gene arrangement in third chromosomes of Drosophila pseudoobscura in western United States. Black columns—samples taken around 1940; hatched columns—1957; stippled columns—1963-1965; black diamonds—absence of samples. (After Dobzhansky, Anderson, and Pavlovsky)

Many changes take place in the natural habitats of the flies; for example, droughts and rainy years occur, or some related and probably competing species of Drosophila become relatively more or less abundant. No correlations between these environmental changes and the genetic changes in *Drosophila pseudoobscura* have thus far been detected. By analogy with industrial melanism in British moths (Chapter 7), it seemed possible that contamination of the natural habitats with traces of DDT and other insecticides may have brought about genetic alterations in Drosophila populations. To date, however, experiments made to test this possibility have given negative results (Anderson, Oshima, et al. 1968). This increases the attractiveness of a hypothesis that is difficult to test experimentally, namely, that, by recombination or by mutation, new and better coadapted chromosomes emerge from time to time and spread in natural populations. Suppose, for example, that a "new" PP chromosome has arisen, the carriers of which have high fitness in Pacific Coast localities where the "old" chromosomes were not well adapted. Balancing natural selection will then favor the spread and increase in frequency of these chromosomes, until they reach their equilibrium frequencies. This is not an ad hoc hypothesis; we know that chromosomes with the same gene arrangement contain different supergenes in geographically more or less remote populations, as shown by the experiments on populations of geographically uniform and mixed origins reviewed in Chapter 8.

The existence of genetically different but cytologically identical chromosomes is instructive also in another respect. We have seen above that environmentally very different though spatially close localities are inhabited by populations with clearly distinct chromosome frequencies. These chromosome frequency differences can reasonably be interpreted, by analogy with the seasonal changes in natural and the changes in experimental populations, as selective responses to the habitats. Is this interpretation equally applicable to macrogeographic frequency differences, such as those shown in Fig. 9.2? For example, is the very high frequency of AR chromosomes in Arizona and other Great Basin states due to greater prevalence there of the habitats in which these chromosomes occur than in California and in Texas, where they are much less frequent? Such an interpretation is not called for; the AR chromosomes in these ecogeographic regions are most probably genetically adjusted in different ways to their respective habitats. This may be a situation analogous to the so-called area effect found among colonies of land snails (see below).

Diversity of Chromosomes and of Environments

We have seen in Chapter 5 that some species of Drosophila are chromosomally monomorphic and others polymorphic; in some the polymorphism is not differentiated geographically, so that one cannot speak of chromosomal races, whereas in others the geographic differentiation is very clear. In certain species of this last category, an interesting relation is revealed between the racial characteristics and the environments in which these races live. *Drosophila willistoni* (da Cunha, Dobzhansky, et al. 1959 and references therein) inhabits an extensive territory, from Florida to Argentina. About 50 different inverted sections have been recorded in all chromosomes of this species; some of the inversions are rare and local, but at least 4 are almost species-wide in distribution. The point of major interest is that near the center of the geographic distribution, in central Brazil, many different inversions occur, and most individuals in natural populations are heterozygous for many inversions. The mean number of inversions per individual in these populations is about 9, and one individual was found heterozygous for 16 inversions, probably the record high for an individual that is not a species hybrid. As one proceeds from the center toward the periphery of the distribution area, the inversion heterozygosis decreases; in Argentina, southernmost Brazil, the West Indies, Florida, Guatemala, and also coastal Equador and northeastern Brazil, the inversion means are between 2 and 1, and even lower. Carson (1959, 1965) and Carson and Heed (1964) found a similar situation in *Drosophila robusta* in the eastern United States—high polymorphism in central and low polymorphism in peripheral parts of the geographic area.

Two different, though complementary rather than exclusive, interpretations of the above findings have been suggested by da Cunha and Dobzhansky and by Carson. As suggested long ago by Vavilov (1926), the center of the distribution area of a species is the territory in which the latter is, by and large, most "at home," and in which it has mastered the greatest variety of environmental opportunities or of available ecological niches. For a species of Drosophila, this may mean that it can subsist on many kinds of fruits and other fermenting substances, that its adaptedness is both wider and greater than that of related competing species, and that it withstands all seasonal vicissitudes and microclimatic contingencies. At the periphery of the distribution, except where an absolute barrier, such as a sea coast, prevents migration, the species has mastered perhaps only a few places in which to live. Genetic diversity—in the species under consideration, the chromosomal diversity—may be the means whereby the mastery of diversified environments is achieved. Of course, if it happens that the geographic center of the species distribution area is ecologically more confining than some more peripheral territory, the genetic diversity may actually be low at the center. This seems to be the case in *Drosophila pseudoobscura*.

Carson (1959) emphasizes that geographically or ecologically peripheral populations are often small and inbred and are dominated by "homoselection," which tends to evolve largely homozygous genotypes specialized to overcome the difficulties of marginal existence. Central populations, on the other hand, are dominated by "heteroselection," which leads to high heterozygosis, heterotic buffering, evolution of supergenes locked in chromosomal inversions, and wider but less plastic genetic adaptive capabilities. Carson supposes further that, although central populations possess a greater immediate adaptive flexibility, the peripheral ones are a more prolific field of genetic experimentation and may give rise to adaptive divergence and to the initiation of new species.

Microgeographic Races in Cepaea Snails

Every living species has the means for active or passive dispersal of its progeny. Because of walking, flying, swimming, and the scattering of seeds, pollen, spores, etc., the birthplaces of the parents and their offspring are separated by some average distance. Individuals within this distance are called sympatric, and those at greater distances allopatric. The distinction between sympatric and allopatric populations, especially in a continuously inhabited territory, is not well defined. It is nevertheless of considerable interest in evolutionary biology; as stated above, races are usually allopatric populations, while polymorphs are variations within a sympatric population. Because the dispersal proficiencies of different living forms vary within a very wide range, races also vary from microgeographic to major geographic.

Since land snails are mostly slow-moving animals, it is not unexpected that populations of the same species living a few steps apart are sometimes genetically different. The extensive studies of Gulick (1905 and earlier) and Welch (1938) on Achatinella in Hawaii, of Crampton (1916, 1932) on Partula in Tahiti and Moorea, and of Diver (1940 and earlier) on Cepaea brought to light numerous examples of geographically very close but phenotypically quite distinctive populations of snails. Since the environments inhabited by these populations appear to be very much the same, it was assumed that the differentiation of the populations is not a result of differential adjustment to the living conditions by natural selection. Random genetic drift seemed a more plausible explanation.

The works of Cain and Sheppard (1950, 1954) and of their numerous disciples in Britain (see Currey, Arnold, and Carter 1964 and Cain, Sheppard, et al. 1968 for references), and of Schnetter (1950) and Sedlmair (1956) in Germany changed the situation (see also Ford 1964 for a review). The common European snail *Cepaea nemoralis* is highly polymorphic. The shell may be yellow (recessive) or brown (dominant). The outside surface of the shell may have five black bands (recessive) or may be unbanded (dominant); some of the bands may be missing or fused together. Working in the neighborhood of Oxford, England, Cain and Sheppard discovered that the incidence of shells of different colors varies, depending on the type of habitat. Figure 9.4 shows that in beechwoods most of the shells are brown and unbanded, whereas on rough herbage and in hedgerows the shells are predominantly yellow and banded. Cain and Sheppard were also able to demonstrate that one of the selective factors responsible for this regularity is predation of the snails by birds, particularly the thrush *Turdus ericetorum*. This bird has the habit of carrying the snails that it finds to "anvils," that is, to nearest convenient stones, where it breaks the shell and consumes the animal inside. The broken shells accumulate near the anvils, and their characteristics can be compared with those of living snails in the vicinity. This permitted Cain and Sheppard to prove that the predation is indeed selective—among the broken shells the proportion of those judged not protectively colored is greater than in the surrounding territory. Thus, in one locality the proportion of banded shells was 264 among 560 living snails, and 486 among 863 predated individuals—a statistically significant difference.

Although the evidence adduced by Cain and Sheppard and their school has established beyond doubt that natural selection is involved in the formation of local races of Cepaea, it does not rule out the possibility that some of the differences may also be due to random drift and to the operation of the founder principle. Lamotte (1951, 1959) in France and Goodhart (1962, 1963) in England did not find the same correlation between the characteristics of the local populations and their habitats that was observed by Cain and Sheppard. Lamotte

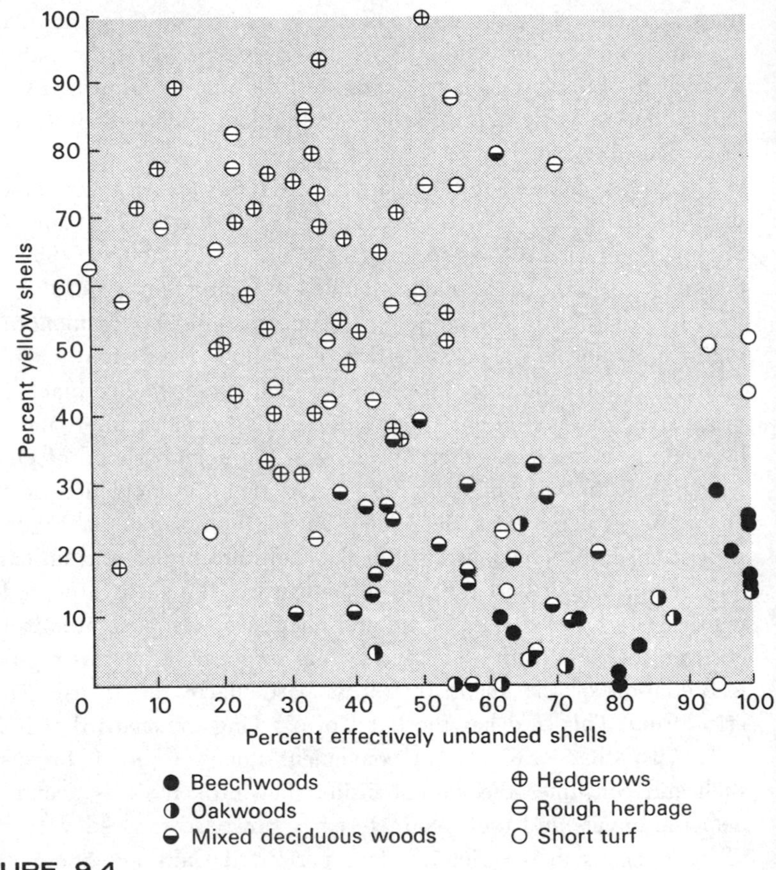

FIGURE 9.4

Shell color and banding in the snail Cepaea nemoralis in different environments. (After Cain and Sheppard)

observed a suggestive relationship between the sizes of the colonies and their diversities, the diversity of small colonies being greater than that of large ones. This is what one would expect if intercolonial diversity is due to random genetic drift in small populations. On the other hand, Lamotte found that snails with differently colored shells differ in temperature preferences and tolerances, so that some selection in addition to drift is probably involved.

Cain and his colleagues (Cain and Currey 1963 and Cain, Sheppard, et al. 1968) also found, in the chalk district of Marlborough Downs not far from Oxford, that the incidence of the color variants of the snails remains uniform over areas vastly larger than that of a panmictic snail population, irrespective of the character of the habitat and the color of the background. This phenomenon, which they have named the "area effect," they believe to be also a product of natural selection, though they have no clue to what factors are responsible for this selection. They observe that "the abruptness of change of gene frequencies in both color and banding might be caused by the change-over from one balanced gene complex to another requiring very different frequencies." If I understand them correctly, this can refer only to epistatic interactions between the genes, or supergenes, which govern the shell color and the rest of the genotype, the latter being different in different populations. If so, their area effect is closely related to the variety of coadaptive relationships observed in chromosomal races of *Drosophila pseudoobscura* (see above).

Stability and Change in Races of Maniola jurtina

Maniola jurtina is a common European butterfly, which has been the subject of brilliant studies by Ford and his collaborators (Ford 1964 and references therein, Creed, Ford, and McWhirter 1964, and Dowdeswell and McWhirter 1967). An inconspicuous but most useful character is the number of spots, varying from none to five, on the underside of the hind wings, particularly in females. A large territory, extending from most of England to southern France, Finland, and Bulgaria, is inhabited by populations with a unimodal spot distribution in females, zero spots being the most frequent class. This "general European stabilization" is maintained despite the pronounced diversity of the habitats where the butterflies occur.

Suddenly in western Devon and Cornwall (southwest England) and on the nearby islands of Scilly, this "stabilization" gives way to remarkable variation. The Cornish race has a bimodal spot distribution in females, zero and two spots being common, and one, three, and more spots rare. On three "large" Scilly islands (682 acres or more) the classes with zero, one, and two spots are about equally common; on "small" islands (40 acres or less) there is a variety of populations with bimodal (zero and two) and with unimodal (zero or two) spot distributions.

The most striking fact (which, owing to the kindness of Professor E. B. Ford, I had the opportunity to observe personally) is that the boundary between the Cornish and the English (general European) races is quite abrupt, a matter of a few meters. Moreover, this boundary often shifts from year to year by several kilometers eastward or westward. The location of the boundary on a given year coincides with no perceptible environmental barriers. The insect concerned is a good flier and can cross the boundary easily. Such sharp boundaries are not unknown elsewhere; for example, White, Carson, and Cheney (1964) found an abrupt border between the chromosomal races of the grasshopper *Moraba viatica* in Australia. Ford and his colleagues have adduced indirect but convincing evidence that the spot distribution in populations of *Maniola jurtina* is subject to strong natural selection. Just what is the selective agency in operation is unknown; it is presumably not the number of spots as such but some pleiotropic effects of the genes responsible for the spotting.

The maintenance of several distinct island races on the Scilly archipelago is explained easily, since the butterfly does not fly across the intervening stretches of sea. The origin of these races poses a more difficult problem. Ford rejects the hypothesis that the racial differences between the islands might have arisen because of the islands having been originally populated by small numbers of founders (the founder principle; see Chapter 8). Ford's argument is that the spotting in some of the island populations has been observed to undergo changes from year to year, provoked by man's interference with the habitats or by weather fluctuations. Thus, the removal of cattle from the small island of Tean has altered its vegetation, and this was followed by changes in the spotting patterns of its population of *Maniola jurtina.* The genetic flexibility of the island populations contrasts, then, with the "stabilization" of the general European race, seemingly impervious to environmental variations.

It is tempting to compare this situation with flexible chromosomal polymorphisms in some populations and species of Drosophila, contrasting with rigid polymorphisms in other populations and species (Dobzhansky 1962). For example, clear-cut seasonal changes in the frequencies of the gene arrangements are observed in some populations of *Drosophila pseudoobscura*, whereas other populations of the same species show no such changes. The gene pool of some populations has reached a degree of adaptedness that makes superfluous genetic reconstructions in response to recurring environmental changes.

Racial Variation of Blood Groups in Man

At the turn of the century, Landsteiner discovered that, when the red blood cells of some persons are placed in the blood sera of some other persons, the cells are clumped or agglutinated. By means of this reaction, people can be divided into four "classical" blood groups: O, A, B, and AB, which are inherited by means of three alleles of a single gene locus—i, I^A, and I^B. The incidence of these blood groups, and of the alleles determining them, varies in populations of different countries. Additional loci, determining immunochemical characteristics of the red blood cells, some of them comprising "systems" of alleles, have been discovered since, and still more are being added almost annually. Some of these are "private" blood groups, so-called because variant antigens have been found only in single families or in single persons, and a vast majority of people test alike with the immunological reagents employed. In addition to the classical OAB system, the following variations of red blood cells give polymorphisms in some or in all human populations:

Rh	at least 15 alleles	Lutheran	2 alleles
MNSs	4 alleles	Kell	2 alleles
Lewis	4 alleles	Duffy	3 alleles
P	2 alleles	Kidd	2 alleles
Diego	2 alleles		

Mourant and his collaborators (Mourant 1954 and Mourant, Kopec, and Domaniewska-Sobzak 1958, and Race and Sanger 1968) have published excellent reviews of the enormous literature dealing with blood groups in man; Mourant's book has 1705 bibliographic references, and numerous works have appeared since 1954. The distribution of the classical blood groups is known best. With the exceptions indicated below, human populations contain persons of the four groups, but in different proportions. The Indians of Central and South America, where they are unmixed with people of European and African descents, are all homozygous for the gene i (O blood group). Blood group O predominates also among the Indians of North America, with the exception of some tribes (such as the Blackfeet) which may have the highest known incidence of A. In the Old World, the gene I^B is, as shown in Fig. 9.5, most frequent in central Asia, Mongolia, Tibet, northern India, and also in a part of Madagascar. Geographic gradients of declining frequencies go westward, eastward, and southward. The allele I^B is rare in Australia and Polynesia. The allele I^A is common in western Europe, South Australia, and, as stated above, in some Indian tribes in North America.

The Rhesus locus is most complex. In fact, there has been much inconclusive controversy between those who believe that only a single gene with many alleles is involved, and those who assume three closely linked genes (*C*, *D*, and *E*) which by rare recombination give rise to a variety of supergenes and of serological phenotypes. This issue need not concern us here; the *CDE* notation is more convenient, and we use it in Table 9.2. The allele, or supergene, *CDe* is common in Asia, among Australian aborigines, and also in Europe; it is infrequent among the Negroid populations of Africa. Conversely, *cDe* is most frequent in Africa, but it occurs in low frequencies throughout the rest of the world. The allele *cde* (Rh-negative) is predominately European,

TABLE 9.2

Percentage frequencies of the Rh blood group alleles in some human populations (After Mourant 1954)

Population	CDE	CDe	CdE	Cde	cDE	cdE	cDe	cde	Tested
English	0.1	43.1	0	0.7	13.6	0.8	2.8	38.9	1038
French	0	41.0	0	0.9	13.0	0.4	3.9	40.8	1672
Germans	0.4	43.9	0	0.6	13.7	1.0	2.6	37.8	2472
Spaniards	1.0	43.0	0	1.3	12.4	0.5	5.4	36.4	2000
Basques	0.9	43.3	0	1.1	4.3	0.2	7.8	42.5	547
Egyptians	0	49.5	0	0	9.0	0	17.3	24.3	184
Nigerians	0	9.9	0	2.2	9.1	0	65.1	13.6	165
Bantus	0	4.7	0	5.8	8.5	0	59.6	21.4	600
Australia (aboriginal)	2.1	56.4	0	12.9	20.1	0	8.5	0	234
Indonesians	1.2	83.4	0	0	10.6	0	4.8	0	247
Bengali	1.6	63.3	0	6.5	7.6	0	3.9	17.7	236
South Chinese	0.5	75.9	0	0	19.5	0	4.1	0	250
Eskimos	3.4	72.5	0	0	22.0	0	2.1	0	158
Navajo	5.8	32.6	0	17.2	35.6	1.6	6.9	0	305

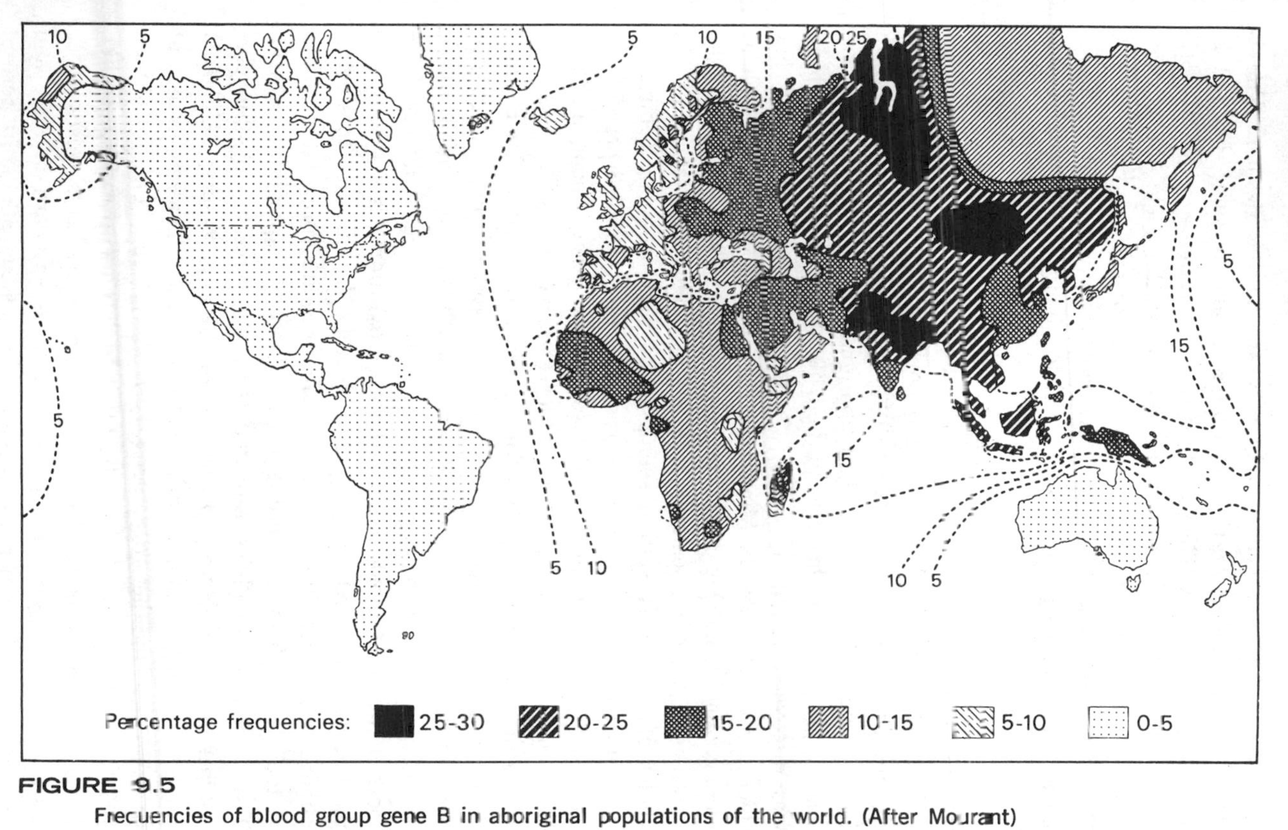

FIGURE 9.5
Frequencies of blood group gene B in aboriginal populations of the world. (After Mourant)

reaching the highest known frequency among the Basques. *Cde* is found mainly among Australian aborigines, and in some, but not in all, American Indians. Thus we have an instructive example of polymorphs that can be used as racial characters: the allele *cDe* is characteristic of Africans, but it occurs also, more or less as a rarity, among Europeans and Asiatics. There is no need to suppose that the Europeans and Asiatics who carry this allele always have some remote Negroid ancestry.

The MN blood group system is geographically less variable than the OAB or the Rh system, since most populations show fairly uniform frequencies of *M* and *N* genes, usually with a slight predominance of the former. However, *M* reaches high frequencies among American Indians, and *N* among Australian aborigines. The other blood group systems are as yet insufficiently studied. The Diego locus polymorphism was discovered originally among South American Indians and has since been found also in Mexico and in North America, as well as among Chinese and Japanese. Race and Sanger (1968) call one of the two Diego alleles "essentially a Mongolian character." One of the alleles of the Duffy system seems to be very frequent (above 90 percent) among west African Negroes, but rare (about 3 percent) in Europeans. An allele in the Kidd system is prevalent among Negroes, less common among Europeans, and least known among Chinese.

The development of methods for the separation of proteins by electrophoresis in starch and other gels made possible exploration of genetic variations of human blood sera (Smithies and Connell 1959). Two serum proteins, haptoglobins and transferrins, have attracted the most attention. Haptoglobins are detected by adding free hemoglobin to aliquots of a blood serum, and subjecting the mixtures to electrophoresis. Haptoglobin-hemoglobin complexes are formed that move in the electric field, depending on the kind of haptoglobin present. Two common gene alleles, Hp^1 *and* Hp^2, give three haptoglobin phenotypes on starch electrophoresis, one of the homozygotes (Hp^1Hp^1) producing a single band, the other (Hp^2Hp^2) three bands at a different position, and the heterozygote (Hp^1Hp^2) all four bands (Giblett 1961). The allele Hp^2 is most frequent in India (up to 90 percent), while the frequencies of Hp^1 show ascending gradients westward to Europe, southwestward to Africa, as well as eastward in Asia, and particularly in the Americas, reaching frequencies above 70 percent in Peru and Chile (See Buettner-Janusch 1966). Giblett (1964) and others discovered at least eleven more haptoglobin phenotypes, which occur in very low frequencies, at least in the populations studied.

Transferrin is an iron-binding protein that plays an important role in the iron metabolism of the body. Smithies and Connell (1959) and Giblett (1961) discovered 3 kinds of transferrins: C, which is the commonest, and B and D, which are less common. About 15 kinds have since been described, some of them apparently restricted to certain human populations. The mode of inheritance is not known precisely, but apparently multiple alleles of one locus are involved.

The technically as well as genetically most complex, yet potentially most interesting, polymorphism has been found by Grubb, Podliachouk, Ropartz, Steinberg, and others (see Steinberg 1962, 1967, and Buettner-Janusch 1966 for reviews) in human gamma globulins. The testing procedure requires using (1) red blood cells from a donor of group O who is homozygous for the allele giving the D antigen of the Rh system (see above), (2) an anti-D serum, and (3) an agglutinating serum, most readily obtained from patients with rheumatoid arthritis. The possibility of detecting various antigens, of which about 20 have been found, depends, of course, on the availability of proper sera, a restriction that makes the data obtained by different authors not always comparable. The inheritance of the variant antigens is best interpreted as due to two genes or supergenes, *Gm* and *Inv*, the former giving rise to at least 17, and the latter to 4, different antigens.

Table 9.3 reproduces Steinberg's list of *Gm* alleles found in populations tested with a battery of antigens: Nos. 1, 2, 3, 5, 6, 13, and 14.

TABLE 9.3

Gn alleles commonly found in certain populations

Population	Alleles Found	Population	Alleles Found
Caucasoid	Gm^{1}, $Gm^{1,2}$, $Gm^{3,5,13,14}$	Bushmen	Gm^{1}, $Gm^{1,5}$, $Gm^{1,13}$, $Gm^{1,5,13,14}$
Negroid	$Gm^{1,5,13,14}$	Pygmy	$Gm^{1,5,13,14}$, $Gm^{1,5,6}$
	$Gm^{1,5,14}$, $Gm^{1,5,6}$	Micronesian	Gm^{1}, $Gm^{1,3,5,13,14}$
Mongoloid	Gm^{1}, $Gm^{1,2}$, $Gm^{1,13}$, $Gm^{1,3,5,13,14}$	Melanesian	Gm^{1}, $Gm^{1,2}$, $Gm^{1,3,5,13,14}$, $Gm^{1,5,13,14}$
Ainu	Gm^{1}, $Gm^{1,2}$, $Gm^{1,13}$, Gm^{2}		

Some alleles (or supergenes) give rise to combinations of several detectable antigens. According to Muir and Steinberg (1967), the common alleles found among Negroes are $Gm^{1,5,13,14}$, $Gm^{1,5,14}$, $Gm^{1,5,6}$, *and* $Gm^{1,5,6,14}$. In Mongoloids (including the Amerindians) the common alleles are Gm^{1}, $Gm^{1,2}$, and $Gm^{1,5}$, which are the same as among Caucasoids. African Bushmen have the same alleles as Negroes and also Gm^{1}, $Gm^{1,13}$, and an edemic allele, $Gm^{1,5}$. Among the Melanesians, an interesting difference is found between the speakers of two groups of languages—the allele $Gm^{1,2}$ is absent among those who use so-called MN languages but present in speakers of NAN languages.

Polymorphism in red blood cell and serum antigens is certainly not confined to the human species. Nonhuman primates have been examined for the antigens of the OAB, MN, Rh, and certain other systems (review in Wiener 1965 and Sullivan and Nute 1968). Chimpanzees are polymorphic for O and A, and orangs, gibbons, and baboons for A, B, and AB or very similar antigens. Antigens related to M and N are both found in chimpanzees, while chimpanzees and gibbons seem to differ with respect to their antigens of the Rh system. Transferrin polymorphisms have been observed in chimpanzees (7 phenotypes), gorillas (4), and gibbons (2), as well as in monkeys of the genus Macaca (11 alleles, a review in Buettner-Janusch 1966).

An impressive variety of red blood cell antigens has been revealed in domestic cattle (Stormont 1959) and sheep (Stansfield, Bradford, et al. 1964). Breeds are often recognizable by the alleles they carry. Protein polymorphisms (hemoglobins, transferrins, lactoglobulins, albumins, esterases, amilases) have been found in cattle, sheep, goats, pigs, horses, and poultry (bibliography in Ashton, Gilmour, et al. 1967). Blood groups and enzyme polymorphisms have been utilized in studies of the genetic population structure in domestic mice (Petras 1967) and in the deer mice, *Peromyscus maniculatus* (Rasmussen 1962). Briles and his colleagues (Briles, Allen, and Mullen 1957 and Briles 1964) have obtained data on blood group polymorphisms in chickens, indicating heterotic superiority in heterozygotes. Ferrell (1966) found significant differences in blood group frequencies in colonies of song sparrows (*Melospiza melodia*) living in various localities around San Francisco Bay in California. The enzyme polymorphisms found in Drosophila, discussed in Chapter 7, are phenomena analogous to the blood polymorphisms considered in the present chapter.

Races of Mankind

Boyd (1950) was the first to utilize the data then available on blood group polymorphisms to characterize human races. His classification was as follows:

1. Hypothetical Early European: the highest known incidence of Rh negative (*cde* in Table 9.2), probably no I^B. Basques are the modern descendants of this race.
2. European or Caucasoid: the next highest incidence of *cde*, a relatively high incidence of *CDe*.
3. African or Negroid: very high frequencies of *cDe*, moderate ones of *cde*.
4. Asiatic or Mongoloid: high frequencies of I^B, little if any *cde*.
5. Amerindian: zero to very high incidence of I^A, probably no I^B or *cde*, low incidence of *N*.
6. Australoid: high incidence of I^A and *N*, no *cde*.

Except for the Hypothetical Early European race, this classification represents no radical departure from the classificatory schemes of anthropologists who use externally visible traits, such as skin color, hair shape, or facial features. Nevertheless, some conservative anthropologists objected to Boyd's essay. Why substitute invisible blood traits for easily visible morphological ones? This objection misses the point. The advantage of blood antigens is that their inheritance is relatively simple, diagnosis by competent technicians is accurate, and differences between populations can be expressed in terms of gene frequencies rather than of visual impressions or at best as means of measurements. Skin and hair colors and shapes are polygenic traits; the mode of their inheritance is uncertain, and they are modifiable by environmental circumstances. As pointed out above, genetically simple traits make it possible to gain an insight into the nature of racial variation. This is not to deny that a racial classification should ideally take cognizance of all genetically variable traits, oligogenic as well as polygenic.

Garn (1965), fully conversant with the evidence of classical anthropology and of genetics, proposes the following classification of 9 "geographical races," some of them further subdivided into "local races":

1. Amerindian, with 4 local races.
2. Australian, with 2 local races.
3. African, with 4 local races.
4. European, with 5 local races.
5. Indian, with 2 local races.
6. Asiatic, with 5 local races.
7. Polynesian.
8. Micronesian.
9. Melanesian.

In addition, he recognizes "puzzling, isolated, numerically small local races": Lapp, Pacific Negrito, African Pygmy, and Eskimo; "long-isolated marginal" local races: Ainu, Bushmen, and Hottentots; and "hybrid, local races of recent origin": North American Colored, South African Colored, Ladino, and Neo-Hawaiian. Since there is no substantial difference between "geographical" and "local" races, this classification may as well be presented as comprising 35 races, some including many millions and others only thousands of persons, some expanding and others in the process of being submerged by hybridization with their more numerous neighbors.

Coon (1965) divides mankind into 5 "subspecies"—Caucasoid (more or less corresponding to Garn's European), Mongoloid (Garn's Asiatic plus Amerindian), Congoid (African), Capoid (Bushmen), and Australoid (Australian plus Melanesian). The Pygmies in Africa and the Negritoes in Australasia he considers, somewhat ambiguously, to be "races."

It should be made unequivocally clear that the number of races or subspecies which one chooses to recognize by giving them vernacular or formal Latin names is largely, though not completely, arbitrary. Every local population of human beings, or of any other sexual species, probably differs from other populations in the incidence of some gene alleles in its gene pool. Every population is, then, racially distinct from every other. But it would serve no useful purpose to give every population a name as a separate race. In a continuously inhabited territory populations have no boundaries; they merge into each other. The incidence of some blood group alleles has repeatedly been shown to be significantly different in the possessors of different surnames (Hatt and Parsons 1965 is a recent example). Possessors of the same surname do not constitute separate Mendelian populations, but the inhabitants of districts in which a given surname is particularly common may be such populations. Moreover, one should not confuse race differences and race names. The former are objectively ascertainable biological phenomena; the latter are matters of convenience. For some purposes it may be expedient to divide mankind into 35, or 9, races; for others, into only 5 named races.

There will always be some populations intermediate between, and failing to fit into, any of the races; some of the inhabitants of western Siberia and Turkestan are examples of the former type, while European Lapps are examples of the latter. Intermediate populations have traditionally been regarded as mixtures, that is, hybrids of the populations between which they are intermediate. To some extent this view is, of course, correct; gene diffusion within a species is normal and ubiquitous, unless geographic barriers intervene. The great plains of Europe and western Asia have few such barriers, the Caucasoid and Mongoloid races or subspecies are connected by many transitional populations, and the gene frequencies (see Fig. 9.5 for an example) show regular gradients. It is unnecessary, however, to suppose that in all cases there was a time in the past when the races were unmixed and the gene frequency gradients were absent.

By contrast, geographic barriers that impede travel also impede gene diffusion and make gene frequency gradients steeper or steplike at the barriers. Examples of this situation in respect to man are the Sahara Desert, separating African and European races, and the Himalaya range, separating Mongoloid from Indian races. It is evidently both an acknowledgment of the objectively ascertainable biological situation and a matter of convenience that race boundaries are drawn, whenever possible, at the breaks or steepenings of the gene frequency gradients. The number and the magnitude of the gene frequency differences impose other restrictions on the arbitrariness whereby populations are included in the same race or are given different racial names. Thus, Mongoloid and Amerindian populations are frequently placed together in a single race, but it would hardly occur to anybody to combine Amerindian and African populations and to contrast them with the Mongoloids. Although some separations and combinations are debatable, others are clearly reasonable, and still others obviously impossible.

Plant Races or Ecotypes

Turesson (1922, 1925, 1930) pioneered experimental studies on the genetic adaptedness of local populations to their habitats. Working with as many as 31 different plant species, he demonstrated that local populations found in different environments are often genetically different, and argued that the distinctions are adaptive in the respective environments. His method was to transplant individuals from different local populations and to grow them under uniform conditions in his experimental garden. The transplants, though often modified as compared to their appearance in their native habitats, nevertheless did not become identical, proving that they were genetically distinct. Thus, the hawkweed *Hieracium umbellatum* transplanted from populations on sandy seashores in southern Sweden was prostrate, with long shoots and narrow, hairy leaves; from rocky localities it had short shoots and broader, shiny leaves; and plants from inland localities were tall and erect. Turesson regarded species (or ecospecies, as he called them) as mosaics of local populations, which evolve different ecotypes in different habitats. He defined an ecotype as "the product arising as a result of the genotypical response of an ecospecies to a particular habitat."

Clausen, Keck, and Hiesey (1940, 1948), Clausen (1951), and Clausen and Hiesey (1958) carried the experimental study of ecotypes much further. Using species in which an individual can be cut into several parts, they replanted the parts in different environments in three transplant gardens in California—one close to sea level at Stanford, and the others at 4600- and 10,000-foot elevations in the Sierra Nevada Mountains. Their most extensively studied materials were species of yarrow (Achillea) and cinquefoil (Potentilla), and their ecotypes were geographic races from various altitudinal and climatic zones in California and from foreign countries. The findings were not only that the races were genetically distinct, but also that their norms of reaction (Chapter 2) were adaptively circumscribed to fit the environments to which they were native. Individuals of different races planted together, and also parts of the same individual replanted in different environments, were distinguishable.

Yarrows (*Achillea lanulosa*), native to different elevations in the Sierra Nevada and transplanted into the garden at Stanford, can be seen in Fig. 9.6. Frequency distributions and the mean heights of about 60 plants from each locality as they develop at Stanford are shown. There are clear gradients of diminishing plant heights, as well as changes in the foliage, as one moves from lower to higher elevations. In addition, many plants native to high elevations failed to produce flowers at Stanford. Races from higher elevations showed winter

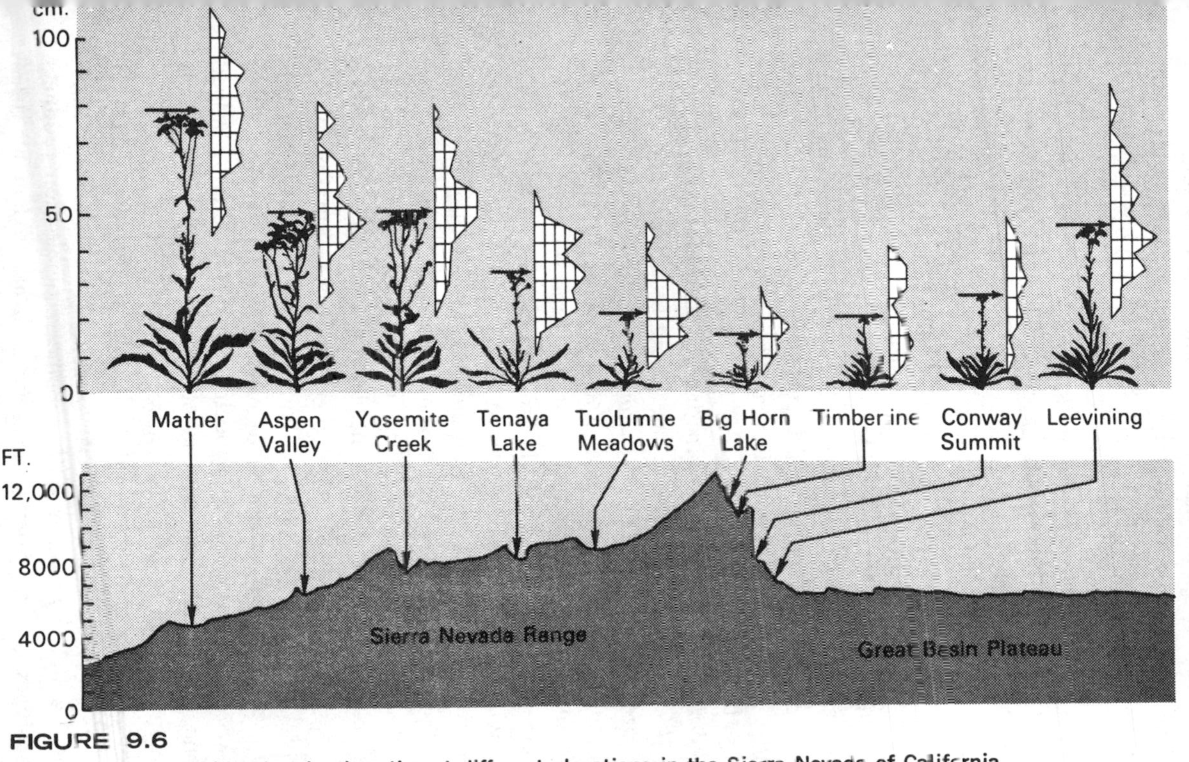

FIGURE 9.6

Yarrows (Achillea lanulosa), native at different elevations in the Sierra Nevada of California, grown under uniform environment in an experimental garden. The distribution curves of the height of the plants are shown to the right of the drawings of the average plants from the respective populations. (After Clausen, Keck, and Hiesey)

dormancy (i.e., did not grow in winter) at Stanford, whereas races from low elevations either had no dormancy or had summer dormancy. The climatic conditions at high elevations make winter dormancy essential for survival, but at Stanford winter dormancy is unadaptive.

Specimens of *Potentilla glandulosa* from different elevations were planted in the three experimental gardens. The race native at low elevations developed best in the Stanford garden, did reasonably well in the garden at 4600 feet, but was killed at 10,000 feet. The mid-elevation race was robust and abundantly flowering at 4600 feet; at Stanford it was less tall, was winter dormant, and produced few flowers; it hardly survived and seldom matured seed at 10,000 feet. The alpine race is a dwarf, but it was the only one that consistently matured seed at 10,000 feet; at 4600 feet it grew somewhat taller than in its native habitat; at Stanford it was winter dormant, very weak and hardly surviving.

Very interesting hybridization experiments on the altitudinal races of *Pontentilla glandulosa* have been carried out by Clausen and Hiesey (1958). Three generations (F_1–F_3) were raised of hybrids between an alpine and a low-elevation race. For a total of 19 distinguishing traits, the authors estimate that at least 100 gene differences are involved; some trait differences are due to a single pair or to two pairs of alleles, while others are polygenic. With so many gene differences, hardly any two individuals in the F_2 generation are identical in appearance or in physiological properties. Some of the F_2 segregants were "cloned," that is, cut into three parts, which were replanted in the three experimental gardens at different elevations in California. Among the 542 segregants so treated, some survived for the duration of the experimental period (5–9 years) at one or more transplant gardens (long survival), whereas others were weak and died usually within 1 or 2 years (short survival). As expected, some individuals survived long in one garden but soon died in others. The numbers of plants with different survival records are shown in Table 9.4.

Some of the segregants survived well only at the low, others at the low and the intermediate, or at the intermediate and the high, or only at the high, transplant station. Some recombination products of the genotypes of the parental races were ill adapted at all three stations—a phenomenon of hybrid breakdown that will be discussed in the next chapter. However, the class most abundantly represented

TABLE 9.4

Duration of survival (long or short) of cloned F_2 segregants from a cross of an alpine and a low-altitude race of Potentilla glandulosa (After Clausen and Hiesey 1958)

Elevation (feet) of Transplant Garden			Number of Segre-gants	Elevation (feet) of Transplant Garden			Number of Segre-gants
90	4600	10,000		90	4600	10,000	
Short	Short	Long	26	Long	Long	Long	148
Short	Long	Long	56	Long	Short	Long	25
Short	Long	Short	68	Long	Long	Short	97
Short	Short	Short	58	Long	Short	Short	64

among the segregants consisted of plants that survived at all three stations, perhaps indicating the formation of genotypes with a norm of reaction more permissive than the norms in the parental races crossed. As might have been expected, most of the plants that survived best at 10,000 feet resembled the native alpine race also in their structural traits, and the plants surviving best at the lowermost station resembled the low-altitude race. But, interestingly enough, there were exceptions; some of the "synthetic alpines" contained a number of traits inherited from the foothill parental race. The six segregants that were notably successful at all three stations tended morphologically to resemble the foothill parent. A caveat must be entered at this point: since the segregants were transplanted artificially into well-tended experimental gardens and were not exposed to competition with native vegetation, their degree of adaptedness in the various environments is probably overestimated rather than underestimated.

Is there a difference between a geographic race and an ecotype? To a large extent the two terms are synonymous. Zoologists, however, seldom speak of ecotypes, and Mayr (1963) concludes, "It would appear safer to describe the result of studies in experimental taxonomy in less rigorous terms, such as climatic race or edaphic race, than in terms of the ecotype with its typological implications." The outstanding botanist and evolutionist Grant (1963) does not mention ecotypes in his book. It may be advisable to restrict the term ecotype to races that occur mosaic-fashion in a quasi-continuously inhabited territory, wherever a certain type of environment (e.g., sand, clay, or calcareous soils) appears. If so, the correlation between the colors and patterns of Cepaea snails and their habitats discovered by Cain, Sheppard, and their colleagues (see above) may be an example of ecotypic variation in an animal species. In any case, the experiments of Clausen and his colleagues can be regarded as shedding much needed light on the genetics of racial variation.

The Possible Adaptive Significance of Racial Variation in Man

One might think that the adaptive significance of the race differences in man, as well as of the intrapopulational polymorphisms, would be comprehensively investigated and known in detail. On the contrary, at best only more or less plausible speculations exist concerning some characteristics, and not even conjectures concerning the rest. Reference was made above to the opinion of Hooton, a leading anthropologist of a generation ago, that race differences are "principally non-adaptive." This view has probably been shared by a majority of anthropologists. At the opposite pole is the dictum that neutral genetic traits are nonexistent or extremely rare, so that if some trait seems to be neutral it means only that its adaptive significance has not yet been discovered.

In the particular case of the human species the difficulty is compounded. The environments in which people now live are radically different from those in which their ancestors lived 1000, or even 100 or fewer, generations ago. Therefore, the adaptive function of a genetic difference at our time level may be entirely unlike its function in the past, when the human species and its subdivisions were in the process of formation. A best-seller has called man "The Naked Ape," but he usually goes clothed and is not an ape. Hairiness or lack of hair, skin pigmentation, or visual acuity may have been important for survival in the past; at present these traits may be nearly neutral or may affect fitness via differential mating success rather than differential survival.

The frequent mention of blood groups as examples of adaptively neutral traits goaded many investigators to discover physiological correlates of the blood groups that are not neutral. Starting with the pioneering work of Aird, Bentall, and Roberts (1953), the burgeoning literature has been ably summarized by, among others, Clarke (1961) and Reed (1961). A relationship between the classical O, A, B, and

AB blood groups and duodenal and gastric ulcers is probably the most thoroughly established. The percentages of persons with different blood groups among patients with these ulcers and in a control population are as follows:

	Blood group			
	O	A	B	AB
Duodenal ulcer	58.7	30.6	7.2	3.5
Gastric ulcer	49.1	40.0	8.2	2.7
Controls	49.0	39.1	9.4	2.5

The incidence of the O group is higher than in the controls among sufferers from duodenal but not from gastric ulcers, while among patients with gastric ulcers (and with cancers of the stomach) a higher frequency of blood group A is sometimes observed. Pernicious anemia has shown a statistically significant association with blood group A in some populations but not in others.

When the blood group of a fetus is unlike that of its mother, an immunological incompatibility may arise, and a statistically significant excess of fetal or neonatal death may be observed. This is established beyond doubt for Rh-positive children of Rh-negative mothers (i.e., *D* versus *d*; see above). Similar incompatibilities for OAB and MN blood systems are less securely verified. Another association that has been claimed is that between blood group A and carcinoma of the cervix.

The taste threshold for solutions of phenylthiocarbamide (PTC) is determined largely by a single gene; specific solutions of this substance taste intensely bitter to some, and seem to have no taste at all to other persons. Only a minority of people are intermediate in their taste thresholds. The incidence of the dominant allele for tasting this substance, and of the recessive allele for taste-blindness to it, are racially variable. Several investigators (Kitchin, Evans, et al. 1959) found that among patients with adenomatous goiters nontasters are significantly more frequent than they are in the control population.

The above observations have shown that polymorphisms like those for blood group genes or for PTC tasting are not neutral, in the sense that they are significantly correlated with certain pathological conditions. Only a small percentage of persons with blood group O have duodenal ulcers, but among those who have such ulcers people with O blood are more frequent than among those who do not. Whether this explains the race differences in the incidence of the blood groups is a quite different question. It happens that the associations thus far discovered are with diseases affecting mainly people of postreproductive ages, so that the Darwinian fitness is but little changed. Moreover, it is not evident how these associations explain the observed racial differences. Do the American Indian tribes that have only O blood suffer from duodenal ulcers more frequently than do the inhabitants of the interior of Asia, among whom O blood is less prevalent? Or are we to suppose that environments in America are less ulcerogenic than those in Asia? There are no data for or against either conjecture.

A very different hypothesis—and one which, if confirmed, might explain the racial differences in the incidence of O, A, and B blood groups—was advanced by Vogel, Pettenkofer, and Helmbold (1961). They attempted to correlate the prevalence of these blood groups with epidemic diseases that ravaged the populations of various countries. Thus, blood group B is common in parts of the world subjected to epidemics of plague, while A and O may possibly be related to the prevalence of smallpox and syphilis. The evidence that these diseases are less dangerous to the carriers of some than of other blood groups is, however, far from convincing (a detailed and critical review in Otten 1967).

Although the rule is not without exceptions, dark-skinned populations live most in countries with abundant sunshine, and light-skinned ones where sunshine is at least seasonally deficient. The notion that dark pigmentation is protective against sunburn is very old; the idea that light skin may facilitate synthesis of vitamin D is more recent. Both notions are plausible, but the supporting evidence is not nearly as complete as could be wished. A summary of the extensive literature has been provided by Blum (1959, 1961). There is no doubt that ultraviolet wavelengths shorter than 3200 Ångstrom units cause erythema (reddening) and sunburn of unpigmented skin, but an acquired tan in otherwise light-skinned races gives protection as effective as the pigmentation of darker races. There is good evidence of higher incidence of skin cancers in lightly pigmented than in darker individuals, with presumably comparable amounts of sun exposure. Loomis (1967) argues that, whereas light pigmentation is adaptive because it facilitates the formation of vitamin D, dark pigmentation is protective against excessive and toxic doses of the substance.

Hamilton and Heppner (1967) found that zebra finches exposed to artificial sunlight used almost 23 percent less metabolic energy after they were dyed black; the authors conjecture that "dark human skin coloration may maximize the absorption of solar radiation in situations where energy must be expended to maintain body temperature, as at dawn and dusk in otherwise hot climates." It has also been surmised that dark skin may function as protective coloration for an individual stalking game or escaping from predators.

Body size, body build, and pigmentation vary widely in different countries (maps in Lundman 1967). It is very probable that these variations help (or helped in the past) human populations to adapt to climatic and other conditions in the countries they inhabit. The evidence in favor of this hypothesis is neither as complete nor as convincing, however, as could be wished. Heat conservation is assisted by minimizing the body surface in relation to mass, whereas heat dissipation, on the contrary, is helped by maximizing the radiating body surface. According to Allen's rule (see the next section), one expects cold-dwelling populations to be as nearly globular as possible—thickset, with relatively short arms and legs. Conversely, inhabitants of hot countries profit by a linear body build, with relatively long arms and legs. Bergmann's rule (see the next section) states that races of the same species of animal that live in cold climates are, on the average, larger than those in hot climates.

Attempts to apply these rules to human races have revealed a general agreement and also not a few exceptions (Hulse 1963, Schreider 1964b, Garn 1965). Scholander, Walters, et al. (1950) challenged the applicability of these rules to man on the ground that human beings deal with cold and heat stress by appropriate clothing and types of dwellings, rather than by body build. This is doubtless so, but in evolution several different adaptive contrivances are often used convergently to respond to an environmental challenge. Summaries and reviews of pertinent literature may be found in Barnicot (1959), Harrison (1961), Schreider (1963), and Baker and Weiner (1966).

Riggs and Sargent (1964) investigated "simulated survival" under conditions of humid heat in 19 matched pairs of white and Negro young male volunteers. Twenty-two "heat casualties" were observed—irregular pulse, anhidrosis, hypohidrosis, heat exhaustion; only one was a Negro. In other studies Baker and his colleagues (Baker 1966 and Baker, Buskirk, et al. 1967) compared physiological reactions to cold on the Andean Altiplano in native Quechua Indians and in partially acclimatized whites. During exposures for 2 hours to a temperature of

10°C, the Quechuas had higher rectal temperatures, skin temperatures, and toe temperatures, but lower mean oxygen consumption and lower cumulative heat exchange than the whites. To what extent this situation could be changed by lifelong cold acclimatization of the whites is difficult to tell. A similar question arises in connection with the findings by Mason, Jacob, and Balakrishnan (1964) that basal metabolic rates are lower among Indian than among European women living in Bombay.

It would be unprofitable to attempt here even a cursory review of the confused and confusing literature on alleged genetic differences between races of man in intelligence and other psychic traits (see Berelson and Steiner 1964, Pettigrew 1964, and Jensen 1969 for references). It is frequently found that the mean IQ scores for Negroes tend to be lower than those of whites living in the same cities or states, but what part of this difference is conditioned genetically and what part is due to social discrimination and inequality of opportunity cannot readily be determined. Even if some portion were shown to be genetic, this would still not justify racist attempts to deprive certain races and social classes of their human rights and dignity. Indeed not even a racist, unless he is wholly ignorant or dishonest, can deny that human races overlap so broadly in measurable psychic traits that every race contains many individuals who score both higher and lower than the means of other races. Schreider (1964a) found a positive correlation between stature and intelligence in the French population. Here, again, it is uncertain to what extent this correlation is genetically or environmentally caused. Whatever the answer may be, would anybody be prepared to argue that tall people are entitled to social preferment over short ones?

Ecogeographical Rules

Adaptive significance of some traits can be inferred from indirect but still convincing evidence. Structural and physiological peculiarities of desert, alpine, prairie, tundra, tropical forest, and other types of vegetation have been known since pre-Darwinian times. Reduction of the evaporating leaf surface, transformation of leaves into spines, development of pubescence or waxy covering on the epidermis, presence of chlorophyll in the surface layers of the bark, and fleshiness of the leaves, twigs, and stems are well-known examples of the adaptations in desert plants. Astonishing resemblances between American cacti and African euphorbias provide spectacular illustrations of evolutionary convergence.

When the distribution area of a species extends over territories with different climatic, edaphic, and other environmental conditions, the races inhabiting these territories often reflect their differential adaptedness in their structural and physiological traits. Some rules, statistically valid but admitting of exceptions caused by special conditions, were noted by nineteenth-century naturalists. The modern formulation and statistical verification of these rules are due chiefly to the work of Rensch and his numerous students (critical reviews in Rensch 1929, 1960a, b, and Mayr 1942, 1963).

Gloger's rule states that, in mammals and birds, races inhabiting warm and humid regions have more melanin pigmentation than races of the same species in cooler and drier regions; arid regions are characterized by the accumulation of yellow and reddish brown phaeomelanin pigmentation. Among insects, the pigmentation increases in humid and cool and decreases in dry and hot climates, the humidity being apparently more effective than temperature (Zimmerman 1931 in wasps, Dobzhansky 1933 in ladybird beetles, and Watt 1968 in Colias butterflies). For the ladybirds, eastern Asia (Siberia and Japan) is the center of heavily pigmented races; to the southwest and southeast of this region, lighter and lighter races are encountered, until centers of very pale races are reached in California in the Western, and in Turkestan in the Eastern, hemisphere.

In mammals and birds, races that live in cooler climates are larger in body size than races of the same species in warmer climates (*Bergmann's rule*). Petersen (1949) finds, however, that in Scandinavian butterflies the wing length decreases northward. In warm-blooded animals, races inhabiting cooler regions have relatively shorter tails, legs, ears, and beaks than those from warmer climates (*Allen's rule*). Petersen finds this rule to apply to some, although not to all, European butterfly races. According to Rensch, Mediterranean races of species of Carabus beetles tend to be smaller and more elongate and to have relatively longer antennae and legs than those from central and northern Europe. In birds, races with relatively narrower and more acuminate wings tend to occur in colder, and those with broader wings in warmer, climates (*Rensch's rule*). Also, the inhabitants of colder countries deposit more eggs per clutch than those of warm countries. In mammals, shorter but coarser hair and a decrease in the amount of down are observed in warm countries. Fish of cooler waters tend to have a larger number of vertebrae than do races of the same species living in warmer waters; an increase in salinity has the same effect as a decrease in temperature. Forms that inhabit swiftly flowing waters tend to be larger and to have more streamlined body shapes than inhabitants of sluggish or stagnant waters; cyprinid fishes isolated in desert springs tend to lose their pelvic fins (Hubbs 1940).

The adaptive significance of some of the rules is evident; that of others, conjectural. The function of longer pelage and greater amount of wool in mammals of cold lands is heat conservation (Irving 1966). As pointed out above, Bergmann's and Allen's rules are also concerned with problems of temperature regulation. Larger body size means a relatively smaller body surface, and consequently lesser loss of heat. Protruding body parts, such as extremities, tails, and ears, are subject to rapid loss of heat. Increasing the body surface in relation to the mass is therefore unfavorable in cold and desirable in warm climates.

Strange as it may seem, the correlations between racial traits and the environment were used in the past as arguments against the natural selection theory. Racial differentiation was considered a result of direct modification by the environment, perpetuated by a gradual change of the germ plasm in the direction of the phenotypic alteration. This interpretation appeared to be borne out by experiments showing that changes analogous to those observed in a particular geographical race may also be effected in other races of the same species by exposure to environmental agents. Thus, in the classical experiments of Standfuss, exposure to heat treatment produced, from the pupae of central European races, butterflies that resembled varieties from Syria or southern Italy. On the other hand, treatment of the central European races with cold resulted in a resemblance to races from northern Scandinavia. Exposure of a mammal to cold or heat may produce, respectively, increase or decrease of the hair length, a change analogous to the difference distinguishing geographic races from high and low latitudes. Bergmann's, Allen's, and Gloger's rules (see above) have their counterparts among the changes that can be induced in animals by the application of appropriate environmental stimuli, namely, changes in temperature and humidity. The same is true for plants, in which the phenotypic changes wrought by external agents may simulate the characteristics of races and species existing in nature.

The parallelism between genotypic racial variability and phenotypic modifications has, however, a more subtle and significant meaning. As pointed out in Chapter 2, harmony between the organism and its environment involves the possession of genotypes with favorable norms of reaction. A norm of reaction is favorable if frequently recurring environmental stimuli evoke phenotypic modifications that enable the organism to survive and to reproduce. It is no accident, therefore, that mammals inhabiting lands with temperate climates react to temperature changes by growing warmer fur in winter and a lighter coat in summer. On the other hand, the genes for warm fur may be fixed in permanently cold regions and for light fur in permanently warm climates. As pointed out particularly by Schmalhausen (1949), Waddington (1957, 1960), and Levins (1968), development may be canalized to yield a fixed outcome or may be plastic to produce varying phenotypes adaptive in different environments. The term organic selection has been coined to describe the parallelism between racial genotypic and environmental phenotypic variability.

Heterosis and Hybrid Breakdown in Race Crosses

The gene pool of a natural population is a genetic system adapted to the environments that the population inhabits. The genetic systems of different races are adapted to different environments. When these genetic systems diverge in the evolutionary process, the gene recombination in the progeny of race hybrids may yield some genotypes of low fitness in the environments of one or of both parental races. We have seen in this chapter an example of this situation. In the experiments of Clausen and Hiesey (1958), it will be recalled, alpine and foothill races of *Potentilla glandulosa* were crossed; some of the F_2 segregants (Table 9.4) were poorly viable when grown at low or at intermediate or at high elevations, or at all of these. To be sure, some segregants seemed to be able to grow in experimental gardens at all three elevations. If these results are taken at face value, the gene recombination produced well adapted as well as poorly adapted genotypes.

The classical work of Goldschmidt (summarized in 1934 and 1940) on the gypsy moth (*Lymantria dispar*) revealed that its geographical races differ, not only in externally visible traits, but also in response to temperature, in rates of development, and in the length of incubation period needed for the young caterpillars to hatch from the eggs in spring. Hybridization of the races in laboratory experiments leads in some cases to production of intersexual hybrids. Thus, the races from northern Japan (Hokkaido) and from Europe have "weak" sex-determining factors. The sex determiners in races from Turkestan, Manchuria, and southwest Japan are "half-weak" or "neutral." Those from central Japan are "strong." The cross weak ♀ × strong ♂ produces in the F_1 generation normal males and intersexual females. The reciprocal cross, strong ♀ × weak ♂, gives normal females and males in the F_1 generation, but in the F_2 half of the males are intersexual. Denoting the female determiners by F, the male determiners by M, and the weak and strong alleles by the subscripts w and s, the above results have been interpreted by Goldschmidt thus:

F_wM_w (weak ♀) × M_sM_s (strong ♂)
= F_wM_s (intersex) and $F_wM_sM_w$ (♂)

F_sM_s (strong ♀) × M_wM_w (weak ♂)
= F_sM_w (♀) and $F_sM_wM_s$ (♂)

One F is normally sufficient to suppress the effects of a single M and to produce a female; an F_w, however, is not strong enough to overpower an M_s, and hence an F_wM_s individual is an intersex. Likewise, individuals of the constitution $F_sM_wM_w$ that appear in the F_2 from the cross strong ♀ × weak ♂ are not males but intersexes. The degree of intersexuality varies greatly among different crosses. In crosses where the "strengths" of the sex determiners in the parental races differ only slightly, the intersexes are so much like normal females or males that they are fertile. Where the difference between the parental races is greater, however, the intersexes are sterile. Finally, crosses between extremely strong and weak races transform the intersexes into individuals of the sex opposite to that which they should have had according to their chromosomal constitution (XX instead of XY females, and XY instead of XX males).

Sympatric strains have sex determiners of similar "strength." Indeed, a mutation that would appreciably change the "strength" may make its carriers intersexual and may be eliminated by normalizing natural selection. Populations with sex genes different enough to produce sterile hybrids when crossed never inhabit contiguous territories. The "strength" of the sex genes shows geographic gradients. As indicated above, the island of Hokkaido is inhabited by the "weakest" race of *Lymantria dispar*. From there, through Manchuria, Korea, southwestern Japan, and central Japan, progressively "stronger" races are encountered. The "strongest" race lives in the northern parts of the main island of Japan, separated from the "weakest" one in Hokkaido only by the Tsugaru Strait, which is, however, broad enough to prevent mingling of the races living on its opposite shores.

Vetukhiv (1953, 1954, 1956) and Wallace (1955) studied F_1 and F_2 hybrids between populations of the same species of Drosophila having different geographic origins. Larval viability under crowded conditions, fecundity, and longevity were examined. Three species, *Drosophila pseudoobscura*, *D. willistoni*, and *D. melanogaster*, were utilized. The populations of the same species were outwardly indistinguishable and were not regarded as representing different "races." Nevertheless, the F_1 hybrids between the populations were, as a rule, significantly superior, and the F_2 hybrids inferior, to the parental populations.

A synopsis of Vetukhiv's results, prepared by Wallace (1968a), is shown in Table 9.5. In this table, the means of the two parental populations are taken to be unity, and the hybrids are characterized by ratios of their scores divided by the parental means.

The ratios for F_1 hybrids in Table 9.5 are all above unity, indicating that these hybrids are luxuriant or heterotic. Since the parental popu-

TABLE 9.5

Viability, fecundity, and longevity in F_1 and F_2 hybrids between geographic populations of the same species of Drosophila (The mean of the parental populations is taken as unity.) (After Vetukhiv and Wallace)

Trait	F_1	F_2
Viability		
D. pseudoobscura	1.18	0.83
D. willistoni	1.14	0.90
Eggs per female per day		
D. pseudoobscura	1.27	0.94
Longevity		
D. pseudoobscura		
At 16°C	1.25	0.94
At 25°C	1.13	0.78–0.95

lations were not inbred, the origin of this luxuriance is unclear. Perhaps it is due to heterozygosis for greater numbers of genes, and here a comparison with the experiments of Wallace and Mukai on newly induced mutations, described in Chapter 6, may be relevant. The ratios for F_2 hybrids are all below unity; the loss of fitness in the F_2, as compared to the parental, populations is easier to interpret: the gene pool of each population is an integrated system, which is disrupted by hybridization.

Concordant results have been obtained by Anderson (1968) in a study of the body size in populations of *Drosophila pseudoobscura* from different parts of the species distribution area and in their hybrids. The F_1 hybrids were not consistently larger than the mean of the two parents crossed, but the F_2 hybrids were significantly smaller than their F_1 parents. On the other hand, MacFarquhar and Robertson (1963) found neither a luxuriance in F_1 nor a breakdown in F_2 hybrids between geographic populations of a fairly closely related species, *D. subobscura*. The cause of this discrepancy is a matter of speculation.

Study of the integration of the gene pool of a population was carried a step further by Brncic (1954) and Kitagawa (1967) in *Drosophila pseudoobscura* and by Wallace (1955) in *D. melanogaster*. By means of rather complex series of crosses that need not be considered here in detail, the following four classes of flies were obtained under conditions such that their relative viabilities could be measured: (1) those with both chromosomes of a pair derived from the population of a single locality; (2) those with the two chromosomes of a pair from different localities; (3) those with one chromosome the result of recombination (crossing over) between chromosomes from different localities, and the other chromosome unrecombined from one locality; and (4) those with both chromosomes of a pair the products of recombination of chromosomes of different geographic origin. Although the results varied from cross to cross, the most common viability sequence was $2 > 1 > 3 > 4$. In addition to confirming the older findings of Vetukhiv (see above), these experiments suggest that integration of the genotype occurs on both the interchromosomal and the intrachromosomal level. A gamete transporting a haploid set of chromosomes from a given population contains a coadapted system of genes. The coadaptation is partly lost when the chromosomes of a set are of different geographic origins and is further decreased if some of the chromosomes are compounded of sections of different geographic origins.

The fitness of race hybrids in man has been a controversial subject for a long time. Partisans of "race purity" prophesy that the hybridization of races will have dire consequences, but critical reviews of the evidence by Trevor (1953) and Penrose (1955) show that, if anything, race hybrids exhibit traces of heterosis. Hulse (1957) found in the Swiss canton of Ticino that people whose parents came from different villages (the products of "exogamous" marriages) are slightly but significantly taller than those having parents from the same village ("endogamous" marriages). By far the most extensive and careful study in this connection is that of Morton, Chung, and Mi (1967) on interracial crosses in Hawaii. Here hybrids of people of Chinese, Japanese, European, and Polynesian extractions have been studied for prenatal and postnatal mortality, congenital malformations, and morbidity, as well as for anthropometric traits. The results can be summarized in the words of the authors:

> First generation hybrids between races in man are intermediate in size, mortality, and morbidity between the parental groups. At the present time, human populations do not represent coadapted genetic combinations which are disrupted by outcrossing. It remains an untested hypothesis that outcrossing effects might be observed in a more rigorous environment.

Do Races Exist?

This question may seem odd at the conclusion of a chapter on race. However, as stated earlier, some authors contend that races do not exist. A glimpse at history and a reconsideration of the issue may be a good way to summarize the chapter. Linnaeus was interested mainly in species, which he regarded as the primordial created entities. He was too good a naturalist, however, not to see that there are subdivisions within species, which he called varieties. Darwin paid considerable attention to varieties, because he held that species develop from varieties by gradual divergence.

As the years passed, the term variety was stretched to include phenomena as disparate as geographic races, breeds of domesticated animals and plants, and intrapopulational polymorphs. Buffon was among the first to employ "race" in a biological sense in 1749, and the word gradually came to be used in connection with "varieties" of mankind. The importance of geographic isolation, and hence of geographic races, for speciation was stressed by Moritz Wagner in 1868 and 1889, and in a more nearly modern form by Karl Jordan in 1896 and 1905. Esper in 1781 distinguished between "essential varieties," which he called subspecies, and "accidental varieties." The employment of subspecies as a taxonomic category goes back to Schlegel in 1844 (see Mayr 1963 for a more detailed historical account).

The race and subspecies concepts are utilized in anthropology and biology primarily for purposes of classification. It has been known since antiquity that different parts of the world are inhabited by different varieties of men. Geographic explorations from the sixteenth to the nineteenth century called this ineluctable fact to general attention. Mankind is a polytypic species, that is, a species including two or several races or subspecies. Over the years, zoologists and botanists discovered that many, in fact most, of their species are also polytypic. They also found that they had described too many species, since some so-called species can be treated more meaningfully as subspecies constituting polytypic species. Modern systematists, studying better known groups such as birds and mammals, have done important work in revising the older classifications in accord with the polytypic species concept.

Mayr (1963, 1969) gives many examples of the results of this reform. Thus, in 1909 a list of species of birds of the world gave 19,000 names; in 1951 the number was reduced to about 8600 species with some 28,500 subspecies, about 3.3 subspecies per species on the average. Mayr does not believe that this average will be increased materially by future studies. The pocket gophers *Thomomys bottae* and *T. umbrinus* have 213 subspecies in the southwestern United States and Mexico (Anderson 1966). This is probably the record, since not even mankind has been honored by so extensive a subdivision. The biology of these gophers, dependent as they are and isolated by soil and vegetation barriers, makes the extensive race differentiation understandable.

In less intensively studied groups of animals, most species are monotypic, that is, not subdivided into subspecies with Latin names. Botanists use the subspecies category less often than zoologists; the important book of Stebbins (1950) mentions subspecies briefly, and that of Grant (1963) not at all. A species may be or may seem monotypic for various reasons. It may be geographically undifferentiated into genetically distinct populations. Relictual species may be reduced to a single colony. In species with strong migratory propensities the gene exchange between populations may prevent genetic divergence. On the other hand, geographic differentiation may involve physiological, cytological, or biochemical, rather than externally visible, morphological traits, and systematists are disinclined to name subspecies that they cannot distinguish by inspection. Species of Drosophila are excellent examples of this situation. Chromosomal races are known in many species, and evidence is accumulating that racial variation in enzyme polymorphisms is at least as common (see above). The tropical American *Drosophila polymorpha* is, however, one of the very few species with a conspicuous color pattern polymorphism and racial variation (da Cunha 1949, Heed 1963, and Heed and Blake 1963). Racial variation in external morphology is so slight that it is detectable only by careful biometric studies (Zürcher 1963 and Cals-Usciati 1964 in *D. melanogaster*, Monclus 1953 in *D. subobscura*, Sokoloff 1966 and Anderson 1968 in *D. pseudoobscura* and *D. persimilis*).

We have seen that the number of races one recognizes in a polytypic species is to a certain extent arbitrary; mankind may be divided into 5, 9, 35, and other numbers of races. I am not convinced that the 213 subspecies named in western American and Mexican pocket gophers (see above) are so clear-cut that they could not be combined into fewer but more inclusive subspecies (Ingels 1950). It should be stressed that the arbitrariness is not always explainable by insufficient data or remediable by more study. Descriptive and anthropometric observations on human populations have already shown that the geographic distributions of different racial traits are often uncorrelated or weakly correlated (maps in Lundman 1967). Not all peoples with black skins have frizzly hair; both tall and short peoples may have prominent or flat noses. Studies on the geographic gradients of gene frequencies have accentuated this lack of rigorous correlations. Not only do the blood group frequencies not vary in parallel with pigmentation or hair shape, but also the frequency gradients in different blood group systems show no strong correlations.

Opponents of racial or subspecific classifications have argued that one should study the geographic distributions of genes and character frequencies, rather than attempt to delimit races or subspecies. The truth is that both kinds of studies are necessary. Gene and character geography is the basis of the biological phenomenon of racial variation; classification and naming are indispensable for information storage and communication. The fact that races are not always or even not usually discrete, and that they are connected by transitional populations, is in itself biologically meaningful. It is evidence that gene flow between races is not only possible but actually realized. Gene flow between species, however, is limited or prevented altogether. To hold that because races are not rigidly fixed units they do not exist is a throwback to typological thinking of the most misleading kind.

Without endeavoring to produce a series of formal definitions, we may summarize the explication of some of the concepts dealt with in this chapter. A Mendelian population is a community of individuals of a sexually reproducing species within which matings take place. There is a hierarchy of Mendelian populations. The most inclusive Mendelian population is the species. The lowermost member of the hierarchy is a panmictic unit, within which matings take place at random. In a continuously inhabited territory the panmictic units have no boundaries and merge into each other. A local population, including one or several panmictic units, is a deme. A race is a cluster of local populations that differs from other clusters in the frequencies of some gene alleles or chromosomal structures. A subspecies (following Mayr 1969) is a "geographically defined aggregate of local populations which differ taxonomically from other such subdivisions of the species." A subspecies is, then, a race that a taxonomist regards as sufficiently different from other races to bestow upon it a Latin name.

CHAPTER TEN

REPRODUCTIVE ISOLATION

Realized and Potential Genotypes

Desmond (1961) estimates that the number of human beings who ever lived is of the order of 77 billion (7.7×10^{10}), distributed in time as follows:

600,000 B.C. to 6000 B.C.: 12 billion
6000 B.C. to A.D. 1650: 42 billion
A.D. 1650 to A.D. 1962: 23 billion

On the assumption that a human individual is heterozygous, on the average, for some 3200 genes (cf. Chapter 7), the potentially possible number of gametes with different gene constellations is 2^{3200} or 10^{963}. Because of linkage, some of these constellations are much less likely to arise than others. Nevertheless, the probability of two individuals, whether siblings or persons not closely related, having the same genotype is remote indeed. The array of genotypes composing mankind at present will be replaced by a new array in the next generation. Which genotypes will or will not be realized within a panmictic unit, and even within a species, is to a considerable extent a matter of chance. If genius has a genetic basis, a genius may or may not appear in a given generation, or ever.

Consider now the genes in different species. There is no way at present to recombine human genes with those of, for example, mice or Drosophila (see, however, the hybrids of tissue culture cells mentioned in a later section). It is known that the insemination of eggs of one species by the sperm of a remote species fails altogether or gives inviable zygotes (see below). It is also known that gene recombination in the progenies of viable and fertile hybrids between closely related species gives rise to genotypes some of which are of low fitness. The genotypes composing a species are arrays of gene constellations that enable their carriers to survive and reproduce in certain ecological niches. In terms of Wright's model (Chapter 1), most of the realized genotypes cluster on the adaptive peaks. The adaptive peaks are separated by adaptive valleys—a metaphor for the multitude of gene combinations that are not formed at all or are eliminated on account of the low fitness of their carriers.

It is advantageous for populations that occupy different adaptive peaks to steer clear of the production of gene combinations in the progeny that would fall in the adaptive valleys. To be sure, a minority of the gene combinations formed by the hybridization of species might be fit, perhaps fit enough to spread onto as yet unoccupied adaptive peaks (see Chapter 11). A majority, however, would be relatively ill adapted in any environment. If so, transpecific hybridization would weaken the reproductive potentials of the populations engaged in such hybridization and gene exchange. Here, then, is a challenge which the evolution of life has constantly had to face. The sexual process and the resulting gene recombination facilitate adaptation by creating genetic materials for natural selection to work with; yet, unrestricted gene exchange decreases fitness because it yields too many worthless genotypes. The challenge is very old—it has existed ever since the emergence in evolution of various methods of gene exchange: transformation, transduction, parasexuality, and finally sex and meiosis.

Many and varied solutions of the problem arising from this challenge are known. A radical but trivial solution is to do away with gene exchange altogether. Asexual reproduction, by fission or budding, is one such solution. Parthenogenesis (apogamy), which makes the entire progeny genetically identical with the mother, is another. Still a third is substitution of obligatory self-fertilization for outbreeding, observed in hermaphroditic forms, especially among plants. Most interesting, however, are the nontrivial solutions compatible with sexual reproduction and outbreeding.

How do sexual, and obligatorily or at least facultatively outbreeding, species avoid disruption of their genetic systems by interspecific hybridization? There is again a variety of methods, among which two classes are easily distinguishable: geographic and reproductive isolation. When races or species are allopatric, their representatives do not meet and hence do not mate. Geographic isolation alone would, however, offer an only partial solution to the problem of maintenance of organic diversity. In any locality, there could live only one kind of organism, only a single Mendelian population. In reality, there are many, often many thousands, of sympatric animal and plant species. Moreover, unless separated by uninhabitable terrain, the distribution areas of allopatric species may come into contact. Gene exchange and gene diffusion may occur in the contact or "suture" zones (Remington 1969). This, in fact, is what actually happens with races; the results may be geographically and genetically intermediate populations, gene frequency gradients, and finally a merging of the races into a single population.

Reproductive isolation limits or prevents this gene exchange by means intrinsic to the organisms themselves. In principle at least, geographically isolated populations may be genetically identical. Reproductively isolated populations are necessarily different genetically, although the converse is not true (races are genetically different populations that are not reproductively isolated). Sympatric populations of sexual and outbreeding organisms must be isolated reproductively, for otherwise they could not keep apart and would fuse into a single population. Many species may coexist sympatrically, and their sympatry is *prima facie* evidence of their reproductive isolation. Obvious exceptions to this rule are human races. They are in part sympatric, but they are kept genetically apart, at least to some extent, by social instead of reproductive isolation.

Classification of Reproductive Isolating Mechanisms

Not only are there different kinds of reproductive isolation, but also the isolation of species in the same group of organisms is often achieved by different means. Moreover, the isolation of a given pair of related species may result from several isolating mechanisms reinforcing each other's actions. Considered physiologically, different isolating mechanisms have scarcely a common denominator; yet genetically their effect is the same, that is, limitation or prevention of the gene exchange between the populations of different species. Dobzhansky (1937a) proposed "physiological isolating mechanisms" as a general designation for all causes, other than geographic isolation, that impede gene exchange between populations. The word physiological in this context led to misunderstanding, however, and Mayr (1942) changed the term to "reproductive isolating mechanisms."

Several variant classifications of reproductive isolating mechanisms have been proposed (Dobzhansky 1937d, Mayr 1942, 1963, Stebbins 1950, Riley 1952, Grant 1963, and others). The following one seems convenient:

1. *Premating* or *prezygotic* mechanisms prevent the formation of hybrid zygotes.
   - A. *Ecological* or *habitat isolation.* The populations concerned occur in different habitats in the same general region.
   - B. *Seasonal* or *temporal isolation.* Mating or flowering times occur at different seasons.
   - C. *Sexual* or *ethological isolation.* Mutual attraction between the sexes of different species is weak or absent.
   - D. *Mechanical isolation.* Physical noncorrespondence of the genitalia or the flower parts prevents copulation or the transfer of pollen.
   - E. *Isolation by different pollinators.* In flowering plants, related species may be specialized to attract different insects as pollinators.
   - F. *Gametic isolation.* In organisms with external fertilization, female and male gametes may not be attracted to each other. In organisms with internal fertilization, the gametes or gametophytes of one species may be inviable in the sexual ducts or in the styles of other species.
2. *Postmating* or *zygotic* isolating mechanisms reduce the viability or fertility of hybrid zygotes.
   - G. *Hybrid inviability.* Hybrid zygotes have reduced viability or are inviable.
   - H. *Hybrid sterility.* The F_1 hybrids of one sex or of both sexes fail to produce functional gametes.
   - I. *Hybrid breakdown.* The F_2 or backcross hybrids have reduced viability or fertility.

It may be useful to reiterate that any reproductive isolation is caused by genetic differences between populations, whereas with geographic isolation this is not necessarily so. Habitat isolation (see 1A, above) comes closest to bridging the gap, but even there the attachment of different species to different habitats is evidently due to their genetic constitutions.

Habitat Isolation

The literature on reproductive isolating mechanisms is so extensive that a whole book would be needed to review it exhaustively. The examples discussed below were chosen to illustrate the remarkable diversity of the situations encountered in nature. More examples can be found in the books of Stebbins (1950, 1958a), Clausen (1951), Patterson and Stone (1952), Linsley (1958), Grant (1963, 1964b), Mayr (1963), Ford (1964), and others.

Among plants, related species often differ in their soil, sun or shade, or elevation requirements. For several years I had the opportunity to observe *Arctostaphylos mariposa* and *A. patula* in the Yosemite region of the Sierra Nevada of California (Dobzhansky 1953). Both species are large bushes or small trees, but they differ in several clear cut and easily visible traits. *Arctostaphylos mariposa* occurs at relatively low elevations, up to about 4800 feet, and *A. patula* at higher ones, from 4600 feet upward. In the narrow altitudinal belt where both species live, *A. mariposa* occupies drier and exposed situations, and *A. patula* more sheltered sites. Even in that belt at least 90 percent of the trees belong to either one or the other species, and in some localities the altitudinal replacement of the species occurs without any hybrids being produced. In other localities some hybrids can be identified. A great majority of them are F_1 hybrids, but Professor G. L. Stebbins assures me (private communication) that he can detect signs of introgression (cf. Chapter 11) where both species occur side by side.

Muller (1952) has studied instances of habitat isolation of several forms of related species of oaks (Quercus). *Quercus mohriana* grows in Mexico and Texas on limestone or in shallow soils overlying limestone, whereas *Q. havardi* is confined to sandy soils. In some places, however, limestone and sand are juxtaposed, and there hybrids between the two species frequently occur. A third species, *Q. grisea,* is an inhabitant of igneous outcrops. Sometimes it grows also on dolomitic limestones, where it meets *Q. mohriana* and hybridizes with the latter. These species are kept separate mainly by habitat isolation. On the other hand, Muller mentions the existence of numerous related species growing together in the same habitat that do not produce hybrids, evidently because of other forms of isolation.

A curious analogy to this situation in oaks is described by Mori and

Matutani (1953) in caddis flies in Japan. These insects engage in daily swarming during which matings occur; each species, however, has its swarming place in a different kind of habitat. The authors refer to this as a "habitat segregation."

The toads *Bufo americanus* and *B. fowleri* inhabit the eastern and central United States. Blair (1941, 1942), Volpe (1952), and Cory and Manion (1955) showed that where these species occur together viable and fertile hybrids are produced. In Indiana and Michigan, three toad populations are found. One breeds early in spring and is pure *B. americanus*; another breeds late and is *B. fowleri*; the breeding period of the third is intermediate, and it consists of individuals resembling the pure species and the experimentally produced hybrids between them. This temporal isolation is insufficient to keep the species apart; in some localities in New Jersey *B. americanus* is no longer found, and the population consists of *fowleri*-like and hybrid individuals. The best interpretation of the situation is that the hybridization is of recent origin. Initially, *B. americanus* and *B. fowleri* occurred in different natural habitats. Human activities, such as destruction of forests and damming of streams, created new habitats acceptable to both species of toad. The gene exchange and formation of hybrid populations are the consequences.

The three-spined stickleback (*Gasterosteus aculeatus*) is a common fish in cold and temperate parts of the Northern Hemisphere. It has attracted the attention of numerous investigators for more than a century, because it has three morphologically, physiologically, and behaviorally different forms (Heuts 1947, Münzing 1963, and Hagen 1967). The form *leiurus* is a fresh-water fish with 3–8 lateral bony plates and an osmoregulatory capacity that maintains the chlorine content of the blood at a constant level in waters of low salinity. *Trachurus* lives in the sea in winter but migrates to river estuaries to breed in spring and summer. It has 30–36 bony plates and a variable chlorine ion concentration. The third form, *semiarmatus*, is a hybrid between the first two, with a highly variable number of plates (8–29).

In Europe, only *trachurus* is found along the coasts of northern Russia, Norway, Scotland, Ireland, and also the Black Sea, and only *leiurus* in fresh waters of Mediterranean countries; all three forms are found along the coasts of the North Sea and the western Baltic, with *leiurus* rare in the estuaries but common uprivers. Hagen (1967) made a most careful study of the situation in the Little Campbell River in

British Columbia. In the headwaters only *leiurus* is found and is very common there; close to the estuary, where the salinity is still low, *trachurus* immigrates in spring and breeds in large numbers during the summer. Only a 1-mile-long section of the stream is the "hybrid zone," where *leiurus,* hybrids, and *trachurus* occur with frequencies of about 10 : 8 : 1. Hagen correctly regards the parental forms, *Gasterosteus aculeatus* (*leiurus*) and *G. trachurus,* as full species, despite the fact that neither sexual behavior nor sterility barriers between them seem to exist. The gene exchange between the species is nevertheless blocked within 1 mile, no introgression occurs, and the hybrids are severely selected against outside the confines of the hybrid zone with its intermediate habitats.

Temporal Isolation

Gene exchange is impossible between species that breed at different times. Perhaps the most extreme instance of this kind of isolation has been recorded in three species of the orchid Dendrobium (Holttum 1953). Their flowers open at dawn and wither by nightfall. The flowering is brought about by a meteorological stimulus, such as a sudden storm on a hot day. However, one of the species flowers 8 days, another 9 days, and a third 10 or 11 days after receiving the stimulus. Many species and races of plants require long, and others short, days for flowering. This photoperiod dependence is under genetic control, as shown most clearly by Smith (1950).

Countless examples of related animal species that differ in breeding season abound in monographs on the systematics and ecology of particular zoological groups. Tretzel (1955) has published an especially detailed study of the temporal isolation, and also of habitat isolations, among cognate species of spiders in Germany. A perusal of the numerous instances recorded by this author suggests some negative correlation between these isolating mechanisms: species found in the same habitats have different breeding seasons more often than do species that live in different habitats. This correlation is, however, far from complete. There are also instances of species some of which are diurnal and others nocturnal in their activities.

Alexander and Moore (1962) have stated:

The periodical cicadas make up a truly amazing group of animals; since their discovery over 300 years ago, the origin and significance of their extended life cycles have been a continual source of puzzlement to biologists. Their incredible abilty to emerge by the million as noisy, flying, gregarious, photopositive adults within a matter of hours after having spent 13 or 17 years underground as silent, burrowing, solitary, sedentary juveniles is without parallel in the animal kingdom.

There are three pairs of very closely related (sibling) species of Magicicada in the United States, one member of each pair having a 17- and the other a 13-year development time:

17 Years	13 Years
M. septendecim	*M. tredecim*
M. cassini	*M. tredecassini*
M. septendecula	*M. tredecula*

Geographically, the species with 17-year cycles occur together, sympatrically, and so do the species with 13-year cycles; the former have, however, a more northern and the latter a more southern distribution, and only in a few places do the two kinds coexist. In any one place a mass emergence of the cicadas occurs only once in 17 (or in 13) years, but in different places the "broods" emerge in different years. The geography of the broods is rather precisely mapped, and in some places the periodicity is known to have been maintained for over three centuries. There are 13 large broods with 17-year cycles and 5 with 13-year cycles. Remarkably enough, all three sibling species emerge together in a given locality. They differ clearly in their "songs" and, to some extent, in microhabitat preferences.

Alexander and Moore have never found a heterospecific mating pair in nature, although the species do mate in captivity. Such differences in the behavior of species in nature and in experimental environments have also been found in Drosophila (see below). It is not known whether the 13-year and the 17-year sibling species intercross in the localities where they may emerge together, an event that will occur presumably once in 221 years.

The spectacular abundance of the species of Magicicada in the years of mass emergence indicates that they are highly successful insects. Why, then, do they have so dilatory a development (related genera develop much faster), and why do they appear only once in 17 or 13 years in a given locality? Alexander and Moore (1962) and Lloyd and Dybas (1966) suggest that this may be a singular adaptation to escape parasites and predators. No parasite is known with a life cycle of this length, and the predators (chiefly birds) are rapidly satiated for a week or two in the years of mass emergence. The relatively very small numbers of stragglers, cicadas that emerge a year after the mass eruption, are quickly destroyed by predators. This explains why the three sympatric species always emerge together; getting out of step would be selectively disadvantageous. It is less easy to understand the origin of the brood of each species emerging in different years, and the appearance of the 17- and the 13-year cycles.

Ethological Isolation

Ethological isolation is a widespread and probably the most effective kind of reproductive isolation between species of animals, particularly arthropods and higher vertebrates. For the maintenance of a species with separate sexes it is indispensable that its females and males encounter each other and perform the acts that lead to the union of their sex cells. This is achieved in different organisms by a great variety of means, many of them quite ingenious and some extravagantly bizarre. Ethological isolation is effected in correspondingly diverse and often curious ways.

Ethology, the study of animal behavior, has been one of the expanding areas of modern biology in recent years. Sexual behavior has attracted a great deal of attention. Every species uses its own courtship and mating techniques; in some animals these techniques become rituals of extraordinary complexity. Birds are, like human beings, predominantly visual animals, and their activities are easily observable. This is probably the reason why so much modern ethological research has been done on birds (one of the pioneering works was that of Lorenz 1941; for general accounts and references, see Thorpe 1963, Tinbergen 1965, 1966, and Marler and Hamilton 1966). Among birds, species-specific plumage, adornments, courtship displays, and pantomimes, as well as songs, play important roles in species recognition and the avoidance of interspecific matings. Analysis of some courtship rituals has shown them to be recognizable modifications of threat, attack, and appeasement behaviors in the same or related species. When the rituals of different species do not coincide, attempted courtship may result in a genuine attack.

The courtship behavior of bowerbirds (Ptilorhynchidae), about nineteen species of which live in Australia and New Guinea, is most remarkable (Marshall 1954). The males build "bowers," each species in its own way, which they decorate in various styles. For the adornment of the bowers some use brightly colored flowers or fruits, or construct approach avenues of bleached bones, stones, pieces of metal if they can get them, or meadows of moss. One species goes so far as to paint its bower with fruit pulp, for which purpose it uses a tool—a wad of dry grass or spongy bark. Females are enticed to enter the bower, where copulation takes place. Nests are built elsewhere, however, without particular adornments.

In frogs and toads, auditory stimuli play an important role in the congregation of females and males of the same species in their breeding places. During the breeding season, the calls of the males attract the females and also stimulate other males to come and to begin calling, so that "choruses" of conspecific individuals are formed. The response to the call is species specific. The role of auditory stimuli in the reproductive isolation of species has been studied particularly by W. F. Blair and his numerous disciples (see Blair 1958a and 1964 for references).

Table 10.1 gives examples of the characteristics of the calls in eleven species of hylid frogs occurring in Florida. The dominant frequency (in cycles per second) and the average duration (in seconds), as revealed by sound spectograms, are shown. Even in regard to only the three characteristics shown in Table 10.1, no two species have the same call. Actually the differences are more numerous—for example, the repetition rate for the trilling species varies from 9.7 notes per second for *Hyla phaeocrypta* to 50.0 for *H. ocularis*.

Direct experimental evidence that the different calls act as isolating mechanisms is not as ample as could be wished. Blair and Littlejohn (1960) exposed mature females of *Pseudacris streckeri* to recorded calls of males of their own species and of the related *P. ornata*, reproduced by a loudspeaker. The females were attracted by the conspecific calls more often than by the heterospecific ones. Littlejohn and Loftus-Hills (1968) made similar experiments on frogs of the *Hyla ewingi* complex. The specificity in the response of female tree crickets (Oecanthus) to the calling songs of their males has been demonstrated by a similar technique (Walker 1957).

Pheromones are chemical substances that serve as signals eliciting behavioral reactions in individuals of the same species. They are impor-

TABLE 10.1

Average values for dominant frequency and duration of call in eleven species of hylid frogs (After Blair 1958b)

Species	Dominant frequency, cycles per second	Duration, seconds	Trilling
Hyla gratiosa	435	0.16	—
H. crucifer	2467	0.14	—
Pseudacris ornata	2562	0.06	—
Hyla cinerea	3407	0.18	—
H. squirella	3457	0.24	—
Acris gryllus	3914	0.04	—
Hyla versicolor	2444	0.84	+
Pseudacris nigrita	3325	0.39	+
Hyla phaeocrypta	4500	1.99	+
H. femoralis	4800	2.35	+
H. ocularis	7125	0.16	+

tant in species recognition and hence in ethological isolation. Most of the relevant studies deal with insects and mammals, but sex pheromones are widespread in the animal kingdom. The chemical nature of some of the pheromones is becoming known (Jacobson and Beroza 1963, Wilson and Bossert 1963, and Roelofs and Comeau 1969). They are mostly small molecules, with molecular weights in the range 80–300. They are not necessarily species specific, nor are they expected to be so—species that are allopatric, live in different habitats, or have different breeding seasons may well use the same pheromone, or react in experiments to the pheromones of other species. Roelofs and Comeau found, however, that two sympatric sibling species of gelechiid moths are attracted, respectively, to *cis*-9-tetradecenyl acetate and to the *trans* isomer of this substance. Moore (1965a,b) tested olfactory discrimination in the deer mice *Peromyscus maniculatus* and *P. polionotus*. The former species shows greater sexual discrimination than the latter, a difference that may be due to *P. polionotus* being geographically "insular," whereas *P. maniculatus* shares its territory with other species.

Species of Drosophila have been widely used for observations and experiments on ethological isolation. Spieth (1968 and references therein) carried out several painstaking studies of the courtship and mating in some 200 species; Bastock and Manning (1955 and Manning 1959) described in detail the sexual behavior of *D. melanogaster* and *D. similans,* Brown (1964) of *D. pseudoobscura* and its relatives, and Koref-Santibañez (1964) of species related to *D. mesophragmatica.* The courtship patterns involve various permutations of the orientation of a male toward another individual: tapping the latter with his forelegs, circling the object of courtship, vibrating, flicking, waving, scissoring, or fluttering of one or both wings, licking the genitalia of the female, spreading her wings, curling the abdomen, mounting, and intromission, as well as rejection signals on the part of the courted female or another male of the same or different species.

Chemical, visual, tactile, and auditory (more precisely, vibrational) stimuli are involved, playing different roles in different species. Spieth and Hsu (1950) and Grossfield (1966) showed that, whereas some species mate equally efficiently with light and in the dark, others are unable to copulate without light, and still others are intermediate. Closely related species may be unlike in this respect; for example, *Drosophila pseudoobscura* is light independent, but *D. subobscura* fails to mate in the dark. Wallace and Dobzhansky (1947) found *D. subobscura* males to be responsible for this failure; the females of the same species confined in darkness with males of *D. pseudoobscura* are sometimes inseminated.

Some of the species endemic to the Hawaiian archipelago have evolved rituals quite novel for drosophilids, namely, "lek" behavior and defense of territory (Spieth 1968). Males of these species take up positions on a single frond of a fern or on the trunk of some tree, where they engage in sexual displays. Their females, who feed elsewhere, are attracted to come to be courted and inseminated, whereupon the females depart for oviposition. Each male defends a certain area against intrusion by other males, and occasionally by unreceptive females. Interestingly enough, males of some of the species that engage in defense of territory have evolved special modification of the structure of their organs, particularly the proboscis and front legs.

Three different techniques have been used in experiments on ethological isolation in Drosophila. In no-choice experiments, females of one species are confined with males of another, and after the lapse of a certain time the proportions of the females inseminated and left virgins are recorded. In double-choice experiments, two kinds of sexually receptive females and males are placed in an observation chamber, and the matings that take place are recorded by inspection. Finally, the two kinds of females are exposed to males of one of them, and the proportions of the inseminated and virgin females are recorded. This is called, rather misleadingly, the male-choice technique, although it is agreed by most investigators that it is really the female who accepts or rejects the advances of the males, who are well-nigh promiscuous. Bateman (1948) argued very cogently that "undiscriminating eagerness in males and discriminating passivity in females" should be induced by natural selection in all bisexual species, except where strict monogamy combined with a sex ratio of unity eliminates intrasexual selection. Indeed, an excess of male sex cells contrasts, in most organisms, with a relatively limited production of female ones. Female fertility is limited by egg production and by the capacity to feed and rear the young. Male fertility, on the other hand, is largely a question of the number of females an individual inseminates. The "eagerness" and "passivity" of males and females are, thus, consequences of the fact that females produce much fewer gametes than do males.

The three techniques outlined above generally give concordant results. When the species tested are not close relatives (as judged by their structural traits), only homogamic (between likes) and no heterogamic (between unlikes) matings take place. With closely related species or with races or strains of the same species, some heterogamic matings do occur, or the matings may be at random. As an illustration, the matings recorded in observation chambers with various combinations of species of the *obscura* group of the genus Drosophila are shown in Table 10.2. It can be seen that the homogamic matings outnumber the heterogamic ones in all combinations.

Malogolowkin, Solima, and Levene (1965) proposed an isolation coefficient to measure the degree of isolation. It is computed by subtracting the proportion of all matings that are heterogamic from the proportion of homogamic ones. This coefficient is plus one if the isolation is complete, zero if there is no isolation, and negative if the heterogamic matings outnumber the homogamic ones.

Several items of interest should be noted in Table 10.2. Although the isolation is quite pronounced in all combinations, some interspecific matings do occur. This does not necessarily mean that there is gene exchange between these populations in nature. Some of them are allopatric—the strains of *Drosophila bifasciata* and *D. imaii* are from Japan, *D. pseudoobscura, D. persimilis,* and *D. miranda* from North America, and *D. subobscura* from Europe. Most crosses produce no vi-

TABLE 10.2

Numbers of matings recorded and the coefficients of ethological isolation for different combinations of species of the Obscura Group of Drosophila. (After Dobzhansky, Ehrman, and Kastritsis 1968)

Species A	Species B	A♀ × A♂	B♀ × B♂	A♀ × B♂	B♀ × A♂	Isolation
D. bifasciata	D. imaii	229	375	13	9	+0.94 ± 0.01
	D. pseudoobscura	112	198	1	17	+0.89 ± 0.03
	D. persimilis	157	203	1	8	+0.96 ± 0.01
	D. miranda	114	142	1	4	+0.96 ± 0.02
	D. subobscura	166	156	10	11	+0.88 ± 0.03
D. imaii	D. pseudoobscura	102	126	10	53	+0.57 ± 0.05
	D. persimilis	136	112	0	6	+0.96 ± 0.01
	D. miranda	90	94	2	4	+0.94 ± 0.02
	D. subobscura	142	59	5	1	+0.96 ± 0.02
D. pseudoobscura	D. persimilis	56	54	2	3	+0.91 ± 0.04
	D. miranda	111	105	1	0	+0.98 ± 0.01
	D. subobscura	80	52	3	2	+0.92 ± 0.03
D. persimilis	D. miranda	89	97	0	0	+1.00
	D. subobscura	73	43	1	0	+0.98 ± 0.01
D. miranda	D. subobscura	49	43	4	1	+0.90 ± 0.05

able hybrids. Moreover, the isolation in nature tends to be stronger than under laboratory conditions. Thus, although hybrids of *D. pseudoobscura* and *D. persimilis* are obtainable in the laboratory, not a single heterogamic mating was observed among 305 copulating pairs collected in nature in 1951 in a locality where both species were present (Dobzhansky 1951). The avoidance of heterogamic matings may be even less pronounced in combinations that produce no viable hybrids (e.g., *D. imaii* with *D. pseudoobscura*) than in those where such hybrids do appear (e.g., the crosses between *D. pseudoobscura, D. persimilis,* and *D. miranda,* Table 10.2). And, finally, the isolation is often greater in one direction than in the reciprocal; thus the females of *D. bifasciata* and *D. imaii* accept males of *D. pseudoobscura, D. persimilis,* and *D. miranda* less easily than the females of the last three species accept males of the first two.

Strains of the same species of different geographic origins may mate at random. For example, Anderson and Ehrman (1969) tested strains of *Drosophila pseudoobscura* derived from places as remote as British Columbia, California, Texas, and Sonora, Mexico. The isolation indices ranged from $+0.14 \pm 0.19$ to -0.09 ± 0.10, none significantly different from zero. On the other hand, Baker (1947) in *D. arizonensis,* Patterson and Wheeler (1947) in *D. peninsularis,* Miller and Westphal (1967) in *D. athabasca,* Dobzhansky (1944) in *D. sturtevanti,* and Dobzhansky and Streisinger (1944) in *D. prosaltans* all found more or less pronounced preferences for homogamic matings among geographic races of the same species. Dobzhansky, Pavlovsky, and Ehrman (1969) tested 25 combinations of geographic strains of the Transitional race of *D. paulistorum;* they found isolation coefficients ranging from $+0.09 \pm 0.09$, which is not significantly different from zero, to a high value of $+0.84 \pm 0.05$. The existence of such intraspecific variations is important; genetic materials are present within species from which ethological barriers separating them can be built.

Mechanical Isolation

The complex structure of the genitalia in many animals, especially insects, has attracted the attention of morphologists and systematists, because closely related species can often be accurately classified by their genitalia. The "lock-and-key" theory was propounded by Leon Dufour in pre-Darwinian days and later elaborated, especially by K. Jordan (1905). According to Dufour, the female and the male genitalia are so exactly fitted to each other that even slight deviations in the structure of either make copulation impossible. The genitalia of each species are "a lock that can be opened by one key only." Different species are isolated from each other by the noncorrespondence of their genitalia.

Experimental evidence on mechanical isolation is scanty. In laboratory cultures *Drosophila melanogaster* males may attempt copulation with *D. pseudoobscura* females, and their genitalia occasionally become locked together. Variations of body size within a species do not seem to hinder copulation. In Drosophila giant and dwarf mutants, and large- and small-bodied flies produced from well-fed and from starved larvae, cross easily and give offspring. The usefulness of genitalia for distinguishing species does not necessarily mean that they are important in mechanical isolation. The reason for their usefulness is that the complexity of genitalic structures is often so great that species differences are more likely to be manifested in these structures than in the relatively simple external ones.

Pollination Barriers

Since the Cretaceous period, a prime factor in the evolution of the angiosperms, the flowering plants, has been mutualistic relationships with animals that effect the transfer of pollen and hence cross-fertilization. Differences in the flower structure in related species may hinder or prevent interspecific pollen transfer. This is an analogue of mechanical isolation through noncorrespondence of the genitalia among animals. In plants there is an additional variable—flowers of different species, because of their particular odors, structures, or colors, may attract different animals as pollinators. There is voluminous literature on pollination mechanisms and floral and pollination ecology; the outstanding recent studies of the genetic aspects of the situation are those of V. Grant (1963; see also V. Grant and K. A. Grant 1965, and K. A. Grant and V. Grant 1968) and of Ehrlich and Raven (1964).

Some plants have flowers of simple structure and are pollinated "promiscuously" by many different pollen carriers. Others have complex flowers and are specialized for pollination by one or several species of insects, birds, or bats. An excellent example is the species pair of columbines *Aquilegia formosa* and *A. pubescens* (Grant 1963). The former has nodding flowers with relatively short upward-directed spurs; it is pollinated by hummingbirds hovering below the flowers. *Aquilegia pubescens* has erect flowers with long spurs, pollinated by hawk moths inserting their probosces from above. Three species of Penstemon sympatric in California are crossable in experiments, but rarely cross in nature because one of them is pollinated by Xylocopa bees, another by hummingbirds, and a third by wasps (Straw 1956). *Salvia apiana* and *S. mellifera* are incompletely isolated by a difference in flowering seasons, and in addition by attracting pollinating bees of different body size (K. A. Grant and V. Grant 1964).

Probably the extreme of specialization of pollination mechanisms is reached in some orchids. Certain species of orchids attract males of definite species of wasps, apparently by copying the sex-attractant odors and, to some extent, the shapes of the females of the latter. The wasps then engage in "pseudocopulation" with the flower, and in the process effect the transfer of pollen (Ames 1937, Kullenberg 1951, Dodson 1967, and references therein).

The phenomenon of flower constancy has been observed in pollinators as diverse as bees, hawk moths, and hummingbirds. An individual pollinator may visit preferentially, one after another, flowers of a certain species, or even of the same color variety if more than one variety is available. Other individuals "work" preferentially on flowers of different colors or different species (Bateman 1951). This factor may not make the reproductive isolation complete, but it will at least diminish the frequency of contamination (Levin and Kerster 1967a, b, 1968). The advantage conferred by flower-constant behavior has been described as follows: "A flower-feeding animal, once it has learned how to work a given flower mechanism, can thereafter obtain more food in less time by continuing to visit other flowers of the same type" (Grant 1963).

Gametic Isolation

Copulation in animals with internal fertilization, the release of eggs and spermatozoa in forms with external fertilization, or pollen deposition on the stigma of a flower in plants is followed by a chain of reactions that bring about the actual union of the gametes, the fertilization proper. These reactions may be out of harmony in different species, resulting in a hindrance or prevention of the formation of hybrid zygotes.

It has been known since the classical work of Lillie (1921) that, if eggs of the two species of sea urchins, *Strongylocentrotus purpuratus* and *S. franciscanus,* are exposed to mixtures of sperm of both species, homogamic fertilizations greatly outnumber heterogamic ones. The same is true of fishes, in species that discharge their sexual products in water, as well as in forms with internal fertilization (e.g., Zander 1962).

The environment that spermatozoa encounter in the reproductive tracts of females of foreign species may be unfavorable. Patterson and his colleagues have done much work on such phenomena in species of Drosophila (Patterson and Stone 1952, and references therein). For example, cross-insemination between the related species *Drosophila virilis, D. americana, D. montana,* and *D. lacicola* is followed by a rather rapid loss of the mobility of the sperm in the sperm receptacles of foreign species, whereas in conspecific inseminations the mobility is conserved for a long time. A similar situation is found by Koref-Santibañez in *D. pavani* and *D. gaucha* (1964). In some species there is a so-called insemination reaction. A rapid secretion of a fluid into the cavity of the vagina takes place after copulation, causing a swelling of the organ. This swelling persists for some hours after intraspecific copulations, whereupon the vagina returns to its normal condition. Insemination by a male of a foreign species, however, gives a more violent reaction. The vagina remains swollen for days, and sometimes the secretion solidifies and obstructs the passage of eggs, making the female sterile. This does not occur in all Drosophila species: some species and species groups show no insemination reaction either after normal or after interspecific copulations.

Nor does a loss of viability of foreign sperm occur in all species. However, the number of offspring that result from one interspecific copulation between *D. pseudoobscura* and *D. persimilis* is, on the average, less than from one intraspecific mating, although the viability of the sperm in foreign sperm receptacles appears to be unaffected. Interspecific copulation results in the delivery of fewer spermatozoa than intraspecific insemination (Dobzhansky 1947). Also, in crosses of *D. pseudoobscura* with *D. imaii* foreign sperm may be inactivated and expelled from the vagina (Dobzhansky, Ehrman, and Kastritsis 1968).

Species and geographic races of mosquitoes of the *Culex pipiens* group exhibit some remarkable crossing relationships (Laven 1967 and references therein). Crosses may be successful in both directions (i.e., strains A♀ × B♂ and B♀ × A♂), in one direction only, or in neither direction. Thus, female mosquitoes from Hamburg produce hybrids when crossed to males from Oggelshausen (Germany), but the cross Oggelshausen ♀ × Hamburg ♂ gives no progeny. Laven showed that what is involved here is an incompatibility of the Oggelshausen egg cytoplasm with spermatozoa of the Hamburg race. The female hybrids from the compatible cross were outcrossed for fifty generations to Oggelshausen males, thus obtaining a strain with Hamburg cytoplasm and Oggelshausen chromosomal genes. The crossing behavior of this gene substitution strain remained, nevertheless, that of the Hamburg race.

Most impressive studies on reproductive isolation between sibling species of ciliate Protozoa have been made by Sonneborn and his school (Sonneborn 1957 and references therein). As many as sixteen sibling species (which Sonneborn calls "varieties") are lumped under the name *Paramecium aurelia*. Each species has at least two "mating types," which function as different sexes, since conjugation occurs between the mating types and does not ordinarily happen within a mating type. Strains of these infusoria can be propagated asexually by cell fission; the resulting clones may be tested for their ability to form sexual unions. Different strains of the same species can be crossed, and the progenies of the exconjugants sometimes exhibit heterosis (Siegel 1958). By contrast, strains belonging to different species do not, as a rule, conjugate. Some do form pairs, however, which adhere to each other briefly and then fall apart without conjugation. Only a small minority of the species combinations tested can undergo a conjugation process, which rarely if ever results in fit progeny. Death may occur even in the F_1 generation, which then exhibits hybrid inviability. Other species crosses give a viable F_1, but the F_2 generation and the backcrosses are completely or nearly completely inviable. This may be regarded as a hybrid breakdown (see below). *Paramecium caudatum, P. bursaria, Tetrahymena pyriformis,* and *Euplotes patella* are other names applied to groups of sibling species that exhibit reproductive isolation more or less similar to that in *P. aurelia.*

In flowering plants a complex sequence of processes intervenes between the deposition of pollen on the stigma and the formation of a zygote. The pollen must germinate, the pollen tube grow down the style, and the pollen nuclei fuse with the nuclei in the ovules to form the embryo and the endosperm nuclei. This sequence may be disrupted at any point in hybrid crosses.

Crosses of different species of Datura were studied by Blakeslee and his collaborators (review and references in Avery, Satina, and Rietsema 1959). The speed of pollen tube growth is frequently greater in the style of its own species than in that of a foreign species. Moreover, the pollen tubes may burst in the styles of a foreign species before they reach the ovary. Table 10.3 shows the results of 174 attempted crosses between races and species of Gilia (phlox family). Grant (1963) summarizes the results as follows:

The incompatability barrier in the cobwebby gilias is manifested at different stages of flowering and fruiting. A flower pollinated with foreign pollen may fail to set a capsule; the capsule may ripen but contain only or mainly shriveled seeds; a reduced number of plump seeds may develop; or numerous plump seeds may form in the capsule but fail to germinate. From the developmental standpoint, there is evidently not one incompatability barrier but several.

It can only be added that in some instances the incompatibility could be overcome in experiments by the application of certain chemical substances, such as growth hormones (Emsweller and Stuart 1948). The pollen abortion may be under a simple genetic control; for example, Cameron and Moav (1957) found in a species of Nicotiana a gene that causes abortion of pollen not carrying it in hybrids with another species.

TABLE 10.3

Average number of normal seeds per flower, percentage of the crosses that produced any offspring, and number of hybrid individuals per ten flowers pollinated in Gilia (After Grant 1963)

Cross	Seeds per Flower	Percentage of Successful Crosses	Hybrids per Ten Flowers
Within a population	17.8	100	22
Between races	15.2	73	12
Between species of a section	3.7	43	3
Between species of different sections	0.004	2	0.038

The Inviability of Hybrids

Fantastic tales of most disparate animals mating and bringing forth offspring found ready credence in ancient, medieval, and even more recent times. Perhaps the ultimate fable of this type was that the ostrich came from hybridization of a camel and a sparrow. A critical reaction against such absurdities led some biologists to postulate that only crosses within a species produce viable and fertile hybrids. The truth lies in between the two extremes—some closely related species do hybridize in nature to a certain extent, and additional hybrids can be obtained in experiments.

Cells in tissue cultures can be induced to coalesce, and hybrid cells have been obtained carrying chromosomes of mouse and rat, mouse and hamster, and even mouse and man (Ephrussi and Weiss 1965, Weiss and Ephrussi 1966, and Weiss and Green 1967). Of course, a union of cells or gametes of different species does not necessarily lead to adult hybrid progeny; the life of a hybrid zygote may be cut short at any stage. In animals with external fertilization, spermatozoa may enter eggs of representatives of different classes and phyla (echinoderms × mollusks, echinoderms × annelids), but the sperm nucleus or its chromosomes may be eliminated from the cleavage spindle. Eggs of species of fish can be inseminated by sperm of different species, genera, or even families. All sorts of disturbances may, however, occur in the zygotes, from chromosome elimination during cleavage, through arrest of gastrulation or of organ formation, to death of the embryos in advanced stages.

In some instances the weakness of the hybrid can be overcome in experiments. Laibach's (1925) hybrids between species of flax constitute a classical example. In the cross *Linum perenne* ♀ × *L. austriacum* ♂ hybrid seeds fail to germinate if left to their own devices. If the embryos are freed from the seed coat (the seed coat being a maternal tissue), however, germination takes place, and the seedlings give rise to luxuriant hybrid plants that are fertile and produce normal seeds in the F_2 generation. Still greater is the suppression of seed development in the cross *L. austriacum* ♀ × *L. perenne* ♂, and yet if the diminutive embryos are extracted from the seeds and placed in a nutrient solution they continue to grow. After some days they may be transferred to moist paper and allowed to grow. The seedlings are then planted in soil.

Moore's (1949a,b, 1950) studies on isolating mechanisms in North American species of frogs (Rana) are another classic. Table 10.4 summarizes some of the information Moore obtained. The distribution areas of all these species overlap, making them sympatric in parts of their ranges. Because of preferences for different habitats during the breeding season, the species are to some extent, but never completely, isolated ecologically. Some species, however, show complete seasonal isolation. After artificial insemination of the eggs, from 0 to 100 percent of the hybrid embryos develop normally to the adult stage. Moore's estimates of the efficacies of the isolating mechanisms taken separately are shown in Table 10.4. In conjunction, they seem to give complete or nearly complete isolation in all cases.

TABLE 10.4

Estimates of the potency of geographic isolation (G), ecological isolation (E), seasonal isolation (S), and hybrid inviability (D) in species of frogs (After Moore 1949)

Complete isolation = 100; Absence of isolation = 0

Males Females	Rana sylvatica	Rana pipiens	Rana palustris	Rana clamitans	Rana catesbeiana	Rana septentrionalis
Rana sylvatica		G 29 E 70 S 60 D 100	G 61 E 40 S 100 D 100	G 59 E 30 S 100 D 100	G 68 E 70 S 100 D 100	G 80 E 60 S 100 D ?
Rana pipiens	G 59 E 70 S 60 D 100		G 74 E 70 S 40 D 0	G 67 E 70 S 100 D 100	G 62 E 85 S 100 D 100	G 88 E 80 S 100 D 100
Rana palustris	G 13 E 40 S 95 D 100	G 0 E 70 S 40 D 0		G 3 E 60 S 95 D 100	G 23 E 70 S 100 D 100	G 72 E 50 S 100 D ?
Rana clamitans	G 28 E 30 S 100 D 100	G 0 E 70 S 100 D 100	G 24 E 60 S 95 D 100		G 18 E 30 S 50 D 100	G 22 E 20 S 0 D 100
Rana catesbeiana	G 51 E 70 S 100 D 100	G 0 E 85 S 100 D 100	G 47 E 70 S 100 D 100	G 29 E 30 S 50 D 100		G 93 E 30 S 50 D ?
Rana septentrionalis	G 0 E 60 S 100 D ?	G 0 E 80 S 100 D 100	G 37 E 50 S 100 D 100	G 36 E 20 S 0 D 100	G 79 E 30 S 50 D 95	

Rana pipiens is a complex of races or closely related species. Its distribution ranges from southern Canada to Mexico and Costa Rica. Northern races have lower minimum and maximum temperature limits for normal embryonic development than do southern races. Thus, strains from Canada and Vermont have tolerance limits of about 5–28°C, while those from Florida have a range from 11 to 35°C. It should be noted, however, that in Canada this species breeds in May or June, in northern New York in mid-April, and near New York City in early April. These differences in breeding dates provide a "northern environment" for the developing eggs. The breeding season in North Carolina is February-March, while in Georgia and Florida the frogs may breed in any month of the year. Northern frogs develop more rapidly than southern ones at low temperatures, but at high temperatures the difference is diminished or even inverted.

The viability of hybrids between the races of *Rana pipiens* appears to be correlated with their adaptedness to different climatic conditions. Thus, the eggs of the Vermont race give normal development when fertilized with Wisconsin sperm. Fertilization of Vermont eggs by the sperm of New Jersey or Oklahoma races results in normal or slightly retarded development rates and a slight enlargement of the head of the embryo. With Louisiana sperm, the retardation is slight and the head enlargement moderate. With Florida or Texas sperm, a marked retardation and a strong enlargement of the head are observed. With eggs of southern and sperm of northern races, the embryos have markedly reduced heads and retarded developments. In either case, the hybrids between geographically extreme members of the series are inviable. Situations similar to those found by Moore in frogs have also been described by Minamori (1957) in species and races of the fishes *Misgurnus anguillicaudatus* and *Cobitis taenia*, and much earlier by Goldschmidt (1934) in the gypsy moth (*Lymantria dispar*) in northern Eurasia.

Some species crosses produce hybrids of one sex only; either male or female zygotes die, while the viability of the other sex is affected little or not at all. Haldane's rule (1922) states, "When in the F_1 offspring of two different animal races one sex is absent, rare or sterile, that sex is the heterozygous sex." In mammals and most insects, males are heterogametic, and male hybrids are defective more frequently than females. In birds, butterflies, and moths, females are heterogametic, and female hybrids are less viable than males.

A possible mechanism underlying Haldane's rule is as follows (Dobzhansky 1937b). *Drosophila pseudoobscura* and *D. miranda* differ in their gene arrangements, and some genes that lie in one species in the X chromosome lie in the other in the autosomes, and vice versa. The cross *D. miranda* ♀ × *D. pseudoobscura* ♂ produces viable female and abnormal male hybrids; the reciprocal cross gives rise to viable females, but the males die off. Suppose that *D. pseudoobscura* has in its X chromosome a group of genes *A* that lie in the autosomes of *D. miranda,* and that a group of genes *B*, which in *D. miranda* lie in the X chromosome, are located in the autosomes of *D. pseudoobscura;* with respect to these genes, the constitution of the females of both species and also of the female hybrids is alike, *AABB*. Males of *D. pseudoobscura* and the male hybrids from the cross *D. miranda* ♀ × *D. pseudoobscura* ♂ are *ABB*; *D. miranda* males and the male offspring from the cross *D. pseudoobscura* ♀ × *D. miranda* ♂ are *AAB*. The genotypes of the pure species are so adjusted that the constitution *ABB* in *D. pseudoobscura* and *AAB* in *D. miranda* permit the development of "normal" males. The constitution *AABB* is normal for females of either parent and for hybrid females as well. The constitution *ABB* is incompatible with the genotype of *D. miranda,* and *AAB* with that of *D. pseudoobscura*. The hybrid males are poorly viable or lethal because of disruption of the gene balance.

Varieties of Hybrid Sterility

The sterility of hybrids between species has interested people since Aristotle, who discussed at length the sterility of mules. As stated previously, some biologists have attempted to define species as arrays of individuals that fail to produce viable and fertile offspring when crossed. Such definitions are invalid, however, because gene exchange between species may be prevented by any one of the reproductive isolating mechanisms or by a combination of several of them. Viable and fertile hybrids may be obtained in experiments between undoubtedly distinct species that are completely isolated reproductively in nature.

Hybrid sterility is not a result of a general weakness of a hybrid organism, as some authors have surmised. Many hybrids with reduced viability are nevertheless fertile, and some sterile hybrids (e.g., mules) are somatically vigorous. The cause of the sterility typically resides in the reproductive organs, not in the body. Dobzhansky and Beadle (1936) transplanted larval testes of hybrids between *Drosophila pseudoobscura* and *D. persimilis* into larvae of the parental species, and vice versa. Male hybrids between these species have grossly abnormal spermatogenesis and are sterile. If a transplanted testis of a pure species becomes attached to the sexual ducts of a hybrid, the latter becomes fertile, the functional sperm coming, of course, from the implanted testis. Hybrid testes do not develop functional spermatozoa in the bodies of pure species, however, and the gonads of the host are not affected by the presence of a hybrid testis.

In 1913, Federley made the discovery that the chromosomes fail to pair at meiosis in the sterile hybrids between species of the moth Pygaera. The disruption of the chromosome pairing is followed by various abnormalities in the chromosome distribution during the meiotic divisions and by irregularities in spermiogenesis. No functional spermatozoa are formed, and the hybrid males are sterile. Failures of the chromosome pairing at meiosis have since been observed in numerous other sterile hybrids, animals as well as plants (reviews in White 1954 and Stebbins 1958a). A plausible inference is that the disruption of the chromosome pairing at meiosis is the cause of the sterility. As we shall see, this inference is indeed warranted for some, though not for all, forms of sterility.

Some hybrids are sterile despite normal meiosis. The primroses *Primula verticillata* and *P. floribunda* both have nine chromosomes in the haploid set. Their hybrid forms nine bivalents at meiosis, but is nevertheless sterile (Upcott 1939). As stated above, male hybrids between *Drosophila pseudoobscura* and *D. persimilis* are completely sterile, although the chromosome pairing at meiosis varies from completely normal to none at all (Dobzhansky 1934). Within each species, strains are encountered that produce hybrids in which no bivalents are formed at meiosis, and other strains that produce hybrids with bivalents only and no univalents. The meiotic divisions are, however, abnormal in either case: the first division spindle elongates enormously and bends into a ring, the cell body fails to divide, the second meiotic division is absent, and the giant binucleate spermatids degenerate.

In other hybrids, such as those between *Drosophila melanogaster* and *simulans*, the gonads are rudimentary, and, as shown by Kerkis (1933), spermatogenesis and oogenesis do not advance beyond spermatogonia and oogonia. In many sterile hybrids between species of mammals (horse × ass, horse × zebra, domestic cow × yak), as well as of birds and fishes, the degenerative changes in the gametogenesis set in before meiotic pairing should normally begin (see Stebbins 1958a for references). Among some 150 hybrids between species of the plant genus Geum, Gajewski (1957, 1959) found completely sterile, partially fertile, and nearly normally fertile examples. Hybrids between some species fail to flower; others have disturbances in the development of the floral parts, failures of the meiotic chromosome pairing, or, finally, normal chromosome pairing followed by breakdowns in gametophyte development.

Meiosis, like any other physiological process, is controlled by the genotype of the organism. The normal course of meiosis involves a succession of events so delicately balanced that the failure of any one of them, or simply altered timing, causes disruptive changes. Genetic disharmonies brought about by the hybridization of species or of varieties that do not normally cross in nature may alter the process of meiosis at any stage. Chromosome pairing may fail despite the presence of pairs of chromosomes with similar genes arranged in identical linear series. The resulting sterility will be due to the genetic constitution of the organism. This is *genic sterility*. On the other hand, chromosomes may fail to pair and to form bivalents because they have no structurally similar partners. Chromosomes may contain identical genes, but they may be differently arranged. Sterility due to such structural dissimilarities is *chromosomal sterility*.

Müntzing (1961 and earlier), Stebbins (1958a), and Grant (1963) distinguish haplontic and diplontic sterilities. The former occur in plants when the haplonts, that is, the pollen grains and the embryo sacs, "are killed by their own genetic constitution." In the latter, "the constitution of the plant conditions a reduced or inhibited capacity to form progeny through sexual reproduction" (Müntzing 1961). On the basis of these criteria, all hybrid sterility in animals must be diplontic, because animal sex cells, which are haploid, generally function even if they carry grossly abnormal chromosome complements (see the next section). It remains to be added that the chromosomal and genic or the haplontic and diplontic, sterilities are not mutually exclusive and occur together in some hybrids. Furthermore, any one of these mechanisms may cause any degree of sterility, from barely noticeable to complete.

Chromosomal Sterility

Suppose that two chromosomes carry the genes *ABCD* and *EFGHI*. A translocation gives two "new" chromosomes, *ABFE* and *DCGHI*. Meiosis in homozygous normals and in translocation homozygotes produces sex cells that carry every gene once and only once. In a translocation heterozygote six classes of sex cells can be produced: (1) *ABFE*, *DCGHI*; (2) *ABCD*, *EFGHI*; (3) *ABFE*, *EFGHI*; (4) *ABCD*, *DCGHI*; (5) *ABFE*, *ABCD*; and (6) *DCGHI*, *EFGHI*. Classes 1 and 2 carry balanced gene complements, but in classes 3–6 certain genes are deficient and others are present in duplicate; 1 and 2 are termed regular or orthoploid, and 3–6 exceptional or heteroploid gametes.

The fate of heteroploid gametes is different in animals and in plants. In animals, sex cells with grossly unbalanced gene complements retain their functional ability. If a translocation heterozygote is crossed either to a homozygous normal or to a translocation homozygote, however, the part of the progeny coming from the heteroploid gametes will suffer from deficiencies of some genes and duplications of others and will be inviable or crippled. In Drosophila, females heterozygous for a translocation crossed to normal males, or normal females crossed to males heterozygous for a translocation, deposit numerous eggs many of which contain "dominant lethals" and fail to develop. That the heteroploid sex cells nevertheless function can be demonstrated by having the chromosomes involved in the translocation marked with appropriate mutant genes. Females and males heterozygous for the same translocation are then intercrossed. Some heteroploid gametes are complementary, in the sense that one of them is deficient for the genes which the other carries in duplicate, and vice versa (gametes 3 and 4, and 5 and 6, above). The union of complementary gametes gives viable zygotes, which can then be identified by the mutant markers they carry.

In plants, meiotic divisions give rise to gametophytes that undergo several cell divisions before the generative cells, the gametes proper, are produced. Gametophytes with deficiencies and duplications are usually aborted. Only small duplications and, even more rarely, deficiencies pass through the gametophytes in nonpolyploid species. Translocation heterozygotes in plants produce mixtures of good and aborted pollen grains and embryo sacs; they are "semisterile," but the seeds that they mature are viable.

The degree of sterility depends on the proportions of orthoploid and heteroploid gametes; if the six classes of gametes shown above are equally frequent, the translocation heterozygote will produce 66.7 percent inviable progeny (in animals) or aborted pollen (in plants). A heterozygote for two, three, or more translocations may then produce few viable progeny and will be effectively sterile. In reality the proportions of heteroploid gametes vary, depending on the relative sizes of the chromosome sections exchanged and on other factors. In structurally homozygous individuals every chromosome has one and only one homologue with an identical gene arrangement. In translocation heterozygotes, some chromosomes consist of sections that are homologous to parts of two or more chromosomes. In inversion heterozygotes, chromosomes have homologues with the same genes arranged in different linear sequences. Meiotic pairing is due to a mutual attraction between homologous genes rather than between chromosomes as such. In structural heterozygotes, parts of the same chromosome may be pulled in different directions simultaneously. Pairing of some chromosome sections may be delayed or not attained at all. The more extensive the differences in gene arrangement between the chromosomes of the parents, the greater is the competition for pairing at meiosis, and the more frequent are the failures of pairing and disjunction.

A special case of chromosomal sterility is that of cryptic structural hybridity (Stebbins 1950, 1958a, Stephens 1950, and Grant 1963). Translocation of small blocks of genes may give rise to chromosomes that are homologous except for some small segments (e.g., chromosomes *ABCDEF* and *GHIJKLM* giving rise to *ABCDLF* and *GHIJKEM*). A hybrid with such chromosomes is, of course, a translocation heterozygote. And yet, because of the competition for pairing, this hybrid will have apparently normal bivalents at meiosis (*ABCDEF/ABCDLF* and *GHIJKLM/GHIJKEM*). Fifty percent of the gametes of such a hybrid will contain duplications and deficiencies. Accumulation of cryptic structural differences in several, or in all, chromosomes of a set may perhaps give virtually complete sterility, despite regular chromosome pairing at meiosis in the hybrid. The sterility of the primrose hybrids mentioned previously may well be of this kind.

Allopolyploidy as a mode of species formation will be discussed in Chapter 11. Here we are interested in one aspect of the story, since the best available evidence that the sterility of some species hybrids is, in fact, chromosomal comes from studies on allopolyploid plants. The pioneering work of Karpechenko (1928) may still be used as a paradigm, although similar results have since been obtained with many plant species. Radish (*Raphanus sativus*), cabbage (*Brassica oleracea*), and their F_1 hybrids have the same number of chromosomes, namely 18. In the parental species, 9 bivalents are formed regularly at meiosis; pollen and ovules contain 9 chromosomes. Little or no meiotic pairing takes place in the F_1 hybrids; the 18 univalents are distributed at random to the daughter cells at the first meiotic division, these cells come to contain varying numbers of chromosomes, mostly 6–12, no viable pollen and ovules are formed, and the hybrids are sterile. In some cells, however, the first meiotic division is abortive, viable pollen and ovules with 18 chromosomes are produced, and their union gives rise to F_2 plants with 36 chromosomes. These tetraploid plants, called radocabbage or Raphanobrassica, are quite vigorous, combine in their appearance the traits of both radish and cabbage, and, what is most interesting, are fully fertile.

The lack of chromosome pairing at meiosis in the diploid hybrid may be caused either by differences in the gene arrangements in the chromosomes of raddish and cabbage, or by disharmonies in the hybrid genetic constitution. The first possibility is upheld and the second refuted by the normal chromosome pairing and bivalent formation in the tetraploid Raphanobrassica. In this tetraploid hybrid every chromosome has one and only one partner with a similar gene arrangement. The sterility of the diploid hybrid is chromosomal, and fertility is restored in the tetraploid.

Genic sterility is not relieved by chromosome doubling. A genetic constitution that causes disruption of the meiotic pairing in a diploid has the same effect in a tetraploid hybrid. This is demonstrated by observations on tetraploid spermatocytes occasionally found in the testes of the sterile hybrids between *Drosophila pseudoobscura* and *D. persimilis* (see the next section). The meiotic chromosome pairing is no different in the tetraploid from that in the diploid spermatocytes; the meiotic divisions are equally abnormal, and the hybrids with tetraploid cells in their testes are completely sterile.

Comparative studies of chromosome behavior in fairly numerous diploid plant hybrids, and in the tetraploids obtained from them, led Darlington (1937) to formulate the following rule, applicable to hybrids whose sterility is chromosomal: Sterile diploid hybrids with little or no chromosome pairing at meiosis give allopolyploids that are fertile and display mostly or only bivalents at meiotic division. Conversely, the allopolyploids derived from diploids with many bivalents show an irregular chromosome pairing. The fertility of allopolyploids tends to be inversely proportional to that of their diploid ancestors. Chromosome pairing in a diploid hybrid shows that the gene arrangement in the chromosomes of the parents is similar enough for some or all chromosomes of one species to find approximate homologues among those of the other. The doubling of the chromosome complement gives to each chromosome three potential mates more or less similar to it. In the competition for pairing, the pairing of the chromosomes of the same species is disrupted by the presence of the partial homologues. As a result, bivalents, trivalents, quadrivalents, and univalents are formed in varying proportions; gametes with unbalanced chromosome complements are produced; and the hybrid is more or less sterile. When the chromosomes of the parental species fail to pair in the diploid because of extensive dissimilarities in the gene arrangement, however, every chromosome in the allotetradiploid has only one mate, with which it can pair with little or no interference from the chromosomes of the other species. Hence, only bivalents are produced, meiosis is regular, and fertility is restored.

Genic Sterility in the Hybrids of Drosophila pseudoobscura × D. persimilis

Drosophila pseudoobscura and *D. persimilis* are sibling species. The cross *D. persimilis* ♀ × *D. pseudoobscura* ♂ gives F_1 hybrid males with small testes; the reciprocal cross, *D. pseudoobscura* ♀ × *D. persimilis* ♂, produces hybrid males with testes of normal size; in either case the males are completely sterile (Lancefield 1929). Backcrosses of the F_1 hybrid females to *D. persimilis* or *D. pseudoobscura* males give sons with testes of variable size, ranging from normal to very small. Males with small testes are always sterile; those with large ones are sometimes fertile. The sterility is due to a profound modification of the process of spermatogenesis. The meiotic chromosome pairing is variable; no univalents, some univalents, or only univalents may be present at the first meiotic division. Irrespective of the numbers of bivalents and univalents formed, only a single, very abnormal meiotic division takes place. The spermatids degenerate (Dobzhansky 1934). The disturbances in spermatogenesis are, in general, greater the smaller are the testes in a hybrid. Testis size is, therefore, a measure of the degree of departure from the normal course of the spermatogenesis. The disturbances in the hybrids are confined to the gonads; the rest of the reproductive system (the sexual ducts and external genitalia) is normal.

The hypothesis that the sterility of hybrids between *D. pseudoobscura* and *D. persimilis* is genic was tested and verified (Dobzhansky 1936). Backcrosses were made of the F_1 hybrid females to males of both parental species. In the progenies of such backcrosses individuals appear that carry all possible combinations of the chromosomes of the parents (Figs. 10.1 and 10.2). Experiments were so arranged that, by having the parental chromosomes marked by mutant genes with easily visible external effects, the combination of the chromosomes present in a given male was recognizable by his phenotype.

The mean sizes of the testes in the backcross males with different chromosomes are shown in Figs. 10.1 and 10.2. Backcross males with an X chromosome and the autosomes of the same species have testes as large as the males of the parental species. Such males are usually fertile, whereas the males with small testes are all sterile. The more dissimilar the X chromosome and the autosomes become in species origin, the smaller is the testis size. The smallest testes are observed in males with the X chromosome of one species and all autosomes of the other. With a single exception, all chromosomes act alike, and their action is additive. Thus, individuals carrying the *D. pseudoobscura* X chromosome and *D. persimilis* autosomes have very small testes (class 16, Fig. 10.1). The introduction of one fourth or one third or one second chromosome of *D. pseudoobscura* increases the testis size (classes 13–15, Fig. 10.1). Simultaneous introduction of the fourth and third (class 12), or the second and third (class 10), or the second and fourth (class 11), or the second, third, and fourth (class 9) chromosomes of *D. pseudoobscura* increases the testis size more than does each of these chromosomes alone. Males having all chromosomes of the same species or having one third or one fourth chromosome of the other species are sometimes fertile (classes 1, 2,

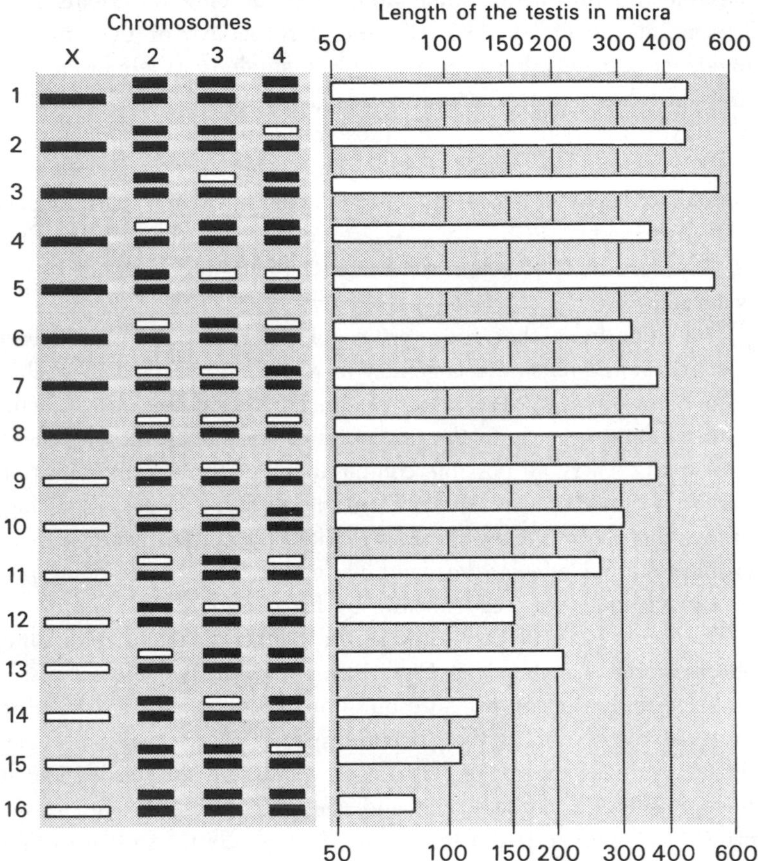

FIGURE 10.1

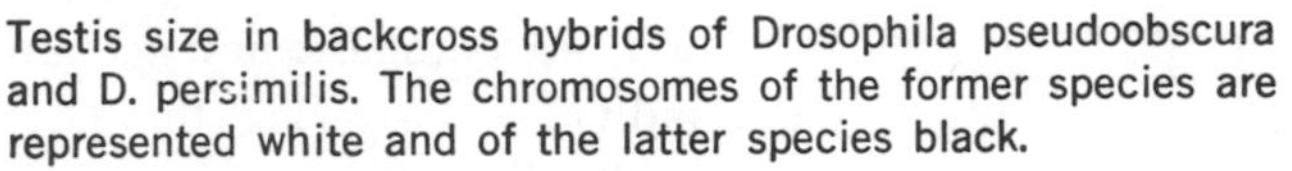

Testis size in backcross hybrids of Drosophila pseudoobscura and D. persimilis. The chromosomes of the former species are represented white and of the latter species black.

and 3). If both the third and the fourth chromosomes disagree in origin with the rest of the complement, the male is sterile. The exception mentioned above is that, in backcrosses to *D. persimilis* males, sons having the *D. pseudoobscura* third chromosome and the *D. persimilis* X chromosome have larger testes than their brothers homozygous for the *D. persimilis* third chromosome.

We can conclude that all the tested chromosomes of *D. pseudoobscura* and *D. persimilis* carry genes concerned with the fertility of

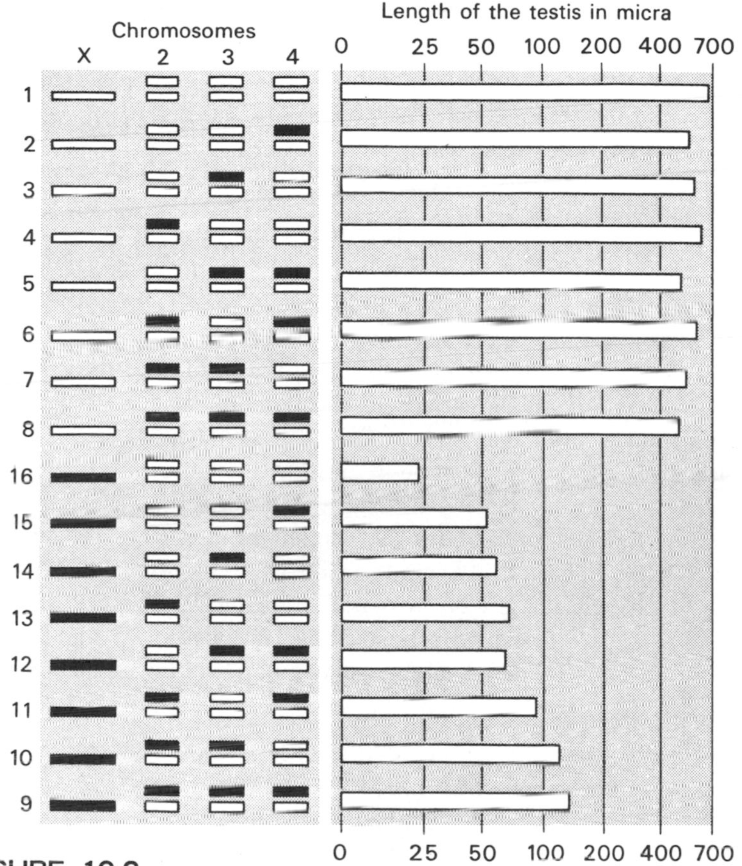

FIGURE 10.2

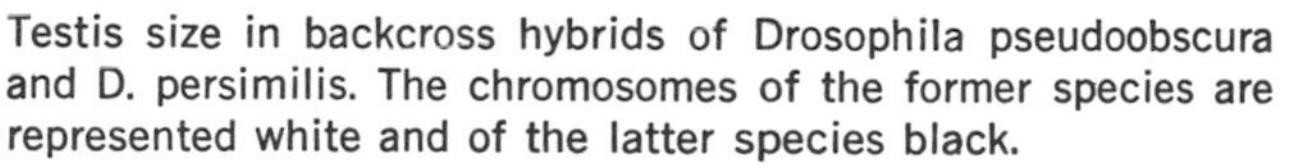
Testis size in backcross hybrids of Drosophila pseudoobscura and D. persimilis. The chromosomes of the former species are represented white and of the latter species black.

males of the same species and the sterility of hybrid males. Moreover, the X, the second, the third, and the fourth chromosomes have at least two such genes apiece. The minimum number of genes concerned with the sterility is, therefore, eight, but the actual number is almost certainly greater. The sterility of hybrids between animal species is most often genic, whereas chromosomal sterility seems prevalent in the plant kingdom, at least among flowering plants. However, Stebbins (1958a) lists several botanical examples of interspecific hybrids, the sterility of which may plausibly be ascribed to their genetic constitution.

Genic Sterility in the Hybrids Drosophila virilis × D. americana

The sterility in the hybrids between the Oriental species *Drosophila virilis* and its American analogues *D. americana* and *D. texana* has been analyzed by Patterson and his colleagues (review in Patterson and Stone 1952). Crosses between *D. virilis* and *D. americana* or *D. texana* produce hybrids of both sexes; the fertility of these hybrids varies, depending on the strains of the parental species used, but on the whole it is lower than that of pure species. Hybrids in the F_2 and in the backcrosses are likewise semisterile. If *D. virilis* females are crossed to *D. texana* males, and the male hybrids are backcrossed to *D. virilis* females, only about 25 percent of the male progeny are fertile; if the initial cross is made using *D. texana* females, however, some 82 percent of the backcross males are fertile. The two series of crosses may be represented schematically as follows (X^v and X^t, and Y^v and Y^t, being X and Y chromosomes of *D. virilis* and *D. texana*, respectively):

P: *virilis* (X^vX^v)♀ × *texana* (X^tY^t)♂
Backcross: *virilis* (X^vX^v)♀ × X^vY^t♂
X^vY^t males (25% fertile)
P: *texana* (X^tX^t)♀ × *virilis* X^vY^v♂
Backcross: *virilis* (X^vX^v)♀ × (X^tY^v)♂
X^vY^v males (82.5% fertile)

The backcross males in both series are similar in that they carry the *Drosophila virilis* X chromosome, but in the series represented first they have the *D. texana* Y, and in the second series the *D. virilis* Y chromosome. Since the chromosomes of *D. virilis* and *D. texana* are distinguishable cytologically because of inversions and translocations, it has been possible to identify the constitution of the fertile backcross males by examining the chromosomes in their offspring. The fertile males with the *D. texana* Y chromosome invariably also possess the second and the fifth chromosomes of the same species. The Y, second, and fifth chromosomes of *D. texana* carry complementary genes that must be simultaneously present to enable a male to be fertile. Whether a single gene or several genes in each of these chromosomes are concerned is not known.

Sterility and Nucleo-Cytoplasmic Imbalance

A hybrid inherits its chromosomes from both parents, but its cytoplasm chiefly or entirely from its mother. Although the genetic information is transmitted mainly through the nuclei and their chromosomes, some of it is also carried in the cytoplasm. The genome and the plasmon of a hybrid can be distinguished. The sterility of some hybrids, especially among plants, is due to genome-plasmon incompatibilities.

Michaelis (1954 and references therein) made most painstaking studies on hybrids between species of Epilobium. The cross *E. luteum* ♀ × *E. hirsutum* ♂ (i.e., flowers of the former species pollinated by the latter) gives healthy hybrid plants, with normal flowers and 15–20 percent good pollen. The reciprocal cross (*E. hirsutum* ♀ × *E. luteum* ♂) yields dwarfish plants, with underdeveloped flowers and no good pollen. The hybrids (*E. luteum* ♀ × *E. hirsutum* ♂) were pollinated by *E. hirsutum* pollen in several successive generations. In this way, plants were obtained that have acquired the *E. hirsutum* genes but have retained the *E. luteum* plasmon. Do such plants behave in crosses to *E. luteum* as the pure *E. hirsutum* did? The answer is no—the hybrids produced develop normally and have some 20 percent good pollen. The plasmon of *E. luteum* has retained its properties despite having carried an *E. hirsutum* genome for several generations.

Much work has been done by Japanese investigators (Kihara 1959, 1967, and references therein) on cytoplasmic male sterility in crosses between some species of wheat, Triticum, and of the related genus Aegilops. For example, hybrids between the bread wheat, *T. vulgare*, and *Ae. caudata* produce no good pollen but their ovules function normally. By a series of backcrosses to *T. vulgare* it is possible to obtain "substitution lines" having Triticum nuclei in Aegilops cytoplasm; they are female fertile but male sterile. Some strains of wheat, however, carry "fertility restorer genes," which make plants with Aegilops cytoplasm and Triticum genomes male fertile. These fertility restorers are specific for the cytoplasms of each species of Aegilops that induces sterility. Grun and Aubertin (1965) described cytoplasmic male sterilities in crosses between species of Solanum that closely resemble those in wheats.

The properties of an egg cell cytoplasm may be determined by its own self-replicating constituents, or by the chromosomes that were present in the egg before meiosis and fertilization. The former mechanism is cytoplasmic inheritance in the strict sense; the latter is known as maternal effect or predetermination of the cytoplasm by the maternal chromosomes. The elegant studies of Ehrman (1960, 1962) brought to light remarkably diverse sterility mechanisms in *Drosophila paulistorum.* This superspecies consists of at least five semispecies (see above and Chapter 11) which yield, when crossed, fertile F_1 females and sterile males with grossly abnormal spermatogenesis. The F_1 females, backcrossed to males of either parental semispecies, give progenies consisting again of fertile females and males all of which are sterile.

The absence of any fertile males in the backcross progenies seemed puzzling, since some of them should carry the genes of only one semispecies. (*Drosophila paulistorum* has three pairs of chromosomes, which in the hybrids between the semispecies tend to be inherited as units, on account of heterozygosis for inverted sections. Therefore, one might expect about one-eighth of the males resulting from the first backcross to have all their chromosomes like their fathers, and consequently to be fertile.) Making use of mutant gene markers, Ehrman showed that all males coming from the eggs deposited by a hybrid female are sterile, regardless of their own chromosomal constitution. In the particular crosses that she studied, any one of the three semispecies-foreign chromosomes present in a female makes all her male progeny sterile. However, in the backcrosses some females appear that no longer carry "foreign" chromosomes; when crossed to males of the same semispecies, they produce sons all of which are fertile. The chromosomal constitution of a mother, then, influences her sons but not necessarily her grandsons.

The hybrid sterility in *D. paulistorum* is, thus, of two kinds: that of the F_1 hybrid males, due to their own genetic constitution, and that of the backcross males, induced by the genetic constitution of their mothers. Williamson and Ehrman (1967 and Ehrman and Williamson 1969) found a third kind of sterility, possibly related to the second. Certain strains, which we may denote as A and B, give sterile F_1 males when crossed. These sterile males were ground and centrifuged, and the supernatant injected into A females, which were then crossed to A males. The progeny is evidently genetically identical with the A strain and is not hybrid at all; nevertheless the males coming from the eggs deposited by these females a week or longer after the injection are sterile. Neither A nor B males injected with the same supernatant nor their pure-bred male progenies, however, are sterile. The "infectious sterility" is induced in the eggs of the injected mothers, when these eggs develop into male individuals.

These results suggest that one or both of the strains crossed carry some symbiotic microorganisms, to which these strains are genetically adjusted, and the presence of which does not interfere with fertility. Hybridization has a disruptive effect on this adjustment, and the sterility is a consequence. Kernaghan and Ehrman (1970) have described what appears to be Mycoplasma-like symbionts in the testes of sterile males.

Hybrid Breakdown

Consider species that are sympatric, between which matings or cross-pollinations occasionally occur and result in viable and fertile hybrids. Nevertheless, no gene exchange may take place between the populations of these species if the recombination products of their genotypes are discriminated against by natural selection. Moreover, as will be argued in the next chapter, the loss of fitness resulting from hybridization of genetically diverging populations may well be a challenge to which these populations could respond by developing reproductive isolation.

The often cited sibling species *Drosophila pseudoobscura* and *D. persimilis* are crossable in laboratory experiments, although no hybrids have been found in nature. The F_1 hybrids appear to be as vigorous as the parental species; the hybrid males are sterile, but the females produce numerous eggs when backcrossed to males of either parent. The viability of the backcross progenies, however, is dramatically reduced, as the following data clearly show (Dobzhansky 1936 and Weisbrot 1963).

Drosophila pseudoobscura females homozygous for the sex-linked genes beaded (*bd*), yellow (*y*), short (*s*), Bare (*Ba*, second chromosome), and purple (*pr*, third chromosome) were crossed to *D. persimilis* males homozygous for the recessive orange (*or*, third chromosome) and heterozygous for the dominant Curly (*Cy*, fourth chromosome). The F_1 generation was as follows:

Ba ♀♀	432	845	*bd y s Ba* ♂♂	401	786
Ba Cy ♀♀	413		*bd y s Ba Cy* ♂♂	385	

Males are somewhat less numerous than females, the discrepancy being due to a decrease of viability caused by the mutants *bd, y,* and *s.* The *Ba Cy* hybrid females were backcrossed to *D. pseudoobscura* males homozygous for purple (*pr*) and orange (*or*). The females have, then, every chromosome, except the small fifth, marked with at least one mutant gene. In the backcross progeny the genetic constitution of every male may be ascertained from its appearance. If the crossing over in the X and in the third chromosomes is disregarded, sixteen classes of males are expected to appear in equal numbers, carrying the combinations of the chromosomes of the parental species represented diagrammatically in Figs. 10.1 and 10.2. Only eight classes of females are distinguishable (since the sex-linked recessive genes do not manifest themselves in the heterozygous females).

The results actually obtained are summarized in Table 10.5; the column headed "Class Number" refers to the Figs. 10.1 and 10.2. Table 10.5 shows that males are fewer than females, and that representatives of the different classes are far from equally numerous. The yield of adult backcross individuals per mother is very small; a majority of the backcross individuals evidently die. A feature at first sight paradoxical is that class 1 of the males, which consists of individuals having only *Drosophila pseudoobscura* chromosomes, is almost obliterated. If the decrease in viability were due only to mixing the chromosomes of the two species, class 1 would be expected to be the most viable category. A closer examination of Table 10.5 shows that the number of individuals of a given class recovered in this backcross is inversely proportional to the number of mutant genes this class carries. All the classes carrying *bd, y,* and *s* have few survivors. The gene *Ba* also depresses viability greatly, *pr* follows next, while *or* and *Cy* are relatively innocuous. The same mutant genes produce no drastic effects on the viability of pure species and of F_1 hybrids. The results indicate that the eggs deposited by F_1 hybrid females give individuals

TABLE 10.5
(See explanation in text.)

Class Number	Males Phenotype	Males Observed	Females Phenotype	Females Observed
1	bd y s Ba pr	2	Ba pr	41
2	bd y s Ba pr Cy	. . .	Ba pr Cy	32
3	bd y s Ba or	4	Ba or	92
4	bd y s pr	7	pr	190
5	bd y s Ba or Cy	1	Ba or Cy	89
6	bd y s pr Cy	7	pr Cy	372
7	bd y s or	14	or	140
8	bd y s or Cy	13	or Cy	336
9	or Cy	147		
10	or	143		
11	pr Cy	62		
12	Ba or Cy	17		
13	pr	58		
14	Ba or	21		
15	Ba pr Cy	14		
16	Ba pr	6		
	Crossovers 121		Crossovers 311	
	Total 637		Total 1603	

afflicted with a general constitutional weakness. Mutant genes that do not impair greatly the viability of pure species or of F_1 hybrids are semilethal in individuals developing from the eggs deposited by hybrid females.

To test this hypothesis further, experiments were so arranged that the class of backcross progeny identical in constitution with *D. pseudoobscura* (corresponding to class 1 in Table 10.5) was free from mutant genes, and the class having hybrid autosomes (corresponding to class 9 in Table 10.5) carried several mutants. The result was the opposite of that observed in the first experiment: class 9 was depressed in frequency, whereas individuals in class 1 survived.

Some diagrammatically clear cases of hybrid breakdown have also been described in species of cottons (Stephens 1949, 1950, and references therein). *Gossypium hirsutum, G. barbadense*, and *G. tomentosum* intercross easily and give fertile and vigorous F_1 hybrids. Only in the F_2 generation do types of low viability make their appearance. For example, among 110 F_2 seeds from the cross *G. hirsutum* var. *punctatum* × *G. tomentosum*, there were found:

7 seeds with small embryos that failed to germinate.
36 seeds with apparently normal embryos that failed to germinate.
9 seedlings that failed to expand their cotyledons.
22 seedlings that died within 3 weeks.
16 unthrifty seedlings at 3 weeks old.
20 strong seedlings at 3 weeks old.

Among many further examples that could be cited are the hybrids between *Zauschneria cana* and *Z. septentrionalis*, *Layia gaillardioides* and *L. hieracioides* (Clausen 1951), and *Lycopersicon esculentum* and *L. chilense* (Rick 1963a). The hybrids of Zauschneria and of Layia species are vigorous, perhaps even heterotic, show normal meiotic chromosome pairing, and are quite fertile. Yet among 2100 Zauschneria F_2 hybrids planted, only 250 survived, and even these were attacked by mildew and rust. Among the F_2 hybrids between Layia species many died early, and the survivors grew less rapidly than the parental species. This slow development is unadaptive in the environments of the parental species. In Clausen's words:

In *Layia*, lack of resistance against the drought of the California summer is compensated for by speedy development which enables the species to bloom early during the moist spring. The interchange of genes in the hybrid of *gaillardiodes* and *hieracioides* resulted in the development of lateness, a new character for *Layia* but an undesirable one for survival in the California climate, because lateness was not accompanied by development of protection against drought, as it is in other genera of the *Madiinae*.

CHAPTER ELEVEN

PATTERNS OF SPECIES FORMATION

Historical and Philosophical Antecedents

Man's reactions to the endless diversity and changeability of what he perceives are ambivalent. On the one hand, the diversity is esthetically enthralling. On the other, it overtaxes his memory and impedes setting his experience in order. The oldest and universal means for bridling the runaway multiformity of nature is human language. A word, a name, applies not to a single object but to an array of individually distinct entities. A need is felt, however, to reconcile the variegated experience with the semantic simplification. Parmenides (around 500 B.C.) made the earliest known and the most radical proposal—all variety and change are mere illusions, the true existence is one and immutable. Plato, a century later, erected a more sophisticated theory of immutable essences or archetypes (*Éidos*). As mentioned in Chapter 1 and elsewhere, what is real according to Plato are the eternal and changeless ideal Man, Horse, Pine, and Drosophila; the men, horses, pines and drosophilae that we actually see are only pale shadows of their perfect and ineffably beautiful *Éidos*.

Aristotle (fourth century B.C.) thought that the essences are not stored somewhere in heaven beyond human reach. They are embodied and expressed in the things we see and in the individuals we meet. It should be obvious, however, that to grasp the essence is a greater achievement than to behold its evanescent expression. Platonic and Aristotelian doctrines of archetypes or essences have strongly influenced medieval as well as some modern philosophies. Typological thinking has become habitual, not only among biologists and other scientists, but among the general public as well. Learned treatises are written about the one essential, unitary, and uniform Human Nature, which men always and everywhere are supposed to possess (a specimen of this genre is Jonas 1966). Opinions about this Nature are expressed casually in everyday conversation. Now, if there is an archetype for

Man, is every biological species also the embodiment of its own *Eidos*? The problem of species became embroiled in the philosophical dispute between the so-called realists and the nominalists; the former affirmed and the latter denied the reality of universal essences of Man, Horse, Dog, and other species. To a nominalist only individuals are real. A name applied to a group of individuals does not imply the existence of any supraindividual entity so designated.

In the eighteenth century, Linnaeus classified all living things (and minerals as well) into species, which he grouped into genera, orders, and classes. Linnaeus was an Aristotelian realist, and he maintained that biological species are real entities, whose essences were created by God. "There are," he stated, "as many species as produced in the beginning by the Infinite Being." Linnaeus was just as firmly convinced that genera are also primordial entities: "Every genus is natural, directly created at the beginning. Hence, it cannot be subdivided or combined gratuitously, or according to some theory." In contrast, he put no store in varieties: "A botanist need not bother about varieties. . . . There are as many varieties as are different plans grown from the seeds of the same species." This implies that the archetypes are more real than individual "varieties," which are, after all, only more or less successful imitations of the archetypes. According to this view, a biologist tries to descry the essence of a species or a genus through the murky exterior of individual variation. Probably no biologist will at present subscribe to these statements, and yet typology is implicit in the thinking of many practitioners of our science.

Darwin's goal was to demonstrate that "species are only strongly marked and permanent varieties, and that each species first existed as a variety." It is not surprising that at times he seemed to take a nominalist stance, seemingly denying the reality of species. Thus, in the concluding pages of "Origin of Species" we read: "In short, we shall have to treat species in the same manner as those naturalists treat genera, who admit that genera are merely artificial combinations made for convenience." Yet, on the very next page, Darwin states, "Our classifications will come to be, as far as they can be so made, genealogies; and will then truly give what may be called the plan of creation." Now, genealogies cannot be "made for convenience"; Darwin evidently does not argue that biological classifications are arbitrary; they reflect the objectively ascertainable clusters of organic forms, which Darwin ascribes to a community of descent. Classification by descent cannot be invented by biologists; it can only be discovered.

Taxonomic categories are arbitrary in a quite different sense. The number of supraspecific categories now used is fairly large: subgenus, genus, section, tribe, subfamily, family, superfamily, etc. The recognition that the units comprising a given complex of living forms are related by propinquity of descent is not arbitrary. The evaluation of a complex as a subgenus, genus, tribe, or family is a matter of convenience. Most of the Linnaean genera of insects are now treated as families. A classifier can exercise his choice, within the bounds of convenience and consistency. For example, a decade ago the genus Drosophila had some 750 described species in 8 subgenera (Wheeler 1959). One could just as well make these full genera and raise the species groups within them to the status of subgenera.

As another example, some paleoanthropologists have seen fit to give specific and generic names to every scrap of bone of fossil hominids discovered. The only telling argument against this practice is, however, that littering the scientific nomenclature with useless names breeds confusion. Many authors now assume only two hominid genera: Australopithecus with two species, *Australopithecus africanus* and *A. robustus,* and Homo also with two species, *Homo erectus* and *H. sapiens* (e.g., Campbell 1966). Yet Robinson (1967) puts the species *africanus* in the genus Homo, and *robustus* in the genus Paranthropus. What kind of evidence will decide which genera are the valid ones? Perhaps finding more fossils will indicate which classification is more convenient.

Emergence of the Biological Species Concept

A taxon is defined by Mayr (1969) as "a taxonomic group of any rank that is sufficiently distinct to be worthy of being assigned to a definite category." We have seen that Linnaeus regarded his taxa, species as well as genera, as primordially created entities. Darwin preferred to view species and genera as "artificial combinations made for convenience." At present, species are considered biologically more meaningful entities than genera. A species is, of course, a taxon, like a genus or a family; but, more importantly, a species is a supraindividual biological system. In the latter sense, a species is more than a group concept. A species is composed of individuals as an individual is composed of cells, or as a termite or an ant colony is composed of fertile and sterile members. A biological species is an inclusive Mendelian population; it is integrated by the bonds of sexual reproduction and parentage. By contrast, the species of a genus or the genera of a family or the components of other taxa have no such bonds, although they are related by common descent, usually many generations back.

The biological species concept is a product of modern understanding of the genetic structure of Mendelian populations. Nevertheless, it was foreshadowed before Darwin and even before Linnaeus. Thus John Ray wrote in 1686:

> After a long and considerable investigation, no surer criterion for determining species has occurred to me than the distinguishing features that perpetuate themselves in propagation from seed. Thus, no matter what variations occur in the individual or the species, if they spring from the seed of one and the same plant, they are accidental variations and not such as to distinguish a species . . . (quoted in Mayr 1963).

Cuvier in 1815 defined a species as "the reunion of individuals descended from one another, or from common parents, or from such as resemble them as strongly as they resemble each other." Similar ideas were expressed by several other early authors who were not evolutionists (see Mayr 1957, 1969, Greene 1961, and Zavadsky 1968 for references). Recognition that a species is not only a taxon but also a reproductive community is not incompatible with a belief that it is a manifestation of its unchanging *Éidos.*

Perhaps the most compelling evidence that species evolve from races or "varieties" was obtained in studies on the variation of species in space. Although Linnaeus had in his collections some animals and plants from remote countries, most of his materials came from his native Sweden. He worked mainly on sympatric rather than allopatric forms of life. Now, sympatric species of sexually reproducing organisms are discrete breeding communities, and usually these are also discrete in their outward characteristics. A student of the fauna or the flora of a single reasonably small territory generally encounters few difficulties in delimiting species.

With allopatric forms, however, problems arise. The eighteenth and nineteenth centuries saw rapid progress in the geographic exploration of the world. Biological museums received materials from diverse countries, and zoologists and botanists had to describe and to classify them. The problem of how to distinguish divergent races from closely related species sometimes defied solution, since races that inhabit remote territories may be about as distinct as sympatric species. Imagine an extraterrestrial zoologist who has seen two specimens each of Swedes, Bushmen, and Eskimos. He might well conclude that they represent three different species of primates. We know that this is not so, because we are familiar with numerous geographically and structurally intermediate populations, all of which exchange genes freely with at least their geographic neighbors. In any case, the typological and nonevolutionary species concept of the pioneer taxonomists floundered when it had to face the phenomenon of geographic races. The discomfiture of the taxonomists proved, however, a boon to the biologists.

Lamarck, who had ample experience in classifying plants and animals, saw that new evidence called for a new explanation. By the time of Darwin the evidence had grown to be overwhelming. The new paradigm was an ancestral species transforming into a derived one, or splitting into two or more derived ones. Darwin, although he was quite familiar with the phenomenon of geographic races, did not differentiate them from nongeographic "varieties" and polymorphisms. It remained for Wagner (1889), K. Jordan (1905), D. S. Jordan (1905), Semenov-Tian-Shansky (1910), Rensch (1929, 1960a), and Mayr (1942, 1963) to develop the theory of allopatric species formation and of polytypic species. Geographic isolation of allopatric races or subspecies is a usual, or even necessary, antecedent of species formation. The views of the early authors (up to Rensch 1929, but not Rensch 1960a!) were tinged with Lamarckism: allopatric populations become genetically different because they are changed by the different environments in the countries they inhabit. It is only fair to note that, while genetics was only groping for its fundamental concepts, the view that organisms are changed by their environments represented no more than a restatement in ambiguous terms of the observed facts.

We saw in Chapter 9 that most animal and plant species are complexes of local populations which differ from each other genetically to some extent (see Mayr 1963 and Grant 1963 for a more thorough treatment of this topic). Only some relictual species confined to very small territories are likely to be single panmictic populations. The difference between the populations may be sufficiently large to make most individuals recognizable as belonging to a certain population or to a group of populations. A taxonomist may then give them racial or subspecific names. Species that comprise two or more subspecies are polytypic; mankind is a prime example of a polytypic species. No race or subspecies of a polytypic species represents the essence or the archetype of that species (although the subspecies named first is technically known as the typical or the nominate subspecies).

The realization that many or most species are polytypic led to a better understanding of the biological nature of species, as well as to considerable simplification of taxonomy. At the beginning of the current century, some taxonomists succumbed to the temptation of assigning species names to every local race distinct enough for most specimens of it to receive determination labels. This occurred mainly in well studied groups, such as mammals, birds, and some genera of insects, in which most species had already been described, prompting specialists to overestimate intraspecific differences. The pandemonium of specific and generic splitting in paleoanthropology has been mentioned. A salutary reaction set in with the introduction of the polytypic species concept (see Mayr 1942, 1963, and 1969 for particulars). For example, the check list of birds had in 1910 some 19,000 species; although some additional ones have been described since, the number of recognized species has been reduced to about 8600.

Races of sexual and normally outbreeding organisms remain genetically distinct because they are usually allopatric; geographic isolation keeps the gene exchange infrequent enough to prevent swamping of the interpopulational differences. The example of human races attests to what happens when races become sympatric—they tend to merge into a single, variable population. In contrast to races, species are able to maintain their genetic integrity despite sympatric coexistence. For a long time biologists groped for an explanation of this fact.

Reference was made in Chapter 10 to the widespread but erroneous belief that viable and fertile hybrids result only from intraspecific, not from interspecific, crosses. Also, some entomologists assumed that differences in the genitalia mark species but not races. There are so many exceptions to these rules that they cannot be used as definitions of species. Lotsy (1931) regarded the "synagameon," which he defined as "a habitually interbreeding community of individuals," as the fundamental unit. This view has the merit of directing attention to Mendelian populations as biological realities, but the definition does not differentiate between local populations or demes and species.

Dobzhansky (1937a) pointed out that the process of species formation, in contrast to race formation, involves the development of reproductive isolating mechanisms. An ancestral species is transformed into two or more derived species when an array of interbreeding Mendelian populations becomes segregated into two or more reproductively isolated arrays. Species are, accordingly, systems of populations; the gene exchange between these systems is limited or prevented in nature by a reproductive isolating mechanism or perhaps by a combination of several such mechanisms. In short, a species is the most inclusive Mendelian population. Mayr (1969) has rephrased the definition thus: "Species are groups of interbreeding natural populations that are reproductively isolated from other such groups." Grant's (1963) definition is as follows: "The sum total of the races that interbreed frequently or occasionally with one another, and that intergrade more or less continuously in their phenotypic characters, is the species."

Some Difficulties of the Biological Species Concept

The biological species concept expressed in these variant definitions is accepted by many biologists and criticized by others. The perennial controversy about species continues. Consideration of some of the objections is in order here.

First of all, it is sheer miscomprehension to allege that our species criterion is "intersterility." This term is an ambiguous locution, which may mean hybrid inviability, hybrid sterility, or both. Reproductive isolation subsumes these as well as the other isolating mechanisms discussed in Chapter 10. What is more, lack of "intersterility" in captivity or in an experimental garden does not rule out the possibility that reproductive isolation may be present in nature.

Another objection raised is that to talk about reproductive isolation is meaningless in asexual, parthenogenetic, or obligatorily self-pollinating forms, and yet systematists name species everywhere. Of course, the criterion of reproductive isolation is applicable only where there are Mendelian populations. Rejection of the species criterion on this ground, however, overlooks the duality of the species concept.

Species is not only a category of classification but also a form of supra-individual biological integration. In the former sense, any taxon may be called a species if it is convenient to use this designation. Species then become as arbitrary as subgenera, genera, and other categories. The virtue of the biological species concept is precisely that it makes the species a category that betokens a biologically highly significant fact. Having achieved reproductive isolation, a Mendelian population henceforth becomes a biological system evolving independently from other such systems.

Some logicians (e.g., Gregg 1954) see no difference between biological taxonomy and the classification of inanimate objects. This is a refusal to take evolution seriously; organisms resemble each other because they are descendants of common ancestors. This statement is not applicable to books or automobiles, except as a loose metaphor. When applied to species of sexually reproducing forms of life, extreme nominalism becomes ludicrous. The species mankind is not an invention of a taxonomist but a biological (as well as sociological and existential) reality. The development in recent years of phenetic, numerical, or computer taxonomies has led to lively disputes, some of which are tangential to the species problem (Sokal and Sneath 1963, Sokal 1965, Sokal and Camin 1965, Mayr 1965a,b, 1969, and others). In brief, phenetic taxonomists classify organisms according to their overall similarities rather than their descent (phylogeny). Modern computers permit what otherwise would be prohibitively laborious calculations of the degree of similarity of taxa with respect to as many of their characteristics as possible. The resulting "phenograms" look like classical phylogenetic trees, but show statistical estimates of the numbers of similarities and differences in the selected characters between any two taxa chosen for study. Critics have pointed out that treating randomly selected characters as equivalent is liable to result in misestimating the genetic similarities of the organisms classified. As Mayr (1969) caustically remarks, "Some users of electronic data processing have suggested that thinking and theory become unnecessary if we merely entrust our fate to the computer." Moreover, sibling species (see the next section) are not perceived by numerical taxonomies, whereas some race differences may loom unduly large.

It is not true that species defined as reproductively isolated Mendelian populations are operationally unusable because working taxonomists can only rarely obtain the pertinent information. Examination of the morphology and the geographic origin of an adequate sample of specimens of a given kind of animals or plants yields indirect but usually reliable evidence concerning the genetic limits of species populations. Classical taxonomists have grasped the existence of biological species intuitively, but correctly in a majority of cases. Moreover, the kinds of animals and plants that some preliterate peoples recognize as distinct, and on which they bestow names, quite often correspond to what zoologists and botanists recognize as species (trees in Amazonian forests, according to Pires, Dobzhansky, and Black 1953, and vertebrates in New Guinea, according to Diamond 1966). To be sure, other primitive peoples are less perspicacious taxonomists (Berlin, Breedlove, and Raven 1966).

The definition of species as reproductively isolated groups of populations is not intended to provide an infallible yardstick, which would always indicate whether two samples of specimens represent one or more species or only races. Rather, the value of this definition lies in its substitution of analytical judgments for less communicable intuitions. The incipient species, the species in *statu nascendi*, will always involve a residue of borderline cases for which the decision will be arbitrary (see below). This difficulty comes from the species being not fixed but evolving. The nonexistence of borderline cases could mean only that evolution has run its course and is no longer happening.

The approaches to recognition of sympatric species are different from those used with allopatric ones, as discussed in detail by Mayr (1942, 1963), Stebbins (1950), and Grant (1963). Two or more Mendelian populations can be sympatric, and can coexist indefinitely in the same territory, only if they are reproductively isolated, at least to the extent that the gene exchange between them is kept under control by natural selection. The genetic gaps between sympatric species are, as a rule, absolute. Since genetic differences are usually reflected in the morphology, greater or lesser morphological hiatus is usually found between species. "If a taxonomist receives a series of specimens from a particular locality, he is almost never in doubt as to whether they belong to one or to several species."

By contrast, "the gaps between allopatric species are often gradual and relative, as they should be, on the basis of the principle of geographic speciation" (Mayr). Allopatric populations do not directly exchange genes, simply because they are allopatric. Whether such populations are also isolated reproductively, so that they could maintain their genetic differences if they were to become sympatric, is sometimes a moot point. The problem can often be resolved by observing that the populations in question are united by a continuous chain of intermediate populations in the geographically intervening localities. The presence of such a chain of intermediate populations is prima facie evidence that gene exchange is possible between the populations by diffusion through the intervening space. This is clearly the case with races of the human species.

Is the criterion of reproductive isolation applicable to forms that were not contemporaneous? Obviously, nobody can make hybridization experiments of *Australopithecus africanus* with *Homo erectus*, and of *H. erectus* with *H. sapiens*. Simpson (1943, 1961) has rightly called this "only a pseudoproblem." His admirably lucid analysis can best be stated in his own words: "A taxonomic species is an inference as to the most probable limits of the morphological species from which a given series of specimens has been drawn." A morphological species, which in turn is an inference as to the most probable limits of the biological (genetic) species, is "a group of individuals that resemble each other in most of their visible characters, sex for sex and variety for variety, and such that adjacent local populations within the group differ only in variable characters that intergrade marginally." Species succeeding each other in time "should be so defined as to make the morphological difference between them at least as great as sequential differences among contemporaneous species of the same group or closely allied groups."

It is not true that the interbreeding versus reproductive isolation criteria are never applicable to fossil forms. We are reasonably certain that the Neanderthal man was a race of *Homo sapiens*, and not a separate species. One reason for this belief is that *H. sapiens neanderthalensis* varied in space as well as in time, and some variants were intermediate between the "classic" Neanderthal and *H. sapiens sapiens*. Also the population that left its remains on Mount Carmel in Palestine is like populations regularly formed in modern species where the territory of one subspecies abuts on that of another. It is rather misleading to call such populations hybrids between subspecies (see Mayr 1963 for a discussion of primary and secondary intergradation).

Gene Differences between Species

De Vries, the founder of the mutation theory, believed that a simple mutation gives rise to a new species (see Chapter 2). Goldschmidt claimed a special category, systemic mutations, which generate not only species but genera and families as well; Lamprecht maintained that there exist special genes differentiating species, and others responsible for intraspecific variation (see Chapter 3). These views have few or no adherents at present. The only known kind of mutation that may at once bring a new species into being is a doubling of the chromosomal complement in a hybrid of two pre-existing species (allopolyploidy; see below). Otherwise genetic differences between species are compounded gradually of many genic and chromosomal alterations, each change having arisen ultimately by mutation.

Just how many gene and chromosome changes differentiate species is difficult to determine. The classical Mendelian method of observing segregation in progenies of crosses has its applicability severely limited by the inviability or sterility of most interspecific hybrids. Where F_2 hybrids between species can be obtained, the variability may be so great that any two individuals are visibly different. For example, Baur (1930) studied the hybrids between species of snapdragons *Antirrhinum majus* and *A. molle*. The F_1 hybrids are, on the whole, intermediate between the parents and no more variable than the latter. In the F_2, however, the variability is spectacular. Most individuals show various recombinations of the parental traits, but few or none can be mistaken for pure *A. majus* or pure *A. molle*. Some individuals possess characteristics present in neither parent, but found in other species of Antirrhinum or other genera of the family Scrophulariaceae. One such segregant was described as a "new species"—*A. rhinanthoides*, because it had certain attributes of the genus Rhinanthus. Baur estimated the number of gene differences between *A. majus* and *A. molle* as more than one hundred. Results similar in principle were obtained by Honing in species of Canna, Wickler in carnations (Dianthus), East in tobacco (Nicotiana), and Clausen in violets (Viola).

The early literature has been reviewed by Renner (1929). Among the more recent works on the genetic analysis of segregations in F_2 of sterile interspecific hybrids in plants, those of Gajewski (1957) on

Geum, Grant (1946) on Gilia, Harland (1936) on cottons (Gossypium), Stubbe (1940) and Mather and Vines (1951) on Antirrihinum, and Rick and Smith (1953) and Tal (1967) on Lycopersicum and Solanum, must be mentioned. The species differences are more or less highly polygenic.

By and large, morphologically distinct species give fertile hybrids less often in animals than in plants. Perhaps this is the reason why animal species proved rather more refractory to genetic analysis. Morgan (1919) wrote, "The slightest familiarity with wild species will suffice to convince any one that they differ from each other generally, not by a single Mendelian difference, but by a number of small differences." Nevertheless, Drosophila geneticists liked to assume that "in general, related species have essentially the same complements of genes" (Sturtevant 1948). The evidence against this assumption accumulated only gradually. *Drosophila melanogaster* is outwardly very close to *D. simulans*. Nevertheless, their hybrids are sterile, and one or the other sex, depending on the direction of the cross, is inviable (Sturtevant 1920–21). Using a sophisticated genetic technique, Pontecorvo (1943) estimated that no fewer than nine genes must be responsible for this inviability. Similarly, *D. pseudoobscura* and *D. persimilis* are morphologically almost indistinguishable, and yet the F_1 hybrid males are sterile and the backcross progenies suffer hybrid breakdown (see Chapter 10). It is evident that the outward similarity of these species pairs is underpinned by rather different systems of genes. This is what Harland (1936) meant by his dictum, "The modifiers really constitute the species."

Irwin and his colleagues have made notable studies of the genetic differences between species of pigeons and doves (Columbidae) in regard to the cellular antigens of their blood corpuscles (Irwin 1953, Irwin and Cumley 1943, Stimpfling and Irwin 1960, and references therein). Some of the species can be crossed and produce fertile hybrids. Comparison of the parental species and their F_1 hybrids shows that a certain proportion of the antigens are shared by both parents, whereas others occur in one species only. The F_1 hybrids usually have all the antigens present in both parents, and rarely new hybrid antigens, the latter being interaction products of the parental ones (Irwin 1966). By backcrossing the hybrids to the parental species, it is possible to isolate and identify the antigens that differentiate the species. It is not possible to determine by this technique how many antigenic substances are shared by the different species, since the genes determining these substances do not segregate in the hybrid offspring. However, fairly large numbers of differences are detected. Thus, nine antigenic characters distinguish the pearneck and ring doves (*Streptopelia chinensis* and *S. risoria*). Only rarely have any of these antigens been found in the other 23 species studied.

The genetic study of species differences entered a new phase when the technique of protein discrimination by electrophoretic mobilities became available (see Chapter 7). This overcomes, at least to some extent, the most serious limitation of the methodology of Mendelian genetics—the taxa compared need not be crossable and capable of giving fertile hybrids. With this technique, Duke and Glassman (1968) compared the enzyme xanthine dehydrogenase in 29 species belonging to 9 species group of the genus Drosophila. The electrophoretic mobilities of the enzyme are distinguishably different in most species; if the mobility in *D. melanogaster* is taken as 100, the enzyme mobilities in other species vary from 57 to 102. What is most remarkable is that the species considered to be related on morphological grounds had enzyme mobilities more similar than the less closely related ones.

The usefulness of protein discrimination studies by means of the relatively simple technique of electrophoresis can be pushed even further. If certain assumptions stated in Chapter 7 are granted, the proportions of the genes represented by similar and by different alleles in the forms examined can be estimated. Hubby and Throckmorton (1965) used this technique to compare 10 species of the *virilis* group of Drosophila. This is a compact array of related species in which, on the basis of cytological evidence (Stone, Guest, and Wilson 1960), two "phylads" of, respectively, 4 and 6 still more closely related species can be distinguished. A summary of the results is shown in Table 11.1. On the average, some 36 proteins have been examined per species. From 2.6 to 28 percent of these were "unique," that is, found in only a single species. The average proportion of such unique proteins in the species studied was about 14 percent. Between 17 and 38 percent of the proteins in a species are shared with the other species of the same phylad, but not with those of the other phylad. And, finally, 43–76 percent are "ancestral" proteins, which are present in species of both phylads.

In a more recent paper, Hubby and Throckmorton (1968) analyzed 9 groups of 3 species each, that is, 27 species, of Drosophila. In each

TABLE 11.1

Numbers of proteins studied in ten species of the virilis group of Drosophila, and percentages of these proteins unique to a given species, restricted to the phylad, and common to the species group (After Hubby and Throckmorton 1965)

Phylad and Species	Number Studied	Percentage Unique to Species	Percentage Common to Phylad	Percentage Common to Group
Virilis phylad				
D. americana	38	5.3	23.7	71.1
D. texana	42	21.4	16.7	61.9
D. novamexicana	38	7.9	21.1	71.1
D. virilis	38	2.6	21.1	76.3
Average for phylad	39.0	9.3	20.7	70.1
Montana phylad				
D. littoralis	39	28.2	25.6	46.2
D. ezoana	35	8.6	29.7	65.7
D. montana	37	18.9	37.8	43.2
D. lacicola	29	20.7	20.7	58.6
D. borealis	42	19.0	28.6	52.4
D. flavomontana	29	10.3	37.9	51.7
Average for phylad	35.2	17.6	29.4	53.0
Grand average	36.6	14.3	25.9	59.8

triplet of species, two are morphologically scarcely distinguishable sibling species, and the third is clearly different but still a relative of the first two. The members of the different triplets are still more distinct; in fact, some of them belong to different subgenera of Drosophila. Remarkably enough, the pairs of sibling species have, on the average, only about 50 percent of their proteins in common, despite their external similarities being so great that a museum taxonomist would find it difficult to distinguish them at all. Moreover, the percentages of proteins shared in pairs of sibling species range from a high of 86 (*D. victoria* and *D. lebanonensis*) to a low of only 23 (*D. willistoni* and *D. paulistorum*). The members of a triplet (i.e., the two siblings and the related nonsibling) share, on the average, about 11.6 percent of the proteins. The authors "interpret these results to indicate that speciation does not *require* a change in a large number of loci."

I believe that their findings warrant the opposite conclusion. On the assumptions that a Drosophila has at least 10,000 gene pairs (Chapter 3), and that the genetic differences detected in the proteins are a fair sample of all gene differences (Chapter 7), the sibling species must differ in thousands of genes. Whether there are distinct reproductively isolated species differing in really small numbers of loci remains to be discovered; certainly the sibling species studied by Hubby and Throckmorton could have been regarded as candidates for such close genic similarity, and yet they proved to have many different genes.

Species as Genetic Systems

There are two ways of looking at the genetic architecture of species differences. First, the genes in which species differ may act in development largely independently of one another. The species difference is then an aggregate or a conglomeration of gene differences. Second, the genes may also interact and cooperate in such ways that development is an emergent product of their actions. The genotypes of different species are, then, organized systems or patterns. In the first case, it is tempting to compare the genes with solo players, and in the second with members of a symphonic orchestra. The two possibilities are evidently not mutually exclusive. The first may be realized more often in lower organisms and the second in higher ones, or the first in plants and the second in animals. Finally, there may be all gradations between gene aggregations and gene systems. Some evidence of gene patterning has already been mentioned. Members of sibling species pairs, such as *Drosophila melanogaster* and *D. simulans,* or *D. pseudoobscura* and *D. persimilis,* seem to be much the same in their morphological and physiological phenotypes. Yet recombinations of their genes produces hybrid breakdown and developmental disharmonies. On the other hand, the hybrid swarms found where some plant species hybridize sometimes consist of individuals apparently as fit as the parental species. Of course, it is possible that the unfavorable combinations of parental genes have been eliminated.

Hollingshead showed as early as 1930 that some strains of *Crepis tectorum* carry a dominant gene that produces no visible effects in the pure species; if, however, a hybrid between *C. tectorum* and *C. capillaris* carries this gene, it does not develop beyond the cotyledon stage. Accordingly, the crosses in which the *C. tectorum* parent is homozygous for the gene in question produce no viable seedlings, whereas 50 percent, or 100 percent, of such seedlings occur in cultures in which the gene is heterozygous or absent. The same gene is lethal for seedlings of the hybrids *C. tectorum* × *C. leontodontoides* and *C. tectorum* × *C.*

bursifolia, but not in the crosses *C. tectorum* × *C. setosa* and *C. tectorum* × *C. taraxacifolia.* The isolation between *C. tectorum* and certain of its congeners would become complete if *C. tectorum* were homozygous for the gene that is lethal in the hybrids.

A remarkably parallel situation, but one that occurs in animal rather than plant species, has been extensively studied by Gordon (1948, 1951, and Gordon and Rosen 1951). Natural populations of the platyfish, *Xiphophorus maculatus,* are often polymorphic. The dominant gene *Sp* produces a spotted pattern consisting of macromelanophores; *N* gives a broad black band on the flanks of the fish; *Sr* forms a series of horizontal lines; *Sd* gives dark spots on the dorsal fin; *Sb* causes a darkening of the ventral parts. If, however, strains of *X. maculatus* carrying any of these genes are crossed to the swordtail, *X. helleri,* the effects of the genes in the hybrids are greatly hypertrophied. The gene *Sp* initiates the development of cutaneous melanomas; *N* gives melanotic tumors anywhere along the black band on the side of the body; *Sd* causes melanotic tumors on the dorsal fin; *Sb* gives melanomas along the midventral line. The effects of *Sr* are exaggerated in F_1 hybrids but no tumors appear; if, however, the F_1 is backcrossed to the swordtail, some individuals in the backcross progeny develop tumors along the flanks. It is evident that the swordtail carries genes which interact with certain apparently useful platyfish genes in such a manner as to make the latter semilethal in the hybrids.

When certain strains of the cotton *Gossypium barbadense* are crossed to some strains of *G. hirsutum,* the F_1 hybrids are weak, have shortened internodes, and have the stem, petiole, and leaf midribs covered with a layer of cork. Hybrids between other strains of the same species, however, are vigorous and free from the "Corky" syndrome. Stephens (1946) has shown that the strains producing Corky hybrids carry complementary alleles, or complementary genes, ck^x (in *G. hirsutum*) and ck^y (in *G. barbadense*). The Corky syndrome is due to the simultaneous presence in the genotype of a plant of these complementary genes. The geographic distribution of the ck^x and ck^y alleles is very interesting. The species areas of *G. barbadense* and *G. hirsutum* overlap in the West Indies and on the northern fringe of South America. The genes that give the Corky syndrome occur almost exclusively in strains of the two species in the zone of the overlap. The Corky hybrids are poorly viable and rarely give F_2 generations in nature. When an F_2 generation is obtained from non-Corky F_1 hybrids, the F_2 hybrids are

TABLE 11.2

Percentages of germinable seeds that developed into vigorous individuals, and of vigorous individuals that were fertile, in different generations of hybrids of Gilia malior and G. modocensis (After Grant 1966a)

Generation	Seeds Giving Vigorous Individuals	Fertility Among Vigorous Individuals
F_1	100	±0
F_2	18	±0
F_3	24	45
F_4	25	83
F_5	22	66
F_6	37	71
F_7	80	96
F_8	97	100
F_9	100	100

deficient in vigor. Thus, the Corky condition eliminates the F_1 hybrids and prevents the production of a degenerate F_2. Gerstel (1954) found a "red lethal" gene, Rl_a, which is viable in *G. hirsutum* but lethal in hybrids with another species, *G. arboreum.*

Rick (1963a) found what he calls "differential zygote lethality" in hybrids of tomato species *Lycopersicon esculentum* and *L. chilense.* The hybrid of these species is fertile, so that F_2 and backcross progenies are obtained. In these progenies, a pronounced differential survival is observed, which favors the constellations of the genes of the parental species and discriminates against mixtures and recombinations.

Grant (1966a, b) combined natural and artificial selection for vigor and fertility in the hybrids between two normally self-pollinating species of desert plants, *Gilia malior* and *G. modocensis.* The artificially obtained F_1 hybrids were somatically vigorous but almost completely sterile. The progress of the selection is shown in Table 11-2. Although the selection was quite successful, it is evident that only a minority of the recombinations of the parental genes gave rise to reasonably harmonious genetic systems.

Incipient Species and the Borderline between Race and Species

Species evolve from races by the accumulation of genetic changes. If the foregoing statement is true, and if the divergence due to accumulation of the gene differences is a gradual process, then instances must be found (and they are found) in which two or more races have diverged so much as to approach, but not to attain completely, the status of reproductively isolated species. The gradualness of the divergence could not be postulated a priori. Goldschmidt (1940) believed that species do not evolve from races but arise through sudden "systemic" mutations. Once the process of speciation is fully consummated, species can no longer be mistaken for races. Groups of Mendelian populations can be recognized as distinct species, or as subspecies of one species, usually without hesitation, provided that sufficient material for study is available. It is for this reason that the category of species has shown a remarkable stability from the time of Linnaeus to our day. The cat and the lion, the horse and the ass, the Norway rat and the black rat belong to different species. The Siamese and the alley cat, the Arabian charger and the draft horse, the maize of Iowa and that of Mexico are distinct races and not distinct species. The same is true of the human races. The claim that there is more than a single living species of Homo can be treated only as an eccentricity or the manifestation of race prejudice.

The reason why biologists spend more time discussing doubtful borderline cases than undoubted species and undoubted races is not that the former are very common. It is rather that borderline cases are interesting to evolutionists: the relative rarity of such instances indicates that, although the process of divergence is a gradual one, speciation in the strict sense, that is, the development of reproductive isolation, is a crisis that is passed relatively rapidly.

The borderline cases recorded in the literature have been so thoroughly examined by Mayr (1963, particularly Chapters 11-16) for animals, and by Grant (1963, Chapters 12-16) for plants, that an attempt to review them here would be supererogatory. Perhaps the most striking are the "rings of races." Sympatric populations that share the same territory without gene exchange or intergradation are distinct species. Yet in some cases they are found to be united by a chain of allopatric races that grade into each other and into the extreme members of the series. Although the terminal links of the chain behave as reproductively isolated species, a gene flow through the connecting links is at least potentially possible.

The salamander *Ensatina eschscholtzi* lives in the mountains encircling the central valley of California, but not in the valley itself.

Six subspecies replace each other along the ring. Although some of these differ quite strikingly in coloration and other traits, the transitions between the subspecies in the intermediate localities are quite gradual. The genes of one subspecies obviously diffuse into the neighboring ones. In the mountains of southern California, however, the most distinct subspecies (*eschscholtzi, croceator,* and *klauberi*) meet without intergradation. The populations behave in southern California as full-fledged species, and yet they are connected by the populations living to the north. They could exchange genes, though not directly but via a circuitous route through the northern subspecies (Stebbins 1949, 1957).

Superspecies and Semispecies

Mayr (1963, 1960) and Amadon (1966) define a superspecies as "a monophyletic group of closely related and largely or entirely allopatric species" or as "a group of entirely or essentially allopatric taxa that were once races of a single species but which now have achieved species status." The components of a superspecies are semispecies or allospecies. Semispecies are "populations which have part way completed the process of speciation. Gene exchange is still possible among semispecies, but not as freely as among conspecific populations" (Mayr 1963). One of the examples of superspecies and semispecies given by Mayr consists of the paradise magpies (Astrapia) of New Guinea; the semispecies of these magnificently colored birds differ strikingly from one another, and yet hybridize where their distribution areas come in contact.

The superspecies *Drosophila paulistorum,* mentioned in Chapter 10, presents a situation different from Astrapia in interesting ways. Dobzhansky and Spassky (1959) first noticed that strains classified as *D. paulistorum* from different parts of South America belong to several groups, which are now considered incipient species or semispecies (Dobzhansky and Pavlovsky 1967 and Dobzhansky, Pavlovsky, and Ehrman 1969 for further references). Crosses between the semispecies occur with difficulty, because of more or less strong ethological isolation; even when the females have no "choice," being confined with males of a different semispecies, most of them remain virgins until they die of old age, although the males continue to court them diligently. The few females that are inseminated produce, however, vigorous hybrid progeny of fertile females and sterile males. As mentioned in Chapter 10, the sterility, at least in the backcrosses, is caused by peculiar predetermination of the cytoplasm by the maternal chromosome complement before meiosis (Ehrman 1960). The genetic basis of the ethological isolation is quite different and not so unusual (Ehrman 1961). Apparently numerous polygenes are involved; the sexual acceptability of an individual, whether female or male, is a function of what semispecies most of its genes came from, and after repeated backcrosses to a particular semispecies the hybrid progeny behaves like this semispecies.

All semispecies are chromosomally highly polymorphic; Kastritsis (1967, 1969b) found 89 different inversions among 115 strains studied. Some inversion polymorphisms are shared by more than a single semispecies, whereas others are restricted to only one. Some inversions are homozygous in, and hence diagnostic of, a given semispecies.

The geographic distribution of the semispecies is shown in Fig. 11.1. Each inhabits an area of its own, but in some places their areas overlap. When two or even three semispecies are sympatric (and they have been attracted to the same banana bait), they rarely if ever cross, and thus behave like full-fledged species. Probably the best evidence of this has been adduced by Malogolowkin, Solima, and Levene (1965) and by Ehrman (1965). The map in Fig. 11.1 shows that the Andean-Brazilian semispecies occupies the most extensive territory in which no other semispecies occurs. On the contrary, the Amazonian and Orinocan semispecies more often occur together, and also with the Andean-Brazilian. One may expect (and the expectation is experimentally verified) that the Andean-Brazilian females will be inseminated by Amazonian and Orinocan males more easily than will the Amazonian or Orinocan females by Andean-Brazilian males.

Ehrman recorded the coefficients of ethological isolation obtained when sympatric strains of two semispecies are placed together; she compared them with the isolation observed between strains of the same semispecies but of allopatric origin. The average isolation coefficient for sympatric strains turned out to be 0.85, and between allopatric strains of the same semispecies 0.67. This can mean only that the pressure of natural selection maintaining the isolation is greater where the populations are exposed to the risk of hybridization than where they are not.

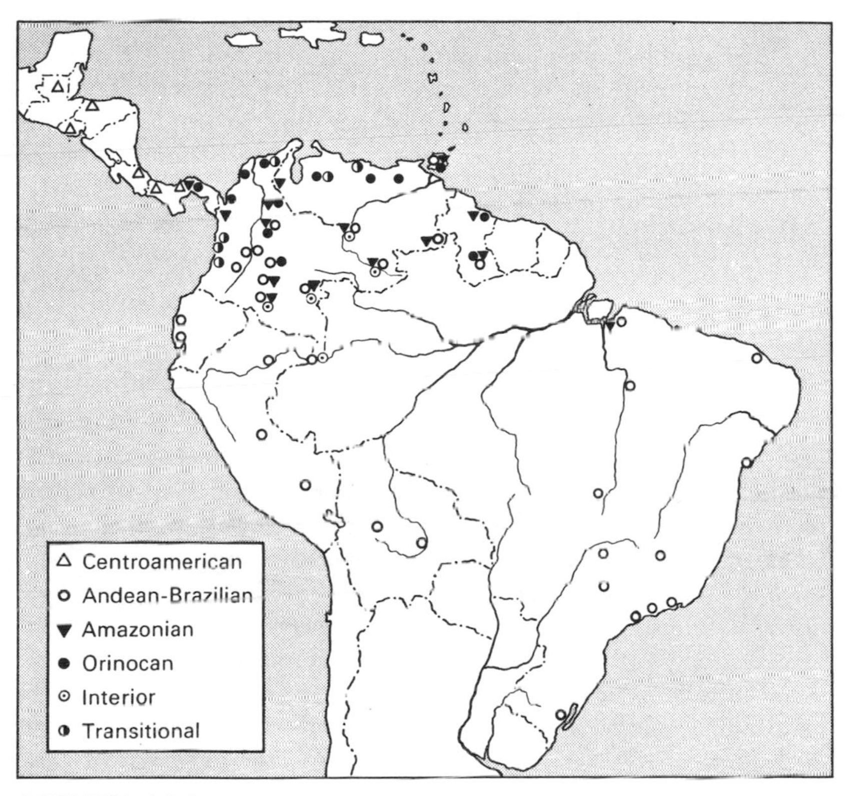

FIGURE 11.1
Known geographic distribution of the semispecies that compose the superspecies Drosophila paulistorum.

In western Colombia and northern Venezuela "Transitional" populations are found (Fig. 11.1.). Different strains, even from the same locality, exhibit a variety of behaviors; some of them, but not all, cross easily and give fertile hybrids with Centro-American or with Andean-Brazilian strains. On the other hand, some Transitional strains show appreciable ethological isolation from each other, and some crosses produce sterile F_1 hybrid males. Both genetical (Dobzhansky, Pavlovsky, and Ehrman 1969) and cytological (Kastritsis 1969b) evidence militates against the supposition that the Transitional is a hybrid of other semispecies; it is more likely to be the survivor of the primitive or ancestral *Drosophila paulistorum*. Because of the fertility of the hybrid females, the possibility of some gene flow between the semispecies nevertheless cannot be ruled out entirely. Furthermore, the findings of Ehrman and Williamson referred to in Chapter 10 suggest that the sterility of the male hybrids may, in a sense, be extraneous to the genotypes of these flies and due, in part, to different symbiotic microorganisms. The same can be inferred from the spontaneous origin of sterility of hybrid males in strains that were formerly fertile (Dobzhansky and Pavlovsky 1967). *Drosophila paulistorum* is, then, an example of a group of species still in *statu nascendi.*

Sibling Species

It is probably a general rule that species differ in numerous genes. Some of these gene differences manifest themselves outwardly in such traits as colors and patterns, proportions and sizes, of various body parts. Others change the physiological, ecological, and behavioral characters of their carriers. Some mutational changes may, perhaps, involve nucleotide substitutions the gene products of which have identical effects in development. What proportions of the gene differences belong to these various classes we do not know. What we do know is that in some groups of organisms (such as pheasants, birds of paradise, and some butterflies) apparently closely related species are spectacularly heterogeneous to our eyes. The judgment of close relationship rests here on the facility with which these species hybridize in experiments or in nature, or on the species having geographic distributions that suggest a subspecific or semispecific (see the preceding section) status. At the opposite extreme, there are species that differ only in some recondite details (such as minor differences in the genitalia among insect species) or that appear altogether identical. These are sibling species—reproductively isolated arrays of populations that show little or no morphological distinctions. Sibling species are not necessarily the same as semispecies or incipient species. Some siblings may have completed the process of speciation and become reproductively isolated without acquiring differences easily apprehended by outward appearance.

Only the recognition of sibling species resolved the longstanding puzzle of why malaria is endemic in some European and Mediterranean countries and yet absent in others where the proven vector, *Anopheles maculipennis,* is commonly found. The simple solution is that at least six sibling species were confused under the name *A. maculipennis*. Whereas some of them feed by preference, or at least occasionally, on man, others are "zoophilous," feeding on other animals. Careful investigation disclosed that, although the adult insects are very hard to distinguish as to species, there are fairly reliable diagnostic traits in the color patterns of the eggs and in the manner in which the eggs are put together in the "egg floats." In addition, the various species differ in ecological preferences—fresh or brackish, flowing or stagnant waters—as well as in the gene arrangements in their chromosomes, in mating habits, and in geographic distribution. A similar complex of six sibling species of Anopheles is found in North America. Sibling species have been discovered also in such important disease vectors as *Anopheles gambiae* and *Aedes* (references to the extensive literature in Bates 1949, Kitzmiller, Frizzi, and Baker 1967, Davidson, Patterson, et al. 1967, and McClelland 1967).

Several complexes of sibling species have been studied in detail in the genus Drosophila. The siblings *D. pseudoobscura* and *D. persimilis* have repeatedly been mentioned in this book, and here we need only briefly summarize the information. The two were believed to be quite indistinguishable morphologically until Rizki (1951) found slight differences in the male genitalia, which make possible determination of the species of single males. Single females cannot be told apart by inspection. The distribution area of *D. pseudoobscura* extends from British Columbia to the highlands of Mexico, Guatemala, and the vicinity of Bogota, Colombia. The area of *D. persimilis* is much smaller and is included within that of its sibling. Where the species are sympatric, they differ in ecological preferences, *D. pseudoobscura* being more abundant in warmer and drier regions and *D. persimilis* in cooler, more humid localities.

The ethological isolation was described in Chapter 10; this is probably the key isolating mechanism preventing gene exchange between the populations of the siblings in nature. The sterility of hybrid males and the viability breakdown in backcross progenies were also discussed in Chapter 10. The most conclusive evidence of lack of gene diffusion between the species is that each of them has its own set of chromosomal polymorphs, not encountered even as exceptions in the other sibling (Dobzhansky and Epling 1944).

The superspecies *Drosophila paulistorum* described in the preceding section is a member of a group of six sibling species—*D. willistoni*,

D. tropicalis, D. insularis, D. equinoxialis, D. paulistorum, and *D. pavlovskiana.* Although the evidence here is not as conclusive as it is for *D. pseudoobscura* and *D. persimilis,* the *willistoni* group of siblings seem also to be reproductively isolated from each other. The claims of fertile hybrids between these species (Winge 1965) could not be confirmed in the very careful experiments of Ehrman and Petit (1968). Excellent recent studies of sibling species of the *D. ananassae* group have been published by Futch (1966), and of the *D. auraria* group by Kurokawa (1960).

In Diptera other than mosquitoes and Drosophila, much work has been done on sibling species of houseflies, Musca (a review in Saccà 1967), on black flies, Simuliidae (Rothfels 1956, and Landau 1962), and on Ceratopogonidae (Nielsen 1951). Examples in other animals are crickets (Fulton 1952, Ohmachi and Mazaki 1964, and Alexander 1968), shrimps Artemia (Halfer-Cervini, Piccinelli, et al., 1968), sea cucumbers Thyonella (Manwell and Baker 1963), oysters (Ostrea, Urosalpinx), and finally infusoria (Sonneborn 1957). For more examples, see Mayr 1963.

The description and classification of species has traditionally been, and to a large extent continues to be, the province of systematists working in museums and herbaria. Sibling species cannot be distinguished, however, by classical museum techniques. A museum taxonomist dislikes being unable to write a species determination label. For example, a pinned and dried female of *Drosophila paulistorum* can only be determined as belonging to the *willistoni* group of siblings. Not unexpectedly, some taxonomists contend that only forms which can be distinguished by their time-honored methods should be considered species. Less understandably, this contention has won support from some geneticists. Species are, however, phenomena of nature that exist regardless of our ability to distinguish them. The techniques of species investigation change with time, and cultural and biochemical tests are now used routinely to classify some microorganisms. Is it really necessary to have Drosophila pinned, dried, and shriveled before classifying them? To demand that modern taxonomists use only the techniques of Linnaeus is about as logical as to direct modern medicine to eschew any methods not utilized in Linnaeus' time.

We saw in Chapter 10 that sibling species in ciliate protozoans are referred to not as species but as "varieties." Sonneborn (1957) proposed the term syngen "for the potentially common gene pool, for organisms capable of 'generating together.'" "Syngen" is a synonym of "species."

The Multiple-Gene Hypothesis of the Origin of Reproductive Isolation

Reproductive isolation has two aspects: (1) the interbreeding of species A with species B is difficult or impossible, whereas (2) individuals of A as well as of B are fully able to breed inter se. The reproductive biology of any species is organized to insure the procreation of a sufficient number of offspring. At the same time, it militates against gene exchange with other species. It is important to visualize how this state of affairs develops. Mutations that alter the sexual behavior, the breeding time, or the structure of the genitalia may occur in any species, but such mutations are not workable isolating mechanisms. Genetic changes that engender reproductive isolation must not only prevent cross-breeding between the mutant and the original type, but also must simultaneously insure normal reproduction of the mutants. Where isolation involves incapacitation of the hybrids, this effect must be restricted to the hybrids and must leave the parental populations unaffected.

Reproductive isolation arising in a single step can hardly become established in sexual and outbreeding forms. Mutants appear in populations at first as heterozygotes, and inviable or sterile heterozygotes will be eliminated by normalizing natural selection, regardless of how fit the corresponding homozygotes might be. This initial disadvantage is mitigated in hermaphrodites capable of self-fertilization and in parthenogenetic and asexual forms. The mutant, if viable, may reproduce and establish a small colony. Cross-fertilization and outbreeding may then be resumed.

In sexual and obligatory cross-fertilizing forms the formation of isolating mechanisms entails, not single mutational steps, but building systems of complementary genes. Assume that a population has the genetic constitution *aabb*, where *a* and *b* are single genes or groups of genes, and that this population is broken up into two allopatric, geographically isolated parts. In one part, *a* mutates to *A* and a local race *AAbb* is formed. In the other part, *b* mutates to *B*, giving rise to a race *aaBB*. Since individuals having constitutions *aabb*, *Aabb*, and *AAbb* interbreed freely, there is no difficulty in establishing the gene *A* in the population. The same is true for the gene or genes *B*, since *aabb*, *aabB*, and *aaBB* also interbreed freely. But the cross *AAbb* × *aaBB* is at a disadvantage, because the interaction of *A* and *B* produces one of the reproductive isolating mechanisms. If the carriers of genotypes *AAbb* and *aaBB* surmount the extrinsic barriers separating them, they are then able to become sympatric, since interbreeding is no longer possible.

Reproductive Isolation as a Product of Genetic Divergence and Natural Selection

To state that races are incipient species is not tantamount to saying that every race is a future species. Race differentiation is reversible; race divergence may be superseded by convergence. This is, in fact, what is happening to the human species. To become species, races must evolve reproductive isolation. The question naturally presents itself, what causes bring about the development of reproductive isolating mechanisms? Two hypothetical answers have been proposed. First, reproductive isolation is a by-product of the accumulation of genetic differences between the diverging races. The same genes that make the races diverge in morphological and physiological traits render them reproductively isolated. Second, the isolation is built up by natural selection, when and if the gene exchange between the diverging populations generates recombination products of low fitness. The establishment of reproductive isolation is a special kind of genetic divergence. These two hypotheses are not mutually exclusive. Needless disputes have arisen because they were mistakenly treated as alternatives.

The first hypothesis has long been implicit in the thinking of systematists (discussion and references in Mayr 1942 and Rensch 1960a), but its genetic formulation is due to Muller (1940, 1942). The gene pool of a population is an integrated system of genes; evolutionary changes are not mere additions or subtractions of unrelated gene elements. The initial advantage of most mutations that arise and become established in a species is slight. As the accumulation of gene differences continues, genes that at one time might have been easily dispensed with become essential constituents of the genotype (Harland 1936 and Schmalhausen 1949). In the course of evolution, the functions of a gene in the development may undergo changes. If in two or more races or species the gene functions diverge, the gene systems may no longer be compatible in hybrids. In Muller's opinion, all isolating mechanisms may arise in this manner:

> Which kind of character becomes affected earliest, and to what degree . . . will depend in part upon its general complexity (which is correlated with the number of genes affecting it), in part on the nicety or instability of the equilibria of processes necessary for its proper functioning, and in part on the accidental circumstances that determined just which incompatible mutations happened to become established first.

The second hypothesis was, according to Grant (1963), suggested as far back as 1889 by A. R. Wallace, and later by Fisher (1930) and Dobzhansky (1940). It starts from the same premise as that of Muller (see above), namely, that the genotype of a species is an integrated system adapted to the ecological niches in which the species lives. Gene recombination in the offspring of species hybrids may lead to the formation of discordant gene patterns that decrease the reproductive potentials of both interbreeding populations. Suppose that incipient species, A and B, are in contact in a certain territory. Mutations arise in either or in both species that make their carriers averse to mating with the other species. The nonmutant individuals of A that cross to B will produce a progeny inferior to the pure species. Since the mutants breed only or mostly within the species, their progeny will be superior in fitness to that of the nonmutants. Consequently, natural selection will favor the spread and establishment of the mutant condition.

Sturtevant (1938) and Bruce Wallace (1968a) have pointed out one of the possible causes that might initiate such a process. Suppose that the gene arrangement *ABCDEFGH* in a chromosome is modified in one race to *AFEDCBGH* and in another race to *ABGFEDCH*. Heterozygotes carrying the ancestral and either of the modified arrangements will produce few or no inviable offspring. In a hybrid carrying the two modified arrangements, crossing over in the section *CDEF* will give chromosomes *AFEDCH* and *ABGFEDCBGH*. Such chromosomes may be inviable. Hence prevention of the interbreeding of carriers of *AFEDCBGH* and *ABGFEDCH* will have a selective advantage.

There is good experimental evidence that selection can build up reproductive isolation. Koopman (1950) made use of the observation that the ethological isolation between *Drosophila pseudoobscura* and *D. persimilis* is weaker at a low (16°c) than at a higher (25°c) temperature. He placed in population cages equal numbers of females and males of the two species, marked by two different recessive mutants. The offspring of matings within and between species are distinguishable by inspection; if the two species are *aaBB* and *AAbb*, the hybrids are wild-type, *AaBb*. In every generation, the hybrids were destroyed, and the populations were continued with equal numbers of the pure species. Therefore, the flies that mated with representatives of their own species had their progenies included among the parents of the next generation, whereas those mating with the other species suffered "genetic death." After only five generations of selection, the proportions of hybrids among the offspring fell to a fraction of the former value.

Working with the same two species as Koopman, Kessler (1966) used a superior technique, observing the matings directly. He selected both for weaker and for stronger ethological isolation. To weaken the isolation, he selected the females and males that mated soonest with individuals of the opposite sex of the foreign species, and subsequently mated them conspecifically. Since the species were marked by recessive mutant genes, the progenies of conspecific and heterospecific matings were distinguishable. To strengthen the isolation, individuals were selected that failed to mate with the other species, and then mated conspecifically. In eighteen generations of selection, Kessler obtained populations with both weaker and stronger isolation, in *D. pseudoobscura* and in *D. persimilis* alike. Although Manning (1961) was able to select *D. melanogaster* for increased and decreased mating speed, Kessler's results were due to changed behavioral responses to individuals of another species, and not simply to greater or lesser eagerness of the flies to mate.

In the experiments of Koopman and Kessler, pre-existing isolating mechanisms were enhanced or reduced. Wallace (1954) and Knight, Robertson, and Waddington (1956) initiated ethological isolation between two strains of *Drosophila melanogaster* that were marked by different recessive genes and that previously showed no mating discrimination. As in Koopman's experiments, the selection was made by discarding the progenies of heterogamic matings. A weak but statistically significant preference for homogamic matings developed over some thirty generations.

In order that natural selection may promote reproductive isolation, there must be a challenge of loss of fitness owing to gene flow between populations. Reduced viability or fertility of hybrid offspring provides such a challenge. This is another way of saying that postmating isolating mechanisms may act as stimuli for the development of premating isolation (cf. Chapter 10). Postmating isolating mechanisms (i.e., hybrid inviability, sterility, breakdown, or combinations of these) are, then, consequences of differential adaptedness of races or species to the conditions of life in their respective distribution areas. They are by-products of genetic divergence, although in experiments they can be enhanced or weakened by selection. Haley, Abplanalp, and Enya (1966) successfully selected strains of the domesticated Japanese quail (Coturnix) to produce viable offspring when artificially inseminated with the sperm of jungle fowl (*Gallus bankiva*) or domestic chickens. The experiments of Grant (1966a,b) on selection for viability and fertility in species hybrids of Gilia have already been mentioned.

Rick (1963b) has studied the wild tomato (*Lycopersicon peruvianum*), which grows in stream valleys along the coast of Peru. Intercrosses of northern strains with southern ones give few viable seeds. Nevertheless, geographically intermediate populations form a "compatibility bridge," and the genetic unity of the species is maintained. Rick states:

> It is not difficult to understand how such barriers might arise gradually in races that have been isolated for long periods of time. Different reaction norms for rates of embryo and endosperm development, osmotic values, and other developmental characteristics might have become fixed by selection while races were adapting to the new environments into which they were migrating.

Similar situations have been observed by, among others, Grant (1954) in Gilia, Stebbins (1957) in Elymus, Kruckeberg (1957) in Streptanthus, Vickery (1959, 1964, 1966) in species of monkey flowers (Mimulus), and Levin and Kerster (1967b) in Phlox.

Hoenigsberg and Koref-Santibañez (1960) found differences in courtship patterns between some laboratory strains of *Drosophila melanogaster,* which result in a preference for homogamic matings. Similar preferences exist, as pointed out previously, in Transitional populations of *D. paulistorum* (Dobzhansky, Pavlovsky, and Ehrman 1969b) and in geographic strains of *D. birchii* (Ayala 1965). It is unlikely that these rudiments of ethological isolation were built by natural selection for their function as isolating mechanisms; on the other hand, they are genetic raw materials from which reproductive isolation may be compounded by natural selection. It is appropriate to mention at this point that not all genetic divergence leads to changed mating preferences. Robertson (1966) selected a population of *D. melanogaster* for adaptedness to a modified diet. Although the adaptation involved multiple, polygenic gene differences, no trace of ethological isolation between the original and the changed strains was found.

There is ample, though of necessity indirect, evidence that selection builds isolating mechanisms in nature. At least the premating isolating mechanisms between closely related species should be enhanced in the geographic areas where hybridization is most likely to occur. The observations of Ehrman (1965) on sympatric and allopatric strains of *Drosophila paulistorum* are among the most elegant verifications of this prediction. The ethological isolation is greater among sympatric than among allopatric strains of the same pairs of semispecies.

In a series of papers Grant (1954b, 1958, 1965, 1966d, a general discussion in 1963) has supplied a demonstration that mechanisms preventing the hybridization of species of Gilia arise by selection under conditions of sympatry. Five related species occur in the foothills and valleys of California and are often found growing side by side. Four other species occur in coastal localities in North and South America; they are completely allopatric with respect to one another, and largely so with respect to the five inland species. Experimentally obtained species hybrids are highly sterile in all combinations tried. Yet the allopatric species can be crossed quite easily, giving 18.1 hybrid seeds per flower on the average. In contrast, the sympatric species are separated by crossability barriers and yield only 0.2 seeds per flower when cross-pollinated artificially.

We saw in Chapter 10 that mating call differences are important isolating factors between related species of anuran amphibians. Littlejohn (1965) has analyzed the mating calls of allopatric and sympatric populations of two species of Australian frogs:

Whereas mating calls of remote allopatric populations of *Hyla ewingi* and *H. verreauxi* are very similar, those of the sympatric populations are quite distinct. . . . It is suggested that the marked differences between sympatric populations have resulted from the direct action of selection for increased reproductive efficiency, i.e., the slight differences present in the allopatric populations have been reinforced in the sympatric populations.

Some evidence of sympatric reinforcement of species differences in mating calls has also been recorded by Blair (1955) in *Microhyla olivacea* and *M. carolinensis* and by Ball and Jameson (1966) in *Hyla regilla* and *H. californiae*. No such reinforcement was found, however, by Michaud (1964) in *Pseudaoris clarki* and *P. nigrita,* or by Blair and Littlejohn (1960) in *P. ornata* and *P. streckeri*. Hubbs and Delco (1960, 1962), working with four species of the fish Gambusia, found that males of sympatric species are better able to distinguish females of their own and foreign species than are males of allopatric species. A similar difference has been recorded by Smith (1965) for sympatric and allopatric strains of mice, *Peromyscus eremicus* and *P. californicus*.

The examples just cited of the reinforcement of premating isolating mechanisms may be viewed as instances of character displacement, which Brown and Wilson (1956) defined as "the situation in which, when two species of animals overlap geographically, the differences between them are accentuated in the zone of sympatry and weakened or lost entirely in the parts of their ranges outside this zone." Habitat, temporal, and ethological isolations are particularly likely to arise in this manner. In addition to the examples mentioned by Brown and Wilson and by Mayr (1963), reference may be made to the works of Brower (1959) on butterflies of the *Papilio glaucus* group, of Schoener (1965) on bill size differences among sympatric species of birds, and of Levin and Kerster (1967b) on Phlox.

Instances of the lack or the weakness of premating isolating mechanisms, where the populations of related species are wholly or largely allopatric, are perhaps as significant as their presence where the species are sympatric. Thus Zaslavsky (1966, 1967) found that hybrids between the ladybird beetle species (or semispecies) *Chilocorus bipustulatus* and *Ch. geminus* are sterile, but detected no ethological isolation at all, at least under experimental conditions. Hybrid belts, formed where the geographic areas of two species come into contact, have been studied to find whether premating isolating mechanisms may be formed there. The most thoroughly investigated cases are those of two species of crows (Corvus) in Europe, and of grackles (Quiscalus) in the United States. The evidence is ambiguous (Mayr 1963, Yang and Selander 1968, and Johnsgard 1967).

Whether postmating isolating mechanisms can be reinforced by natural selection is also an open problem. If the progeny of hybrids is inferior in fitness, it would seem advantageous to the species concerned to prevent hybridization, either by premating isolation or, failing that, by such postmating mechanisms as inviability or sterility of F_1 hybrids. Group selection (see Chapter 12) could, theoretically, bring such a result about. However, because the efficiency of group selection is low relative to the selection of individual genotypes, it is doubtful that isolating mechanisms frequently arise in this way.

Polyploid Species in Plants

We have seen that differences between species involve many genes, and often also chromosomal changes. The transformation of one species into another in time, or the splitting of an ancestral species into two or several derived ones, is, therefore, a slow process. Yet, alongside this gradual method of species formation, new species may also emerge in a single generation, by polyploidy. This may occur, it will be recalled, either through doubling of the chromosome complement in the hybrid between two previously existing species (allopolyploidy), or through multiplication of the chromosomes of a single species (autopolyploidy). In either case, the polyploid possesses, at least initially, all the genes that were present in its ancestors and no new ones. However, the ancestral species may continue to exist side by side with the polyploid; the organic diversity is, therefore, augmented.

Species formation through polyploidy has occurred in all major groups of plants, with the possible exception of fungi, but only rarely in animals. Excellent reviews of the state of the knowledge about plant polyploids can be found in Stebbins (1950), Grant (1963), and Schwanitz (1967). Only a brief consideration is needed here.

Grant (1963) estimates that as many as 47 percent of species of angiosperms are recent or ancient polyploids; the frequency is higher among monocotyledons (58 percent) than among dicotyledons (43 percent). Polyploidy is rare among the gymosperms, although the redwood (*Sequoia sempervirens*) is a polyploid. Manton (1950) and Klekowski and Baker (1966) found polyploidy very common among ferns. Some of the most important cultivated plants, such as wheat, oat, cotton, tobacco, potato, banana, sugar cane, and coffee, are polyploid.

Since most individuals in sexual and outbreeding plants are more or less complex heterozygotes, no sharp distinction can be drawn between auto- and allopolyploids (Stebbins 1950). As a convention, one may take the doubling of the chromosomes within a Mendelian population to produce autopolyploids, while hybrids between reproductively isolated populations are allopolyploids. Müntzing (1961) argues that autopolyploidy is an evolutionary factor of some consequence, and so it is among cultivated plants. Schwanitz (1967) lists tetraploid varieties of the clovers *Trifolium pratense* and *T. hybridum* and *Brassica rapa*, as well as of the ornamental plants *Cyclamen, Primula, Hyacinthus, Petunia, Crocus,* and *Antirrhinum*, that are superior to their diploid ancestors. Most of the wild species believed to be autopolyploid were later shown to be probably or certainly allopolyploid. Mosquin (1967) believes that a subspecies of *Epilobium angustifolium* is tetraploid or hexaploid. That allopolyploidy is a far more widespread method of species formation is, however, generally admitted.

Since allopolyploids possess gene complements of two or even three species, their reaction norms may be intermediate between those of the parents or may combine the properties of both. Sometimes, though not always, an allopolyploid possesses the environmental tolerances of both parents; such allopolyploids are likely to have high adaptive values. There is a considerable and still growing literature concerning the geographic regularities in the distribution of polyploids (Clausen, Keck, and Hiesey 1945, Gustafsson 1947, Stebbins 1950, Löve 1951, 1964, Johnson and Packer 1965, and references therein). The floras of arctic, subarctic, and recently glaciated territories have proportionately more polyploid species than do warmer and geologically more ancient lands. Löve and Löve (1949) give the list of percentages of polyploid species in different floras (arranged from south to north) shown in Table 11.3

The formation of species through polyploidy is a process vastly more rapid than the more ubiquitous race divergence; polyploids are therefore most likely to colonize newly opened lands, such as those recently freed from the continental ice sheets (Babcock and Stebbins 1938 and Stebbins 1950). Also, the floras of high latitudes include

TABLE 11.3

Percentages of polyploid species of angiosperms

Area	Latitude, °N	Polyploids	Area	Latitude, °N	Polyploids
Timbuctoo	17	37	Iceland	63–66	64
Cyclades	37	34	Sweden	55–69	56
Sicily	37	37	Kolguev	69	64
Hungary	46–49	47	Finland	60–70	57
Schleswig-Holstein	54	50	Norway	58–71	58
Denmark	54–58	53	South Greenland	60–71	72
Great Britain	50–61	57	Spitzbergen	77–81	74
Faroes	62	61			

many perennial herbs and relatively few woody species; the incidence of polyploidy is known to be higher among the former than among the latter.

The hypothesis of Babcock and Stebbins is questioned by Löve (1951, 1964), who postulates instead that polyploids possess a selective superiority because of their greater genetic variability. Hutchinson, Silow, and Stephens (1947) pointed out that polyploids may show a greater variety of phenotypes than diploids because of dominant and semidominant genes. A diploid may carry either one or two dominant alleles, while genotypes with one, two, three, and four dominants may be formed in a tetraploid. Since genetic changes that affect polygenic traits are usually neither dominant nor recessive, Hutchinson et al. are of the opinion that the "evolutionary potentialities of a young polyploid will rapidly approach those of a diploid." Löve sees a confirmation of his view in the findings of S. Mangenot and G. Mangenot (1962) that the flora of tropical Africa contains a high proportion of very old polyploids, though a low percentage of young ones.

What makes polyploids of outstanding interest to evolutionists is the possibility, not only of creating new species experimentally, but also of tracing the phylogeny of existing polyploid species, and sometimes of re-creating them. Karpechenko's classical work on "radocabbage" (Raphanobrassica) was discussed in Chapter 10. Radocabbage not only is fully fertile with itself, but also produces sterile triploid hybrids when crossed to its ancestors, radish and cabbage. Another classic in this field is Müntzing's (1932) synthesis of the tetraploid ($2n = 32$) mint species *Galeopsis tetrahit* from its diploid ($2n = 16$) parents, *G. pubescens* and *G. speciosa*. The cross of the two diploid species gives a highly sterile F_1 hybrid with disturbed chromosome pairing at meiosis. Müntzing found a single plant in the F_2 generation, which proved to be triploid (24 chromosomes); this he outcrossed to *G. pubescens*. A single seed in the resulting progeny was a tetraploid (32 chromosomes), which gave rise to a strain of fertile "artificial *tetrahit*," identical in all essentials to the *G. tetrahit* found in nature. The hybrids of artificial and natural *tetrahit* have normal meiosis with 16 bivalents, and at least some of these are fully fertile.

The unraveling of the phylogeny of species of wheat (Triticum) and related grasses (Aegilops, Secale, Agropyron) constitutes one of the greatest success stories of cytogenetics, made possible by the work of many scientists in several countries (for references to the voluminous literature, see Sears 1948, 1956, Unrau 1959, and Kihara 1965). Diploid ($2n = 14$), tetraploid ($4n = 28$), hexaploid ($6n = 42$), and artificially produced octoploid ($8n = 56$) species are known. The analysis is made in terms of "genomes," that is, sets of 7 chromosomes each, differing in gene contents and gene arrangements, derived from different diploid ancestors. To oversimplify the story, if a hexaploid wheat, such as *Triticum vulgare* or *T. spelta*, with 42 chromosomes, is crossed to a tetraploid, such as *T. durum*, with 28 chromosomes, the hybrids have at meiosis up to 14 bivalents and 7 univalents. The inference is that *T. vulgare* has 2 genomes, or a total of 14 chromosomes in its gamete, sufficiently similar to the 2 genomes of *T. durum* to pair and to form bivalents. A cross of *T. durum* to the diploid einkorn wheat, *T. monococcum*, with 14 chromosomes, gives up to 7 bivalents and 7 univalents in the hybrid. Finally, *T. vulgare* crossed to *T. monococcum* gives up to 7 bivalents and 14 univalents. If the genome of *T. monococcum* is denoted as *A*, then the tetraploid wheats have also genome *A* and some other genome, *B*. The hexaploid wheats have genomes *A* and *B*, and also some third genome, *D*.

Genome *D* comes from the diploid grass, *Aegilops squarrosa*, with 14 chromosomes. The hypothesis suggests itself that the hexaploid wheats arose by a doubling of the chromosomes in a hybrid between some tetraploid wheat, and *Ae. squarrosa* or its close relative. This was verified independently by Sears and McFadden in America and by Kihara in Japan. A tetraploid wheat, *Triticum dicoccoides*, was crossed to *Ae. squarrosa*; by doubling the chromosomes in the resulting triploid hybrid, a hexaploid was obtained that proved to be strikingly similar to the existing hexaploid, *T. spelta*.

As so often happens, further studies have disclosed some complications. Presumably all of the seven chromosome genomes have descended in some remote past from some primordial genome, and subsequently became differentiated by gene mutations and rearrangements. Not surprisingly, some genomes still remain so similar that some of their chromosomes became paired at meiosis in the hybrids. One tetraploid wheat species, *Triticum timopheevi,* has the genomes *AG,* instead of *AB* as in *T. durum* and other tetrapoloids. Among the Aegilops species there is a diploid, *Aegilops caudata,* carrying genome *C*; a tetraploid, *Ae. cylindrica,* with genomes *C* and *D*; a diploid, *Ae. comosa,* with genome *M*; and several tetraploid and hexaploid species combining *DM, CM, DCM,* and their derivatives C^uM^o, DC^uM^o, C^uM^t, etc. Genome *B* may have been derived from a species of the grass Agropyron; an allotetraploid hybrid of this with *T. aegilopoides* (a relative of the neolithic cultivated *T. monococcum*) gave the tetraploid *T. dicoccoides,* the ancestor of the modern hard wheat, *T. durum.* In the Neolithic or Bronze Age there arose the hexaploid *T. spelta,* and later the modern cultivated bread wheat, *T. vulgare.* Similar analyses of the allopolyploid descent have been made for species of cottons (Hutchinson, Silow, and Stephens 1947), tobaccos, and some other plants.

Animal Polyploids

The prevalence of polyploids among plants and their scarcity among animals constitute a striking difference between evolutionary patterns in the two kingdoms. Muller (1925) surmised that this is due to the preponderance of hermaphroditism (monoecy) among plants, and the separation of sexes (dioecy) among animals. Where sex is determined by a mechanism like that in Drosophila, polyploidy may result in the production of intersexes and other abnormal and sterile types. If the ratio of the numbers of X chromosomes and of sets of autosomes is intermediate between that in females (1 : 1) and that in males (1 : 2), the individual is a sterile intersex. A tetraploid individual arising by mutation in nature will cross to a normal diploid of the opposite sex, and the progeny will consist of triploid females and intersexes. A part of the progeny of a triploid female crossed to a normal male is also intersexual.

Muller's argument lost its force with the discovery that both in dioecious plants and in mammals the male determining genes are carried in the Y chromosome, and the female determiners in the X. In the plant *Melandrium album*, triploid individuals with two X's and a Y chromosome are fertile males, and not intersexes as they are in Drosophila (Warmke and Blakeslee 1940 and Westergaard 1948). In mice, diploid individuals with a single X and no Y are fertile females, and not sterile males as in Drosophila (Welshons and Russell 1959).

Natural polyploid species in animals occur among hermaphrodites, such as earthworms (Omodeo 1952) and planarians (Aeppli 1952), or in forms with parthenogenetic females, for example, some beetles, moths, sow bugs, shrimps, goldfish, and salamanders (Suomalainen 1947a, 1962, Seiler 1946, Vandel 1941, Cherfas 1966, Uzzell and Goldblatt 1967, and others). Astaurov (1969) obtained fertile allotetraploid hybrids of *Bombyx mori* (the silkworm moth) and the closely related *B. mandarina*. In his experiments Astaurov utilized the technique of artificial induction of parthenogenesis by heat treatment. He obtained triploid hybrids with two chromosome sets of *B. mori* and one of *B. mandarina*. Such hybrids are sterile, but artificial parthenogenesis leads to the development of some mosaic individuals, with triploid and hexaploid tissues, the latter having four sets of *B. mori* and two sets of *B. mandarina* chromosomes. These individuals are weakly fertile; crossed to the ordinary diploid *B. mandarina*, they yield more fertile allotetraploid moths (two chromosome sets from each parental species).

Wide variations in the chromosome numbers in related species of certain groups of animals led some authors to infer that polyploidy has played an evolutionary role in the animal kingdom comparable to that among plants. A critical analysis of these claims (see White 1954 and Matthey 1952, 1964a,b) showed that what is actually involved is either chromosome fusion and fragmentation in forms with diffuse centromeres (see Chapter 5), or changes owing to frequent translocations. This simpler interpretation has been borne out by measurements of the DNA contents in the nuclei of the alleged polyploids. In the fish family Salmonidae the chromosome numbers vary from 58 to 104; Rees (1964) nevertheless found similar amounts of DNA in *Salmo salar* (60 chromosomes) and *S. trutta* (80 chromosomes). Suomalainen (1965) also found uniform DNA contents in species of the geometrid moths Cidaria with chromosome numbers ranging from 12 to 32 (haploid). In polyploids, the amounts of DNA would be expected to form a series of multiples of some basic amount.

A much bolder claim in favor of polyploidy in animal evolution has been advanced by Ohno and Atkin (1966, Atkin and Ohno 1967) and by Taylor (1967). The amounts of DNA per nucleus vary greatly in different vertebrates and prochordates, being generally small in lower forms, which may be surmised to reflect the conditions in the ancestors of the vertebrate phylum. Thus, the ascidian Ciona has approximately 6 percent, the cephalochordate Amphioxus 17 percent, and the cyclostome fishes Lampetra and Eptatretus 38 and 76 percent, respectively, of the amounts in the nuclei of man and most mammals. At the opposite extreme, the lungfish Lepidosiren and some urodele amphibians have very high amounts, up to 35 times the mammalian value. Ohno and his colleagues ascribe these wide variations to a series of polyploidizations having occurred in the evolution of vertebrates and their ancestors. However, since polyploidy cannot be easily established in species with separate sexes and regular outcrossing, they see themselves as forced to relegate the hypothetical polyploidization to great antiquity, 300 million years ago, when "fishes ancestral to the terrestrial vertebrates of today did not have a firmly established chromosomal sex-determining mechanism, and there was no barrier against polyploid evolution then."

All that need be said concerning this highly imaginative speculation is that polyploidy is by no means the only known process whereby the amount of genetic material in the nucleus is changed. That duplications of chromosome sections occur in evolution is amply attested by the presence of numerous "repeat" areas in the salivary gland chromosomes of Drosophila and other polytene chromosomes.

Sympatric, Stasipatric, and Saltational Species Formation

The essence of the process of species formation is the establishment of reproductive isolation between arrays of Mendelian populations. Most often this process occurs gradually, while the genetically diverging races are allopatric, that is, live in different territories. The geographic isolation is antecedent to reproductive isolation. One of the recurrently debated issues is, however, how often the speciation is not gradual but saltational, and occurs while the diverging populations are not allopatric but sympatric. Speciation by way of allopolyploidy is evidently saltational, and no less evidently it occurs sympatrically, since the species ancestral to the allopolyploid must occur in close proximity in order to produce hybrids.

Whether populations can diverge gradually while living in the same territory is a different issue. Mayr (1963 and earlier) has expertly marshaled arguments showing that this is unlikely to occur except under very special circumstances. The hypothesis of sympatric speciation, he states, "is neither necessary nor supported by irrefutable facts. It overlooks the fact that speciation is a problem of populations, not of individuals, and it minimizes the difficulties raised by dispersal and recombination of genes during sexual reproduction." Several authors, most recently Maynard-Smith (1966), Bush (1969), and Pimentel, Smith, and Soans (1967), have nevertheless devised plausible genetic models of how sympatric speciation could occur.

The findings of Lewis and Raven (1958 and Lewis 1966) and of Kyhos (1965) are in a different category. Reference has been made to the work of Grant (1966a,b) who, in nine generations, selected viable and fertile recombination products among poorly viable and semisterile hybrids between *Gilia malior* and *G. modocensis*. According to Lewis and Raven, *Clarkia franciscana* is a narrow endemic growing within the distribution area of *C. rubicunda*, from which it is reproductively isolated, and not far from the distribution margin of *C. amoena*. The authors infer that *C. franciscana* arose "*in situ*" from *C. rubicunda* "as a consequence of a rapid reorganization of the chromosomes due to the presence, at some time, of a genotype conducive to extensive chromosome breakage." Similarly, Kyhos finds that "*Chaenactis glabriuscula* is the living ancestor of *Ch. fremontii* and *Ch. stevioides*," from which it differs in chromosome number ($n = 6$ in the first species and $n = 5$ in *Ch. fremontii* and *Ch. stevioides*), as well as in a series of translocations.

White (1968) and his colleagues have studied the chromosomes in some 160 species and semispecies of flightless Australian grasshoppers of the subfamily Morabinae. At least 34 translocations resulting in chromosome fusion and 20 resulting in dissociation of chromosomes have taken place in the phylogeny of this group. Particularly interesting are the superspecies *Moraba viatica* and *M. scurra*. Semispecies with different karyotypes occupy adjacent territories; hybrids, which are often translocation heterozygotes, are found, nevertheless, only in very narrow zones of overlap, sometimes some hundred meters wide. The translocation heterozygosis reduces the fitness of its carriers. How, then, have the species and semispecies with different chromosome complements established themselves in the first place? White suggests that this may have happened "stasipatrically." A translocation establishes itself at first in a small local colony, either at the periphery of the distribution area of the ancestral species or inside it, by a process of random genetic drift (Chapter 8). If members of this colony possess high fitness, they subsequently spread and displace the ancestral form in a certain area. White believes that the allopatric and stasipatric models of species formation "seem essentially different" but are nevertheless "not entirely antithetical." Perhaps the stasipatric may be regarded as a special case of the allopatric model (Key 1968).

CHAPTER TWELVE

PATTERNS OF EVOLUTION

Concepts of Progressive Evolution

Teilhard de Chardin (1959) saw in organic evolution "only one event, the grand orthogenesis of everything living toward a higher degree of immanent spontaneity." He was using, or misusing, the word orthogenesis in an unusual sense; the theory of orthogenesis proposes that "evolution is in a great measure an unfolding of preexisting rudiments" (Berg 1969). This is at variance with everything that modern biology has learned about evolution. Teilhard de Chardin himself supposed, on the contrary, that the evolutionary process proceeds by "groping"—a term that, to him, meant "pervading everything so as to find everything." This is a splendid, though somewhat impressionistic, characterization of evolution molded by natural selection. The groping leads to evolutionary progress in some lines of descent, extinction in many more lines, and evolutionary stasis in the rest.

That there has been progress in evolution is intuitively evident (Stebbins 1969). In Barbour's (1966) words, "By almost any standard man represents a higher level than primeval mud." And yet attempts to define what constitutes progress have met with only mediocre success. Mere change is not necessarily progress; in fact, the evolution of many parasitic groups can be regarded as regressive. The simplest and most primitive existing forms of life, microorganisms and viruses, show an adaptedness to their ecological niches not manifestly inferior to that of the most complex and advanced forms, including man.

Simpson (1949) finds only one really universal trend in evolution: "a tendency for life to expand, to fill in all the available spaces in the livable environments, including those created by the process of that expansion itself." Now, the "expansion" of life can occur by various means. One is cladogenesis (Rensch 1947) or splitting (Simpson 1953), which leads to diversification of phylogenetic lineages, adaptation to a greater variety of ecological niches, and, as a rule, growth of the biomass taken up by representatives of a lineage. Speciation, the appearance in time of two, several, or many contemporaneous species descended from a single ancestral one, is the most thoroughly studied kind of cladogenesis and is discussed in Chapters 10 and 11. Some evolutionists argue that the emphasis on speciation, started in Darwin's "Origin of Species" and continued in modern genetics, may have gone too far. The proliferation of more and more species may be a sign of the biological success or ascendancy of a lineage, but hundreds or thousands of species of a genus (such as Drosophila) or a family or superfamily (such as parasitic wasps, Ichneumonoidea) may be merely so many variations on the same theme.

The emergence of new organs, the development of novel ways of dealing with the environment, and advances into new adaptive zones are earmarks of anagenesis (Rensch 1947), phyletic evolution (Simpson 1949), or arogenesis (Takhtajan 1966). An anagenetic line may be represented by a single species or by the same number of species at different time levels. Anagenesis and cladogenesis are often mixed in various proportions. Human evolution is an excellent example of anagenesis. There were apparently only two hominid species (*Australopithecus africanus* and *A. robustus*) in the early Pleistocene, and there has been only one from the middle Pleistocene to the present (*Homo erectus* followed by *H. sapiens*). The cladogenetic element was confined to the formation of races or subspecies; some of these may have died out, but most races of *H. erectus* were eventually transformed into races of *H. sapiens*.

Huxley (1942) and Rensch (1947, 1968) arrived independently at fairly similar lists of characteristics of anagenesis or progressive evolution. These are increased complexity and rationalization of structures and functions, especially complexity and rationalization of central nervous systems; open-ended improvement, permitting further improvement, and increasing independence of the environment, making for greater autonomy of the organism. Simpson (1949) pointed out that none of these criteria can be taken as a valid touchstone of progress. There is no denying, however, that each one is applicable to some evolutionary developments that we may choose to call progressive. For example, the transition from unicellular to multicellular organisms clearly involved a structural complication. Consider, however, the evolutionary sequence fish → amphibian → reptile → mammal → man. The sequence is usually taken to be progressive, yet "it would be a brave anatomist who would attempt to prove that recent man is more complicated than a Devonian ostracoderm." Although the development of a variety of sense organs is generally taken as progress, mammals and man lack certain kinds of senses present in other vertebrates—the lateral line organs of fishes, which perceive variations in pressure, or the directional receptors for heat radiation present in pit vipers.

Thoday (1953) defined biological progress as increase in fitness. In his usage, however, "fitness" includes the capacity of species or of other units of evolution not only for immediate but also for future survival in future environments. "The probability that a unit of evolution will survive for a given long period of time, such as 10^8 years, that is to say, will leave descendants after the lapse of that time, is the fitness of the unit." This definition leads to at least two difficulties. First, it is impossible to predict which of the existing organisms will have descendants not only after 10^8 years, but even after periods some orders of magnitude shorter. Simpson (1969) goes so far as to say, "No known actions of man can guarantee or even make probable the *indefinite* survival of our species, which will almost certainly become extinct in due course, whatever we do now." Yet it is our cherished idea that man is the most progressive and fittest product of evolution! Second, if we use retrodiction instead of prediction, the primordial organism from which all the rest have descended must have been the fittest (Ayala 1969c).

There is, nevertheless, no denying that fitness in Thoday's sense, which may perhaps be designated as the durability of a unit of evolution, is a highly significant consideration. Slobodkin (1964, 1968) rightly says, "The animals that are now alive are successful players at the evolutionary game in that . . . extinction represents a kind of losing." The difficulty that a biologist encounters is that "the applicability of the concept of an existential game to evolution in any interesting way is contingent on the logical possibility of a non-intelligent player or an automation being able to be effective in an existential game, without being explicitly programmed with information about the future." Slobodkin then argues that success in the "game" depends on the interaction of behavioral, physiological, ecological, and genetic mechanisms. Their interaction with each other and with the environment should conserve or maximize the "homeostatic ability" of the population. This occurs when organisms "respond to environmental perturbations in such a way that they not only minimize the departure from steady state conditions caused by the perturbation but also maximize their ability to withstand further perturbations."

Kimura (1961) has made an interesting attempt to envisage progressive evolution as an increase in the amount of genetic information. We saw in Chapter 1 that the DNA content of a human chromosome set is about three orders of magnitude greater than that in a bacterial cell. It is reasonable to infer that the amount of genetic information is also greater in man than in a bacterium. Kimura is on less secure ground when he chooses to assume that man's ancestors who lived 500 million years ago, in Cambrian times, carried much smaller amounts of genetic information. However, having made this assumption, and proceeding also on the premise that evolution has occurred at a constant rate during this very long time, he computes the rate of accumulation of genetic information as approximately 0.29 bit per generation. The total amount of genetic information accumulated since the Cambrian in the lineage leading to mammals and man is estimated to be of the order of 10^8 bits. The maximum amount of information that can be stored in the DNA of a human chromosome set is inferred to be about 10^{10} bits. This difference means, according to Kimura, that "either the amount of genetic information which has been accumulated is a small fraction of what can actually be stored in the chromosome set or, more probably, the DNA code itself is highly redundant." Although these numerical estimates can hardly be relied upon, even to the order of magnitude, Kimura's approach may lead to important developments if methods are found to actually measure the amount of accumulated genetic information.

One may well concede that each of the above concepts of evolutionary progress is meaningful, and yet agree with Simpson (1949) that "within the framework of the evolutionary history of life there have been not one but many different sorts of progress." Except in the new (1967) edition of his 1949 book, Simpson eschewed the topic of progressive evolution under this name in his later books (1953, 1961, 1964b, 1969). I feel that a biologist may reasonably speak of evolutionary progress, provided only that he makes clear what kind of progress is meant. My colleague F. J. Ayala has pointed out (unpublished) that the concept of progress, including biological progress, necessarily is axiological, that is, refers to some kind of value in reference to which we choose to consider objects or events. Our choice of values may or may not be determined by whether the valuation can be measured exactly.

The approach of evolution to man is almost inevitably in one's mind in considering the evolutionary history of life on earth. If made explicit (as Teilhard de Chardin 1959 has done), this emphasis is legitimate, although totally inapplicable to the evolution of the plant kingdom. The increase of individuation is another feasible criterion. Clones of billions of individuals, genetically identical except for newly arisen mutations, are easily obtained in many microorganisms. In sexually reproducing and outbreeding species, on the other hand, no two individuals have the same genotype. Moreover, an individual mammal may remain alive for years or even decades, during which it accumulates information about its environment. Cultural transmission in man makes this accumulation different in kind as well as in quantity from that characteristic of other living species. The destruction of billions of individuals is easily supported among bacteria, whereas in vertebrates an individual's life is hedged and sheltered by many physiological and developmental homeostatic mechanisms.

Sex and Genetic Recombination

Sexual reproduction is the most widespread and successful of the mechanisms of recombination of genetic materials. The origin and development of these mechanisms constituted an outstanding achievement of progressive evolution, because it facilitated further progress. This achievement must have been made early in the history of life. The number of organisms in which no gene recombination is known to occur is steadily dwindling; at least some of these organisms are descendants of forms in which gene recombination did exist.

The diversity of organs that serve the functions of mating, pollination, and fertilization in animals and plants is immense. The variety of behavior patterns underpinning these functions is no less impressive. Some of the most fascinating chapters of zoology and botany are concerned with this diversity. The cellular mechanisms of meiosis, indispensable for sexual reproduction, are more standardized, but even these display interesting and sometimes bizarre variations (an excellent review in White 1954). The role of sex in evolution attracted the attention of evolutionists, beginning with Darwin. It was Weismann, however, who pointed out in 1891 that sexual reproduction results in the formation of ever new combinations of hereditary determinants, later called genes. Weismann's idea was developed in genetic terms in the early nineteen thirties by Fisher, Morgan, Muller, and Wright. Among recent authors, Muller (1964), Crow and Kimura (1965), Maynard-Smith (1968b), and Bodmer (1970) endeavored to place the arguments on a quantitative basis.

Just how greatly evolution is speeded up by sex is controversial. The conclusions of various authors differ, depending on what assumptions they choose as plausible in their calculations. The basic consideration is, however, simple and straightforward. Suppose that substitution of alleles A_2, B_2, C_2, etc., for A_1, B_1, C_1, etc., is an adaptive improvement. Mutations from A_1 to A_2, B_1 to B_2, etc., occur infrequently. Under strictly asexual reproduction, these mutations must happen among the descendants of a single individual. With gene recombination, the mutations may occur in different individuals and in different places, and be subsequently joined together. Bodmer (1970) considers a simple model of two genes in a haploid organism. The combination of the favorable mutant alleles A_2 and B_2 in a single individual will be achieved, on the average, in less than one-half as many generations in sexual as in asexual populations. Moreover, the advantage of recombination is strongly dependent on population size. According to Bodmer, the advantage is greater in small than in large populations. Crow and Kimura (1965), using a different model, found that, on the contrary, recombination is more important in large than in small populations.

Full-fledged sexual reproduction involves the union of the nuclei of two cells, gametes, followed sooner or later by meiosis and the formation of new gametes that contain half as many chromosomes as the zygote did. This occurs in eukaryotes, from protozoans to man. A remarkable variety of phenomena of gene recombination exists also, however, in prokaryotes, which lack discrete nuclei. One of the mechanisms is DNA-mediated transformation, described in Chapter 2. First discovered in pneumococci, it has been found in several other microorganisms, among them *Bacillus subtilis* and *Hemophilus influenzae*. How widespread gene recombination by transformation may be in nature is, however, an open question.

The already classical work of Lederberg and Tatum showed that certain strains of *Escherichia coli* containing so-called fertility or *F* factors undergo conjugation, that is, pairwise union of cells, followed by recombination of genetic materials. The conjugating bacterial cells do not fuse, however, to form zygotes. Moreover, the exchange of genetic materials is not equal and mutual, as it is in some eukaryotic protozoans, such as paramecia. Instead, one of the conjugating cells (F^+) is the donor of a chromosome, which is transferred as a whole or in part to the recipient (F^-) cell. Thereupon, recombination of the indigenous and the transferred chromosomes takes place (Jacob and Wollman 1961). Escherichia and probably other bacteria have a single chromosome, which is a closed circle.

Apparently the most widespread and important gene recombination mechanism in prokaryotes is transduction, first studied in Salmonella and Escherichia. This involves the transfer of genetic materials from one cell to another by means of a temperate phage serving as a vector. A temperate phage does not normally cause lysis of the bacterial cell in which it occurs, but reproduces at the same rate as the cell containing it. It may migrate, however, from one cell to another, doing this without the cells coming into physical contact, as in conjugation. The phage carries a part of the genetic material of the bacterial cell, and this material may undergo recombination with the chromosome of the recipient cell. In specialized transduction, a phage transfers only some particular section of the host chromosome, rather than any part of it. Different strains of bacteriophages may also undergo recombination with each other if two or more of them infect simultaneously the same host bacterial cells.

It must be understood, of course, that in prokaryotes, and also in some eukaryotes, gene exchange occurs by no means in every generation. Several or many asexual generations are interposed between the recurrent recombination events. This evolutionary pattern makes sense. Indeed, gene recombination is just as efficient in making new gene constellations as in breaking them. Suppose that a highly fit gene combination A_2B_2 has arisen in a population predominantly A_1B_1. If the genes A and B are not linked, meiosis in an $A_1A_2B_1B_2$ heterozygote will produce only one-quarter of A_2B_2 gametes. By contrast, asexual reproduction of A_2B_2 individuals will yield (barring mutation) all A_2B_2 offspring. By the time the next sexual generation occurs, the fitter genotype is likely to have multiplied in comparison to its frequency in the generation in which it arose.

The Retrogression of Sexuality—Selfing

Viewed in the aspect of the adaptation of a species to its environment, gene recombination amounts to evolutionary experimentation. It produces swarms of new genotypes, some of which may be advantageous in some environments. By the same token, genotypes will arise that will be disadvantageous and will be discriminated against by natural selection. The life cycle of a mammal or a tree may be longer than that of a bacterium by four or more orders of magnitude. In more complex organisms, especially higher animals, fertilization, meiosis, and gene recombination occur in every generation.

Asexual reproduction is found predominantly in microorganisms. As Stebbins (1950) has stated:

> Organisms like bacteria, fungi, and the smaller algae are usually small in size, are destroyed in huge numbers by their enemies, and depend for their survival chiefly on their ability to build up large populations rapidly in favorable medium. During the periods of increase the production of any organism not adapted to this medium is a wasted effort on the part of a population whose very life depends on reproductive efficiency. Such a growing population cannot experiment with new gene combinations; it must sacrifice flexibility to immediate fitness.

The necessity of such sacrifice under some circumstances explains why sexuality, as well as gene recombination, which is its corollary, went into eclipse in many evolutionary lines.

Some invertebrate animals and most higher plants are hermaphrodites. The production of both female and male gametes in the same individual makes self-fertilization possible but by no means necessary. There is a great diversity of contrivances that increase the probability of cross-fertilization in hermaphrodites or even ensure that it will occur. The most radical one is self-sterility, which makes the male gametes unable to fertilize the female gametes of the same individual. Selfing is rare or absent in most hermaphroditic snails and oligochaete worms, but it seems to be the rule in the remarkable hermaphroditic fish, *Rivulus marmoratus* (Harrington and Kallman 1968). Elsewhere self- and cross-pollination or insemination are both possible, and their relative frequencies vary all the way from zero to 100 percent. Self-pollination is obligatory, however, in so-called cleistogamous flowers, which are so constructed that foreign pollen is excluded (e.g., subterranean clover, Morley and Katznelson 1965).

Stebbins (1950, 1957), Grant (1958), Allard (1965), and others have sought to relate the reproductive biology to the ecological characteristics of higher plants. Among 101 species of grasses examined by Stebbins, 26 species are self-incompatible and hence cross-pollinated. All of them are perennial in growth habit, half of them being rhizomatous (underground stems or runners) and the other half cespitose (matted). Among 32 self-pollinating species, 27 are annuals and 5 cespitose perennial. The intermediate group of 43 species, in which both selfing and crossing are frequent, contains 39 cespitose perennials, 1 rhizomatous perennial, and 3 annuals. One of the advantages of selfing is that formation of seeds is assured regardless of possible shortage of insect pollinators and unfavorable weather conditions. This is evidently more important for annual than for perennial plants. Another possible advantage is that a self-pollinating population can withstand occasional reduction to very small size (Moore and Lewis 1965), and that new colonies may start from a single seed.

Selfing is a form of evolutionary opportunism, which sacrifices the evolutionary plasticity given by gene recombination for immediate, and perhaps ephemeral, adaptive advantages. According to Stebbins (1957), it leads into an evolutionary "blind alley," because "it apparently closes the door to the elaboration of radically new adaptive devices. On the other hand, a group of species may travel a long way down this 'alley' by evolving new variations on the theme laid down for them by their cross-fertilizing ancestors."

The studies of Allard and his school (Allard 1965, Harding, Allard, and Smeltzer 1966; Jain and Marshall 1967; Jain 1969, and references therein) have disclosed that the genetic system in at least some self-pollinators preserves more genetic variability than was formerly thought possible. Thus, despite more than 95 percent frequency of selfing in populations of lima beans (*Phaseolus lunatus*), the populations maintain a balanced polymorphism for the alleles S and *s* of a certain gene, the heterozygotes S*s* having a fitness higher than the homozygotes SS and *ss*. Furthermore, the heterotic advantage is frequency-dependent; it is greatest when the heterozygotes are rare in a population. A parallel situation has been found in barley.

The two wild oats species, *Avena fatua* and *A. barbata,* are introduced weeds that are highly successful in California. The former species has only 3–5 percent of outcrossing by cross-pollination, whereas in the latter outcrossing is even less frequent. The wild populations of both show considerable variability, the nature of which is interestingly different in the two species. The variability in *A. barbata* is to a large extent due to a reaction norm permitting adaptive phenotypic plasticity; that in *A. fatua* shows genetic polymorphisms maintained by heterotic balancing selection.

Festuca microstachys is an array of many genetically different lines, reproducing almost exclusively by selfing. Allard and Kannenberg (1968) found no evidence of cross-pollination in some 20,000 individuals. Individual plants here appear to be homozygous, and yet the natural populations are genetically highly diversified, because they are complex mixtures of numerous selfing pure lines. Systematists have subdivided *F. microstachys* into eight named "species." Since they are aggregations of noninterbreeding pure lines, rather than Mendelian populations, giving them one or eight or some other number of Latin names is an arbitrary procedure.

Such arbitrariness has plagued the systematics of other selfing plants as well (Mansfield 1951 and Schwanitz 1967). For example, the name *Triticum aestivum* (or *T. vulgare*) applies to at least 404 varieties of soft wheats. They represent mostly different combinations of a smaller number of genetic characteristics and preserve their distinctiveness when propagated by self-pollination.

The Retrogression of Sexuality—Apomixis

Selfing conserves the outer appearances and the underlying basic processes of sexual reproduction—flowers or genitalia, meiosis, female and male gametes. Only fertilization is, in a sense, frustrated; the uniting gametes are formed in the same individual. Selfing and crossing are often facultative, and both may occur with varying frequencies in the populations of the same species (Rick, 1947, 1950).

Apomixis dispenses with the union of gametes. The variety of apomictic phenomena is quite impressive (see Gustafsson 1947, Stebbins 1950, White 1954, Carson 1967, and Suomalainen 1969). The opportunism of biological evolution is displayed in full view: the sexuality has been allowed to break down in many different ways.

In parthenogenesis, an egg cell develops into an embryo and ultimately an adult without fertilization. In some plants, the embryo sac develops normally, but the meiosis is suppressed, giving rise to an ovule with the diploid chromosome number, which develops without fertilization (diplospory). Instead, the embryo sac may be pushed aside, and the embryo develops from the diploid somatic tissue (apospory, adventitious embryony). Curiously enough, in some apomictic plants (Citrus, Rubus, and others) pollination of the flowers is necessary for the production of germinable seed, although the ovule does not unite with the male gamete (pseudogamy). Still other apomicts no longer flower or mature seeds; they reproduce entirely by bulbs, bulbils, runners, stolons, etc. (e.g., the water weed *Elodea canadensis* in Europe).

Arrhenotokous parthenogenesis is found in hymenopteran and in a few representatives of other insect orders (such as scolytid beetles, Takenouchi and Takagi 1967). The females are diploid and come from eggs fertilized by spermatozoa; the males are haploid and develop from unfertilized eggs. Here the evolutionary advantages of sex are preserved, because gene recombination occurs regularly at meiosis in females. Hymenoptera are perhaps the most progressive order of insects. Thelytokous parthenogenesis is a very different phenomenon, resembling diplospory in plants. Meiosis is usually suppressed, and unfertilized eggs yield only or mostly females so that the populations are then unisexual. An outlandish situation has been discovered, however, in the amazon molly, *Mollienesia formosa* (C. P. Haskins, E. F. Haskins, and Hewitt 1960, Kallman 1962, and Hubbs 1964). Populations of this fish consist of females, which are "sexual parasites." They attract males of related species (*M. latipinna* and *M. sphenops*) to inseminate them; their eggs require for development stimulation by spermatozoa, but the sperm nuclei are eliminated. Hence the progeny of a female is identical with the mother in genotype. In a colony of *M. formosa* studied by Kallman, roughly 80 percent of the individuals belonged to only two clones.

Several, though mostly short-range, evolutionary advantages explain the sacrifice of sex in favor of apomixis. In contrast to self-fertilization, apomixis facilitates the maintenance of heterozygosity and of the resulting hybrid vigor. We saw in Chapters 5 and 7 that the polymorphism maintained by heterotic balancing selection entails a "price"; homozygotes whose fitness is lower than that of the heterozygotes are inevitably produced in every generation. With many polymorphisms the "price" is likely to be too high. Apogamy offers an escape, if all of the offspring have (barring mutation) the same genotype as their mother.

Where apomixis has not displaced fertilization entirely, the high heterozygosity of apomictic strains can be brought to light. Thus, in the blue grass Poa apomictic progenies are usually uniform, whereas sexual progenies evince great variability, sometimes no two individuals being alike. This explains one of the sad pages in the history of genetics. Mendel's experiments on the crossing of hawkweeds (Hieracium) appeared to him to give results contradictory to those he obtained in his classical work on peas. Mendel did not know that most hawkweeds are highly heterozygous apomicts.

Apomixis facilitates the establishment of polyploids, particularly those with odd numbers of chromosome complements, triploids, pentaploids, etc. We saw in Chapter 11 that polyploids are rare in nature in bisexual animals; polyploidy in animals is usually combined with parthenogenesis and unisexual populations. Triploid and tetraploid races and species have long been known, however, in the brine shrimp Artemia, wood lice Trichoniscus, the moth Solenobia, and earthworms Eisenia and Allolobophora (review in White 1954). The extensive studies of Suomalainen (1962, 1969) brought to light many parthenogenetic triploid and tetraploid species and races of weevils (Curculionidae), closely related to species that are bisexual, diploid, and presumably ancestral. Parthenogenetic triploid strains have been discovered even in vertebrates, such as a goldfish (Cherfas 1966) and a lizard (Darevsky 1966).

Apomixis, like polyploidy, is more widespread in the plant than in the animal kingdom. Apomictic species are found in many families, together with sexual ones. Many plants can reproduce both by apomixis and by cross-pollination, thus exploiting the advantages of both types of reproductive biology—generating ever new genotypes by recombination, and guarding the genotypes of proven fitness by the prevention of recombination. Some genetic systems are inherently unstable, however, and can be perpetuated only by asexual means. In this category are polyploidy with odd numbers of chromosome sets (triploids, pentaploids), and aneuploidy with some of the chromosomes triplicated whereas others are diploid. Such bizarre genotypes have, nevertheless, high fitness in some wild and especially in cultivated plants, and are conserved by apomixis (including artificial grafting and propagation by cutting and replanting in cultivated forms). Some garden plants (e.g., bananas) have lost sexual reproduction entirely, although their wild relatives mature seeds.

The transition from normal bisexual crossing to parthenogenesis has been studied experimentally in Drosophila by Stalker (1954) and Carson (1967a,b). In several species of Drosophila a small proportion of eggs can develop without fertilization, and at least one species, *Drosophila mangabeirai*, has parthenogenesis as the normal method of reproduction. In *D. mercatorum*, Carson was able, by selection, to increase the proportion of parthenogenetic eggs from about 0.1 to 0.4 percent. In nature, the selective stimulus comes from the presence of chromosomal structural heterozygosis, which would cause semisterility under cross-fertilization. Such structural heterozygosis has been found in the parthenogenetic fly *Lonchoptera dubia* (Stalker 1956), in parthenogenetic Drosophila (the works of Stalker and Carson, cited above), in grasshoppers (White, Cheney, and Key 1963), and in many plant apomicts.

When apomixis and selfing become the predominant or exclusive methods of reproduction, they make the delimitation of species a perplexing problem. Botanists have long regarded such genera as Rubus, Crataegus, and Hieracium as notoriously "difficult" (Gustafsson 1947, Stebbins 1950, and Hedberg 1955; see also Maslin 1968 on parthenogenetic vertebrates). The opinions of different authorities on what constitutes a species in these genera vary so widely that it is not uncommon to find one investigator uniting under a single specific name a complex of forms divided by other botanists into numerous "species." The subdivision of the mass of clones into the species *Escherichia coli, Salmonella typhosa,* and *S. enteritidis* is also a matter of taste; one might just as well regard all of them as a single species.

This does not mean that the diversity in asexual groups is continuous. On the contrary, aggregations of many more or less clearly distinct genotypes, each of which reproduces its like, are found. Although the carriers of these genotypes are sometimes called elementary species, they are not integrated groups, like the species in the cross-fertilizing forms. The term elementary species is, therefore, misleading and should be discarded. The existing genotypes do not embody all the combinations of genes potentially possible. As in cross-fertilizing organisms, the genotypes in the asexual ones are clustered on "adaptive peaks," while "adaptive valleys" remain uninhabited. Furthermore, the clusters are arranged in an hierarchical order, in a way again analogous to that in sexual forms. Some of the clusters may, then, be designated as species, others as subgenera, still others as genera, and so on. Which one of these ranks is ascribed to a given cluster is, however, a matter only of convenience, and the decision is in this sense arbitrary. The species, as a category less arbitrary than the rest, is lacking in asexual and obligatory self-fertilizing organisms.

A Genetic Tour de Force—Oenothera

Two plant genera, Hieracium and Oenothera, have played curiously opposite roles in the history of genetics. Mendel, as stated in the preceding section, was frustrated by the results of his experiments on Hieracium, which he was unable to interpret since he did not know that these plants are apogamic. Some years later, de Vries developed his mutation theory on the basis of his work on *Oenothera lamarckiana,* also unaware of the peculiarities of the genetic system in his object. Some of his "mutations" were not mutations at all, but rare recombination products. Many ingenious studies by Renner, Cleland, Emerson, Darlington, Catcheside, Oehlkers, and others were needed to decipher the peculiarities of this genetic system (reviews in Cleland 1950, Stebbins 1950, and Grant 1964a).

Species of Oenothera are widespread in North America; some of them (including *Oenothera lamarckiana*) have become successful weeds when introduced into Europe. *Oenothera hookeri* in California and adjacent states is large-flowered and mostly cross-pollinated, and its 14 chromosomes form 7 bivalents at meiosis. A swarm of other "species," especially in the eastern United States, are permanent translocation heterozygotes, exhibiting at meiosis rings of from 4 to all 14 chromosomes, and from 5 to 0 bivalents. *Oenothera lamarckiana* forms a ring of 12 chromosomes and a single bivalent. The meiotic disjunction of the chromosomes is generally regular, with adjacent chromosomes in the ring going to opposite poles and alternate chromosomes to the same pole; the formation of aneuploid gametes is thus avoided (see Chapter 5). The chromosomes involved in the ring form two "complexes"; the members of each complex are transmitted together, as effectively as though they were a single chromosome and the genes contained in them a single linkage group. The ring-forming oenotheras are mostly small-flowered and self-pollinated. Permanent heterozygosity for the two complexes is tenaciously conserved, and the production of homozygous seeds is obviated by a peculiar mechanism of two different nonallelic lethals, one being included in each complex. Either these lethals may act in early zygotes, or else one of them eliminates the pollen and the other the embryo sac that carries it.

What is the evolutionary meaning of this eccentric genetic system? It combines the advantages of heterosis with the assured seed set from selfing. Chromosomal inversions in Drosophila protect from recombination the supergenes contained in these chromosomes, which yield heterosis in the inversion heterozygotes (Chapter 5). *Oenothera lamarckiana* has, in effect, two grand supergenes in the complexes of the chromosomes forming the ring, and merely a single pair of chromosomes left to form a bivalent at meiosis. The occurrence of new translocations or of rare recombinations between the supergenes produces some of the "mutants" that de Vries observed in his classical work.

In addition to Oenothera, some other plants have attempted the same genetic tour de force. *Rhoeo discolor* has all of its 12 chromosomes forming a single ring at meiosis; it belongs to a monotypic genus of the family Commelinaceae. Some species of Clarkia (Onagraceae, the family to which Oenothera also belongs), Trillium (Liliaceae), Paeonia (Paeoniaceae), and others have gone a part of the way in the same direction.

Rudimentation and Regressive Evolution

Evolution is opportunistic; genetic changes are favored by natural selection if they increase the Darwinian fitness of their carriers at a given time and place, regardless of whether they might be favorable or otherwise in the long run. Any adaptive peak, however temporary, is occupied by a population, if it is at all accessible. Adaptation to new environments may decrease the importance of some organs and functions that were vital in past environments; such organs and functions may then become vestigial and disappear. Zoology and botany provide an abundance of examples of the rudimentation and disappearance of organs, especially among obligatory parasites. Thus many internal parasites have lost the alimentary canal and sense organs, such as eyes, that were doubtless present in their free-living ancestors. Some parasitic plants have lost the chlorophyll. If the acquisition of these organs and substances was a result of progressive evolution, then their disappearance can only be designated as regressive evolution.

Some of the best examples of rudimentation are encountered among cave inhabitants. Reduction and disappearance of the eyes and of the pigmentation of the body, as well as development of organs with tactile functions and of specialized behavior patterns, are observed in cave animals, from vertebrates to insects, crustaceans, and flatworms. Excellent reviews of the pertinent evidence are those of Vandel (1964), Barr (1968), and Poulson and White (1969). Instances in which the same or closely related species occur in the caves, as well as out of them, are particularly enlightening, since here the processes of adaptation to subterranean life may be studied. The variability in the structure of such organs as the eyes within the population of a single cave may be striking. Some individuals have fully developed eyes, whereas in others only rudiments are present; some individuals are fully pigmented, and others colorless. Aberrant individuals resembling the cave forms in certain particulars may be found in the surface populations as well, but their frequency outside the caves is small.

Although some of these variations have been proved to be hereditary, no less important is the extraordinary phenotypic plasticity of certain characters of cave animals. Many years ago Kammerer showed that the salamander *Proteus anguinus,* which, when kept in the dark, has vestigial eyes and little or no pigment, develops eyes and a black pigmentation when grown under light. The genotype of an organism is, in general, adjusted to ensure the development of vital organs and functions in the environments likely to be encountered; a deterioration of this adjustment may be one of the first steps of the rudimentation process.

The phenomena of rudimentation were happy hunting grounds for those who believed that acquired modifications are heritable. If such a Lamarckian explanation is ruled out, two hypotheses need be considered. First, mutation pressure may lead to rudimentation if not opposed by natural selection. Mutations that weaken or destroy an organ are more frequent than those that strengthen it; so long as the functioning of an organ is vital to its possessors, selection will discriminate against such mutations. When an organ ceases to be vital, selection is relaxed and the organ may regress. There is little evidence, however, that this actually happens. Post (1962, 1964, 1965) and Wolpoff (1969) compiled data suggesting that relaxation of the selection in human populations against such traits as color blindness and lessened hearing acuity may lead to increases in their frequencies. Wright (1964a) and Prout (1964) noted the extreme improbability of mutation alone causing the disappearance of an organ in a species, although such a view has been urged by Brace (1963).

Second, natural selection may favor the rudimentation of functionless organs. Weismann introduced the idea of a struggle of parts of the body; since it takes energy to build an organ, genetic changes that remove a superfluous organ save energy and thereby acquire positive adaptive value (Barr 1968 traces this idea back to Darwin, St. Hilaire, and Goethe). Evidence in favor of this view has been summarized by, among others, Rensch (1960), Barr (1968), and Byers (1969). In the nature of things, a direct test of the hypothesis is hardly possible; perhaps the best that can be said in its favor is that it does not encounter objections of the kind Wright and Prout raised against the mutation theory.

Reticulate Evolution

Reproductive isolation and gene recombination resulting from sexual processes stand in mutual opposition. Sex generates a diversity of genotypes, among which there may arise some highly fit ones, which will be promoted by natural selection. Reproductive isolation, on the other hand, hinders or prevents gene exchange between populations that occupy different adaptive peaks. It wards off the origination of swarms of disharmonious gene patterns, and thus conserves the historically evolved arrays of genotypes that constitute the biological species. The conflict is resolved by a compromise. The frequency of gene exchange between Mendelian populations is regulated, so that genetic adaptability is maintained at the price of elimination of the smallest possible numbers of genotypes of low fitness. As could be expected, the compromise is reached in different organisms at different points in the gene exchange-reproductive isolation continuum. By and large, the reproductive isolation of animal species is more rigid than that of plant species. In stable environments, gene recombination is less likely to produce improved genotypes than in natural, and especially in man-made, environments that are in flux.

Infiltration of the genes of one species into another is termed introgression (Anderson 1949). There has been some controversy concerning how widespread and how important in evolution introgressive hybridization is. That it sometimes occurs is undeniable, but some authors have succumbed to the temptation of attributing universality to their special discoveries. Judicious reviews of botanical evidence can be found in Stebbins (1950, 1958b), Grant (1963), and Ehrendorfer (1963), and of zoological results in Mayr (1963) and Remington (1968). Some examples of introgression in both animals and plants were given in Chapters 10 and 11, in connection with the discussion of isolating mechanisms.

Riley (1952 and earlier) described a particularly clear example of introgression in the populations of *Iris fulva* and *I. hexagona* var. *gigantocaerulea*, which inhabit the Mississippi Delta region. The former species grows by preference on clay soils and in shade, whereas the latter prefers tidal marshes and full sun. This ecological isolation has suffered a breakdown because of the destruction of the forests and the drainage of the swamps for pastures. It is especially in such man-made habitats that hybrids between these Iris species are found; the F_1 hybrids are partially sterile, but backcrosses to the parental species give rise to populations of *I. hexagona* with some genes derived from *I. fulva*.

An equally convincing case of introgression is that between the sunflowers *Helianthus annuus*, *H. bolanderi*, and *H. petiolaris* (Heiser 1947). Many sunflower populations are now weeds growing on disturbed soils, and they have arisen by introgressive hybridization of two or more ancestral species. Some species of Eucalyptus in Australia have formed hybrid swarms (for examples, see Clifford and Binet 1954), and so have some species of oaks (Muller 1952 and Forde 1962), Liatris (Levin 1968), and certain other plants. Viemeyer (1958) states that species of Penstemon may amalgamate to a common gene pool by hybridization, and similar claims have been made for other plants (e.g., blueberries, Vaccinium) by other authors. The elegant work of Ownbey (1950) has shown that three species of Tragopogon in the northwestern United States have formed hybrids, which by doubling the chromosome complements gave rise to three derived allopolyploid species.

Some of the alleged cases of introgressive hybridization in animals need critical re-examination. According to Mayr (1963), secondary intergradation occurs when populations that were allopatric in the past come into contact again. Remington (1968), has stated, "From a study of the geographic occurrences of contemporary hybridization among North American animals, it has become apparent that most of the hybrids are produced in a few relatively localized zones, with little hybridizing in the vast areas between these zones of mixing." He records six major and seven minor mixing zones, which he calls suture zones, in the United States and Canada.

The hybridization of formerly ecologically isolated species in new habitats disturbed by human activities belongs in a different category. The two toad species, *Bufo americanus* and *B. fowleri*, discussed in Chapter 10, exemplify this situation. Other likely examples are the hybrids of fish species Notropis (Hubbs and Strawn 1956), turtles Pseudemys (Crenshaw 1965a), and birds—red-eyed towhees (Sibley 1954), orioles (Sibley and Short 1964), and flickers (Short 1965). Inferential but fairly convincing evidence of hybridization exists also for butterflies Pieris (Petersen 1955), leafhoppers (Ross 1958), coccinellid beetles Chilocorus in America (Smith 1962, 1966) and in central Asia (Zaslavsky 1966, 1967), and two species of Drosophila (Pipkin 1968). On the other hand, what appeared to be hybridization in nature of two species of the fruit fly Dacus has been rendered questionable by the study of Gibbs (1968).

Why introgression is more frequent in plants than in animals, and in some groups than in others, is not completely clear. Stebbins (1950, 1958b) and Grant (1958, 1963, 1966c) have advanced some interesting hypotheses. Many plants are perennial, long lived, or capable of asexual reproduction. An individual or a clone may persist for centuries or even millennia. Sexual reproduction by seeds serves in such plants to maintain genetic adaptability and evolutionary plasticity. This gives more freedom for wide outcrossing than exists in animals, where sexual union and gene recombination are obligatory in every generation. Furthermore, the open system of growth and the relative simplicity of plant tissues, as contrasted with the closed system of growth and the great complexity of tissues and organ systems in animals, are also relevant. The fitness of an animal genotype depends on the entire constellation of its genes, as an integrated whole, whereas in plants the genes may affect fitness more nearly independently.

DNA Hybridization

The discovery of the so-called *in vitro* DNA hybridization has opened new possibilities for assessing the genetic similarities and differences between organisms that cannot be hybridized. The double-helical structure of DNA in solution is denatured by heating to nearly 100°C. The resulting single strands of DNA can be renatured, made to come together in pairs, when the solution is cooled. The pairing is specific, that is, the new pairs formed consist of strands that have wholly or at least largely complementary sequences of nucleotides. The amount of pairing between DNA strands of different species decreases as the nucleotide sequences become more and more unlike.

There are several techniques to make the assessment quantitative. Suppose that one wishes to compare the DNA's of species A and B. The DNA from species A is heated to separate the polynucleotide strands, mixed with a solution of agar, and cooled quickly to prevent reunion of the strands. The single strands of A are now immobilized in the agar gel. The DNA of species B is marked with a radioactive tracer, broken by mechanical shearing into fragments of some hundreds of nucleotides, and heated to separate the strands. The single-stranded fragments of B are then incubated at 60°C with A strands entrapped in the agar. The fragments are small enough to penetrate the agar gel and to form duplex structures with the immobilized strands, provided that they find sufficiently similar mates. The amounts of reassociated and of left-over DNA are assayed by measurement of the radioactivity.

McCarthy and Bolton (1963) compared the amounts of binding of DNA fragments of *Escherichia coli* with agar-bound DNA's of certain other bacteria and other organisms. If the amount of binding of the *E. coli* fragments with the conspecific DNA in agar is taken as 100, the binding with other DNA's is estimated as follows:

Salmonella typhimurium	71	*Serratia marcescens*	7
Shigella dysenteriae	71	*Pseudomonas aeroginosa*	1
Aerobacter aerogenes	51	T2 bacteriophage	1

Hoyer, McCarthy, and Bolton (1964) and Hoyer, Bolton, et al. (1965) made extensive experiments comparing the hybridizing abilities of DNA's from man, chimpanzee, Old World and New World monkeys, prosimians, mouse, hamster, guinea pig, chicken, and salmon. As expected, the DNA's of animals belonging to the same family are more similar than those of different families of the same order, of the same order more similar than of different orders, and of the same class more similar than of different classes. An attempt was made to correlate the similarity of the DNA's to the amounts of time elapsed since the separation of the phylogenetic branches to which the animals belong, as estimated by paleontologists. The result is shown in Fig. 12.1. A tolerably good linear relationship is obtained between the logarithms of the DNA similarity and the times of divergence.

A somewhat different technique was used by Laird and McCarthy (1968a) to compare the DNA's of three species of Drosophila. Denatured DNA of one species was entrapped on a nitrocellulose mem-

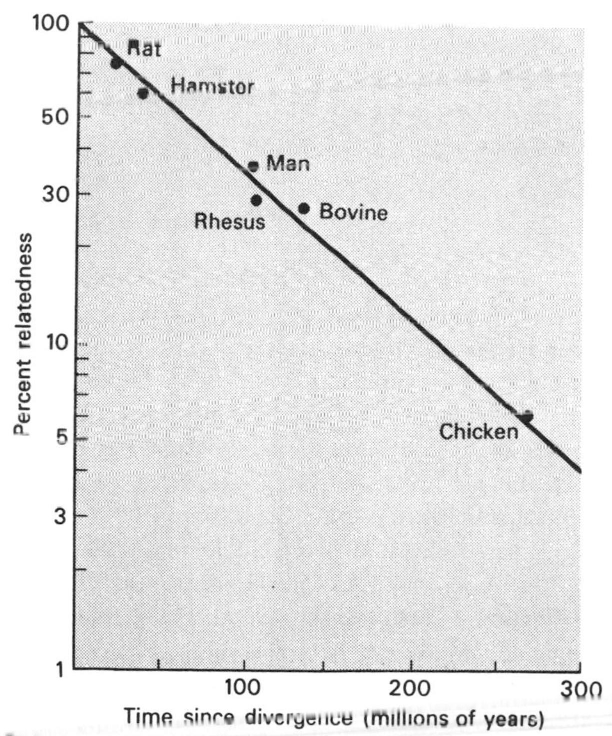

FIGURE 12.1

The relatedness, measured by the reaction of DNA fragments of different vertebrates with those of the house mouse, and its correlation with the time elapsed since the presumed separation of the respective evolutionary lines. (From Hoyer, McCarthy, and Bolton)

brane filter. The DNA of two species, labeled with a hydrogen isotope, was fragmented by shearing and denatured by heating to 95°C. The fragments of the DNA's were then made to compete for reunion with the DNA strands of one of them immobilized in the filter. The more similar are the DNA's, the greater proportions of the fragments find their homologues among the strands in the filter.

The magnitude of the differences found between species of the same genus is considered surprising by the authors. *Drosophila simulans* and *D. melanogaster* have about 80 percent of DNA sequences in common, and 20 percent "too distantly related to form stable duplexes." These species are related closely enough to give some viable, though completely sterile, hybrids. On the other hand, *D. funebris,* which belongs to a different subgenus and is not crossable to the other two, has only about 25 percent of common sequences. Contrary to the opinions of some geneticists (see Chapter 11), species formation entails genotypic reconstructions more profound than the substitution of a few "lucky" mutant alleles.

Redundancy in Genetic Materials

To interpret correctly the evolutionary implications of the results just described, it must be kept in mind that the "common" sequences in the DNA fragments do not necessarily contain identical genes. Gene alleles that differ in several mutational stages by nucleotide substitutions may well be carried in "common" sequences. The proportions of the gene differences between the species arrived at from DNA hybridization experiments are more likely, therefore, to be underestimates than overestimates. On the other hand, the genetic endowment of the same species and individual may contain redundancies, that is, some genes may be represented by numerous copies. This fact was mentioned in Chapter 1. Britten and Kohne (1968) found that in eukaryotic, though apparently not in prokaryotic, organisms, from 15 to 80 percent of the DNA consists of sequences repeated from 100 to 1,000,000 times. The remaining 85 to 20 percent appears to be unique sequences, although the technique used would probably not distinguish between unique sequences and those repeated two or even several times.

The estimates are deduced from the speed of the reassociation of the DNA fragments with DNA strands immobilized in agar or on filters.

The more copies of a sequence that are available, the sooner a fragment is likely to encounter its mate. Conversely, unique sequences may only come into contact with similar ones after a considerable lapse of time, or may never chance to encounter them. In some animals (calf) Britten and Kohne found an apparent discontinuity between a fraction of some 40 percent of DNA, which reassociates rapidly, and one that does so very slowly. By contrast, many degrees of repetition are present in the DNA of a fish (salmon sperm), varying continuously from perhaps 100 to 100,000 copies. In *Drosophila melanogaster* the estimated proportion of partially redundant sequences is only 10 percent (Laird and McCarthy 1968b).

Reduplication of genes has long been recognized as an important evolutionary process. On the assumption that primordial life was represented by a single gene, the thousands of different genes now found in the same gamete in most organisms must be the diverged descendants of the primordial gene. Visible evidence of gene multiplication in the phylogeny can be seen under the microscope in the giant chromosomes of the salivary glands of Drosophila and other flies. The so-called repeats are groups of from two to ten similar stainable discs, which are constantly present at different locations in the same chromosome set. Their genetic homologies are attested to not only by a visible similarity but also by a tendency to somatic pairing. The repeats consist of genes homologous because of common descent. These homologous genes may gradually diverge in structure, however, as a result of the occurrence and fixation of different mutations. Fitch and Margoliash (1970) distinguish two kinds of homologous genes: orthologous genes remain similar in function, as the different hemoglobin genes of the vertebrates; paralogous genes diverge to assume different functions, as the hemoglobin and myoglobin genes.

Redundant and unique genes within the same genotypes may, however, play different roles. Wallace (1963a), Stebbins (1969), and Britten and Davidson (1969) have independently arrived at similar hypotheses that are relevant at this point. Their substance is that most structural genes are found among the unique genetic materials, while the redundant fraction of the genetic endowment may consist mainly of regulatory or controlling genetic elements. The highly complex developmental processes in higher organisms may require a minority of relatively stable structural genes, outnumbered by more labile controlling genes.

According to Wallace, his hypothetical model

does not require variation of structural genes; the stability of DNA and of the amino acid sequences this material encodes does not conflict with our model in any way. This remarkable stability is entirely compatible, however, with genetic variation and with the series of isoalleles revealed by experiments of population geneticists. This variation lies in what may be called the "physiological" as opposed to the "structural" gene and its controlling element.

Britten and Davidson point out that repeated DNA sequences are interspersed with unique sequences, and state, "This is precisely the pattern required in our model if repeated sequences are usually or often regulatory in function." The above authors, as well as Stebbins, surmise that in the evolution of higher organisms genetic changes in the regulatory mechanisms, rather than in the structural genes, play the leading role. Stebbins, in particular, sees here a likely genetic matrix of progressive evolution. It is probably premature to attempt to transform these interesting hypotheses into a more elaborate theoretical structure. Further studies in this field must be awaited with intense interest.

Protein Phylogenies

In the fullness of time, the sequences of the genetic "letters," the nucleotides, in all genes of all organisms may imaginably, become known. It would then be possible to quantify the similarities and the differences between organisms and to erect a classification of living things, based on precise information about their genetic endowments. Although this utopia is very remote, it is becoming possible to determine and to compare the amino acid sequences in the orthologous and paralogous (see the preceding section) proteins of the same and of different species. Different amino acids are often found in similar positions in these linear sequences. Each amino acid in a protein, as we have seen, corresponds to a triplet of nucleotides in the DNA of the gene that codes for this protein. Consequently, it is possible to compute the mutation distances between homologous proteins; the mutation distance is defined "as the minimal number of nucleotides that would need to be altered in order for the gene for one protein to code for the other" (Fitch and Margoliash 1967b). Mutation distances between cytochromes c of organisms ranging from man to yeasts are shown in Table 12.1. A single mutation could transform human cytochrome c into that of a monkey, or vice versa. Some 66 mutations are needed, however, to effect a transformation of human and a yeast (Candida) cytochrome into each other.

Some aspects of protein evolution were considered in Chapters 2 and 8. The point that is essential for us here is that the amino acid sequences in certain proteins of different organisms are far too similar for these likenesses to be ascribed to chance. It is only reasonable to suppose that these proteins are homologous, and that the genes coding for them have descended by gradual accumulation of mutational changes from the same gene in a common ancestor (Ingram 1963, Zuckerkandl and Pauling 1965, Jukes 1966, Margoliash 1963, and others). Fig. 2.2 shows that most of the sites in different hemoglobins are occupied by the same amino acids, and even the sequence of amino acids in the myoglobin is too close to be attributed to chance. In most vertebrate animals, the alpha chains of the hemoglobins consist of 141 amino acids. In Table 8.4 it can be seen that related species (man and gorilla, horse and donkey) differ in 1 or 2 amino acids, representatives of different mammalian orders in about 20, and mammals and fish (carp) in 60–75. In addition to the hemoglobins, the most extensive comparative studies have been made on cytochrome c. The cytochromes c of representatives of different orders of mammals or birds differ in 2–17 amino acids, classes of vertebrates in 7–38, vertebrates and insects in 23–41, and animals and molds and yeasts in 56–72 (Table 12.1).

The mutation distances between homologous proteins have been used by Fitch and Margoliash (1967a,b, 1970) to construct phylogenetic trees of these proteins. Figure 12.2 shows a phylogeny of the cytochromes c. In superficial appearance, this phylogeny looks like classical phylogenetic trees of animal or plant groups constructed on the basis of morphological and paleontological information. The branches of the protein phylogenies show, however, estimates of the mutation distances between the proteins, in the present instance the cytochromes c, of any two species of organisms studied. These estimates are arrived at by statistical techniques that are simple in principle but involve calculations necessitating the use of electronic computers. Somewhat comparable, but not identical, statistical techniques were used by Cavalli-Sforza 1969, Cavalli-Sforza, Barrai, and Edwards 1964, and Cavalli-Sforza and Edwards 1964; to study the relationships

TABLE 12.1

Minimum numbers of mutations required to interrelate pairs of cytochromes c in the organisms studied (After Fitch and Margoliash)

Organism	No.	Number of Mutations Required for Transformation																		
Man	1																			
Monkey (Macaca)	2	1																		
Dog	3	13	12																	
Horse	4	17	16	10																
Donkey	5	16	15	8	1															
Pig	6	13	12	4	5	4														
Rabbit	7	12	11	6	11	10	6													
Kangaroo	8	12	13	7	11	12	7	7												
Duck	9	17	16	12	16	15	13	10	14											
Pigeon	10	16	15	12	16	15	13	8	14	3										
Chicken	11	18	17	14	16	15	13	11	15	3	4									
Penguin (Aptenodytes)	12	18	17	14	17	16	14	11	13	3	4	2								
Turtle (Chelydra)	13	19	18	13	16	15	13	11	14	7	8	8	8							
Rattlesnake (Crotalus)	14	20	21	30	32	31	30	25	30	24	24	28	28	30						
Fish (Tuna)	15	31	32	29	27	26	25	26	27	26	27	26	27	27	38					
Fly (Haematobia)	16	33	32	24	24	25	26	23	26	25	26	26	28	30	40	34				
Moth (Samia)	17	36	35	28	33	32	31	29	31	29	30	31	30	33	41	41	16			
Mold (Neurospora)	18	63	62	64	64	64	64	62	66	61	59	61	62	65	61	72	58	59		
Yeast (Saccharomyces)	19	56	57	61	60	59	59	59	58	62	62	62	61	64	61	66	63	60	57	
Yeast (Candida)	20	66	65	66	68	67	67	67	68	66	66	66	65	67	69	69	65	61	61	41
	No.	1	2	3	4	5	6	7	8	9	10	11	12	13	14	15	16	17	18	19

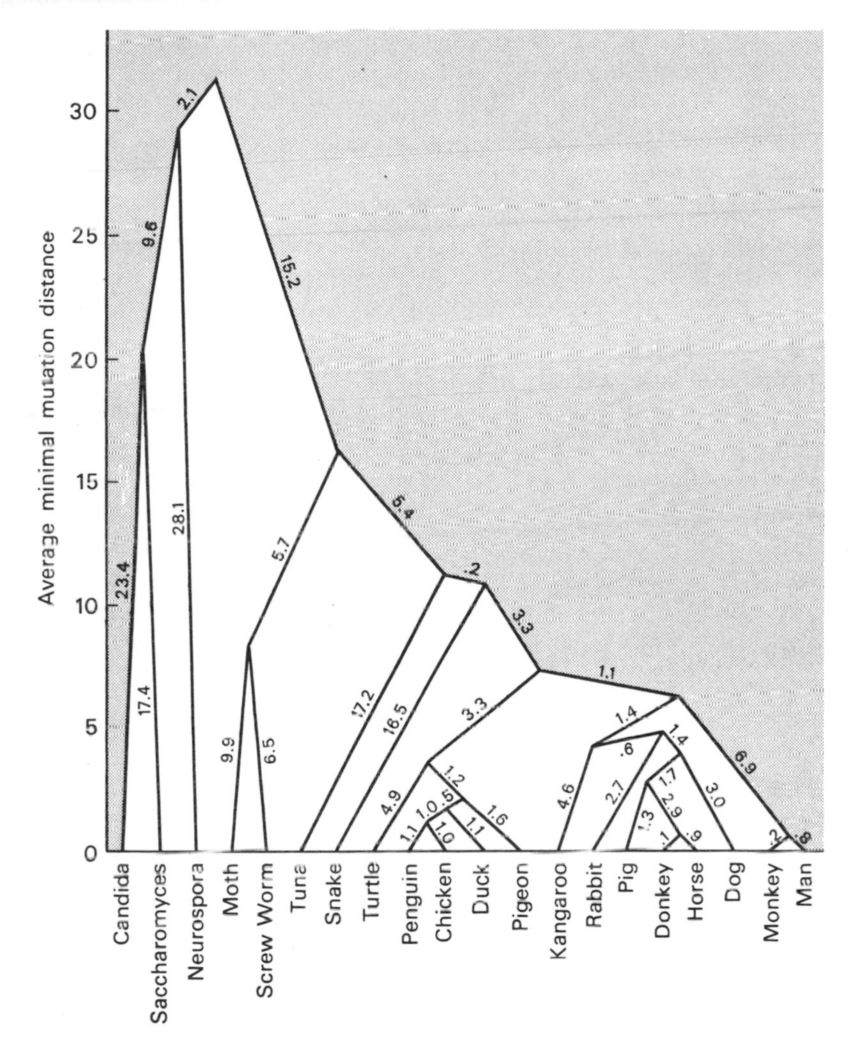

FIGURE 12.2

Phylogenetic relationships of the cytochromes c of different organisms. Further explanation in text. (From Fitch and Margoliash)

of human populations examined genetically and morphologically (see Chapter 8).

Without in the least underestimating the ingenuity and the promise of the method, some inconsistencies in the phylogeny in Fig. 12.2 must be indicated (they are recognized by Fitch and Margoliash 1970, themselves). The turtle turns out to be one of the birds, and quite remote from the only other reptile studied, the rattlesnake. The primates (man and monkey) branch off from the mammalian stem before the marsupial (kangaroo) does. Such inconsistencies cause no surprise to students of phylogenies; they are bound to occur if only a single trait is examined, even so sophisticated a one as the structure of a kind of protein.

A phylogeny of the protein chains composing the hemoglobins and the myoglobins has been constructed with the aid of similar methods (Zuckerkandl 1965 and Fitch and Margoliash 1967b). The kind of hemoglobin found most abundantly in the human adult consists of two alpha and two beta chains, of 141 and 146 amino acids, respectively. In the embryo and in the newborn, however, the prevalent hemoglobin consists of two alpha and two gamma chains. In addition, in the adult there is a relatively small amount of hemoglobin with beta replaced by delta chains. Myoglobin molecules are single chains of 153 amino acids each. All these kinds of molecules (or, rather, the genes coding for them) are descendants of a common gene that was present in a remote ancestor, perhaps a primitive vertebrate or an even earlier form. This gene underwent four duplications, giving rise to five genes, which diverged by accumulating different mutational changes. The earliest duplication, dated conjecturally about 650 million years ago, gave rise to genes coding for myoglobins, on the one hand, and to those coding for hemoglobins, on the other. The second duplication happened in the hemoglobin gene, perhaps 380 million years ago, and differentiated the genes that code for alpha chains from those for the other chains. The third duplication, say 150 million years ago, differentiated the embryonic gamma from the adult beta and delta chains. The last-named ones are the most similar in the amino acid alignments, and their genes may have started to diverge only 35 million years ago (Zuckerkandl and Pauling 1965 and Zuckerkandl 1965).

Rates of Evolution and Evolutionary "Clocks"

If genetic differences were to accumulate at identical rates in all phylogenetic lines, the amount of genetic differentiation observed would be directly proportional to the time elapsed since the separation of any two lines. It is well known that this is not the case; the paleontological record shows that some groups evolve rapidly and others slowly. The assessment of the evolutionary rates is based, of course, on comparative morphological studies. The morphological, as well as the physiological and ethological, differentiation may, however, be taken to reflect, at least roughly, the amount of genetic differences. Exceptions, such as sibling species and convergent evolutionary lines, do not invalidate the general rule.

Paleontologists, most notably Simpson (1949, 1953, 1961, and other works), have firmly established that the rates of origination and extinction (in other words the "longevity") of species, genera, and families are much higher in some groups than in others. Thus, according to Simpson, the mean "survivorship" of a genus of bivalve molluscs is 78 million years, whereas for a genus of carnivore mammals it is only about 8 million years, approximately one order of magnitude difference. The extremes differ, of course, far more than the averages. The oldest, or at any rate one of the oldest, surviving species is the crustacean *Triops cancriformes,* known from the Jurassic period, some 170 million years ago. The brachiopod genus Lingula existed in the Ordovician, 400 million years ago, but the living Lingula is supposed to be specifically distinct from its fossil predecessor. Among mammals, opossums lived in the Cretaceous, 60 million years ago. Another case of extremely slow, bradytelic evolution (or of evolution arrested) is the coelacanth fish, Latimeria, the survivor of a once widespread and abundant group. Among plants, a species of Plantago has endured for about 20 million years (Stebbins and Day 1967). Man, on the contrary, is an example of very rapid, tachytelic evolution. *Homo habilis* (also classified as *Australopithecus africanus habilis*) lived only some 1.75 million years ago (Tobias 1965). *Homo sapiens presapiens* is perhaps 200,000 years old.

The cause of the differences between evolutionary rates is a matter of speculation (Simpson 1949, Schmalhausen 1949, and Rensch 1960a). Although very little is known about the mutation rates in bradytelic and tachytelic groups, mutation is quite unlikely to be the limiting factor in evolutionary transformations. Drosophila is, compared to man, evolving slowly, and yet the best estimates of the average mutation rates per generation are about the same. The mutation rates per unit of time are consequently lower in the rapidly developing hominid line. The evolutionary rate is a complex phenomenon, having to do more with selectional than with mutational processes. Environmental changes generally speed up evolution, and constancy slows it down. As a creative process (see below), evolution has an unpredictable element (at least at the present level of our knowledge).

The term quantum evolution (Simpson 1953) refers to evolutionary events that involve all-or-none transitions: "They are changes of adaptive zone such that transitional forms between the old zone and the new cannot, or at any rate do not, persist." What is involved is the adoption of novel ways of life, and hence of novel adaptive needs and possibilities. The emergence of man from his animal ancestors is a prime example of quantum evolution. Man adapts primarily by culture and to culture, and culture does not exist, except as minuscule rudiments, in nonhuman animals.

Studies on the evolution of proteins have led to new approaches to the problem of evolutionary rates. Given the numbers of amino acid differences in a group of homologous proteins, one can estimate the mutation distances between them (see the preceding section). These estimates can then be compared to those of the times elapsed since the separation of the phyletic lines, as appraised by paleontologists. King and Jukes (1969) give the following average rates of evolution of nine kinds of proteins in mammals, expressed in terms of 10^{-10} substitution per triplet codon of DNA per year:

Insulin A and B	3.3	Immunoglobulin light chain	33.2
Cytochrome c	4.2	Fibrinopeptide A	42.9
Hemoglobin alpha	9.9	Bovine hemoglobin fetal chain	22.9
Hemoglobin beta	10.3	Guinea pig insulin	53.1
Ribonuclease	25.3		

Different proteins in the same group of animals evolve at clearly different rates. We saw in Chapter 8, however, that the amino acids at some positions within a protein chain are rigidly conserved throughout the evolutionary transformations, whereas other portions of the same chains vary more or less freely. It is reasonable to infer that the amino acids in the conservative portions are indispensable for the function which the protein serves. Mutations are probably as frequent in the conservative codons as in the variable ones. The mutants in the former act as lethals or at least as subvitals, and are eliminated by the normalizing natural selection. By contrast, the mutants in the variable codons are close to neutral. If so, they are tolerated by natural selection and are allowed to drift in the gene pool of a population. Most of them are eventually lost, but some reach fixation by random walk processes. The variable portions of the proteins are, then, assumed to evolve mainly by drift, while the constant ones change very slowly, if at all, by natural selection. The assumption of a dichotomy of constant versus variable parts is alleged to be confirmed by statistical tests. The numbers of the substitutions in the variable codons fit the expectations calculated with the aid of the Poisson distribution formula (see Chapter 8). A possible source of bias in such calculations is the deliberate exclusion from consideration of codons that do not fit the assumption of randomness.

A logical next step, made by Zuckerkandl and Pauling (1965), is to suppose that substitutions of the amino acids in protein chains, or at least in the adaptively neutral parts of those chains, occur at uniform rates in time. If so, comparative studies of the protein structure in different organisms can be used as an evolutionary "clock." Knowing the mutational distances between the proteins of representatives of two or several groups of organisms, and having an estimate of the substitution rates, one can arrive at an approximate date for the separation of the ancestral phylogenetic lines of these organisms.

Perhaps the boldest essay to read this "clock" is that of Sarich and Wilson (1967, and Wilson and Sarich 1969). The phylogeny of the higher primates, including man, has several incertitudes. One of these concerns the time of divergence of the three lines, those leading to the now living great apes, to man, and to the Old World monkeys (Cercopithecidae). There is apparently good paleontological evidence which leads some authors to conclude that the three lines separated more or less simultaneously, perhaps as long as 30 million years ago. Other authorities suppose that the separation of the human and the ape stems is much more recent, perhaps 4–5 million years ago, while the divergence of these two lines from the monkeys stem occurred in greater antiquity.

The gist of Sarich and Wilson's arguments is that the serum albumins and transferrins (studied by immunological techniques), the hemoglobins (studied by amino acid sequencing), and the DNA's (studied by "hybridization"; see above) are more similar in man and the apes (chimpanzee and gorilla) than in either of these and the monkeys. The mutational distance between man and chimpanzee hemoglobins is zero, man and gorilla 2, and monkey and man or monkey and ape 15–17. The average rate of amino acid replacement in mammalian hemoglobins is 1 per some 3.5 million years. Therefore, Sarich and Wilson state, "If the probability model of protein evolution . . . is applied to this case, then it can be calculated with the Poisson distribution that there is less than one chance in 10^5 that a sequence difference of zero to two residues could result when the divergence time was 30 million years ago. . . ." The evidence adduced by these authors is certainly impressive, but their argumentation skirts close to the edge of circularity.

Adaptive Radiation and Group Selection

At the beginning of this chapter reference was made to the two evolutionary modes: cladogenesis or splitting, and anagenesis or phyletic evolution. These two modes are not mutually exclusive and, in fact, are combined in various proportions in most lines of descent. Simpson (1953) has said, "It is strikingly noticeable from the fossil record and from its results in the world around us that some time after a rather distinctive new adaptive type has developed it often becomes highly diversified." Anagenesis and cladogenesis together result in adaptive radiation, that is, occupation and subdivision of more and more numerous adaptive zones and ecological niches. "There has been tremendous increase in the number of broad adaptive zones and in the fineness and multiplicity of subdivision of these in the course of geologic time" (Simpson 1953).

Comparison of the diversity of species and genera in different geographic regions and different ecological circumstances is one of the possible approaches to the analytic study of adaptive radiation. It has been known for a long time that there is a greater diversity of both plants and animals in the tropics than in temperate climates, and in the latter than in cold climates. Although one does not find many species of penguins or polar bears on the Equator, the rule has considerable generality. Moreover, it applies to recent as well as to fossil, and marine as well as to land, organisms (Stehli 1968 and Stehli, Douglas, and Newell 1969). Fig. 12.3 shows that the greatest diversity of mosquito genera is found in tropical America, Asia, and Africa, and the least diversity in subarctic and subantarctic lands. It must be understood, of course, that the diversity of taxa may be the reverse of the

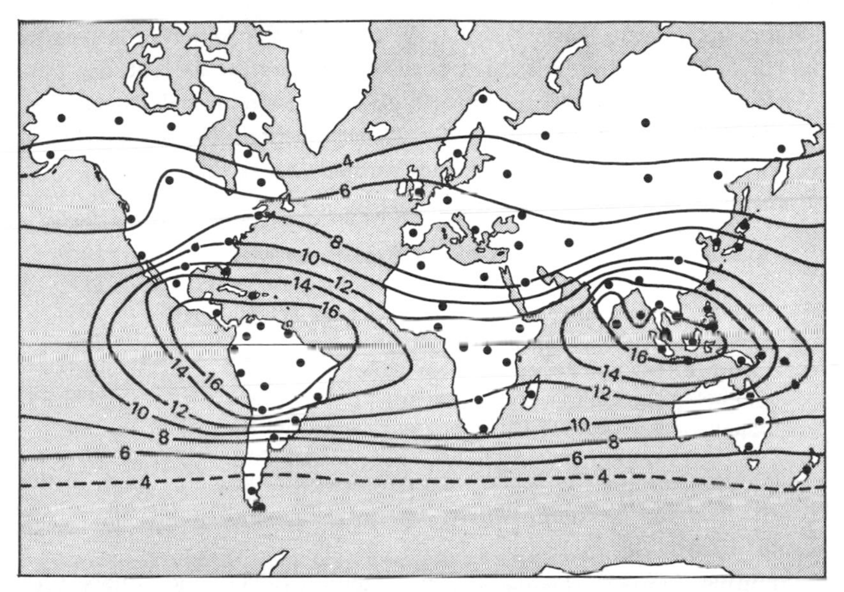

FIGURE 12.3

Numbers of genera of mosquitoes found in different territories. The places sampled indicated by dots. (From Stehli)

abundance of individuals. Although there may be clouds of mosquitoes in summer in northern Canada, Alaska, and Siberia, they are likely to belong to a single or a few species. In Amazonia, on the other hand, each mosquito you see is often specifically different from the last one.

Simpson (1964c) has produced probably the most careful and detailed study of the species diversity of mammals in North and Central America. He subdivided the continental area (ignoring islands) into squares 150 miles on each side and counted the numbers of species recorded in each square. An interesting and complex picture emerged. Species densities varied from 13 on the Melville peninsula (latitude 70°) to 163 in southern Costa Rica (latitude 10°). The north-south gradient is, however, far from regular. The species numbers increase from the extreme north to about the Canada-United States border, and then the diversity stays approximately constant down to the latitude 30°. The increasing trend is then resumed through Mexico to the climax in Costa Rica. A strong dependence on the topography emerges clearly. In the eastern United States the greatest diversity is found in the Appalachians, and in the West in the Rocky

Mountains and the Sierra Nevada. By contrast, the region of low relief in Yucatan and Honduras has low species counts for its latitude.

Baker (1967) records the species diversity of mammals specifically on the lowlands along the Pacific Coasts of both North and South America. His numbers for the zones studied are as follows:

Tundran (Alaska)	25
Coniferan (southeastern Alaska to San Francisco Bay)	92
North Eremian (southern California to Sonora)	114
North Subtropical (Mexico from Sinaloa to Tehuantepec)	109
Tropical (Chiapas to northern Peru)	267
South Eremian (Peru to northern Chile)	18
Nothofagian (middle and southern Chile)	35
Patagonian (southern Patagonia and Tierra del Fuego)	10

The only irregularity is the relatively low number of species in the South Eremian section, which has few lowlands, and these are harsh deserts.

Simpson (1964c), Pianka (1966), Stehli (1968), and others (see the papers just mentioned for further references) have attempted to formulate hypothetical explanations of the above observations. Pianka lists six such explanations, most of which are not mutually exclusive, and all of which may be true in some situations.

1. All biotic communities diversify as time goes on; older communities have more species than younger ones. Because of glaciations, the cold-temperate and cold zone communities are the newest whereas the tropical and subtropical ones are the most ancient.

2. The environmental complexity grows as one proceeds toward the tropics. It also increases from lowlands to territories with complex topographic relief. Life responds to environmental complexity by increasing adaptive diversity.

3. Natural selection in the temperate and cold zones depends mainly on physical factors, and in the subtropics and tropics on biotic factors. There are more ecological niches in the tropics, and they are more finely subdivided there.

4. There are more predators and parasites in the tropics. This supposedly holds the prey populations to lower density levels, decreases the competition, and makes genetic changes more permissive.

5. Regions with stable climates allow the evolution of more specialized adaptedness.

6. Territories with greater energy production support a greater diversity of inhabitants.

Wright (1931, 1932) in his classical papers expounded the view that territorial subdivision of a species into numerous semiisolated colonies, especially colonies with moderate to small genetically effective population sizes, gives the most favorable conditions for evolutionary progress. A species so subdivided has, at it were, numerous scouting parties which "explore" the field of gene combinations. One or more of these parties may "discover" new, previously unoccupied, adaptive peaks (see Chapter 1), or superior genetic endowments for the higher levels of the old adaptive peak. A great majority of the parties, of course, discover nothing new. They eventually die out or become swamped by migrants from the more successful parties. This is interpopulational or group selection, natural selection that promotes the perpetuation and spread of genetically better endowed populations. Differential reproduction, the spread or the extinction, of competing species is a form of group selection. Species are reproductively, rather than only territorially, isolated Mendelian populations.

The effectiveness of group selection and its role in evolution are controversial matters. Among recent authors, Wynne-Edwards (1962) believes that group selection is responsible for the evolution of physiological and behavioral mechanisms that lower the reproductive rates of populations reaching or exceeding optimal densities. These mechanisms prevent the populations from overtaxing and eventually destroying the resources of their habitats.

At the opposite extreme, Williams (1966) uses the principle of Occam's razor to argue that group selection should not be assumed to exist if a form of adaptedness can conceivably be ascribed to intrapopulational selection. He wields this razor with some abandon, leaving only the polymorphism at the *t* locus in the house mouse (Dunn 1964 and Lewontin 1962) plausibly ascribed to group selection. Several mutant alleles at this locus distort the segregation in heterozygous males; a considerable majority of the functional spermatozoa carry the mutant, and only a minority the normal allele. This confers upon the mutant alleles sizable selective advantages, which cause them to spread in mice populations. The mutant alleles are, however, lethal or cause sterility when homozygous, so that a population evidently cannot be homozygous for them. This case, by the way, is a diagrammatically clear demonstration that natural selection does not necessarily increase the adaptedness of a population. Group selection solves this predicament. Mouse colonies with high frequencies of lethal *t* alleles are replaced by migrants from colonies where *t* is absent or infrequent. Levin, Petras, and Rasmussen (1963) have questioned, however, whether group selection is a valid explanation of this situation.

It is easy to see that interpopulational or group selection per se is a process much less efficient than intrapopulational selection. The turnover of generations within a colony is obviously much more rapid than the extinction and replacement of colonies or of species. A 10 percent advantage of the carriers of an allele, A_1, over the carriers of A_2 would increase the frequency of A_1 in a population within a a number of generations much smaller than would be required for the extinction and replacement of colonies. Suppose that a genetic variant decreases the reproductive rate of its carriers when a population has outgrown the resources of its environment. How can such a variant spread in a species? Since natural selection has no foresight, it cannot avoid overpopulation, ending in a catastrophe of extinction. Selection cannot sacrifice the immediate for long-range benefits. One possibility is group selection: most populations commit suicide by excessive growth and destruction of their resources; only the populations that restrain their reproductive potentials survive. I agree with Williams (1966) that a more probable course is selection favoring individual variants within populations which "act on the assumption that reproduction is not worth the effort in crowded situations." In other words, selection favors individual variants that reproduce only when the environment is favorable for the survival of their offspring.

A closely related problem is the development in evolution of various forms of "altruistic" behavior. Individuals may expose themselves to risks of injury or death for the benefit of conspecific individuals. Human altruism presupposes conscious and free choice between alternative courses of action and is seemingly confined to a single species, man. There are, however, countless examples not only of parents defending their progeny, but also of some members of a herd or a flock (such as the strongest adult males) interposing themselves between a predator and the weaker, particularly the immature, members. For our purposes, altruism may be defined as behavior that is likely to bring injury or death to individuals so behaving but that benefits other members of the species.

Haldane long ago (1932) and Wright more recently (1949) discussed the possibility of genes for altruism being promoted by natural selection. This is possible in a large undivided population only if the individuals who benefit are close relatives of the altruist, and therefore are likely to be carriers of the same genes. To say the same thing in another way, the number of genes preserved in the relatives should exceed that lost in the altruist if he perishes. Assuming a single dominant gene, the self-sacrifice of an altruist should save more than two of his children or siblings, or even greater numbers of cousins or more remote relatives. The requirements are progressively relaxed, of course, if the altruist is part of the way through his reproductive age. Post-reproductive individuals should be altruists even if they save only a single relative. With group selection, tribes containing many altruists may well have an advantage over those the members of which fend only for themselves.

The converse is true for genes predisposing toward egotism, aggression, or "criminal" behavior, which benefit the individuals behaving in these ways but injure conspecific individuals. Without group selection such genes will be favored, unless they do much damage to close relatives. Neel and Schull (1968) discuss, however, a model of frequency-dependent selection for such genes:

> A primitive, polygynous society in which each male was highly aggressive might so decimate itself in the struggle for leadership that it was non-viable. At high frequencies of aggressiveness, the non-aggressive male who could keep aloof from the sanguine struggle might stand a better chance of survival and reproduction than the more aggressive, with his chance a function of the amount of aggressiveness in the group. But at low frequencies of the same traits, the aggressive male who assumed leadership (and multiple wives) would be the object of positive selection, and the more passive the group, the greater his reproductive potential. There would thus be an intermediate frequency at which this phenotype (and its genetic basis) would tend to be stabilized.

That natural selection can bring about sacrifice of individuals for the benefit of the species is incontestably shown by social insects. Termite, ant, bee, and wasp societies consist of numerous "workers" that do not reproduce, and a minority, often a single pair that belong to a reproductive caste. Natural selection acts to make the workers serve most efficiently the welfare of this reproductive caste. It is the group, the society, that is the unit of selection. The reproductive caste is analogous to the sex cells, and the workers to the somatic cells of a multicellular body. The origin of the situation need not be envisaged as group selection in the sense of Wynne-Edwards (see above). In a colony, the workers are siblings of the individuals of the reproductive caste that will be the founders of the next generation of colonies. Therefore, the workers are serving to transmit genes identical with their own, although borne in other individuals. The process is selection promoting the survival and perpetuation of groups of genetically related individuals.

The Irreversibility of Evolution

There is considerable literature, much of it polemical, concerning the so-called Dollo's "law" of irreversibility of evolution, first formulated in 1896 or even earlier (for discussions, see Simpson 1953 and Rensch 1960a). Simply stated, the law (or rule) asserts that evolution does not double back on itself to return to ancestral states. Irrevocability of evolution is a complementary principle: the influence of ancestral conditions is not wholly lost in descendent groups.

The validity of these propositions obviously depends on the level of the evolutionary changes considered. A mutation is, in principle, reversible; a substitution of a nucleotide somewhere in a DNA chain may be reversed by restoration of the old nucleotide. A bacterial strain made resistant to an antibiotic by a mutation may revert to sensitivity by another mutation. Mutations restoring an old phenotype do not always restore the old genotype as well; suppressor mutations are by no means rare, and the apparent revertant may actually differ from the ancestral condition not in a single gene change but in two. Starting with a Drosophila population that is, on the average, neutral to light and to gravity, one can select photo-and geopositive and negative strains (see Chapter 7). Relaxation of the artificial selection, or selection in the opposite direction, causes reversion to photo- and geo-neutrality. Whether the reverted neutrals are really genetically identical to the original population is an open question.

More and more genetic changes accumulate as evolution proceeds;

a reversal grows less and less probable, and soon it becomes impossible. An evolutionary event, such as a mutation that is temporarily favored by selection, is reversible; an evolutionary history of many successive gene substitutions is practically irreversible. An analogy with human history suggests itself—an electoral victory of a political party may be reversed at the next election, but the Roman Empire as it was under Augustus cannot possibly be restored, even if this were desirable. Land-dwelling vertebrates are descendants of water-living ancestors; dolphins and whales returned to live in water, but they have not become fishes again.

A loose distinction is often drawn between microevolution and macroevolution, according to the magnitude of the morphological, physiological, or genetic changes. One may say that microevolutionary changes are possibly reversible, whereas macroevolutionary ones are not. This does not mean that there are two kinds of evolution, micro- and macro-, as was contended by some authors, principally believers in orthogenetic "evolution from within." The distinction is quantitative, and any boundary can only be arbitrary.

Certain microevolutionary events can be induced, or reproduced, in experiments. The origin of allopolyploid species is an example (Chapter 11). Some beginnings of macroevolutionary (or, if one prefers, mesoevolutionary) changes are, however, on record. The operation of the founder principle in chromosomally polymorphic experimental populations of *Drosophila pseudoobscura* was described in Chapter 8. An interesting variability, verging on the indeterminacy of the selection results, was observed. Intimations of the irreversibility, or the irrevocability, of evolution can be seen also in the experiments of Strickberger (1963) on chromosomal polymorphism in the same species. The environments in which Drosophila is cultured in laboratories are patently different from the outdoor environments. Not unexpectedly, natural selection, unpremeditated by the experimenter, tends to increase the genetic adaptedness of laboratory populations to the particular environments in which they are kept.

Strikeberger began his experiments with two kinds of populations of *D. pseudoobscura* homokaryotypic for AR and for CH gene arrangements in their third chromosomes (see Chapter 5); some populations had been maintained for many generations in laboratory culture bottles, and others in laboratory population cages. They diverged genetically in response to these different environments, each becoming more efficient in its own environment. Strickberger then set up new polymorphic experimental populations with chromosomes that had different selectional histories. All populations were started by founders with 20 percent AR and 80 percent CH chromosomes. Within about a year (some 10-12 generations) the populations reached new equilibrium frequencies of the chromosomes. The equilibrium depended, however, on the previous selectional histories of the chromosomes. When AR chromosomes came from culture bottles and CH chromosomes from cages, the equilibrium was between 53 and 58 percent AR. In the reverse combination, AR from cages and CH from bottles, the equilibrium was 90–97 percent AR. Most remarkably, these populations showed no trend toward convergence to a common equilibrium level; 1065 days (some 32 generations) from the start, the high equilibrium populations had 92 percent AR chromosomes.

Evolution as a Creative Process

Evolution is a creative process, in precisely the same sense in which composing a poem or a symphony, carving a statue, or painting a picture are creative acts. An art work is novel, unique, and unrepeatable; a copy of a painting, however exact, is only a reproduction of somebody's creation. Some creativity is possible, however, in playing music, the performer adding his own to that of the composer. The evolution of every phyletic line yields a novelty that never existed before and is a unique, unrepeatable, and irreversible proceeding. As we have seen above, only a most elementary component of the evolutionary process, a mutation followed by selection of a single gene allele, may be recurrent. An evolutionary history is a unique chain of events.

The objection has been raised that, after all, different biological species are only different combinations of a more limited assortment of mutational gene variants, something like different figures in the same kaleidoscope. This misses the point entirely. Following this line of reasoning, one would have to say that mosaic pictures have nothing new in them, because they consist of different combinations of the same kinds of colored stones. The stones are the raw materials; the creative act occurs when the stones are put together to compose a picture. Nucleotide substitutions in DNA chains provide gene variants, which are components of the genetic endowments of organisms. Mutation is a chemical (or, if you wish a molecular) process; selection is a biological or "compositionist" process (Simpson 1964a, b; see also Chapter 1).

How can selection, a process devoid of foresight, be the composer of biological symphonies? This idea is preposterous if selection is misrepresented as only a preserver of a lucky minority of random mutations. Selection is, however, much more than a sieve retaining the lucky and losing the unlucky mutants (see Chapter 7). The meaning of the adjective random as applied to mutations was explained in Chapter 3. Colors may be mixed at random in a pile of stones, but the artist composing a mosaic picture does not use them at random. Selection also creates order out of randomness. It renders possible formation of living systems that would otherwise be infinitely improbable. Nothing can be simpler and more ingenious than its mode of operation: gene constellations that fit the environment survive better and reproduce more often than those that fit less well. As the aesthetic satisfaction of the artist and the beholder is the guide in artistic creation (except, perhaps, in some forms of modern "art"), so the supremacy of life over death is the guiding principle of biological creation.

Unfortunately, a creative process is liable to end in frustration and failure. Extinction is the commonest fate of phyletic lines. The present inhabitants of the globe are the descendants of a minority of the denizens of past epochs, and of smaller and smaller minorities as we look further and further back. The extinct forms are miscarriages of natural selection, consequences of its lack of foresight. Man's foresight has not, however, enabled him to avoid miscarriages of his social and political history! Anyway, the living world not only persists but also contains a greater variety of forms and more complex, sophisticated, or "progressive" ones than in the past. Teilhard de Chardin (1959) has said that evolution is "pervading everything so as to find everything." This is an overstatement, since the potentially possible variety of gene patterns is vastly greater than the variety ever realized. Yet he is right in essence: natural selection has tried out an immense number of possibilities and has discovered many wonderful ones. Among which, to date, the most wonderful is man.

BIBLIOGRAPHY

The following abbreviations of the names of periodicals are used:

A.N.	*American Naturalist*
Ch.	*Chromosoma* (Berlin)
C.S.H.	*Cold Spring Harbor Symposia on Quantitative Biology*
E.	*Evolution*
E.B.	*Evolutionary Biology* (Th. Dobzhansky, M. K. Hecht, and Wm. C. Steere, Eds.), Appleton-Century-Crofts, New York
Ga.	*Genetika* (Moscow)
G.R.	*Genetical Research* (Cambridge)
Gs.	*Genetics*
Hs.	*Hereditas*
Hy.	*Heredity*
N.	*Nature*
P.N.A.S.	*Proceedings of the National Academy of Sciences (U.S.A.)*
P. No.I.C.G.	*Proceedings of the Nth International Congress of Genetics*
Q.R.B.	*Quarterly Review of Biology*
S.	*Science*
S.Z.	*Systematic Zoology*
U.T.P.	*University of Texas Publications*
Z.i.A.V.	*Zeitschrift für induktive Abstammungs- und Vererbungslehre*

Abbott, U. K., V. S. Asmundson, and K. R. Shortridge. 1962. Nutritional studies with scaleless chickens. I. The sulphur-containing amino acids. *G.R.*, 3:181-195.

Abplanalp, H., D. C. Lowry, I. M. Lerner, and E. R. Dempster. 1964. Selection for egg number with X-ray induced variation. *Gs.*, 50:1083-1100.

Aeppli, E. 1952. Natürliche Polyploidie beiden Planarien *Dendrocoelum lacteum* (Müller) und *Dendrocoelum infernale* (Steinman). *Z.i.A.V.*, 84:182-212.

Aird, I., H. H. Bentall, and J. A. F. Roberts. 1953. A relationship between cancer of the stomach and the ABO blood groups. *Brit. Med. J.*, 1:799-801.

Alexander, R. D. 1968. Life cycle origins, speciation, and related phenomena in crickets (*Orthoptera, Gryllidae*). *Q.R.B.*, 43:1-41.

Alexander, R. D., and Th. E. Moore. 1962. The evolutionary relationship of 17-year and 13-year cicadas, and three new species. *Misc. Publ. Museum Zool. Univ. Mich.*, 121:1-57.

Ali, M. A. K. 1950. Genetics of resistance to common bean mosaic virus (bean virus 1) in the bean (*Phaseolus vulgaris L.*). *Phytopathology*, 40:69-79.

Allard, R. W. 1965. Genetic systems associated with colonizing ability in predominantly self-pollinating species. Pp. 50-76 *in*: H. G. Baker and G. L. Stebbins (Eds.), "The Genetics of Colonizing Species," Academic Press, New York.

——, and L. W. Kannenberg. 1968. Population studies in predominantly self-pollinating species. XI. Genetic divergence among the members of the *Festuca microstachys* complex. *E.*, 22:517-528.

Allen, A. C. 1966. The effects of recombination on quasi-normal second and third chromosomes of *Drosophila melanogaster* from a natural population. *Gs.*, 54: 1409-1422.

Allfrey, V. G., V. C. Littau, and A. E. Mirsky. 1963. On the role of histories in regulating ribonucleic acid synthesis in cell nucleus. *P.N.A.S.*, 49:414-421.

Allison, A. C. 1955. Aspects of polymorphism in man. *C.S.H.*, 20: 239-255.

——. 1964. Polymorphism and natural selection in human populations. *C.S.H.*, 29:137-150.

Alström, C. H., and R. Lindelius. 1966. A study of the population movement in nine Swedish populations in 1800-1849 from the genetic statistical viewpoint. *Acta Gen. Stat. Medica,* 16:1-44.

Amadon, D. 1966. The superspecies concept. *S.Z.*, 15:245-249.

Ames, O. 1937. Pollination of orchid through pseudo-copulation. *Harvard Univ. Bot. Museum Leaflet* 5.

Anders, G. 1955. Untersuchungen über das pleiotrope Manifestations-muster der Mutante lozenge-clawless ($1_z{}^{cl}$) *von Drosophila melanogaster. Z.i.A.V.*, 87: 113-186.

Anderson, E. 1949. Introgressive Hybridization. John Wiley, New York.

Anderson, S. 1966. Taxonomy of gophers, especially *Thomomys* in Chihuahua, Mexico. *S.Z.*, 15:189-198.

Anderson, W. W. 1968. Further evidence for coadaptation in crosses between geographic populations of *Drosophila pseudoobscura. G.R.*, 12:317-330.

——, and L. Ehrman. 1969. Mating choice in crosses between geographic populations of *Drosophila pseudoobscura. Amer. Midland Natur.*, 81:47-53.

——, C. Oshima, T. Watanabe, Th. Dobzhansky, and O. Pavlovsky. 1968. Genetics of natural populations. XXXIX. A test of the possible influence of two insecticides on the chromosomal polymorphism in *Drosophila pseudoobscura. Gs.*, 58:423-434.

Andrewartha, H. G. 1952. Diapause in relation to the ecology of insects. *Biol. Rev.* 27:50-107.

——, and L. C. Birch. 1954. The Distribution and Abundance of Animals. Univ. Chicago Press, Chicago.

Anonymous. 1965. The Use of Induced Mutations in Plant Breeding. Report of the Meeting of FAO of the United Nations and the IAEA. Pergamon Press, New York.

Arnheim, N., and C. E. Taylor. 1969. Non-Darwinian evolution consequences for neutral allelic variation. *N.*, 223:900-903.

Ashton, G. C., D. G. Gilmour, C. A. Kiddy, and F. K. Kristjansson. 1967. Proposals on nomenclature of protein polymorphism in farm livestock. *Gs.*, 56: 353-362.

Astaurov, B. L. 1969. Experimental polyploidy in animals with special reference to the hypothesis of an indirect origin of natural polyploidy in bisexual animals. *Ga.*, 5, No. 7:129-149.

Aston, J. L., and A. D. Bradshaw. 1966. Evolution in closely adjacent plant populations. II. *Agrostis stolonifera* in maritime habitats. *Hy.*, 21:649-664.

Atkin, N. B., and S. Ohno. 1967. DNA values of four primitive chordates. *Ch.*, 23:10-13.

Auerbach, Ch. 1949. Chemical mutagenesis. *Biol. Rev.*, 24:355-391.

——. 1965. Past achievements and future tasks of research in chemical mutagenesis. *P.* XI *I.C.G.*, 2:275-284.

Auerbach, C., and D. Ramsay. 1968. The influence of treatment conditions on the selective mutagenic action of diepoxybutane in Neurospora. *Japan. J. Genetics*, 43:1-8.

Avery, A. G., S. Satina, and J. Rietsema, 1959. The Genus Datura. Ronald Press, New York.

Aya, T., Y. Kuroki, T. Kajii, K. Oikawa, and T. Miki. 1967. A familial survey of a B/C chromosome translocation. *J. Fac. Sci. Hokkaido Univ. Zool.*, 16:148-157.

Ayala, F. J. 1965. Sibling species of the *Drosophila serrata* group. *E.*, 19:538-545.

——. 1966. Evolution of fitness. I. Improvement in the productivity and size of irradiated populations of *Drosophila serrata* and *Drosophila birchii*. *Gs.*, 53: 883-895.

——. 1967. Evolution of fitness. III. Improvement of fitness in irradiated populations of *Drosophila serrata*. *P.N.A.S.*, 58:1919-1923.

——. 1968. Biology as an autonomous science. *Amer. Scientist*, 56:207-221.

——. 1969a. Genetic polymorphism and interspecific competitive ability in Drosophila. *G.R.*, 14:95-102.

——. 1969b. Evoluton of fitness. V. Rate of evolution of irradiated populations of Drosophila. *P.N.A.S.*, 63:790-793.

——. 1969c. An evolutionary dilemma: Fitness of genotypes versus fitness of populations. *Canad. J. Gen. Cytol.*, 11:439-456.

Babcock, E. B., and G. L. Stebbins, Jr. 1938. The American species of *Crepis*. *Carnegie Inst. Washington Publ.* 504.

Baglioni, C. 1967. Molecular evolution in man. Pp. 317-337 *in:* J. F. Crow and J. V. Neel (Eds.), "Proceedings of the 3rd International Congress on Human Genetics," Johns Hopkins Press, Baltimore.

Bailey, D. W. 1959. Rates of subline divergence in highly inbred strains of mice. *J. Heredity*, 50:26-30.

Baker, H. G. 1965. Characteristics and modes of origin of weeds. Pp. 147-172 *in:* H. G. Baker and G. L. Stebbins (Eds.), "The Genetics of Colonizing Species," Academic Press, New York.

Baker, P. T. 1966. Human biological variation as an adaptive response to the environment. *Eugen. Quart.*, 13:81-91.

Baker, P. T., G. R. Buskirk, J. Kollias, and R. B. Mazess. 1967. Temperature regulation at high altitude. Quechua Indians and U. S. whites during total body cold exposure. *Human Biol.*, 39:155-169.

Baker, P., and J. Weiner. 1966. Biology of Human Adaptability, Oxford Univ. Press, New York.

Baker, R. H. 1967. Distribution of recent mammals along the Pacific coastal lowlands of the Western Hemisphere. S.Z., 16:28-37.

Baker, W. 1947. A study of isolating mechanisms found in *Drosophila arizonensis* and *Drosophila mojavensis*. *U.T.P.*, 4720:126-136.

Ball, R. W., and D. L. Jameson. 1966. Premating isolating mechanisms in sympatric and allopatric *Hyla regilla* and *Hyla californiae*. *E.*, 20:533-551.

Band, H. T. 1963. Genetic structure of populations. II. Viabilities and variances of heterozygotes in constant and fluctuating environments. *E.*, 17:198-215.

——. 1964. Genetic structure of populations. III. Natural selection and concealed genetic variability in a natural population of *Drosophila melanogaster*. *E.*, 18: 384-404.

——, and P. T. Ives. 1963. Genetic structure of populations. I. On the nature of the genetic load in the South Amherst population of *Drosophila melanogaster*. *E.*, 17:198-215.

——, and P. T. Ives. 1968. Genetic structure of populations. IV. Summer environmental variables and lethal and semilethal frequencies in a natural population of *Drosophila melanogaster*. *E.*, 22:633-641.

Bandlow, G. 1951. Mutationsversuche an Kulturpflanzen. II. Züchterisch wertvolle Mutanten bei Sommer- und Wintergersten. *Der Züchter*, 21:357-363.

Barbour, I. G. 1966. Issues in Science and Religion. Prentice-Hall, Englewood Cliffs.

Barnicot, N. A. 1959. Climatic factors in the evolution of human populations. *C.S.H.*, 24:115-129.

Barr, Th. C. 1968. Cave ecology and evolution of troglobites *E.B.*, 2:35-102.

Bartolos, M., and Th. Baramki. 1967. Medical Cytogenetics. Williams & Wilkins, Baltimore.

Bastock, M. 1956 A gene mutation which changes a behavior pattern. *E.*, 10: 421-439.

——, and A. Manning. 1955. The courtship of *Drosophila melanogaster*. *Behaviour*, 8:85-111.

Bateman, A. J. 1948. Intra-sexual selection in Drosophila. *Hy.*, 2:349-368.

——. 1951. The taxonomic discrimination of bees. *Hy.*, 5:271-278.

Bates, Marston. 1949. The Natural History of Mosquitoes. Macmillan, New York.

Bateson, W. 1894. Materials for the Study of Variation. Macmillan, London.

Battaglia, B. 1958. Balanced polymorphism in *Tisbe reticulata*, a marine copepod. *E.*, 12:358-364.

——. 1964. Advances and problems of ecological genetics of marine animals. *P. XI I.C.G.*, 2:451-463.

——, Lazzaretto, I., and L. Malesani-Tajoli. 1966. Attivita sessuale degli omo ed eterozigoti per i geni delle pigmentazione nel copepode marino *Tisbe reticulata*. *Arch. Oceanogr. Limnol.*, 14:359-364.

——, and H. Smith. 1961. The Darwinian fitness of polymorphic and monomorphic populations of *Drosophila pseudoobscura* at 16°C. *Hy.*, 16:475-484.

Baur, E. 1924. Untersuchungen über das Wesen, die Entstehung und die Vererbung von Rassenunterschieden bei *Antirrhinum majus*. *Bibliotheca Genetica*, 4:1-170.

——. 1930. Einführung in die Verenbungslehre. Borntraeger, Berlin.

Beadle, G. W., and M. Beadle. 1966. The Language of Life. Doubleday, New York.

Beardmore, J. A., Th. Dobzhansky, and O. Pavlovsky. 1960. An attempt to compare the fitness of polymorphic and monomorphic experimental populations of *Drosophila pseudoobscura*. *Hy.*, 14:19-33.

Beermann, W. 1955a. Cytologische Analyse eines *Camptochironomus* Artbastards. I. *Ch.*, 7:198-259.

——. 1955b. Geschlechtsbestimmung und Evolution der genetischen Y Chromosomen bei *Chironomus*. *Biol. Zentral.*, 74:525-544.

——. 1956. Inversion-Heterozygotie und Fertilität der Männchen von *Chironomus*. *Ch.*, 8:1-11.

Bell, G. 1969. I-DNA: Its packaging into I-somes and its relation to protein synthesis during differentiation. *N.*, 224:326-328.

Belyaev, D. R., and V. I. Evsikov. 1967. Genetics of fertility of animals. The effect of the fur color on the fertility of mink (*Lutreola vison* Brisson). *Ga.*, 1967, No. 2:21-34.

Bender, H. A., and R. E. Gaensslen. 1967. Physiological genetics. Pp. 487-504 *in:* J. W. Wright and R. Pal (Eds.), "Genetics of Insect Vectors of Disease," Elsevier, Amsterdam.

Benoit, J., P. Leroy, R. Vendrely, and C. Vendrely. 1960. Experiments on Pekin ducks treated with DNA from Khaki Campbell ducks. XXIX *Trans. New York Acad. Sci.*, 22:494-503.

Benzer, S. 1957. The elementary units of heredity. Pp. 70-93 *in:* W. D. McElroy and B. Glass (Eds.), "A Symposium on the Chemical Basis of Heredity," Johns Hopkins Press, Baltimore.

Berelson, B., and G. A. Steiner. 1964. Human Behavior: An Inventory of Scientific Findings. Harcourt, Brace, & World, New York.

Berg, L. S. (1922) 1969. Nomogenesis, or Evolution Determined by Law. M.I.T. Press, Cambridge, Mass.

Berg, R. L. 1960. The ecological significance of correlation pleiades. *E.*, 14:171-180.

Berlin, B., D. E. Breedlove, and P. H. Raven. 1966. Folk taxonomies and biological classification. *S.*, 154:273-275.

Bernal, J. D. 1967. The Origin of Life. Weidenfeld & Nicholson, London.

Bianchi, N. O., and O. Molina. 1966. Autosomal polymorphism in a laboratory strain of rat. *J. Heredity*, 57:231-232.

Birch, L. C. 1955. Selection in *Drosophila pseudoobscura* in relation to crowding. *E.*, 9:389-399.

Birdsell, J. B. 1950. Some implications of the genetical concept of race in terms of spatial analysis. *C.S.H.*, 15:259-314.

Blair, A. P. 1941. Variation, isolating mechanisms, and hybridization in certain toads. *Gs.*, 26:398-417.

——. 1942. Isolating mechanisms in a complex of four toad species. *Biol. Symposia*, 6:235-249.

Blair, W. F. 1955. Mating call and stage of speciation in the *Microhyla olivacea-M. carolinensis* complex. *E.*, 9:469-480.

——. 1958a. Mating call in the speciation of anuran amphibians. *A.N.*, 92:27-51.

——. 1958b. Call differences as an isolating mechanism in Florida species of hylid frogs. *Quart. J. Florida Acad. Sci.*, 21:32-48.

——. 1964. Isolating mechanisms and interspecies interactions in anuran amphibians. *Q.R.B.*, 39:334-344.

——, and M. J. Littlejohn. 1960. Stage of speciation of two allopatric populations of chorus frogs (*Pseudacris*). *E.*, 14:82-87.

Blakeslee, A. F., A. D. Bergner, and A. G. Avery. 1937. Geographical distribution of chromosomal prime types in *Datura stramonium*. *Cytologia*, Jubilee Volume: 1070-1093.

Blaylock, B. G. 1966. Chromosomal polymorphism in irradiated natural populations of *Chironomus*. *Gs.*, 53:131-136.

Bloom, B. S. 1964. Stability and Change in Human Characteristics. John Wiley, New York.

Blum, H. F. 1959. Cancerogenesis by Ultraviolet Light. Princeton Univ. Press, Princeton.

——. 1961. Does the melanin pigment of human skin have adaptive value? *Q.R.B.*, 36:50-63.

Böcher, T. W. 1951. Studies on morphological progression and evolution in the vegetable kingdom. *Dan. Biol. Medd.*, 18, No. 13: 1-51.

Bodmer, W. F. 1970. The evolutionary significance of recombination in prokaryotes. *In* 20th Symposium of the Society for General Microbiology (in press).

Bodmer, W. F., and J. Felsenstein. 1967. Linkage and selection: Theoretical analysis of the deterministic two locus random mating model. *Gs.*, 57:237-265.

Boesiger, E. 1958. Influence de l'hétérosis sur la vigueur des mâles de *Drosophila melanogaster*. *C. R. Acad. Sci.*, 246:489-491.

——. 1962. Sur le degré d'hétérozygotie des populations naturelles de *Drosophila melanogaster* et son maintien par la sélection sexuelle. *Bull. Biol. France Belgique,* 96:3-122.

Bonner, J. 1950. The role of toxic substances in the interactions of higher plants. *Botan. Rev.*, 1950:51-65.

——. 1965. The Molecular Biology of Development, Oxford Univ. Press, New York.

Bonnier, G., and U. B. Jonsson. 1957. Studies on X-ray induced detrimentals in the second chromosome of *Drosophila melanogaster*. *Hs.*, 43:441-461.

Böök, J. H. 1953. A genetic and neuropsychiatric investigation of a North-Swedish population. *Acta Gen. Stat. Medica,* 4:1-139, 345-414.

Borisov, A. I. 1969. The adaptive significance of the chromosomal polymorphism. *Ga.*, 5, No. 5:119-122.

Boyd, W. C. 1950. Genetics and the Races of Man. Little, Brown, Boston.

Boynton, J. E. 1966. Chlorophyll-deficient mutants in tomato requiring vitamin B_1. *Hs.*, 56:171-199, 238-254.

Brace, C. L. 1963. Structural reduction in evolution. *A.N.*, 97:39-49.

Bradshaw, A. D. 1965. Evolutionary significance of phenotypic plasticity in plants. *Adv. Genetics,* 13:115-155.

Bradshaw, A. D., T. S. McNeilly, and R. P. G. Gregory. 1965. Industrialization, evolution, and the development of heavy metal tolerance in plants. *Brit. Ecol. Soc. Symposium,* 5:327-343.

Brenner, S., L. Barnett, F. H. C. Crick, and A. Orgel. 1961. The theory of mutagenesis. *J. Mol. Biol.*, 3:121-124.

Bretscher, M. S. 1968. How repressor molecules function. *N.*, 127:509-511.

Brewbaker, J. L. 1964. Agricultural Genetics. Prentice-Hall, Englewood Cliffs.

Briles, W. E. 1964. Current status of blood groups in domestic birds. *Zeit. Tierzucht. Züchtungsbiol.*, 79:377-391.

——, C. P. Allen, and T. W. Mullen. 1957. The B blood group system of chickens. I. Heterozygosity in closed populations. *Gs.*, 42:631-648.

Britten, R. J., and E. H. Davidson. 1969. Gene regulation for higher cells; a theory. *S.*, 165:349-357.

——, and D. E. Kohne. 1968. Repeated sequences in DNA. *S.* 161:529-540.

Brncic, D. 1954. Heterosis and the integration of the genotype in geographic populations of *Drosophila pseudoobscura*. *Gs.*, 39:77-88.

——. 1961. Non-random association of inversions in *Drosophila pavani*. *Gs.*, 46: 401-406.

——. 1966. Ecological and cytogenetic studies of *Drosophila flavopilosa,* a neotropical species living in *Cestrum* flowers. *E.*, 20:16-29.

——, and S. Koref-Santibañez. 1964. Mating activity of homo- and heterokaryotypes in *Drosophila pavani*. *Gs.*, 49:585-591.

——, and S. Koref-Santibañez. 1965. Geographical variation of chromosomal structure in *Drosophila gasici*. *Ch.*, 16:47-57.

——, S. Koref-Santibañez, M. Budnik, and M. Lamborot. 1969. Rate of development and inversion polymorphism in *Drosophila pavani*. *Gs.*, 61:471-478.

Brooks, W. K. 1899. The Foundations of Zoology. Columbia Univ. Press, New York.

Brower, L. P. 1959. Speciation in butterflies of the *Papilio glaucus* group. II. Ecological relationships and interspecific sexual behavior. *E.*, 13:212-228.

Brown, A. W. A. 1967. Genetics of insecticide resistance in insect vectors. Pp. 505-552 *in:* J. W. Wright and R. Pal (Eds.), "Genetics of Insect Vectors of Disease," Elsevier, Amsterdam.

Brown, R. G. B. 1964. Courtship behaviour in the *Drosophila obscura* group. I. *D. pseudoobscura*. *Behaviour,* 23:61-106.

——. 1965. Courtship behaviour in the *Drosophila obscura* group. II. Comparative studies. *Behaviour,* 25:281-323.

Brown, W. L., and E. O. Wilson. 1956. Character displacement. *S.Z.*, 5:49-64.

Brues, A. M. 1964. The cost of evolution vs. the cost of not evolving. *E.*, 18: 379-383.

Brust, R. A., and W. R. Horsfall. 1965. Thermal stress and anomalous development of mosquitoes (*Diptera:Culicidae*). IV. *Aedes communis*. *Canad. J. Zool.*, 43:17-53.

Buettner-Janusch, J. 1966. Origins of Man. John Wiley, New York.

Buri, P. 1956. Gene frequency in small populations of mutant *Drosophila*. *E.*, 10:367-402.

Burla, H., A. B. da Cunha, A. G. L. Cavalcanti, Th. Dobzhansky, and C. Pavan. 1950. Population density and dispersal rates in Brazilian *Drosophila willistoni*. *Ecology,* 31:393-404.

——, and M. Greuter 1959. Vergleich der Migrationsverhalten von *Drosophila subobscura* und *Drosophila obscura*. Rev. Suisse Zool., 66:272-279.

Burnet, B. 1962. Manifestation of eight lethals in *Coelopa frigida*. *G.R.*, 3:405-416.

Bush, G L. 1969. Sympatric host race formation and speciation in frugivorous flies of the genus *Rhagoletis*. *E.*, 23:237-251.

Bushland, R. C., and D. E. Hopkins. 1951. Experiments with screw-worm flies sterilized by X-rays. *J. Econ. Entom.*, 44:725-731.

Byers, G. W. 1969. Evolution of wing reduction in crane flies (*Diptera:Tipulidae*). *E.*, 23:346-354.

Cain, A. J., and J. D. Currey. 1963. Area effects in *Cepaea*. *Phil. Trans.*, B, 246: 1-81.

——, and P. M. Sheppard 1950. Selection in the polymorphic land snail *Cepaea nemoralis*. *Hy.*, 4:275-294.

——, and P. M. Sheppard. 1954. Natural selection in *Cepaea*. *Gs.*, 39:89-116.

——, P. M. Sheppard, J. M. B. King, M. A. Carter, J. D. Currey, B. Clarke, C. Diver, J. Murray, and R. W. Arnold. 1968. Studies on *Cepaea*. *Phil. Trans.*, B, 253:383-595.

Cals-Usciati, J. 1964. Étude comparative de caractères biométriques en fonction de l'origine géographique de diverses souches de *Drosophila melanogaster*. *Ann. Genetique*, 7:56-66.

Cameron, D. R., and R. M. Moav. 1957. Inheritance in *Nicotiana tabacum*. XXVII. Pollen killer, an alien genetic locus inducing abortion of microspores not carrying it. *Gs.*, 42:326-335.

Campbell, B. G. 1966. Human Evolution. Aldine, Chicago.

Cannon, W. B. 1932. The Wisdom of the Body. Norton, New York.

Carlquist, S. 1966. The biota of long-distance dispersal. I. Principles of dispersal and evolution. *Q.R.B*, 41:247-270.

Carlson, E. A., R. Sederoff, and M. Cogan. 1967. Evidence favoring a frame-shift mechanism for ICR-170 induced mutations in *Drosophila melanogaster*. *Gs.*, 55:295-313.

Carson, H. L. 1951. Breeding sites of *Drosophila pseudoobscura* and *Drosophila persimilis* in the Transition Zone of the Sierra Nevada. *E.*, 5:91-96.

——. 1959. Genetic conditions which promote or retard the formation of species. *C.S.H.*, 24:87-105.

——. 1965. Chromosomal morphism in geographically widespread species of *Drosophila*. Pp. 508-531 *in:* H. G. Baker and G. L. Stebbins (Eds.), "The Genetics of Colonizing Species," Academic Press, New York.

——. 1967a. Selection for parthenogenesis in *Drosophila mercatorum*. *Gs.*, 55: 157-171.

——. 1967b. Permanent heterozygosity. *E.B.*, 1:143-168.

——, F. E. Clayton, and H. D. Stalker. 1967. Karyotypic stability and speciation in Hawaiian *Drosophila*. *P.N.A.S.*, 57:1280-1285.

——, and W. B. Heed. 1964. Structural homozygosity in marginal populations of neoarctic and neotropical species of *Drosophila* in Florida. *P.N.A.S.*, 52:427-430.

——, and H. D. Stalker. 1968. Polytene chromosome relationships in Hawaiian species of *Drosophila*. I. The *D. grimshawi* subgroup. *U.T.P.*, 6818::335-354.

Caskey, C. T., A. Beaudet, and M. Nierenberg. 1968. RNA codons and protein synthesis. 15. Dissimilar response of mammalian and bacterial transfer. RNA fractions to messenger RNA codons. J. Mol. Biol., 37:99-118.

Cattell, R. B., H. Boutourline Young, and J. D. Hundleby. 1964. Blood groups and personality traits. *Amer. J. Human Genetics,* 16:397-402.

Cavalli-Sforza, L. L. 1969. "Genetic drift" in an Italian population. *Sci. American,* 221:30-37.

——, I. Barrai, and A. W. F. Edwards. 1964. Analysis of human evolution under random genetic drift. *C.S.H.*, 29:9-20.

——, and A. W. F. Edwards. 1964. Analysis of human evolution. *P.* XI *I.C.G.*: 923-933.

Cherfas, N. B. 1966. Natural triploidy in females of the unisexual form of the goldfish (*Carassius auratus gibelio* Bloch). *Ga.*, 5:16-24.

Clarke, B. 1962. Balanced polymorphism and the diversity of sympatric species. *Systematics Assoc. Publ.*, 4:47-70.

Clarke, C. A. 1961. Blood group and disease. *Progr. Med. Genetics,* 1:81-119.

——, and P. M. Sheppard. 1960. The evolution of mimicry in the butterfly *Papilio dardanus*. *Hy.*, 14:163-173.

——, and P. M. Sheppard. 1962. Disruptive selection and its effects on a metric character in the butterfly *Papilio dardanus*. *E.*, 16:214-226.

——, and P. M. Sheppard. 1966. A local survey of the distribution of industrial melanic forms in the moth *Biston betularia* and estimates of the selective values of these in an industrial environment. *Proc. Royal Soc.*, B, 165:424-439.

——, P. M. Sheppard, and I. W. B. Thornton. 1968. The genetics of the mimetic butterfly *Papilio memnon* L. *Phil. Trans. Royal Soc. London,* B, 254:37-89.

Clausen, J. 1951. Stages in the Evolution of Plant Species. Cornell Univ. Press, Ithaca.

——, and W. M. Hiesey. 1958. Experimental studies on the nature of species. IV. Genetic structure of ecological races. *Carnegie Inst. Washington Publ.* 615: 1-312.

——, D. D. Keck, and W. M. Hiesey. 1940. Experimental studies on the nature of species. I. Effects of varied environments on western North American plants. *Carnegie Inst. Washington Publ.* 520:1-452.

——, D. D. Keck, and W. M. Hiesey. 1945. Experimental studies on the nature of species. II. Plant evolution through amphiploidy and autoploidy, with examples from the *Madiinae*. *Carnegie Inst. Washington Publ.* 564.

——, D. D. Keck, and W. M. Hiesey. 1948. Experimental studies on the nature of species. III. Environmental responses of climatic races of *Achillea*. *Carnegie Inst. Washington Publ.* 581:1-129.

Clayton, G. A., G. R. Knight, J. A. Morris, and A. Robertson. 1957. An experimental check on quantitative genetical theory. III. Correlated responses. *J. Genetics,* 55:171-180.

——, J. A. Morris, and A. Robertson. 1957. An experimental check on quantitative genetical theory. I. Short-term responses to selection. *J. Genetics,* 55:131-151.

Cleland, R. E. 1950. Studies on *Oenothera* cytogenetics and phylogeny. *Indiana Univ. Publ. Sci.* 16:1-348.

Clifford, H. T., and F. E. Binet. 1954. A quantitative study of a presumed hybrid swarm between *Eucalyptus elaeophora* and *E. goniocalyx*. *Austral. J. Bot.,* 2:325-336.

Cloud, P. E. 1968. Premetazoan evolution and the origins of the metazoa. Pp. 1-72 *in:* E. T. Drake (Ed.), "Evolution and Environment," Yale Univ. Press, New Haven.

Cohen, M. M., M. W. Shaw, and J. W. MacCluer. 1966. Racial differences in the length of the human Y chromosome. *Cytogenetics* 5:34-52.

Colwell, R. N. 1951. The use of radioactive isotopes in determining spore distribution patterns. *Amer. J. Bot.,* 38:511-523.

Commoner, B. 1964. Deoxyribonucleic acid and the molecular basis of self-duplication. *N.,* 203:486-491.

Coon, C. S. 1965. The Living Races of Man. Knopf, New York.

Cordeiro, A. R. 1952. Experiments on the effects in heterozygous condition of second chromosomes for natural populations of *Drosophila willistoni*. *P.N.A.S.,* 38:471-478.

Correns, C. 1909. Vererbungsversuche mit blass (gelb) grünen und buntblättrigen Sippen bei *Mirabilis, Urtica,* und *Linaria*. *Z.i.A.V.,* 1:291-329.

Corrigan, J. J. 1969. D-amino acids in animals. *S.,* 164:142-149.

Cory, L., and J. J. Manion. 1955. Ecology and hybridization in the genus *Bufo* in the Michigan-Indiana region. *E.,* 9:42-51.

Cotter, W. B. 1967. Mating behavior and fitness as a function of single allele differences in *Ephestia kühniella* Z. *E.,* 21:275-284.

Count, G. W. 1950. This is Race. Henry Schuman, New York.

Court-Brown, W. M. 1967. Human Population Cytogenetics. North-Holland, Amsterdam.

Craig, G. B., and W. A. Hickey. 1967. Genetics of *Aedes aegypti*. Pp. 67-131 *in:* J. W. Wright and R. Pal (Eds.), "Genetics of Insect Vectors of Disease," Elsevier, Amsterdam.

Crampton, H. E. 1916. Studies on the variation, distribution, and evolution of the genus *Partula*. The species inhabiting Tahiti. *Carnegie Inst. Washington Publ.* 228:1-311.

——. 1932. Studies on the variation, distribution and evolution of the genus *Partula*. The species inhabiting Moorea. *Carnegie Inst. Washington Publ.* 410: 1-335.

Creed, E. R., E. B. Ford, and K. McWhirter. 1964. Evolutionary studies on *Maniola jurtina:* the isles of Scilly 1958-1959. *Hy.*, 19:471-488.

Crenshaw, J. W. 1965a. Serum protein variation in an interspecies hybrid swarm of turtles of the genus *Pseudemys. E.*, 19:1-15.

——. 1965b. Radiation-induced increase in fitness in the flour beetle *Tribolium confusum.* S., 149:426-427.

Crick, F. H. C. 1967. The genetic code. *Proc. Royal Soc.*, B, 167:331-347.

Crow, J. F. 1957. Genetics of insect resistance to chemicals. *Ann. Review Entom.*, 2:227-246.

——. 1958. Some possibilities for measuring selection intensities in man. Pp. 1-13 *in:* J. N. Spuhler (Ed.), "Natural Selection in Man," Wayne Univ. Press, Detroit.

——. 1968. Some analysis of hidden variability in *Drosophila populations.* Pp. 71-86 *in:* R. C. Lewontin (Ed.), "Population Biology and Evolution," Syracuse Univ. Press, Syracuse.

——. 1969. Molecular genetics and population genetics. *P.* XII *I.C.G.*, 3:105-113.

——, and M. Kimura. 1963. The theory of genetic loads. *P.* XI *I.C.G.*, 3:495-506.

——, and M. Kimura. 1965. Evolution in sexual and asexual populations. *A.N.*, 99:439-450.

——, and N. E. Morton. 1955. Measurement of gene frequency drift in small populations. *E.*, 9:202-214.

——, and R. G. Temin. 1964. Evidence for the partial dominance of recessive lethal genes in natural populations of *Drosophila. A.N.*, 98:21-33.

Crumpacker, D. W. 1967. Genetic loads in maize (*Zea mays* L.) and other cross-fertilized plants and animals. *E.B.*, 1:306-424.

Cunha, A. B. da. 1949. Genetic analysis of the polymorphism of color pattern in *Drosophila polymorpha. E.*, 3:239-251.

——. 1955. Chromosomal polymorphism in the *Diptera. Adv. Genetics,* 7:93-138.

——, Th. Dobzhansky, O. Pavlovsky, and B. Spassky. 1959. Genetics of natural populations. XXVIII. Supplementary data on the chromosomal polymorphism in *Drosophila willistoni* in its relation to its environment. *E.*, 13:389-404.

Currey, J. D., R. W. Arnold, and M. A. Carter. 1964. Further examples of variation of populations of *Cepaea nemoralis* with habitat. *E.*, 18:111-117.

Darevsky, I. S. 1966. Natural parthenogenesis in a polymorphic group of Caucasian rock lizards related to *Lacerta saxicola* Eversmann. *J. Ohio Herpet. Soc.*, 5:115-152.

Darlington, C. D. 1937. Recent Advances in Cytology. 2nd Ed. Blakiston, Philadelphia.

Davidson, E. H. 1968. Gene Activity in Early Development. Academic Press, New York.

——, M. Crippa, and A. E. Mirsky. 1968. Evidence for the appearance of novel gene products during amphibian blastulation. *P.N.A.S.*, 60:152-159.

Davidson, G., H. E. Patterson, M. Coluzzi, G. F. Mason, and D. W. Micks. 1967. The *Anopheles gambiae* complex. Pp. 211-250 *in:* J. W. Wright and R. Pal Eds.), "Genetics of Insect Vectors of Disease," Elsevier, Amsterdam.

Dawood, M. M., and M. W. Strickberger. 1964. The effect of larval interaction in viability in *Drosophila melanogaster.* I. Changes in heterozygosity. *Gs.*, 50: 999-1007.

Dawson, P. S. 1966. Correlated responses to selection for developmental rate in *Tribolium. Ga.*, 37:63-77.

——, and I. M. Lerner. 1966. The founder principle and competitive ability of *Tribolium. P.N.A.S.*, 55:1114-1117.

Dayhoff, M. O., and R. V. Eck. 1968. Atlas of Protein Sequence and Structure 1967-68. National Biomedical Research Foundation, Silver Spring, Maryland.

Demerec, M. 1936. Frequency of "cell-lethals" among lethals obtained at random in the X chromosome of *Drosophila melanogaster*. *P.N.A.S.*, 22:350-354.

——. 1948. Origin of bacterial resistance to antibiotics. J. Bacteriology, 56:63-74.

——. 1955. What is a gene? Twenty years later. *A.N.*, 89:5-20.

——, and U. Fano. 1945. Bacteriophage resistant mutants in *Escherichia coli*. *Gs.*, 30:119-136.

Dempster, E. R. 1955. Maintenance of genetic heterogeneity. *C.S.H.*, 20:25-32.

Desmond, A. 1964. How many people have ever lived on earth? Pp. 27-46 *in:* S. Mudd (Ed.), "The Population Crisis and the Use of World Resources," Indiana Univ. Press, Bloomington.

Diamond, J. D. 1966. Zoological classification system of a primitive people. *S.*, 151:1102-1104.

Dickson, R. C. 1940. Inheritance of resistance to hydrocyanic acid fumigation in the California red scale. *Hilgardia:* 515-522.

Diver, C. 1939. Aspects of the study of variation in snails. *J. Conchol.*, 21:91-141.

——. 1940. The problem of closely related species living in the same area. Pg. 303-328 *in:* J. S. Huxley, "New Systematics," Clarendon, Oxford.

Dobzhansky, Th. 1927. Studies on manifold effect of certain genes in *Drosophila melanogaster*. *Z.i.A.V.*, 43:330-388.

——. 1933. Geographical variation in lady-beetles. *A.N.*, 67:97-126.

——. 1934. Studies on hybrid sterility. I. Spermatogenesis in pure and hybrid *Drosophila pseudoobscura*. *Zeit. Zef. mikr. Anat.*, 21:169-223.

——. 1936. Studies on hybrid sterility. II. Localization of sterility factors in *Drosophila pseudoobscura* hybrids. *Gs.*, 21:113-135.

——. 1937a. Genetic nature of species differences. *A.N.*, 71:404-420.

——. 1937b. Further data on *Drosophila miranda* and its hybrid with *Drosophila pseudoobscura*. *J. Genetics*, 34:135-151.

——. 1937c. Further data on the variation of the Y chromosome in *Drosophila pseudoobscura*. *Gs.*, 22:340-346.

——. 1937d, 1941, 1951. Genetics and the Origin of Species. 1st, 2nd, and 3rd Eds. Columbia Univ. Press, New York.

——. 1940. Speciation as a stage in evolutionary divergence. *A.N.*, 74:312-321.

——. 1944. Experiments on sexual isolation in *Drosophila*. *P.N.A.S.*, 30:335-339.

——. 1946. Genetics of natural populations. XIII. Recombination and variability in population of *Drosophila pseudoobscura*. *Gs.*, 31:269-290.

——. 1947. Genetics of natural populations. XIV. A response of certain gene arrangements in the third chromosome of *Drosophila pseudoobscura* to natural selection. *Gs.*, 32:142-160.

——. 1948. Genetics of natural populations. XVI. Altitudinal and seasonal changes produced by natural selection in certain populations of *Drosophila pseudoobscura* and *Drosophila persimilis*. *Gs.*, 33:158-176.

——. 1951. Experiments on sexual isolation in *Drosophila*. X. Reproductive isolation between *Drosophila pseudoobscura* and *D. persimilis* under natural and under laboratory conditions. *P.N.A.S.*, 37:792-796.

——. 1953. Natural hybrids of two species of *Arctostaphylos* in the Yosemite region of California. *Hy.*, 7:73-79.

——. 1956a. What is an adaptive trait? *A.N.*, 90:337-347.

——. 1956b. Genetics of natural populations. XXV. Genetic changes in populations of *Drosophila pseudoobscura* and *Drosophila persimilis* in some locations in California. *E.*, 10:82-92.

——. 1960. Evolution and environment. Pp. 403-428 *in:* Sol Tax (Ed.), "Evolution after Darwin," Vol. 1, Chicago Univ. Press, Chicago.

——. 1961. On the dynamics of chromosomal polymorphism in *Drosophila.* Pp. 30-42 *in:* J. S. Kennedy (Ed.), "Insect Polymorphism," Royal Entomological Society, London.

——. 1962. Rigid vs. flexible chromosomal polymorphisms in *Drosophila. A.N.*, 96:321-328.

——. 1964a. How do the genetic loads affect the fitness of their carriers in *Drosophila* population? *A.N.*, 98:151-166.

——. 1964b. Genetic diversity and fitness. *P.* XI *I.C.G.*, 3:541-552.

——. 1968a. On some fundamental concepts of Darwinian biology. *E.B.*, 2:1-34.

——. 1968b. Adaptedness and fitness. Pp. 109-121 *in:* R. C. Lewontin (Ed.), "Population Biology and Evolution," Syracuse Univ. Press, Syracuse.

——, W. W. Anderson, and O. Pavlovsky. 1966. Genetics of natural populations. XXXVIII. Continuity and change in populations of *Drosophila pseudoobscura* in western United States. *E.*, 20:418-427.

——, and G. W. Beadle. 1936. Studies on hybrid sterility. IV. Transplanted testes in *Drosophila pseudoobscura. Gs.*, 21:832-840.

——, L. Ehrman, and P. A. Kastritsis. 1968. Ethological isolation between sympatric and allopatric species of the *obscura* group of *Drosophila. Animal Behaviour,* 16:79-87.

——, and C. Epling. 1944. Contributions to the genetics, taxonomy, and ecology of *Drosophila pseudoobscura* and its relatives. *Carnegie Inst. Washington Publ.* 554:1-183.

——, and A. M. Holz. 1943. A re-examination of the problem of manifold effects of genes in *Drosophila melanogaster. Gs.*, 28:295-303.

——, and H. Levene. 1948. Genetics of natural populations. XVII. Proof of operation of natural selection in wild populations of *Drosophila pseudoobscura. Gs.*, 33:537-547.

——, H. Levene, B. Spassky, and N. Spassky. 1959. Release of genetic variability through recombination. III. *Drosophila prosaltans. Gs.*, 44:75-92.

——, R. C. Lewontin, and O. Pavlovsky. 1964. The capacity for increase in chromosomally polymorphic and monomorphic populations of *Drosophila pseudoobscura. Hy.*, 19:597-614.

——, and O. Pavlovsky. 1953. Indeterminate outcome of certain experiments on *Drosophila* populations. *E.*, 7:198-210.

——, and O. Pavlovsky. 1957. An experimental study of interaction between genetic drift and natural selection. *E.*, 11:311-319.

——, and O. Pavlovsky. 1958. Interracial hybridization and breakdown of coadapted gene complexes in *Drosophila paulistorum* and *Drosophila willistoni. P.N.A.S.*, 44:622-629.

——, and O. Pavlovsky. 1967. Experiments on the incipient species of the *Drosophila paulistorum* complex. *Gs.*, 55:141-156.

——, O. Pavlovsky, and L. Ehrman. 1969. Transitional populations of *Drosophila paulistorum. E.*, 23:482-492.

——, and B. Spassky. 1941. Intersexes in *Drosophila pseudoobscura. P.N.A.S.*, 27:556-562.

——, and B. Spassky. 1947. Evolutionary changes in laboratory cultures of *Drosophila pseudoobscura. E.*, 1:191-216.

——, and B. Spassky. 1953. Genetics of natural populations. XXI. Concealed variability in two sympatric species of *Drosophila*. *Gs.*, 38:471-484.

——, and B. Spassky. 1959. *Drosophila paulistorum,* a cluster of species in *statu nascendi. P.N.A.S.*, 45:419-428.

——, and B. Spassky. 1960. Release of genetic variability through recombination. V. Breakup of synthetic lethals by crossing over in *Drosophila pseudoobscura. Zool. Jahrb. Abt. Syst.*, 88:57-66.

——, and B. Spassky. 1963. Genetics of natural populations. XXXIV. Adaptive norm, genetic load and genetic elite in *Drosophila pseudoobscura. Gs.*, 48:1467-1485.

——, and B. Spassky. 1967a. Effects of selection and migration on geotactic and phototactic behaviour of *Drosophila*. I. *Proc. Royal Soc.*, B, 168:27-47.

——, and B. Spassky. 1967b. An experiment on migration and simultaneous selection for several traits in *Drosophila pseudoobscura. Gs.*, 55:723-734.

——, and B. Spassky. 1968. Genetics of natural populations. XL. Heterotic and deleterious effect of recessive lethals in populations of *Drosophila pseudoobscura. Gs.*, 59:411-425.

——, and B. Spassky. 1969. Artificial and natural selection for two behavioral traits in *Drosophila pseudoobscura. P.N.A.S.*, 62:75-80.

——, B. Spassky, and N. Spassky. 1952. A comparative study of mutation rates in two ecologically diverse species of *Drosophila*. *Gs.*, 39:472-487.

——, B. Spassky, and N. Spassky. 1954. Rates of spontaneous mutation in the second chromosomes of the sibling species, *Drosophila pseudoobscura* and *Drosophila persimilis. Gs.*, 39:899-907.

——, B. Spassky, and T. Tidwell. 1963. Genetics of natural populations. XXXII. Inbreeding and the mutational and balanced genetic loads in natural populations of *Drosophila pseudoobscura. Gs.*, 48:361-373.

——, and N. P. Spassky. 1962. Genetic drift and natural selection in experimental populations of *Drosophila pseudoobscura. P.N.A.S.*, 48:148-156.

——, and G. Streisinger. 1944. Experiments on sexual isolation in *Drosophila*. II. *P.N.A.S.*, 30:340-345.

——, and A. H. Sturtevant. 1938. Inversions in the chromosomes of *Drosophila pseudoobscura. Gs.*, 23:28-64.

——, and C. C. Tan. 1936. Studies on hybrid sterility. III. A comparison of the gene arrangement in two species, *Drosophila pseudoobscura* and *Drosophila miranda. Z.i.A.V.*, 72:88-114.

——, and S. Wright. 1941. Genetics of natural populations. V. Relations between mutation rate and accumulation of lethals in populations of *Drosophila pseudoobscura. Gs.*, 26:23-51.

——, and S. Wright. 1943. Genetics of natural populations. X. Dispersion rates in *Drosophila pseudoobscura. Gs.*, 28:304-340.

——, and S. Wright. 1947. Genetics of natural populations. XV. Rate of diffusion of a mutant gene through a population of *Drosophila pseudoobscura. Gs.*, 32:303-324.

Dodson, C. H. 1967. Relationships between pollinators and orchid flowers. *Atlas Simp. Biota Amazonica*, 5:1-72 (Rio de Janeiro).

Dowdeswell, W. H., and K. McWhirter. 1967. Stability of spot-distribution in *Maniola jurtina* throughout its range. *Hy.*, 22:187-210.

Drescher, W. 1964. The sex-limited genetic load in natural populations of *Drosophila melanogaster. A.N.*, 98:167-171.

Druger, M. 1967. Selection and the effect of temperature on scutallar bristle number in *Drosophila*. *Gs.*, 56:39-47.

Dubinin, N. P. 1931. Genetico-automatical processes and their bearing on the mechanism of organic evolution. *J. Exp. Biol.*, 7:463-479. (Russian).

——. 1961. Problems of Radiation Genetics. Gosatomizdat, Moscow.

——. 1964. Problems of Radiation Genetics. Oliver and Boyd, Edinburgh.

——. 1966. Evolution of Populations and Radiation. Atomizdat, Moscow.

——, and D. D. Romaschoff. 1932. Die genetische Struktur der Art und ihre Evolution. *Biol. Zhurnal*, 1:52-95.

——, D. D. Romashov [Romaschoff], M. A. Heptner, and Z. A. Demidova. 1937. Aberrant polymorphism in *Drosophila fasciata* Meig (Syn. - *melanogaster Meig*). *Biol. Zhurnal*, 6:311-354.

——, and G. G. Tiniakov. 1946. Inversion gradients and natural selection in ecological races of *Drosophila funebris*. *Gs.*, 31:537-545.

——, and fourteen collaborators. 1934. Experimental study of the ecogenotypes of *Drosophila melanogaster*. *Biol. Zhurnal*, 3:166-216

Dubos, R. 1965. Man Adapting. Yale Univ. Press, New Haven.

Duke, E. J., and E. Glassman. 1968. Evolution of xanthine dehydrogenase in *Drosophila*. *Gs.*, 58:101-112.

Dun, R. B., and A. S. Fraser. 1959. Selection for an invariant character, vibrissa number, in the house mouse. *Austral. J. Biol. Sci.*, 12:506-523.

Dunn, L. C. 1964. Abnormalities associated with a chromosome region in the mouse. *S.*, 144:260-263.

Edwards, A. J., and J. F. Cauley (Eds.) 1964. Physiological determinants of behavior. *Kansas Studies in Education*, 14: No. 3.

Ehrendorfer, F. 1963. Cytologie, Taxonomie und Evolution bei Samenpflanzen. *Vistas Bot.*, 4:99-186.

Ehrlich, P. R., and P. H. Raven. 1964. Butterflies and plants, a study in coevolution. *E.*, 18:586-608.

Ehrman, L. 1960. The genetics of hybrid sterility in *Drosophila paulistorum*. *E.*, 14:212-223.

——. 1961. The genetics of sexual isolation in *Drosophila paulistorum*. *Gs.*, 46: 1025-1038.

——. 1962. Hybrid sterility as an isolating mechanism in the genus *Drosophila*. *Q.R.B.*, 37:279-302.

——. 1965. Direct observation of sexual isolation between allopatric and between sympatric strains of the different *Drosophila paulistorum* races. *E.*, 19:459-464.

——. 1967. Further studies on genotype frequency and mating success in *Drosophila*. *A.N.*, 101:415-424.

——, and C. Petit. 1968. Genotype frequency and mating success in the *willistoni* species group of *Drosophila*. *E.*, 22:649-658.

——, B. Spassky, O. Pavlovsky, and Th. Dobzhansky. 1965. Sexual selection, geotaxis, and chromosomal polymorphism in experimental populations of *Drosophila pseudoobscura*. *E.*, 19:337-346.

——, and D. L. Williamson. 1969. On the etiology of the sterility of hybrids between certain strains of *Drosophila paulistorum*. *Gs.*, 62:193-199.

Eibl-Eibesfeldt, I. 1965. Angeborenes und Erworbenes in Verhalten einiger Säuger. *Zeit. Tierpsychol.*, 20:705-754.

Eiche, V. 1955. Spontaneous chlorophyll mutations in Scots pine (*Pinus silvestris* L.). *Medd. Stat. Skogsforskningsinst.*, 45, No. 13:1-69.

Eiseley, L. 1959. Charles Darwin, Edward Blyth, and the theory of natural selection. Proc. Amer. Philos. Soc., 103:94-158.

Emerson, S. 1939. A preliminary survey of the *Oenothera organensis* population. *Gs.*, 24:524-537.

Emsweller, S. L., and N. W. Stuart. 1948. Use of growth regulating substances to overcome incompatibilities. in *Lilium. Proc. Amer. Soc. Hort. Sci.*, 51:581-589.

Ephrussi, B., and M. C. Weiss. 1965. Interspecific hybridization of somatic cells. *P.N.A.S.*, 53:1040-1042.

Ergene, S. 1951. Wählen Heuschrecken ein homochromes Milieu? *Deutsche Zool. Zeit.*, 1:123-133, 187-195.

Erlenmeyer-Kimling, L., and W. Paradowski. 1966. Selection and schizophrenia. *A.N.*, 100:651-665.

Falconer, D. S. 1960a. Introduction to Quantitative Genetics. Ronald Press, New York.

——. 1960b. Selection of mice for growth on high and low planes of nutrition. *G.R.*, 1:91-113.

Falk, R. 1961. Are induced mutations in *Drosophila* overdominant? II. Experimental results. *Gs.*, 46:737-757.

——, and N. Ben-Zeev. 1966. Viability of heterozygotes for induced mutations in *Drosophila melanogaster*. II. Mean effects in irradiated autosomes. *Gs.*, 53:65-77.

Feller, W. 1967. On fitness and the cost of natural selection. *G.R.*, 9:1-15.

Fenner, F. 1959. Myxomatosis in Australian wild rabbit—evolutionary changes in an infectious disease. *Harvey Lectures, 1957-58:* 25-55.

——. 1965. Myxoma virus and *Oryctolagus cuniculus,* two colonizing species. Pp. 485-501 *in:* H. G. Baker and G. L. Stebbins (Eds.), "The Genetics of Colonizing Species," Academic Press, New York.

Ferguson, C. W. 1968. Bristlecone pine: science and esthetics. S., 159:839-846.

Ferrell, G. T. 1966. Variation in blood group frequencies in populations of song sparrows of the San Francisco bay region. *E.*, 20:369-382.

Fisher, R. A. 1928. The possible modification of the response of the wild type to recurrent mutations. *A.N.*, 62:115-126.

——. 1930. The Genetical Theory of Natural Selection. Clarendon Press, Oxford.

Fitch, W. M., and E. Margoliash. 1967a. Construction of phylogenetic trees. S., 155:279-284.

——, and E. Margoliash. 1967b. A method for estimating the number of invariant amino acid coding positions in a gene using cytochrome c as a model case. *Biochem. Genetics,* 1:65-71.

——, and E. Margoliash. 1970. The usefulness of amino acid and nucleotid sequences in evolutionary studies. *E.B.*, 4: (in press).

Foerster, R. E. 1968. The sockeye salmon *Oncorhynchus nerka. Fisheries Res. Canada Publ.* 162, Ottawa.

Ford, E. B. 1964. Ecological Genetics. Methuen, London.

——. 1965. Genetic Polymorphism. M.I.T. Press, Cambridge.

Forde, M. M. 1962. Effect of introgression on the serpentine endemism of *Quercus durata*. *E.*, 16:338-347.

Fox, A. S., and S. B. Yoon. 1966. Specific genetic effects of DNA in *Drosophila melanogaster*. *Gs.*, 53:897-911.

Fox, S. W. (Ed.). 1965. The Origin of Prebiological Systems. Academic Press, New York.

Fraenkel-Conrat, H., and B. Singer. 1957. Virus reconstruction. II. Combination of protein and nucleic acid from different strains. *Biochem. Biophys. Acta,* 24: 540-548.

——, and R. C. Williams. 1955. Reconstitution of active tobacco mosaic virus from its inactive protein and nucleic acid components. *P.N.A.S.,* 41:690-698.

Frahm, R. R., and K. I. Kojima. 1966. Comparison of selection responses of body weight under divergent larval density conditions in *Drosophila pseudoobscura. Gs.,* 54:625-637.

Fraser, A., and D. Burnell. 1967. Simulation of genetic systems. XII. Models of inversion polymorphisms. *Gs.,* 57:267-282.

Fraser, G. R. 1962. Our genetical "load." A review of some aspects of genetical variation. *Ann. Human Genetics,* 25:387-415.

Freese, E. 1965. The influence of DNA structure and base composition on mutagenesis. *P.* XI *I.C.G.,* 2:297-306.

Friedman, L. D. 1964. X-ray induced sex-linked lethal and detrimental mutations and their effects on the viability of *Drosophila melanogaster. Gs.,* 49:689-699.

Fulton, B. B. 1952. Speciation in the field cricket. *E.,* 6:283-295.

Futch, D. G. 1966. A study of speciation in South Pacific populations of *Drosophila ananassae. U.T.P.* 6615:79-120.

Gajewski, W. 1957. A cytogenetic study on the genus *Geum* L. *Monogr. Botanicae (Warszawa),* 4:1-415.

——. 1959. Evolution in the genus *Geum. E.,* 13:378-388.

Garen, A. 1968. Sense and nonsense in the genetic code. *S.,* 160:149-159.

Garn, S. M. 1965. Human Races. 2nd Ed. Charles C Thomas, Springfield, Ill.

Garnjobst, L., and E. L. Tatum. 1967. A survey of new morphological mutants in *Neurospora crassa. Gs.,* 57:579-604.

Gause, G. F. 1941. Optical Activity and Living Matter. Biodynamica, Normandy, Mo.

Geitler, L. 1938. Weitere cytogenetische Untersuchungen an natürlichen Populationen von *Paris quadrifolia. Z.i.A.V.,* 75:161-190.

Gerassimova, H. 1939. New experimentally produced strains of *Crepis tectorum* which are physiologically isolated from the original form owing to reciprocal translocation. *C. R. Acad. Sci. URSS,* 25:148-154.

Gerstel, D. U. 1954. A new lethal combination in interspecific cotton hybrids. *Gs.,* 39:628-639.

Gibbs, G. W. 1968. The frequency of interbreeding between two sibling species of *Dacus (Diptera)* in wild populations. *E.,* 22:667-683.

Giblett, G. 1961. Haptoglobins and transferrins. Pp. 132-158 *in:* B. S. Blumberg (Ed.), "Proceedings of the Conference on Genetic Polymorphisms and Geographic Variations in Disease," Grune & Stratton, New York.

——. 1964. Variant haptoglobin phenotypes. *C.S.H.,* 29:321-326.

Gibson, J. B., and J. M. Thoday. 1962. Effects of disruptive selection. VI. A second chromosome polymorphism. *Hy.,* 17:1-26.

——, and J. M. Thoday. 1963. Effects of disruptive selection. VIII. Imposed quasi-random mating. *Hy.,* 18:513-524.

Gilbert, J. J. 1966. Rotifer ecology and embryological induction. *S.,* 151:1234-1237.

Giles, E., R. J. Walsh, and M. A. Bradley. 1966. Micro-evolution in New Guinea: The role of genetic drift. *Ann. New York Acad. Sci.,* 134:655-665.
Service, 38:33.

Gill, K. S. 1963. A mutation causing abnormal mating behavior. *Drosophila Inf.*

Glass, B. 1954. Genetic changes in human populations, especially those due to gene flow and genetic drift. *Adv. Genetics,* 6:95-139.
——. 1955. A comparative study of induced mutation in the oocytes and spermatozoa of *Drosophila melanogaster. Gs.,* 40:252-267, 281-296.
——, M. S. Sacks, E. J. Jahn, and C. Hess. 1952. Genetic drift in a religious isolate. An analysis of the causes of variation in blood group and other gene frequencies in a small population. *A.N.,* 86:145-160.
Goldschmidt, R. 1934. *Lymantria. Bibliogr. Genetica,* 11:1-186.
——. 1938. Physiological Genetics. McGraw-Hill, New York.
——. 1940. The Material Basis of Evolution. Yale Univ. Press, New Haven.
Golubovsky, M. D. 1969. The viability of heterozygotes for lethal mutations characterized by different concentrations in a natural population of *Drosophila melanogaster. Ga.,* 5, No. 8:116-126.
Goodhart, C. B. 1962. Variation in a colony of the snail *Cepaea nemoralis* (L.). *J. Animal Ecol.,* 31:207-237.
——. 1963. "Area effects" and non-adaptive variation between populations of *Cepaea. Hy.,* 18:459-465.
Gordon, M. 1948. Effects of five primary genes on the site of melanomas in fishes and the influence of two color genes on their pigmentation. *In:* "The Biology of Melanomas," *Spec. Publ. New York Acad. Sci.,* 4:216-268.
——. 1951. Genetic and correlated studies of normal and atypical pigment cell growth. *Growth Symposium,* 10:153-219.
——, and D. E. Rosen. 1951. Genetics of species differences in the morphology of the male genitalia of xiphophorin fishes. *Bull. Amer. Museum Nat. Hist.,* 95:409-464.
Gottesman, I.I. 1968. A sampler of human behavioral genetics. E.B., 2:276-320.
Grant, K. A., and V. Grant. 1964. Mechanical isolation of *Salvia apiana* and *Salvia mellifera (Labiatae). E.,* 18:196-212.
——, and V. Grant. 1968. Hummingbirds and Their Flowers. Columbia Univ. Press, New York.
Grant, V. 1954a. Genetics and taxonomic studies in *Gilia.* VI. Interspecific relationships in the leafy-stemmed *Gilias. El Aliso,* 3:35-49.
——. 1954b. Genetic and taxonomic studies in *Gilia.* VII. The woodland *Gilias. El Aliso,* 3:59-91.
——. 1958. The regulation of recombination in plants. *C.S.H.,* 23:337-363.
——. 1963. The Origin of Adaptations. Columbia Univ. Press, New York.
——. 1964a. The Architecture of Germplasm. John Wiley, New York.
——. 1964b. The biological composition of a taxonomic species in *Gilia. Adv. Genetics,* 12:281-328.
——. 1965. Evidence for the selective origin of incompatibility barriers in the leafy-stemmed *Gilias. P.N.A.S.,* 54:1567-1571.
——. 1966a. Selection for vigor and fertility in the progeny of a highly sterile species hybrid in *Gilia. Gs.,* 53:757-775.
——. 1966b. Linkage between viability and fertility in a species cross in *Gilia. Gs.,* 54:867-880.
——. 1966c. The selective origin of incompatibility barriers in the plant genus *Gilia. A.N.,* 100:00-118.
——, and K. A. Grant. 1965. Pollination in the Phlox Family. Columbia Univ. Press, New York.
Greenberg, J. 1964. A locus for radiation resistance in *Escherichia coli. Gs.,* 49:771-778.

Greenberg, R., and J. F. Crow. 1960. Comparison of the effect of lethal and detrimental chromosomes from *Drosophila* populations. *Gs.*, 45:1153-1168.

Greene, J. C. 1961. The Death of Adam. New American Library, New York.

Gregg, J. R. 1954. The Language of Taxonomy. Columbia Univ. Press, New York.

Grinchuk, P. M. 1967. A study of the polymorphism of polytene chromosomes of the black fly *Prosimulium hirtipes (Diptera, Simulidae)* indigenous to the Leningrad region. *Ga.*, 1967, 165-172.

Gripenberg, U. 1964. Size variation and orientation of the human Y chromosome. *Ch.*, 15:618-629.

Grossfield, J. 1966. The influence of light on the mating behavior of Drosophila. *U.T.P.*, 6615:147-176.

Grun, P., and M. Aubertin. 1965. Evolutionary pathways of cytoplasmic male sterility in *Solanum*. *Gs.*, 51:399-409.

Grüneberg, H. 1938. An analysis of the "pleiotropic" effects of a new lethal mutation in the rat (*Mus norvegicus*). *Proc. Royal Soc.*, B, 125:123-144.

——. 1943. Congenital hydrocephaly in the mouse, a case of spurious pleiotropism. *J. Genetics*, 45:1-21.

Gulick, J. T. 1905. Evolution, racial and habitudinal. *Carnegie Inst. Washington Publ.* 25:1-269.

Gunn, D. L., and Ph. Hunter-Jones. 1952. Laboratory experiments on phase differences in locusts. *Anti-Locust Bull.* (London), 12.

Gustafsson, Å. 1941. Preliminary yield experiments with ten induced mutations in barley. *Hs.*, 27:337-359.

——. 1946. The effect of heterozygosity on viability and vigour. *Hs.*, 32:263-286.

——. 1947. Apomixis in higher plants. *Lunds Univ. Arsskrift*, N.F., 43:71-178.

——. 1951. Induction of changes in genes and chromosomes. II. Mutations, environment and evolution. *C.S.H.*, 16:263-281.

——. 1953. The cooperation of genotypes in barley. *Hs.*, 39:1-18.

——. 1963a. Productive mutations induced in barley by ionizing radiations and chemical mutagens. *Hs.*, 50:211-263.

——. 1963b. Mutations and the concept of viability. Pp. 89-104 *in:* E. Akerberg, A. Hagberg, et. al. (Eds.), Recent Plant Breeding Research, John Wiley, New York.

——, and N. Nybom. 1950. The viability reaction of some induced and spontaneous mutations in barley. *Hs.*, 36:113-133.

——, N. Nybom, and U. Wettstein. 1950. Chlorophyll factors and heterosis in barley. *Hs.*, 36:383-392.

Guthrie, G. D., and R. L. Sinsheimer. 1960. Infection of protoplasts of *Escherichia coli* by subviral particles of bacteriophage $\phi\chi$ 174. *J. Mol. Biol.*, 2:297-305.

Hadorn, E. 1951. Developmental action of lethal factors in *Drosophila. Adv. Genetics*, 4:53-85.

——. 1956. Patterns of biochemical and developmental pleiotropy. *C.S.H.*, 21: 363-373.

——. 1961. Developmental Genetics and Lethal Factors. Methuen, London.

——, and H. Niggli. 1946. Mutations in *Drosophila* after chemical treatment of gonads *in vitro*. *N.*, 157-162.

——, S. Rosin, and G. Bertani. 1949. Ergebnisse der Mutationsversuche mit chemischer Behandlung von *Drosophila in vitro*. *P.* VIII *I.C.G.*:256-266.

Haga, T. 1956. Genome and polyploidy in the genus *Trillium*. VI. Hybridization and speciation by chromosome doubling in nature. *Hy.*, 10:85-98.

——, and M. Kurabayashi. 1954. Chromosomal variation in natural populations of *Trillium kamtschaticum* Pall. *Mem. Fac. Sci. Kyushu Univ.*, E, 1:159-185.

Hagedoorn, A. L., and A. C. Hagedoorn. 1921. The Relative Value of the Processes Causing Evolution. Martius Nijhoff, The Hague.

Hagen, D. W. 1967. Isolating mechanisms in three-spine sticklebacks. (*Gasterosteus*). *J. Fish. Res. Bd. Canada*, 24:1637-1692.

Håkansson, A. 1942. Zytologische Studien an Rassen und Rassenbastarden von *Godetia whitneyi* und verwandten Arten. *Lunds Univ. Arsskrift*, N.F., 38, No. 5:1-70.

Haldane, J. B. S. 1932. The Causes of Evolution. Harper, London and New York.

——. 1933. The part played by recurrent mutation in evolution. *A.N.*, 67:5-19.

——. 1937. The effect of variation on fitness. *A.N.*, 71:337-349.

——. 1957. The cost of natural selection. *J. Genetics*, 55:511-524. Reprinted *in:* E. B. Spiess (Ed.), "Papers in Animal Population Genetics," Little, Brown, Boston (1962).

——. 1964. A defense of beanbag genetics. *Persp. Biol. Med.*, 7:343-350.

——, and S. D. Jayakar. 1963a. Polymorphism due to selection of varying direction. *J. Genetics*, 58:237-242.

——, and S. D. Jayakar. 1963b. Polymorphism due to selection depending on the composition of a population. *J. Genetics*, 58:318-323.

Haley, L. E., H. Abplanalp, and K. Enya. 1966. Selection for increased fertility of female quail when mated to male chickens. *E.*, 20:72-81.

Halfer-Cervini, A. M., M. Piccinelli, P. Prosdocimi, and L. Baratelli-Zambruni. 1968. Sibling species in *Artemia (Crustacea: Branchiopoda)*. *E.*, 22:373-381.

Hamilton, W. J., and F. Heppner. 1967. Radiant solar energy and the function of black homeotherm pigmentation: an hypothesis. *S.*, 155:196-197.

Harding, J., R. W. Allard, and D. G. Smeltzer. 1966. Population studies in predominantly self-pollinated species. IX. Frequency-dependent selection in *Phaseolus lunatus*. *P.N.A.S.*, 56:99-104.

Hardy, G. H. 1908. Mendelian proportions in a mixed population. *S.*, 28:49-50.

Harland, S. C. 1936. The genetical conception of the species. *Biol. Rev.*, 11:83-112.

Harrington, R. W., and K. D. Kallman. 1968. The homozygosity of clones of the self-fertilizing hermaphroditic fish *Rivulus marmoratus* Poey (*Cyprinodontidae, Atheriniformes*). *A.N.*, 102:337-343.

Harris, H. 1966. Enzyme polymorphisms in man. *Proc. Royal Soc.*, B, 164:298-316.

——. 1967. Enzyme variation in man: some general aspects. Pp. 207-214 *in:* J. F. Crow and J. V. Neel (Eds.), "Proceedings of the 3rd International Congress on Human Genetics," Johns Hopkins Press, Baltimore.

Harrison, G. A. (Ed.). 1961. Genetical Variation in Human Populations. Pergamon Press, Oxford.

Haskins, C. P., E. F. Haskins, and R. E. Hewitt. 1960. Pseudogamy as an evolutionary factor in the poeciliid fish *Mollienesia formosa*. *E.*, 14:473-483.

Hatt, D., and P. A. Parsons. 1965. Association between surnames and blood groups in the Australian population. *Acta Genetica*, 15:309-318.

Hauschteck, E., G. Mürset, A. Prader and E. Bühler. 1966. Siblings with different types of chromosomal aberrations due to D/F translocation of the mother. *Cytogenetics*, 5:281-294.

Hayes, C. 1951. The Ape in Our House. Harper & Row, New York.

Hedberg, O. 1955. Some taxonomic problems concerning the Afro-alpine flora. *Webbia*, 11:471-487.

Heed, W. B. 1963. Density and distribution of *Drosophila polymorpha* and its color alleles in South America. *E.*, 17:502-518.

——, and P. R. Blake. 1963. A new color allele at the *e* locus of *Drosophila polymorpha* from northern South America. *Gs.*, 48:217-234.

Heiser, C. B. 1947. Hybridization between the sunflower species *Helianthus annus* and *H. petiolaris*. *E.*, 1:249-262.

Helfer, R. G. 1941. A comparison of X-ray induced and naturally occurring chromosomal variations in *Drosophila pseudoobscura*. *Gs.*, 26:1-22.

Héritier, Ph. L'., and G. Teissier. 1934. Une experience de sélection naturelle. Courbe d'élimination du gene "Bar" dans une population de *Drosophiles* en équilibre. *C. R. Soc. Biol.*, 117:1049-1051.

Herskowitz, I. 1949a. Hexaptera, a homoeotic mutant in *Drosophila melanogaster*. *Gs.*, 34:10-25.

——. 1949. Tests for chemical mutagens in Drosophila. *Proc. Soc. Exp. Biol.*, 70:601-607.

Heuts, M. J. 1947. Experimental studies in adaptive evolution in *Gasterosteus aculeatus* L. *E.*, 1:89-102.

Hildreth, P. E. 1956. The problem of synthetic lethals in *Drosophila melanogaster*. *Gs.*, 41:729-742.

——. 1965. Doublesex, a recessive gene that transforms both males and females of *Drosophila* into intersexes. *Gs.*, 51:659-678.

Hiraizumi, Y., and J. F. Crow. 1960. Heterozygous effects on viability, fertility, rate of development and longevity of *Drosophila* chromosomes that are lethal when homozygous. *Gs.*, 45:1071-1083.

Hochman, B. 1961. On fourth chromosome lethals from a natural population of *Drosophila melanogaster*. *A.N.*, 95:375-382.

Hoenigsberg, H. F., and S. Koref-Santibañez. 1960. Courtship and sensory preferences in inbred lines of *Drosophila melanogaster*. *E.*, 14:1-7.

Holland, J. J., M. W. Taylor, and C. A. Buck. 1967. Chromatographic differences between tyrosyl transfer RNA from different mammalian cells. *P.N.A.S.*, 58: 2437-2444.

Hollingshead, L. 1930. A lethal factor in *Crepis* effective only in interspecific hybrids. *Gs.*, 15:114-140.

Holttum, R. E. 1953. Evolutionary trends in an equatorial climate. *Symposia Soc. Exp. Biol.*, 7:159-173.

Horton, I. H. 1939. A comparison of the salivary gland chromosomes of *Drosophila melanogaster* and *D. simulans*. *Gs.*, 24:234-243.

Hosino, Y. 1940. Genetical studies on the pattern types of the lady bird beetle, *Harmonia axyridis* Pallas. *J. Genetics*, 40:215-228.

Hoyer, B. H., E. T. Bolton, B. J. McCarthy, and R. B. Roberts. 1965. The evolution of polynucleotides. Pp. 581-590 *in:* V. Bryson and H. J. Vogel (Eds.), "Evolving Genes and Proteins," Academic Press, New York.

——, B. J. McCarthy and E. T. Bolton. 1964. A molecular approach in the systematics of higher organisms. S., 144:959-967.

Hrishi, N. J., and A. Müntzing. 1960. Structural heterozygosity in *Secale kuprijanovii*. *Hs.*, 46:745-752.

Hubbs, C. 1964. Interactions between a bisexual fish species and its gynogenetic sexual parasite. *Bull. Texas Mem. Museum*, 8:1-72.

——, and E. A. Delco. 1960. Mate preference in males of four species of gambusiine fishes. *E.*, 14:145-152.

——, and E. A. Delco. 1962. Courtship preferences of *Gambusia affinis* associated with the sympatry of the parental populations. *Copeia,* 1962:396-400.

——, and K. Strawn. 1956. Interfertility between two sympatric fishes, *Notropis lutrensis* and *Notropis venustus.* E., 10:341-344.

Hubbs, C. L. 1940. Speciation of fishes. A.N. 74:198-211.

Hubby, J. L., and R. C. Lewontin. 1966. A molecular approach to the study of genic heterozygosity in natural populations. I. The number of alleles at different loci in *Drosophila pseudoobscura. Gs.,* 54:577-594.

——, and L H. Throckmorton. 1965. Protein differences in *Drosophila.* II. Comparative species genetics and evolutionary problems. *Gs.,* 52:203-215.

——, and L. H. Throckmorton. 1968. Protein differences in *Drosophila.* IV. A study of sibling species. *A.N.,* 102:193-205.

Hulse, F. S. 1957. Exogamie et hétérosis. *Arch. Suisse Anthrop. Gén.* 22:103-125.

——. 1963. The Human Species. Random House, New York.

Hutchinson, J. B., J. B. Silow, and J. G. Stephens. 1947. The evolution of *Gossypium.* Oxford Univ. Press, Oxford.

Huxley, J. S. (1942) 1963. Evolution, the Modern Synthesis. Rev. Ed. Harper, New York.

Ingels, L. G. 1950. Pigmental variations in populations of pocket gophers. *E.,* 4:353-357.

Ingram, V. M. 1963. The Hemoglobins in Genetics and Evolution. Columbia Univ. Press, New York.

Irving, L. 1966. Adaptation to cold. *Sci. American,* 214:94-101.

Irwin, M. R. 1953. Evolutionary patterns of antigenic substances of the blood corpuscles in *Columbidae. E.,* 7:31-50.

——. 1966. Interaction of nonallelic genes on cellular antigens in species hybrids of *Columbidae.* II. Identification of interacting genes. *P.N.A.S.,* 55:34-40.

——, and R. W. Cumley. 1943. Interrelationships of the cellular characters of several species of *Columba. Gs.,* 28:9-28.

Ives, P. T. 1945. The genetic structure of American populations of *Drosophila melanogaster. Gs.,* 30:167-196.

——. 1950. The importance of mutation rate genes in evolution. *E.,* 4:236-252.

Jacob, F., and J. Monod. 1961. Genetic regulatory mechanisms in the synthesis of proteins. *J. Mol. Biol.,* 3:318-356.

——, and E. L. Wollman. 1961. Sexuality and Genetics of Bacteria. Academic Press, New York.

Jacobson, M., and M. Beroza. 1963. Chemical insect attractants. S., 140:1367-1372.

Jain, S. K. 1969. Comparative ecogenetics of two *Avena* species occurring in central California. *E.B.,* 3:73-118.

——, and A. D. Bradshaw. 1966. Evolutionary divergence among adjacent populations. I. The evidence and its theoretical analysis. *Hy.,* 21:407-441.

——, and D. R. Marshall. 1967. Population studies on predominantly self-pollinating species. X. Variation in natural populations of *Avena fatua* and *A. barbata. A.N.,* 101:19-33.

Jenkins, J. B. 1967. Mutagenesis at a complex locus in *Drosophila* with the monofunctional alkylating agent, ethyl methanesulfonate. *Gs.,* 57:783-793.

Jensen, A. R. 1969. How much can we boost I.Q. and scholastic achievement? *Harvard Educ. Rev.,* 39:1-123.

Jensen, N. 1965. Multiple superiority in cereals. *Crop. Sci.,* 5:566-568.

Jinks, J. L. 1964. Extrachromosomal Inheritance. Prentice-Hall, Englewood Cliffs.

Johannsen, W. 1909 (1926). Elemente der exakten Erblichkeitslehre. Gustav Fischer, Jena.

John, B., and K. R. Lewis. 1958. Studies on *Periplaneta americana*. III. Selection for heterozygosity. *Hy.*, 12:185-197.

——, and K. R. Lewis. Selection for interchange heterozygosity in an inbred culture of *Blaberus discoidalis* (Serv.) *Gs.*, 44: 251-267.

Johnsgard, P. A. 1967. Sympatry changes and hybridization incidence in mallards and black ducks. *Amer. Midland Natur.*, 77:51-63.

Johnson, A. W., and J. G. Packer. 1965. Polyploidy and environment in arctic Alaska. S., 148:237-239.

Johnson, F. M., C. G. Kanapi, R. H. Richardson, M. R. Wheeler, and W. S. Stone. 1968. An analysis of polymorphisms among isozyme loci in dark and light *Drosophila ananassae* strains from American and Western Samoa. *P.N.A.S.*, 56:119-125.

Johnson, M. P., L. G. Mason, and P. H. Raven. 1968. Ecological parameters and plant species diversity. *A.N.*, 102:297-306.

Jonas, H. 1966. The Phenomenon of Life. Harper & Row, New York.

Jones, H. A., J. C. Walker, T. M. Little, and R. M. Larson. 1946. Relation of color-inhibiting factor to smudge resistance in onion. *J. Agr. Res.*, 72:259-264.

Jordan, D. S. 1905. The origin of species through isolation. S., 22:545-562.

Jordan, K. 1905. Der Gegensatz zwischen geographischer und nichtgeographischer Variation. *Zeit. wiss. Zool.*, 83:151-210.

Jukes, Th. H. 1966. Molecules and Evolution. Columbia Univ. Press, New York.

Käfer, E. 1952. Vitalitätsmutationen, ausgelöst durch Röntgenstrahlen beim *Drosophila melanogaster*. Z.i.A.V., 84:508-535.

Kallman, K. D. 1962. Population genetics of the gynogenetic teleost, *Mollienesia formosa* (Girard). *E.*, 16:497-504.

Kaplan, W. D. 1948. Formaldehyde as a mutagen in *Drosophila*. S., 108:43.

Karpechenko, G. D. 1928. Polyploid hybrids of *Raphanus sativus* L. × *Brassica oleracea* L. *Z.i.A.V.*, 48:1-85.

Kastritsis, C. D. 1966. Cytological studies on some species of the *tripunctata* group of *Drosophila*. *U.T.P.*, 6615:413-474.

——. 1967. A comparative study of the chromosomal polymorphs in the incipient species of the *Drosophila paulistorum* complex. *Ch.*, 23:180-202.

——. 1969a. The chromosomes of some species of the *guarani* group of *Drosophila*. *J. Heredity*, 60:51-57.

——. 1969b. A cytological study of some recently collected strains of *Drosophila paulistorum*. *E.*, 23:663-675.

Kaul, D., and P. A. Parsons. 1965. The genotype control of mating speed and duration of copulation in *Drosophila pseudoobscura*. *Hy.*, 20:381-392.

Keiding, J. 1967. Persistence of resistant populations after the relaxation of the selection pressure. *World Rev. Pest Control*, 6:115-130.

Kennedy, W. P. 1967. Epidemiologic aspects of the problem of congenital malformations. Birth Defects Original Series, Vol. 3, No. 2.

Kerkis, J. 1933. Development of gonads in hybrids between *Drosophila melanogaster* and *Drosophila simulans*. *J. Exp. Zool.*, 66:477-509.

——. 1936. Chromosome configuration in hybrids between *Drosophila melanogaster* and *Drosophila simulans*. *A.N.*, 70:81-86.

——. 1938. The frequency of mutations affecting viability. *Bull. Acad. Sci. USSR (Biol.)*, 1938:75-90.

Kernaghan, R. P., and L. Ehrman. 1970. An electron microscopic study of the etiology of hybrid sterility in *Drosophila paulistorum*. I. Mycoplasma-like inclusions in the testes of sterile males. Chromosoma 29:291-304.

Kerr, W. E., and L. Kerr. 1952. Concealed variabilty in the X chromosome. *A.N.*, 96:405-407.

——, and S. Wright. 1954. Experimental studies of the distribution of gene frequencies in very small populations of *Drosophila melanogaster*. *E.*, 8:172-177, 225-240, 293-302.

Kerster, H. W., and D. A. Levin. 1968. Neighborhood size in *Lithospermum carolinense*. *Gs.*, 60:577-583.

Kessler, S. 1966. Selection for and against ethological isolation between *Drosophila pseudoobscura* and *Drosophila persimilis*. *E.*, 20:634-645.

Kettlewell, H. B. D. 1955. Selection experiments on industrial melanism in the *Lepidoptera*. *Hy.*, 9:323-342.

——. 1961. The phenomenon of industrial melanism in the *Lepidoptera*. *Ann. Rev. Entom.*, 6:245-262.

——. 1965. Insect survival and selection for pattern. S., 148:1290-1296.

Key, K. H. L. 1950. A critique of the phase theory of locusts. *Q.R.B.*, 25:363-407.

——. 1957. Kentromorphic phases in three species of *Phasmatodea*. *Austral. J. Zool.*, 5:247-284.

——. 1968. The concept of stasipatric speciation. *S.Z.*, 17:14-22.

Keyl, H. G. 1961. Chromosomenevolution bei *Chironomus*. I. Strukturabwandlungen an Speicheldrüsen Chromosomen. *Ch.*, 12:26-47.

——. 1962. Chromosomenevolution bei *Chironomus*. II. Chromosomenumbauten und phylogenetische Beziehung der Arten. *Ch.*, 13:464-514.

——. 1965. Duplikationen von Untereinheiten der chromosomalen DNS während der Evolution von *Chironomus thummi*. *Ch.*, 17:139-180.

Kihara, H. 1959. Fertility and morphological variation in the substitution and restoration backcrosses of the hybrids, *Triticum vulgare* × *Aegilops caudata*. *P. X I.C.G.*, 1:142-171.

——. 1965. The origin of wheat in the light of comparative genetics. *Japan. J. Genetics*, 40:45-54.

——. 1967. Cytoplasmic male sterility in relation to hybrid wheat breeding. *Der Züchter*, 37:86-93.

Kimura, M. 1954. Process leading to quasi-fixation of genes in natural populations due to random fluctuations of selection intensities. *Gs.*, 39:280-295.

——. 1960. Optimum mutation rate and degree of dominance as determined by the principle of minimum genetic load. *J. Genetics*, 57:21-34.

——. 1961. Natural selection as the process of accumulating genetic information in adaptive evolution. *G.R.*, 2:127-140.

——. 1968. Genetic variability maintained in a finite population due to mutational production of neutral and nearly neutral isoalleles. *G. R.*, 11:246-269.

——. 1969. The number of heterozygous nucleotide sites maintained in a finite population due to steady flux of mutations. *Gs.*, 61:893-903.

——, and J. F. Crow. 1963. The measurement of effective population number. *E.*, 17:279-288.

——, and J. F. Crow. 1964. The number of alleles that can be maintained in a finite population. *Gs.*, 49:725-738.

——, and J. F. Crow. 1969. Natural selection and gene substitution. *G.R.*, 13:127-141.

——, T. Maruyama, and J. F. Crow. 1963. The mutation load in small populations. *Gs.*, 48:1303-1312.

——, and T. Ohta. 1969. The average number of generations until fixation of a mutant gene in a finite population. *Gs.*, 61:763-771.

——, and G. H. Weiss. 1964. The stepping-stone model of population structure and the decrease of genetic correlation with distance. *Gs.*, 49:561-576.

Kindred, B. 1967. Selection for an invariant character, vibrissa number in the house mouse. V. Selection on non-Tabby segregants from Tabby selection lines. *Gs.*, 55:365-373.

King, J. L. 1966. The gene interaction component of the genetic load. *Gs.*, 53:403-413.

——. 1967. Continuously distributed factors affecting fitness. *Gs.*, 55:483-492.

——, and Th. H. Jukes. 1969. Non-Darwinian evolution. S., 164:788-798.

Kitagawa, O. 1967. Genetic divergence in M. Vetukhiv's experimental populations of *Drosophila pseudoobscura*. *G.R.*, 10:303-312.

Kitchin, F. D., W. H. Evans, C. A. Clarke, R. B. McConnell, and P. M. Sheppard. 1959. PTC taste response and thyroid disease. *Brit. Med. J.*, 1:1069-1074.

Kitzmiller, J. B. 1967. Mosquito cytogenetics. Pp. 133-150 *in*: J. W. Wright and R. Pal (Eds.), "Genetics of Insect Vectors of Disease," Elsevier, Amsterdam.

——, G. Frizzi, and R. H. Baker. 1967. Evolution and speciation within the *maculipennis* complex of the genus *Anopheles*. Pp. 151-210 *in*: J. W. Wright and R. Pal (Eds.), "Genetics of Insect Vectors of Disease," Elsevier, Amsterdam.

Klekowski, E. J., and H. G. Baker. 1966. Evolutionary significance of polyploidy in *Pteridophyta*. S., 153:305-307.

Knight, G. R., A. Robertson, and C. H. Waddington. 1956. Selection for sexual isolation within a species. *E.*, 10:14-22.

Knipling, E. F. 1955. Possibilities of insect control or eradication through the use of sexually sterile males. *J. Econ. Entom.*, 48:459-462.

——. 1967. Sterile technique—principles involved, current application, limitations, and future application. Pp. 587-616 *in*: J. W. Wright and R. Pal (Eds.), "Genetics of Insect Vectors of Disease," Elsevier, Amsterdam.

Koestler, A. 1967. The Ghost in the Machine. Macmillan, New York.

Kojima, R. I., and T. M. Kelleher. 1962. Survival of mutant genes. *A.N.*, 96:329-346.

——, and T. M. Kelleher. 1963. Selection studies of quantitative traits with laboratory animals. *Stat. Genetics Plant Breeding NAS-NRC*, 982:395-422.

Komai, T. 1954. An actual instance of microevolution observed in an insect population. *Proc. Japan Acad.*, 30:970-975.

——, M. Chino, and Y. Hosino. 1950. Contributions to the evolutionary genetics of the lady bettle, Harmonia. I. *Gs.*, 35:589-601.

——, and Y. Hosino. 1951. Contributions to the evolutionary genetics of the lady beetle, Harmonia. II. Microgeographic variations. *Gs.*, 36:382-390.

Kondakova, A. A. 1935. Einfluss des Jods auf das Auftreten letaler mutationen im III. Chromosom von *Drosophila melanogaster*. *Biol. Zhurnal*, 4:721-726.

Konigsberg, W., R. G. Huntsman, F. Wadia, and H. Lehmann. 1965. Haemoglobin $D_{\beta Punjab}$ in an East Anglian Family. *J. Royal Anthrop. Inst.*, 95:295-306.

Koopman, K. F. 1950. Natural selection for reproductive isolation between *Drosophila pseudoobscura* and *Drosophila persimilis*. *E.*, 4:135-145.

Koref-Santibañez, S. 1964. Reproductive isolation between the sibling species *Drosophila pavani* and *Drosophila gaucha*. *E.*, 18:245-251.

Kornberg, A. 1962 Enzymatic Synthesis of DNA. John Wiley, New York.

Korzhinsky, S. I. 1899. Heterogenesis and Evolution. Academy of Science, St. Petersburg (Russian).

Kosiupa, D. E. 1936. The effect of sublimate on the occurrence of lethal mutations in *Drosophila melanogaster*. *Bull. Biol. Med. Exp. URSS*, 2:87-89.

Kozhevnikov, B. Th. 1936. Experimentally produced karyotypical isolation. *Biol. Zhurnal*, 5:727-752.

Kraczkiewicz, Z. 1950. Recherches cytologiques sur les chromosomes de *Lasioptera rubi* Heeg. (*Cecidomyidae*). *Zool. Poloniae*, 5:73-115.

Krimbas, C. B. 1960. Synthetic sterility in *Drosophila willistoni*. *P.N.A.S.*, 46: 832-833.

——. 1961. Release of genetic variability through recombination. VI. *Drosophila willistoni*. *Gs.*, 46:1323-1334.

——. 1964. The genetics of *Drosophila subobscura* populations. I. Inversion polymorphism in populations of southern Greece. *E.*, 18:541-552.

Kruckeberg, A. R. 1957. Variation in fertility of hybrids between isolated populations of serpentine species, *Streptanthus glandulosus* Hook. *E.*, 11:185-211.

Kullenberg, B. 1951. *Ophrys insectifera* L. et les insects. *Oikos*, 3:53-70.

Kurokawa, H. 1960. Sexual isolation among the three races, A, B, and C of *Drosophila auraria*. *Japan. J. Genetics*, 35:161-166.

Kyhos, D. W. 1965. The independent aneuplod origin of two species of *Chaenactis (Compositae)* from a common ancestor. *E.*, 19:26-43.

LaChance, L. E. 1967. The induction of dominant lethal mutations in insects by ionizing radiation and chemicals as related to the sterile-male technique of insect control. Pp. 617-650 *in:* J. W. Wright and R. Pal (Eds.), "Genetics of Insect Vectors of Disease," Elsevier, Amsterdam.

Laibach, F. 1925. Das Taubwerden von Bastardsamen und die Künstliche Aufzucht früh absterbender Bastardembryonen. *Zeit. Botanik*, 17:417-459.

Laidlaw, H. H., E. P. Gomes, and W. E. Kerr. 1956. Estimation of the number of lethal alleles in a panmictic population of *Apis mellifera* L. *Gs.*, 41:179-188.

——, J. G. Reiman, and D. E. Hopkins. 1964. A reciprocal translocation in *Cochliomyia hominivorax (Diptera, Calliphoridae)*. *Gs.*, 49:959-972.

Laird, Ch. D., and B. J. McCarthy. 1968a. Magnitude of interspecific nucleotide sequence variability in *Drosophila*. *Gs.*, 60:303-322.

——, and B. J. McCarthy. 1968b. Nucleotide sequence homology within the genome of *Drosophila melanogaster*. *Gs.*, 60:323-334.

Lamotte, M. 1951. Recherches sur la structure génétique des populations naturelles de *Cepaea nemoralis* L. *Bull. Biol. France*, Suppl., 35:1-239.

——. 1959. Polymorphism of natural populations of *Cepaea nemoralis*. *C.S.H.*, 24:65-84.

Lamprecht, H. 1964. Species concept and the origin of species. The two categories of genes—intra- and interspecific ones. *Agri Hortique Genetica*, 22:272-280.

——. 1966. Die Entstehung der Arten und höheren Kategorien. Springer, Vienna and New York.

Lancefield, D. E. 1929. A genetic study of two races or physiological species in *Drosophila obscura*. *Z.i.A.V.*, 52:287-317.

Landau, R. 1962. Four forms of *Simulium tuberosum* Lundstr. in southern Ontario: Salivary gland chromosome study. *Canad. J. Zool.*, 40:921-939.

Landauer, W. 1948. Hereditary abnormalities and their chemically induced phenocopies. *Growth Symposia*, 12:171-200.

Larson, C. A. 1961. Phenylketonuria. *Genetica Medica*, 4:121-140.

Laugé, G. 1966. Étude comparative des effets d'un traitement thermique sur le développement des gonades et de divers charactères sexuels primaires chez les intersexués triploides de *Drosophila melanogaster*. *Bull. Soc. Zool. France,* 91: 661-686.

Laven, H. 1967. Speciation and evolution in *Culex pipiens*. Pp. 251-275 *in:* J. W. Wright and R. Pal (Eds.), "Genetics of Insect Vectors of Disease," Elsevier, Amsterdam.

Law, L. W. 1938. The effects of chemicals on the lethal mutation rate in *Drosophila melanogaster*. *P.N.A.S.*, 24:546-550.

Lea, A. J. 1955. Association of susceptibility to poliomyelitis with eye and hair color. *S.*, 121:608.

Lea, D. E. 1955. Action of Radiations on Living Cells. 2nd Ed. Cambridge Univ. Press, Cambridge.

Lécher, P. 1967. Cytogénétique de l'hybridation expérimentale et naturelle chez l'isopode *Jaera albifrons syei* Bocquet. *Arch. Zool. Exp. Gen.*, 108:633-698.

Lederberg, J. 1965. Signs of life. Criterion-system of exobiology. *N.*, 207:9-13.

Lejeune, J. 1969. Human cytogenetics. *P.* XII *I.C.G.*, 3:379-387.

——, R. Turpin, and M. Gautier. 1959. Le mongolisme, premier example d'aberration autosomique humaine. *Ann. Génét. Hum.*, 1:41-49.

Lerner, I. M. 1950. Population Genetics and Animal Improvement. Cambridge Univ. Press, Cambridge.

——. 1954. Genetic Homeostasis. Oliver & Boyd, Edinburgh.

——. 1958. The Genetic Basis of Selection. John Wiley, New York.

——. 1968. Heredity, Evolution and Society. Freeman, San Francisco.

——, and H. P. Donald. 1966. Modern Developments in Animal Breeding. Academic Press, London and New York.

Levene, H. 1953. Genetic equilibrium when more than one ecological niche is available. *A.N.*, 87:331-333.

——. 1963. Inbred genetic loads and the determination of population structure. *P.N.A.S.*, 50:587-592.

——, I. M. Lerner, A. Sokoloff, F. K. Ho, and I. R. Franklin. 1965. Genetic loads in *Tribolium*. *P.N.A.S.*, 53:1042-1050.

——, O. Pavlovsky, and Th. Dobzhansky. 1958. Dependence of the adaptive values of certain genotypes of *Drosophila pseudoobscura* on the composition of the gene pool. *E.*, 12:18-23.

Levin, B. R., M. L. Petras, and D. I. Rasmussen. 1969. The effect of migration on the maintenance of a lethal polymorphism in the house mouse. *A.N.*, 103: 647-661.

Levin, D. A. 1968. The structure of a polyspecies hybrid swarm in *Liatris*. *E.*, 22:352-372.

——, and H. W. Kerster. 1967a. An analysis of interspecific pollen exchange in *Phlox*. *A.N.*, 101:387-399.

——, and H. W. Kerster. 1967b. Natural selection for reproductive isolation in *Phlox*. *E.*, 21:679-687.

——, and H. W. Kerster. 1968. Local gene dispersal in *Phlox*. *E.*, 22:130-139.

Levins, R. 1968. Evolution in Changing Environments. Princeton Univ. Press, Princeton.

——, and R. MacArthur. 1966. The maintenance of genetic polymorphism in a spatially heterogeneous environment: variations on a theme by Howard Levene. *A.N.*, 100:585-589.

Levisohn, R., and S. Spiegelman. 1969. Further extracellular Darwinian experiments with replicating RNA molecules. Diverse variants isolated under different selective conditions. *P.N.A.S.*, 63:805-811.

Levitan, M., H. L. Carson, and H. D. Stalker 1954. Triads of overlapping inversions in *Drosophila robusta*. *A.N.*, 88:113-114.

Lewis, E. B. 1967. Genes and gene complexes. Pp. 17-47 *in:* R. A. Brink and E. D. Styles (Eds.), "Heritage from Mendel," Univ. Wisconsin Press, Madison.

Lewis, H. 1966. Speciation in flowering plants. S., 152:167-172.

——, and P. H. Raven. 1958. Rapid evolution in *Clarkia*. *E.*, 12:319-336.

Lewis, K. R., and B. John. 1964. Spontaneous interchange in *Chorthippus brunneus*. *Ch.*, 14:618-637.

Lewontin, R. C. 1955. The effects of population density and composition on viability in *Drosophila melanogaster*. *E.*, 9:27-41.

——. 1962. Interdeme selection controlling a polymorphism in the house mouse. *A.N.*, 96:65-78.

——. 1967a. An estimate of average heterozygosity in man. *Amer. J. Human Genetics,* 19:681-685.

——. 1967b. The genetics of complex systems. *Proc. 5th Berkeley Symposium Math. Stat. Probability,* 4:439-455.

——. 1968. The concept of evolution. Pp. 202-210 *in:* D. L. Sills (Ed.), International Encyclopedia of Social Science, Vol. 5.

——, and J. L. Hubby. 1966. A molecular approach to the study of genic heterozygosity in natural populations. II. Amount of variation and degree of heterozygosity in natural populations of *Drosophila pseudoobscura*. *Gs.*, 54:595-609.

——, and Y. Matsuo. 1963. Interaction of genotypes determining viability in *Drosophila busckii*. *P.N.A.S.*, 49:270-278.

——, and M. J. D. White. 1960. Interaction between inversion polymorphisms of two chromosome pairs in the grasshopper *Moraba scurra*. *E.*, 14:116-129.

Li, C. C. 1955a. Population Genetics. Univ. Chicago Press, Chicago.

——. 1955b. The stability of an equilibrium and the average fitness of a population. *A.N.*, 89:281-296.

——. 1963. The way the load ratio works. *Amer. J. Human Genetics,* 15:315-321.

Lillie, F. R. 1921. Studies of fertilization. VIII. *Biol Bull.*, 40:1-22.

Lindsley, D. L., C. W. Edington, and E. S. von Halle. 1960. Sex-linked recessive lethals in *Drosophila* whose expression is suppressed by the Y chromosome. *Gs.*, 45:1649-70.

——, and E. H. Grell. 1968. Genetic variations of *Drosophila melanogaster*. *Carnegie Inst. Washington Publ.* 627:1-472.

Linsley, E. G. 1958. The ecology of solitary bees. *Hilgardia,* 27:543-599.

Littlejohn, M. J. 1965. Premating isolation in the *Hyla ewingi* complex (*Anura: Hylidae*). *E.*, 19:234-243.

——, and J. J. Loftus-Hills. 1968. An experimental evaluation of premating isolation in the *Hyla ewingi* complex. *E.*, 22:659-663.

Livingstone, F. B. 1964. The distribution of the abnormal hemoglobin genes and their significance for human evolution. *E.*, 18:685-699.

——. 1967. Abnormal Hemoglobins in Human Populations. Aldine, Chicago.

Lloyd, M., and H. S. Dybas. 1966. The periodical cicada problem. II. Evolution. *E.*, 20:466-505.

Lobashov, M. E. 1935. Uber die Natur der Einwirkung der chemischen Agentien auf den Mutationsprocess bei *Drosophila melanogaster*, *Ga.*, 19:200-241.

——, and F. Smirnov. 1934. On the nature of the action of chemical agents on the mutational process in *Drosophila melanogaster*. II. The effect of ammonia on the occurrence of lethal transgenations. *C. R. Acad. Sci. URSS*, 3:174-176.

Loomis, W. F. 1967. Skin-pigment regulation of vitamin-D biosynthesis in man. *S.*, 157:501-506.

Lorenz, K. 1941. Vergleichende Bewegungsstudien an Anatinen. *J. Ornithol.*, 89: 194-294.

Lotsy J. P. 1931. On the species of the taxonomist in its relation to evolution. *Ga.*, 13:1-16.

Löve, A. 1951. Taxonomical evaluation of polyploids. *Caryologia*, 3:263-284.

——. 1964. The biological species concept and its evolutionary structure. *Taxon*, 13:33-45.

——, and D. Löve. 1949. The geobotanical significance of polyploidy. I. Polyploidy and latitude. *Portugeliae Acta Biol.*, Goldschmidt Volumen:273-352.

Lucchesi, J. C. 1968. Synthetic lethality and semi-lethality among functionally related mutants of *Drosophila melanogaster*. *Gs.*, 59:37-44.

Ludwig, W. 1950. Zur Theorie der Konkurrenz. Die Annidation (Einnischung) als fünfter Evolutionsfaktor. *Neue Ergeb. Probleme Zool.*, Klatt-Festschrift 1950:516-537.

Lundman, B. 1967. Geographische Anthropologie. Gustav Fischer, Stuttgart.

Luria, S. E., and M. Delbrück. 1943. Mutations of bacteria from virus sensitivity to virus resistance. *Gs.*, 28:491-511.

McCarthy B. J., and E. T. Bolton. 1963. An approach to the measurement of genetic relatedness among organisms. *P.N.A.S.*, 50:156-164.

McClelland, G. A. H. 1967. Speciation and evolution in *Aedes*. Pp. 277-311 *in:* J. W. Wright and R. Pal (Eds.), "Genetics of Insect Vectors of Disease," Elsevier, Amsterdam.

MacFarquhar, A. M., and F. W. Robertson. 1963. The lack of evidence for coadaptation in crosses between geographical races of *Drosophila subobscura* Coll. *G.R.*, 4:104-131.

McKusick, V. A. 1966. Mendelian Inheritance in Man. Johns Hopkins Press, Baltimore.

McNeilly, T. 1968. Evolution in closely adjacent plant populations. III. *Agrostis tenuis* on a small copper mine. *Hy.*, 23:99-108.

——, and A. D. Bradshaw. 1968. Evolutionary processes in populations of copper tolerant *Agrostis tenuis*. *E.*, 22:108-118.

Magalhães, L. E., A. B. da Cunha, J. S. de Toledo, S. P. de Toledo, H. L. Souza, H. J. Targa, V. Setzer, and C. Pavan. 1965. On lethals and their suppressors in experimental populations of *Drosophila willistoni*. *Mutation Res.*, 2:45-54.

Magalhães, L. E., J. S. de Toledo, and A. B. da Cunha. 1965. The nature of lethals in *Drosophila willistoni*. *Gs.*, 52:599-608.

Magrzhikovskaja, K. V. 1938. The effect of $CuSO_4$ on the mutation process in *Drosophila melanogaster*. *Biol. Zhurnal*, 7:635-642.

Mainx, F., J. Fiala, and E. V. Kogerer. 1956. Die geographische Verbreitung der chromosomalen Strukturtypen von *Liriomyza urophorina* Mill. *Ch.*, 8:18-29.

Makino, S. 1951. An Atlas of the Chromosome Number in Animals. Iowa State College Press, Ames.

——, M. S. Sasaki, K. Yamada, and T. Kajii. 1963. A long Y chromosome in man. *Ch.*, 14:154-161.

——, and N. Takagi. 1965. Some morphological aspects of the abnormal human Y chromosome. *Cytologia*, 30:274-292.

Malécot, G. 1948. Les mathématiques de l'hérédité. Masson, Paris.

——. 1959. Les modèles stochastiques en génétique des populations. *Publ. Inst. Statistique Univ. Paris*, 8:173-210.

Malogolowkin-Cohen, Ch., H. Levene, N. P. Dobzhansky, and A. S. Simmons. 1964. Inbreeding and the mutational and balanced loads in natural populations of *Drosophila willistoni*. *Gs.*, 50:1299-1311.

——, A. S. Solima, and H. Levene. 1965. A study of sexual isolation between certain strains of *Drosophila paulistorum*. *E.*, 19:95-103.

Maly, R. 1951. Cytomorphologische Studien an strahleninduzierten, konstant abweichenden Plastidenfermen bei Farnprothalien. *Z.i.A.V.*, 33:447-478.

Mangenot, S., and G. Mangenot. 1962. Enquête sur les nombres chromosomiques dans une collection d'espèces tropicales. *Rev. Cytol. Biol. Végét.*, 25:411-447.

Manna, G. R., and S. G. Smith. 1959. Chromosomal polymorphism and interrelationships among bark weevils of genus *Pissodes* Germar. *Nucleus*, 2:179-208.

Manning, A. 1959. The sexual behavior of two sibling *Drosophila* species. *Behaviour*, 15:123-145.

——. 1961. The effects of artificial selection for mating speed in *Drosophila melanogaster*. *Animal Behavior*, 9:82-92.

Mansfeld, R. 1951. Das morphologische System des Saatweizens, *Triticum aestivum* L. *Der Züchter*, 21-41-66.

Manton, I. 1950. Problems of Cytology and Evolution in the *Pteridophyta*. Cambridge Univ. Press, Cambridge.

Manwell, C., and C. M. Ann Baker. 1963. A sibling species of sea cucumber discovered by starch gel electrophoresis. *Comp. Biochem. Physiol.*, 10:39-53.

Margoliash, E. 1963. Primary structure and evolution of cytochrome c. *P.N.A.S.*, 50:672-679.

——, and E. L. Smith. 1965. Structural and functional aspects of cytochrome c in relation to evolution. Pp. 221-242 *in:* V. Bryson and H. J. Vogel (Eds.), "Evolving Genes and Proteins," Academic Press, New York.

Marien, D. 1958. Selection for developmental rate in *Drosophila pseudoobscura*. *Gs.*, 43:3-15.

Marinkovic, D. 1967. Genetic loads affecting fertility in natural populations of *Drosophila pseudoobscura*. *Gs.*, 57:701-709.

Marler, P. R., and W. J. Hamilton. 1966. Mechanisms of Animal Behavior. John Wiley, New York.

Marquardt, H. 1952. Uber die spontanen Aberrationen in der Anaphase der Meiosis von *Paeonia tenuifolia*. *Ch.*, 5:81-112.

Marshall, A. J. 1954. Bowerbirds, Their Displays and Breeding Cycles. Clarendon, Oxford.

Marshall, R. E., C. T. Caskey, and M. Nirenberg. 1967. Fine structure of RNA code words recognized by bacterial, amphibian and mammalian transfer RNA. S., 155:820-826.

Martin, J. 1965. Interrelation of inversion systems in the midge *Chironomus intertinctus*. II. A nonrandom association of linked inversions. *Gs.*, 52:371-383.

Maslin, T. P. 1968. Taxonomic problems in parthenogenetic vertebrates. S.Z., 17:219-231.

Mason, E. D., M. Jacob, and V. Balakrishnan. 1964. Racial group differences in the basal metabolism and body composition of Indian and European women in Bombay. *Human Biol.*, 36:374-396.

Mather, K. 1941. Variation and selection of polygenic characters. *J. Genetics*, 41:159-193.

——. 1953. The genetical structure of populations. *Symposium Soc. Exp. Biol.*, 7:66-95.

——. 1955. Polymorphism as an outcome of disruptive selection. *E.*, 9:52-61.

——, and B. J. Harrison. 1949. The manifold effect of selection. *Hy.*, 3:1-52, 131-162.

——, and A. Vines. 1951. Species crosses in *Antirrhinum*. *Hy.*, 5:195-214.

Matthey, R. 1952. Chromosomes de *Muridae*. *Ch.*, 5:113-138.

——. 1963. Cytologie comparée et polymorphisme chromosomique chez des *Mus* africains appartenant aux groupes *bufotriton* et *minutoides*. *Cytogenetics*, 2:290-322.

——. 1964a. La signification des mutations chromosomiques dans les processus de spéciation. Etude cytogénétique du sous-genre *Leggada* Gray. *Arch. Biol.*, 75:169-206.

——. 1964b. Evolution chromosomique et spéciation chez les *Mus* du sous-genre *Leggada* Gray 1837. *Experientia*, 20:1-9.

——. 1966a. Le polymorphisme chromotomique des *Mus* africains du sous-genre *Leggada*. *Rev. Suisse Zool.*, 73:585-607.

——. 1966b. Une inversion péricentrique à l'origine d'un polymorphisme chromosomique non-robertsonien dans une population de *Mastomys* (*Rodentia-Murinae*). *Ch.*, 18:188-200.

Maynard-Smith, J. 1966. Sympatric speciation. *A.N.*, 100:637-650.

——. 1968a. "Haldane's dilemma" and the rate of evolution. *N.*, 219:1114-1116.

——. 1968b. Evolution in sexual and asexual populations. *A.N.*, 102:469-473.

Mayr, E. 1942. Systematics and the Origin of Species. Columbia Univ. Press, New York.

——. 1954. Change of genetic environment and evolution. Pp. 157-180. *In:* Huxley, J., and others, "Evolution as a Process," Macmillan, New York.

——. 1957. Species concepts and definitions. Pp. 1-22 *in:* E. Mayr (Ed.), "The Species Problem," American Associaton for the Advancement of Science, Washington.

——. 1963. Animal Species and Evolution. Belknap, Cambridge.

——. 1965a. Classification and phylogeny. *Amer. Zoologist*, 5:165-174.

——. 1965b. Numerical phenetics and taxonomic theory. *S.Z.*, 14:73-97.

——. 1969. Principles of Systematic Zoology. McGraw-Hill, New York.

Menzel, M. Y. 1964. Preferential chromosome pairing in allotetraploid *Lycopersicon esculentum-Solanum lycopersicoides*. *Gs.*, 50:855-862.

Merrell, D. J. 1968. A comparison of the estimated size and the "effective size" of breeding populations of the leopard frog, *Rana pipiens*. *E.*, 22:274-283.

——, and Ch. F. Rodell. 1968. Seasonal selection in the leopard frog, *Rana pipiens*. *E.*, 22:284-288.

Meselson, M., and F. W. Stahl. 1958. The replication of DNA in *Escherichia coli*. *P.N.A.S.*, 44:671-682.

Mettler, L. E., S. E. Moyer, and K. Kojima. 1966. Genetic loads in cage populations of *Drosophila*. *Gs.*, 54:887-898.

Michaud, T. C. 1964. Vocal variation in two species of chorus frogs, *Pseudacris nigrita* and *Pseudacris clarki* in Texas. *E.*, 18:498-506.

Michaelis, P. 1954. Cytoplasmic inheritance in *Epilobium* and its theoretical significance. *Adv. Genetics*, 6:287-401.

Milani, R. 1967. The genetics of *Musca domestica* and of other muscoid flies. Pp. 315-369 *in:* J. W. Wright and R. Pal (Eds.), "Genetics of Insect Vectors of Disease," Elsevier, Amsterdam.

Milkman, R. D. 1962. The genetic basis of natural variation. IV. On the natural distribution of *cve* polygenes of *Drosophila melanogaster*. *Gs.*, 47:261-272.

——. 1965. The genetic basis of natural variation. VII. The individuality of polygenic combinations in *Drosophila. Gs.*, 52:789-799.

——. 1966. The genetic basis of natural variation. VIII. Synthesis of *cve* polygenic combinations from laboratory strains of *Drosophila melanogaster. Gs.*, 53:863-874.

Miller, D. D., and R. A. Voelker. 1968. Salivary gland chromosome variation in the *Drosophila affinis* subgroup. *J. Heredity*, 59:86-98.

——, and N. J. Westphal. 1967. Further evidence on sexual isolation within *Drosophila athabasca. E.*, 21:479-492.

Millicent, E., and J. M. Thoday. 1961. Effects of disruptive selection. *Hy.*, 16:199-217.

Minamori, S. 1957. Physiological isolation in *Cobitidae*. VI. Temperature adaptation and hybrid inviability. *J. Sci. Hiroshima Univ.*, B, 17:65-119.

——, and Y. Saito. 1964. Local and seasonal variations of lethal frequencies in natural populations of *Drosophila melanogaster. Japan. J. Genetics*, 38:290-304.

Mirsky, A. E., and H. Ris. 1951. The desoxyribonucleic and acid content of animal cells and its evolutionary significance. *J. Gen. Physiol.*, 34:451-462.

Monclus, M. 1953. Variacion geographica de los peines tarsales de los machos de *D. subobscura. Genetica Iberica,* 5:101-114.

Montagu, A. (Ed.). 1964. The Concept of Race. Free Press, Glencoe.

Montalenti, G. 1965. Human population genetics. Synthesis. *Genetics Today. P. XI I.C.G.*, 3:965-972.

Moore, D. M., and H. Lewis. 1965. The evolution of self-pollination in *Clarkia xantiana. E.*, 19:104-114.

Moore, J. A. 1949a. Geographic variation of adaptive characters in *Rana pipiens* Schreber. *E.*, 3:1-24.

——. 1949b. Patterns of evolution in the genus *Rana. Genetics, Paleon., Evolution:* 315-338.

——. 1950. Further studies on *Rana pipiens* racial hybrids. *A.N.*, 84:247-254.

Moore, R. E. 1965a. Olfactory discrimination as an isolating mechanism between *Peromyscus maniculatus* and *Peromyscus polionotus. Amer. Midland Natur.*, 73:85-100.

——. 1965b. Ethological isolation between *Peromyscus maniculatus* and *Peromyscus polionotus. Amer. Midland Natur.*, 74:341-349.

Mooring, J. 1958. A cytogentic study of *Clarkia unguiculata.* I. Translocations. *Amer. J. Botany*, 45:233-242.

——. 1961. The evolutionary role of translocations in *Clarkia unguiculata* (*Onagraceae*). *Recent Adv. Botany, Biosystem.*:853-858.

Moos, J. R. 1955. Comparative physiology of some chromosomal types in *Drosophila pseudoobscura. E.*, 9:141-151.

Morgan, T. H. 1911. The origin of five mutations in eye color in *Drosophila* and their modes of inheritance. S., 33:534-537.

——. 1919. The Physical Basis of Heredity. Lippincott, Philadelphia.

——, A. H. Sturtevant, H. J. Muller, and C. B. Bridges. 1915. The Mechanism of Mendelian Heredity. Holt, New York.

Mori, S., and Matutani, K. 1953. Daily swarming of some caddis fly adults and their habitat segregations. *Dobutsugaku Zasshi*, 62:191-198.

Morley, F. H. W., and J. Katznelson. 1965. Colonization in Australia by *Trifolium subterraneum* L. Pp. 269-285 *in:* H. G. Baker and G. L. Stebbins (Eds.), "The Genetics of Colonization Species," Academic Press, New York.

Morris, D. 1952. Homosexuality in the ten-spined stickleback (*Pygosteus pungitius* L.). *Behavior*, 4:233-261.

Morton, N. E. 1960. The mutational load due to detrimental genes in man. *Amer. J. Human Genetics*, 12:348-364.

——. 1964. Models and evidence in human population genetics. *P.* XI *I.C.G.*:935-951.

——, and C. S. Chung. 1959. Are the MN blood groups maintained by selection? *Amer. J. Human Genetics*, 11:237-251.

——, C. S. Chung, and L. D. Friedman. 1968. Relation between homozygous viability and average dominance in *Drosophila melanogaster*. *Gs.*, 60:601-614.

——, C. S. Chung, and M. P. Mi. 1967. Genetics of Interracial Crosses in Hawaii. Karger, Basel and New York.

——, J. F. Crow, and H. J. Muller. 1956. An estimate of the mutational damage in man from data on consanguineous marriages. *P.N.A.S.*, 42:855-863.

——, H. Krieger, and M. P. Mi. 1966. Natural selection on polymorphisms in northeastern Brazil. *Amer. J. Human Genetics*, 18:153-171.

Mosquin, Th. 1967. Evidence for autopolyploidy in *Epilobium angustifolium (Onagraceae)*. *E.*, 21:713-719.

Mourad, A. K. 1964. Lethal and semilethal chromosomes in irradiated experimental populations of *Drosophila pseudoobscura*. *Gs.*, 50:1279-1287.

Mourant, A. E. 1954. The Distribution of the Human Blood Groups. Blackwell, Oxford.

——, A. C. Kopec, and K. Domaniewska-Sobczak. 1958. The ABO Blood Groups. Charles C Thomas, Springfield, Ill.

Muir, A., and A. G. Steinberg. 1967. On the genetics of the human allotypes, *Gm* and *Inv. Seminars Hematol.*, 4:156-173.

Mukai, T. 1964. The genetic structure of natural populations of *Drosophila melanogaster*. I. Spontaneous mutation rate of polygenes controlling viability. *Gs.*, 50:1-19.

——. 1967. A study of the genetic structure of natural populations in *Drosophila melanogaster* by means of spontaneous polygenic mutation rates. I and II. *Proc. Japan Acad.*, 38:741-746, 747-752.

——. 1969a. The genetic structure of natural populations of *Drosophila melanogaster*. VI. Further studies on the optimum heterozygosity hypothesis. *Gs.*, 61:479-495.

——. 1969b. Maintenance of polygenic and isoallelic variation in populations. *P.* XII *I.C.G.*, 3:293-308.

——, and A. B. Burdick. 1959. Single gene heterosis associated with a second chromosome recessive lethal in *Drosophila melanogaster*. *Gs.*, 44:211-232.

——, and A. B. Burdick. 1961. Examination of the closely linked dominant adaptive gene heterosis as an alternative to single gene heterosis associated with l (2) 55i in *Drosophila melanogaster*. *Japan. J. Genetics*, 36:97-104.

——, S. Chigusa, and I. Yoshikawa. 1964. The genetic structure of natural populations of *Drosophila melanogaster*. II. Overdominance of spontaneous mutant polygenes controlling viability in homozygous genetic background. *Gs.*, 50:711-715.

——, S. Chigusa, and I. Yoshikawa. 1965. The genetic structure of natural populations of *Drosophila melanogaster*. III. Dominance effect of spontaneous mutant polygenes controlling viability in heterozygous genetic backgrounds. *Gs.*, 52:493-501.

——, and T. Yamazaki. 1964. Position effect of spontaneous mutant polygenes controlling viability in *Drosophila melanogaster*. *Proc. Japan Acad.*, 40:840-845.

——, and T. Yamazaki. 1968. The genetic structure of natural populations of *Drosophila melanogaster*. V. Coupling-repulsion effect of spontaneous mutant polygenes controlling viability. *Gs.*, 59:513-535.

——, I. Yoshikawa, and K. Sano. 1966. The genetic structure of natural populations of *Drosophila melanogaster*. IV. Heterozygous effects of radiation-induced mutations on viability in various genetic backgrounds. *Gs.*, 53:513-527.

Muller, C. H. 1952. Ecological control of hybridization in *Quercus*: A factor in the mechanism of evolution. *E.*, 6:147-161.

Muller, H. J. 1925. Why polyploidy is rarer in animals than in plants. *A.N.*, 59:346-353.

——. 1927. Artificial transmutation of the gene. *S.*, 66:84-87.

——. 1928a. The problem of genic modification. *Verh. V Intern. Kongr.*, 1:234-260.

——. 1928b. The measurement of gene mutation rate in *Drosophila*, its high variability, and its dependence upon temperature. *Gs.*, 13:279-357.

——. 1940. Bearing of the "Drosophila" work on systematics. Pp. 185-268 *in:* J. S. Huxley, "New Systematics," Clarendon, Oxford.

——. 1942. Isolating mechanisms, evolution, and temperature. *Biol. Symposia*, 6:71-125.

——. 1950. Our load of mutations. *Amer. J. Human Genetics*, 2:111-176.

——. 1964. The relation of recombination to mutational advance. *Mutation Res.*, 1:2-19.

——. 1965. Means and aims in human genetic betterment. Pp. 100-122 *in:* T. M. Sonneborn (Ed.), "The Control of Human Heredity and Evolution," Macmillan, New York.

——. 1967. What genetic course will man steer? Pp. 521-543 *in:* J. F. Crow and J. V. Neel (Eds.), "Proceedings of the International Congress on Human Genetics," Johns Hopkins Press, Baltimore.

——, and W. D. Kaplan. 1966. The dosage compensation of *Drosophila* and mammals as showing the accuracy of the normal type. *G. R.*, 8:41-59.

Müntzing, A. 1932. Cytogenetic investigations on synthetic *Galeopsis tetrahit*. *Hs.*, 16:105-154.

——. 1961. Genetics, Basic and Applied. LTs Förlag, Stockholm.

——. 1963. A case of preserved heterozygosity in rye in spite of long-continued inbreeding. *Hs.*, 50:377-413.

Münzing, J. 1963. The evolution of variation and distributional patterns in European populations of the three-spined stickleback, *Gasterosteus aculeatus*. *E.*, 17:320-332.

Myrianthopoulos, N. C., and S. M. Aronson. 1966. Population dynamics of Tay-Sachs disease. I. Reproductive fitness and selection. *Amer. J. Human Genetics*, 18:313-327.

Nagel, G. 1961. The Structure of Science. Harcourt, Brace, & World, New York.

Naumenko, V. A. 1936. Lethal mutations in *Drosophila melanogaster* induced by potassium permanganate. *Bull. Biol. Med. Exp. USSR*, 1:204-206.

Neel, J. V. 1958. A study of major congenital defects in Japanese infants. *Amer. J. Human Genetics*, 10:398-445.

——, S. S. Fajans, J. W. Conn, and R. T. Davidson. 1965. Diabetes mellitus. Pp. 105-132 *in:* J. V. Neel, M. W. Shaw, and W. J. Schull (Eds.), "Genetics and Epidemiology of Chronic Diseases," Government Printing Office, Washington.

——, and W. J. Schull. 1962. The effects of inbreeding on mortality and morbidity in two Japanese cities. *P.N.A.S.*, 48:573-582.

——, and W. J. Schull. 1968. On some trends in understanding the genetics of men. *Persp. Biol. Med.*, 11:565-602.

Newman, L. J. 1966. Bridge and fragment aberrations in *Podophyllum peltatum*. *Gs.*, 53:55-63.

Nicoletti, B., and C. Giardina. 1964. Genetic load and reproductive efficiency in natural and laboratory populations of *Drosophila melanogaster*. *Riv. Biol.* 57:209-236.

Nielsen, A. 1951. Contributions to the metamorphosis and biology of the genus *Atrichopogon* Kieffer (*Diptera, Ceratopogonidae*). *Biol. Skrift. Kongr. Dan. Viden. Selskab*, 6, No. 6:1-95.

Nienstaedt, H., and A. H. Graves. 1955. Blight-resistant chestnuts. *Conn. Agr. Exp. Stat. Circ.* 192.

Nilsson, S. E. 1962. Genetic and Constitutional Aspects of Diabetes Mellitus. Almquist & Wiksell, Stockholm.

Nirenberg, M. W., and J. H. Matthei. 1961. The dependence of cell-free protein synthesis in *E. coli* upon naturally occurring or synthetic polyribonucleotides. *P.N.A.S.*,47:1588-1602.

O'Brien, S. J., and R. J. MacIntyre. 1969. An analysis of gene-enzyme variability in natural populations of *Drosophila melanogaster and D. simulans*. *A.N.*, 103: 97-113.

Ohba, Sh. 1967. Chromosomal polymorphism and capacity for increase under near optimal conditions. *Hy.*, 22:169-189.

Ohmachi, F., and S. Mazaki. 1964. Interspecific crossing and development of hybrids between the Japanese species of *Teleogryllus* (*Orthoptera, Gryllidae*).*E.*, 18:405-416.

Ohno, S., and N. B. Atkin. 1966. Comparative DNA values and chromosome complements of eight species of fishes. *Ch.*, 18:455-466.

——, C. Weiler, J. Poole, L. Christian, and C. Steinus. 1966. Autosomal polymorphism due to pericentric inversions in the deer mouse (*Peromyscus maniculatus*) and some evidence of somatic segregation. *Ch.*, 18:177-183.

Oldroyd, H. 1964. The Natural History of Flies. Norton, New York.

Olenov, J. M. 1958. On the increase of the resistance of insects to the action of DDT. *Rev. Entom. URSS.*, 37:520-537.

Omodeo, P. 1952. Cariologia dei *Lumbricidae*. *Caryologia*, 4:173-275.

Oparin, A. I. 1964. The Chemical Origin of Life. Charles C Thomas, Springfield.

Oshima, C. 1967. Persistence of some recessive lethal genes in natural populations of *Drosophila melanogaster*. *Ciencia e Cultura*, 19:102-110.

——, and T. Hiroyoshi. 1956. Genetic studies of resistance to DDT and nicotine sulphate in *Drosophila virilis*. *Botyu Kagaku*, 21:65-70.

Otten, Ch. M. 1967. On pestilence, diet, natural selection, and the distribution of microbial and human blood group antigens and antibodies. *Current Anthrop.*, 8:209-226.

Ownbey, M. 1950. Natural hybridization and amphiploidy in the genus *Tragopogon*. *Amer. J. Bot.*, 37:487-499.

Paik, Y. K. 1960. Genetic variability in Korean populations. *E.*, 14:293-303.

Park, Y. I., C. T. Hansen, C. S. Chung, and A. B. Chapman. 1966. Influence of feeding regime on the effects of selection for postweaning gain in the rat. *Gs.*, 54:1315-1327.

Parsons, P. A., and M. M. Green. 1959. Pleiotropy and competition at the vermilion locus in *Drosophila melanogaster*. *P.N.A.S.*, 45:993-996

Pasternak, J. 1964. Chromosome polymorphism in the blackfly *Simulium vittatum* (Zett). *Canad. J. Zool.*, 42:135-158.

Pätau, K. 1935. Chromosomenmorphologie bei *Drosophila melanogaster* und *Drosophila simulans* und ihre genetische Bedeutung. *Naturwiss.*, 23:537-543.

Patterson, J. T., and W. S. Stone. 1952. Evolution in the Genus *Drosophila*. Macmillan, New York.

——, and M. R. Wheeler. 1947. Two strains of *Drosophila peninsularis* with incipient reproductive isolation. *U.T.P.*, 4720:116-125.

Pauling, L., H. A. Itano, S. J. Singer, and I. C. Wells. 1949. Sickle cell anemia, a molecular disease. *S.*, 110:543-548.

Pavan, C., A. R. Cordeiro, N. Dobzhansky, Th. Dobzhansky, C. Malogolowkin, B. Spassky, and M. Wedel. 1951. Concealed genic variability in Brazilian populations of *Drosophila willistoni*. *Gs.*, 36:13-30.

——, and E. N. Knapp. 1954. The genetic population structure of Brazilian *Drosophila willistoni*. *E.*, 8:303-313.

Pavlovsky, O., and Th. Dobzhansky. 1966. Genetics of natural populations. XXXVII. The coadapted system of chromosomal variants in a population of *Drosophila pseudoobscura*. *Gs.*, 53:843-854.

Paxman, G. J. 1957. A study of spontaneous mutation in *Drosophila melanogaster*. *Ga.*, 29:39-57.

Penrose, L. S. 1955. Evidence of heterosis in man. *Proc. Royal Soc.*, B, 144:203-213.

Pentzos, A. D., E. Boesiger, and A. Kanellis. 1967. Frequences de genes mutants dans plusieurs populations naturelles de *Drosophila subobscura* de Grèce. *Ann. Fac. Sci. Univ. Thessaloniki*, 10:133-152.

Perutz, M. F., and H. Lehman. 1968. Molecular pathology of human haemoglobin. *N.*, 219-902-909.

Petersen, B. 1949. Studies on geographic variation of allometry in some European Lepidoptera. Zool. Bidrag. Uppsala, 29:1-38.

——, 1955. Geographische Variation von *Pieris* (*napi*)*bryoniae* durch Bastardierung mit *Pieris napi*. *Zool. Bidrag Uppsala*, 30:355-397.

Petit, C. 1958. Le déterminisme génétique et psycho-physiologique de la compétition sexuelle chez *Drosophila melanogaster*. *Bull. Biol. France Belgique*, 92: 248-329.

——, and L. Ehrman. 1969. Sexual selection in *Drosophila*. *E. B.*, 3:177-223.

Petras, J. W., and J. E. Curtis. 1968. The current literature on social class and mental disease in America. Critique and bibliography. *Behav. Sci.*, 13:382-398.

Petras, M. L. 1967. Studies of natural populations of *Mus*. I. Biochemical polymorphisms and their bearing on breeding structure. *E.*, 21:259-274.

Pettigrew, Th. F. 1964. Race, mental illness, and intelligence: A social psychological view. *Eugen. Quart.*, 11:189-215.

Philiptschenko, J. 1927. Variabilität und Variation. Borntraeger, Berlin.

Pianka, E. R. 1966. Latitudinal gradients in species diversity: A review of concepts. *A.N.*, 100:33-46.

Pimental, D., G. J. C. Smith, and J. Soans. 1967. A population model of sympatric speciation. *A.N.*, 101:493-504.

Pipkin, S. B. 1968. Introgression between closely related species of *Drosophila* in Panama. *E.*, 22:140-156.

Pires, J. M., Th. Dobzhansky, and G. A. Black. 1953. An estimate of the number of species of trees in an Amazonian forest community. *Bot. Gazette*, 114:467-477.

Polani, P. E. 1969. Abnormal sex chromosomes and mental disorder. *N.*, 223:680-686.

Polanyi, N. 1968. Life's irreducible structure. *S.*, 160:1308-1312.

Polivanov, S. 1964. Selection in experimental populations of *Drosophila melanogaster* with different genetic backgrounds. *Gs.*, 50:81-100.

Pollard, E. C. 1965. The fine structure of the bacterial cell and the possibility of its artificial synthesis. *Amer. Scientist*, 53:437-463.

Ponomarev, V. P. 1937-1938. The effect of lead nitrate on mutation in *Drosophila melanogaster. Biol. Zhurnal*, 7:619-634.

Pontecorvo, G. 1943. Viability interactions between chromosomes of *Drosophila melanogaster* and *Drosophila simulans. J. Genetics*, 43:51-66.

Post, R. H. 1962. Population differences in red and green color vision deficiency: A review and a query on selection relaxation. *Eugen. Quart.*, 9:131-146.

——. 1964. Hearing acuity variation among negroes and whites. *Eugen. Quart.*, 11:65-81.

——. 1965. Notes on relaxed selection in man. *Anthrop. Anz.*, 29:186-195.

Poulson, Th. L., and W. B. White. 1969. The cave environment. *S.*, 165:971-981.

Prakash, S., and R. C. Lewontin. 1968. A molecular approach to the study of genic heterozygosity in natural populations. III. Direct evidence of coadaptation in gene arrangements of *Drosophila. P.N.A.S.*, 59:398-405.

——, R. C. Lewontin, and J. L. Hubby. 1969. A molecular approach to the study of genic heterozygosity in natural populations. IV. Patterns of genic variation in central, marginal and isolated populations of *Drosophila pseudoobscura. Gs.*, 61:841-858.

Pratt, R. T. C. 1967. The Genetics of Neurological Disorders. Oxford Univ. Press, London.

Prevosti, A. 1964. Chromosomal polymorphism in *Drosophila subobscura* populations from Barcelona. *G.R.* 5:27-38.

Prout, T. 1954. Genetic drift in irradiated experimental populations of *Drosophila melanogaster. Gs.*, 39:529-545.

——. 1962. The effects of stabilizing selection on the time of development in *Drosophila melanogaster. G.R.*, 3:364-382.

——. 1964. Observations on structural reduction in evolution. *A.N.*, 98:239-249.

——. 1968. Sufficient conditions for multiple niche polymorphism. *A.N.*, 102:493-496.

Purdom, C. E. 1963. Genetic Effects of Radiations. Academic Press, New York.

Quayle, H. J. 1938. The development of resistance in certain scale insects to hydrocyanic gas. *Hilgardia*, 11:183-225.

Race, R. R., and R. Sanger. 1968. Blood Groups in Man. 5th Ed. Davis, Philadelphia.

Rana, R. S. 1965. Induced interchange heterozygosity in diploid *Chrysanthemum. Ch.*, 16:477-485.

——, and H. K. Jain. 1965. Adaptive role of interchange heterozygosity in the annual *Chrysanthemum. Hy.*, 20:21-29.

Rapoport, J. A. 1939. Specific morphoses induced in *Drosophila* by chemicals. *Bull. Exp. Biol. Med. (Moscow)*, 7:424-426.

——. 1946. Carbonyl compounds and the chemical mechanism of mutation. *C. R. Acad. Sci. URSS*, 54:65-67.

Rasmussen, D. I. 1962. Blood group polymorphism and inbreeding in natural populations of the deer mouse *Peromyscus maniculatus gracilis*. *E.*, 18:219-229.

Reed, T. E. 1961. Polymorphism and natural selection in blood groups. Pp. 80-101 *in:* B. S. Blumberg (Ed.), "Genetic Polymorphisms and Geographic Variation in Disease," Grune & Stratton, New York.

Rees, H. 1961. The consequences of interchange. *E.*, 15:145-152.

——. 1964. The question of polyploidy in the *Salmonidae*. *Ch.*, 15:275-279.

——, and S. Sun. 1965. Chiasma frequency and the disjunction of interchange associations in rye. *Ch.*, 16:500-510.

Reitalu, J. 1968. Chromosome studies in connection with sex chromosomal deviations in man. *Hs.*, 59:1-48.

Remington, Ch. L. 1968. Suture-zones of hybrid interaction between recently joined biotas, *E.B.*, 2:321-438.

Renner, O. 1929. Artbastarde bei Pflanzen. Borntraeger, Berlin.

Rensch, B. 1929. Das Prinzip geographischer Rassenkreise und das Problem der Artbildung. Borntraeger, Berlin.

——. 1947 (1960a). Evolution above the Species Level. Columbia Univ. Press, New York.

——. 1960b. The laws of evolution. Pp. 95-116 *in:* S. Tax (Ed.), "Evolution after Darwin," Chicago Univ. Press, Chicago.

——. 1968. Biophilosophie. Gustav Fischer, Stuttgart.

Rhoades, M. M. 1946. Plastid mutations. *C.S.H.*, 11:202-207.

Richardson, R. H., and K. I. Kojima. 1965. The kinds of genetic variability in relation to selection responses in *Drosophila* fecundity. *Gs.*, 52:583-598.

Richmond, R. C. 1970. Non-Darwinian evolution: a critique. *N.* (in press).

Rick, C. M. 1947. Partial suppression of hair development indirectly affecting fruitfulness and the proportion of cross-pollination in a tomato mutant. *A.N.*, 81:185-202.

——. 1950. Pollination in relation of *Lycopersicon esculentum* in native and foreign habitats. *E.*, 4:110-122.

——. 1963a. Differential zygotic lethality in a tomato species hybrid. *Gs.*, 48: 1497-1507.

——. 1963b. Barriers to interbreeding in *Lycopersicon peruvianum*. *E.*, 17:216-232.

——, and P G. Smith. 1953. Novel variation in tomato species hybrids. *A.N.*, 87:359-373.

Riggs, S. K., and F. Sargent. 1964. Physiological regulation in moist heat by young American Negro and white males. *Human Biol.*, 36:339-353.

Riley, H. P. 1952. Ecological barriers. *A.N.*, 86:23-32.

Rizki, M. T. M. 1951. Morphological differences between two sibling species, *Drosophila pseudoobscura* and *Drosophila persimilis*. *P.N.A.S.*, 156-159.

——. 1952. Ontogenetic distribution of genetic lethality in *Drosophila willistoni*. *J. Exp. Zool.*, 121:327-350.

Robertson, A. 1960. A theory of limits in artificial selection. *Proc. Royal Soc.*, B, 153:234-249.

——. 1962. Selection for heterozygotes in small populations. *Gs.*, 47:1291-1300.

Robertson, F. W. 1955. Selection response and the properties of genetic variation. *C.S.H.*, 20:166-177.

——. 1957. Studies in quantitative inheritance. XI. Genic and environmental correlation between body size and egg production in *Drosophila melanogaster*. *J. Genetics*, 55:428-443.

——. 1966. A test of sexual isolation in *Drosophila*. *G.R.*, 8:181-187.

Robinson, J. T. 1967. Variation and the taxonomy of the early hominids. *E.B.*, 1:69-100.

Roelofs, W. L., and A. Comeau. 1969. Sex pheromone specificity: Taxonomic and evolutionary aspects in *Lepidoptera*. *S.*, 165:398-399.

Ross, H. H. 1958. Evidence suggesting a hybrid origin for certain leafhopper species. *E.*, 12:337-346.

Rothe, H. 1951. Morphologisch-Entwicklungsgeschichtliche und genetische Analyse einer sich variabel manifestierenden Mutation von *Antirrhinum majus* L. *Z.i.A.V.*, 84:74-132.

Rothfels, K. H. 1956. Black flies: sibling, sex, and species grouping. *J. Heredity*, 47:113-122.

Roy, S. K. 1960. Interaction between rice varieties. *J. Genetics*, 57:137-152.

Sabrosky, C. W. 1952. How many insects are there? Yearbook of Agriculture, Separate No. 2290, Washington.

Saccà, G. 1967. Speciation in *Musca*. Pp. 385-415 *in:* J. W. Wright and R. Pal (Eds.), "Genetics of Insect Vectors of Disease," Elsevier, Amsterdam.

Sacharov, W. W. 1936. Iod als chemischer Factor, der auf den Mutationsprozess von *Drosophila melanogaster* wirkt. *Ga.*, 18:193-216.

Salceda, V. M. 1967. Recessive lethals in second chromosomes of *Drosophila melanogaster* with radiation histories. *Gs.*, 57:691-699.

Samjatina, N. D., and O. T. Popova. 1934. Der Einfluss von Iod auf die Entstehung von Mutationen bei *Drosophila melanogaster*. *Biol. Zhurnal*, 3:679-693.

Sankaranarayanan, R. 1964. Genetic loads in irradiated experimental populations of *Drosophila melanogaster*. *Gs.*, 50:131-150.

——. 1965. Further data on the genetic loads in irradiated experimental populations of *Drosophila melanogaster*. *Gs.*, 52:153-164.

——. 1966. Some components of the genetic loads in irradiated experimental populations of *Drosophila melanogaster*. *Gs.*, 54:121-130.

Sarich, V. M., and A. C. Wilson. 1967. Rates of albumin evolution in primates. *P.N.A.S.*, 58:142-148.

Sarkar, I. 1955. A translocation heterozygote in the grasshopper *Gesonula punctifrons*. *J. Heredity*, 46:157-160.

Scharloo, W. 1964. The effect of disruptive and stabilizing selection on the expression of a cubitus interruptus mutant in *Drosophila*. *Gs.*, 50:553-562.

Schatz, G. 1951. Über die Formbilding des Flügel bei Hitzemodificationen und Mutationen von *Drosophila melanogaster*. *Biol. Zentral.*, 70:305-353.

Schlager, G., and M. M. Dickie. 1966. Spontaneous mutation rates at five coatcolor loci in mice. *S.*, 151:205-206.

Schmalhausen, I. I. 1949. Factors of Evolution. Blakiston, Philadelphia.

Schnetter, M. 1950. Veränderungen der genetischen Konstitution in natürlichen Populationen der polymorphen Bänderschnecken. *Verh. Deutsch. Zool. Marburg*, 1950, 192-206.

Schoener, Th. W. 1965. The evolution of bill size differences among sympatric congeneric species of birds. *E.*, 19:189-213.

Scholander, P. F., V. Walters, R. Hock, and L. Irving. 1950. Heat regulation *Biol. Bull.*, 99:225-271.

Scholz, F., and Ch. O. Lehmann. 1962. Die Gaterslebener Mutanten der Saatgerste in Berziehung zur Formenmannigfaltigkeit der Art *Hordeum vulgare* L. *Kulturpflanze*, 10:312-334.

Schreider, E. 1963. Physiological anthropology and climatic variations. Pp. 37-73 *in:* "Proceedings of the Lucknow Symposium (UNESCO) on Environmental Physiology and Psychology in Arid Conditions."

——. 1964a. Recherches sur la stratification sociale des charactères biologiques. *Biotypologie,* 26:105-135.

——. 1964b. Ecological rules, body-heat regulation, and human evolution. *E.,* 18:1-9.

——. 1967. Un mécanisme sélectif possible de la différenciation sociale des caractères biologiques. *Biom. Humaine,* 11:67-83.

Schull, W. J., and J. V. Neel. 1965. The Effects of Inbreeding on Japanese Children. Harper & Row, New York.

——, and J. V. Neel. 1966. Some further observations on the effect of inbreeding on mortality in Kure, Japan. *Amer. J. Human Genetics,* 18:144-152.

Schwab, J. J. 1940. A study of the effects of a random group of genes on shape of spermatheca in *Drosophila melanogaster. Gs.,* 25:157-177.

Schwanitz, F. 1967. Die Evolution der Kulturpflanzen. Bayerischer Landwirtschaftsverlag, München.

Scossiroli, R. E. 1954. Effectiveness of artificial selection under irradiation of plateaued populations of *Drosophila melanogaster. IUBS Symposium Genetics Population Structure,* B15:42-60.

——. 1965. Value of induced mutations for quantitative characters in plant breeding. Pp. 443-450 *in:* "The Use of Induced Mutations in Plant Breeding," Pergamon Press, Oxford.

Scott, J. P. 1967. The evolution of social behavior in dogs and wolves. *Amer. Zoologist,* 7:373-381.

——. 1968. Evolution and domestication of the dog. *E.B.,* 2:243-275.

——, and J. L. Fuller. 1965. Genetics and the Social Behavior of the Dog. Univ. Chicago Press, Chicago.

Scriver, Ch. R. 1967. Treatment in medical genetics. Pp. 45-56 *in:* J. F. Crow and J. V. Neel (Eds.), "Proceedings of the 3rd International Congress on Human Genetics," Johns Hopkins Press, Baltimore.

Sears, E. R. 1948. The cytology and genetics of the wheats and their relatives. *Adv. Genetics,* 2:240-270.

——. 1956. The systematics cytology and genetics of wheat. *Handbuch Pflanzenzüchtung,* 2:164-187.

Sedlmair, H. 1956. Verhaltens-, Resistenz-, und Gehäuseunterschiede bei den polymorphen Bänderschnecken *Cepaea hortensis* (Müll.) und *Cepaea nemoralis* (L.). *Biol. Zentral.,* 75:281-313.

Selander, R. K., W. G. Hunt, and S. Y. Yang. 1969. Protein polymorphism and genic heterozygosity in two European subspecies of the house mouse. *E.,* 23: 379-390.

Semenov-Tian-Shansky, A. P. 1910. Die taxonomische grenzen der Art und ihrer Unterabfteilungen. Friedlander, Berlin.

de Serres, F. J. 1964. Mutagenesis and chromosome structure. *J. Cell. Comp. Physiol.,* 64 (Suppl.):33-42.

Sheppard, P. M. 1969. Evolutionary genetics of animal populations; the study of natural populations. *P. XII I.C.G.,* 3:261-279.

Shields, J. 1968. Summary of the genetic evidence for the transmission of schizophrenia. *J. Psychiat. Res.,* 6 (Suppl.): 95-126.

Shkvarnikov, P. K. 1966. Modern research problems of utilization of experimentally induced mutations in plants. *Ga.,* No. 2:7-19.

Short, L. L. 1965. Hybridization in the flickers (*Colaptes*) of North America. *Bull. Amer. Museum Nat. Hist.*, 129:307-428.

Shultz, F. T. 1953. Concurrent inbreeding and selection in the domestic fowl. *Hy.*, 7:1-21.

Sibley, C. G. 1954. Hybridization in the red-eyed towhees of Mexico. *E.*, 8:252-290.

——, and L. L. Short. 1964. Hybridization in the orioles of the Great Plains. *Condor*, 66:130-150.

Siegel, R. W. 1958. Hybrid vigor, heterosis, and evolution in *Paramecium aurelia*. *E.*, 12:402-416.

Siegel, S. M., and C. Giumarro. 1966. On the culture of a microorganism similar to the precambrian microfossil *Kakabekia umbellata* Barghoorn in NH_3-rich atmospheres. *P.N.A.S.*, 55:349-353.

Sinsheimer, R. L. 1959. A single-stranded deoxyribonucleic acid from bacteriophage 174. *J. Mol. Biol.*, 1:43-53.

Simpson, G. G. 1943. Criteria for genera, species, and subspecies in zoology and paleontology. *Ann. New York Acad. Sci.*, 44:145-178.

——. 1949 (1967). The Meaning of Evolution. Rev. Ed. Yale Univ. Press, New Haven.

——. 1953. The Major Features of Evolution. Columbia Univ. Press, New York.

——. 1961. Principles of Animal Taxonomy. Columbia Univ. Press, New York.

——. 1964a. Organisms and molecules in evolution. S., 146:1535-1538.

——. 1964b. This View of Life. Harcourt, Brace, & World, New York.

——. 1964c. Species density of North American recent mammals. S.Z., 13:57-73.

——. 1969. Biology and Man. Harcourt, Brace, & World, New York.

Singleton, R. 1951. Inheritance of corn grass, a macromutation in maize, and its possible significance as an ancestral type. *A.N.*, 85:81-96.

Skinner, F. A. 1968. The limits of microbial existence. *Proc. Royal Soc.*, B, 171: 77-89.

Slobodkin, L. B. 1964. The strategy of evolution. *Amer. Scientist*, 52:342-557.

——. 1968. Towards a predictive theory of evolution. Pp. 187-205 *in:* R. C. Lewontin (Ed.), "Population Biology and Evolution," Syracuse Univ. Press, Syracuse.

Smith, H. H. 1950. Differential photoperiod response from interspecific gene transfers. *J. Heredity*, 41:199-203.

Smith, M. H. 1965. Behavioral discrimination shown by allopatric and sympatric males of *Peromyscus eremicus* and *Peromyscus californicus* between females of the same two species. *E.*, 19:430-435.

Smith, S. G. 1962. Cytogenetic pathways in beetle speciation. *Canad. Entom.*, 94:941-955.

——. 1966. Natural hybridization in the coccinellid genus *Chilocorus*. *Ch.*, 18: 380-406.

Smithies, O., and G. E. Connell. 1959. Biochemical Aspects of the Inherited Variations in Human Serum Haptoglobins and Transferrins. Ciba Foundation Symposium, Biochemistry of Human Genetics, Churchill, London.

Snow, R., and M. P. Dunford. 1961. A study of interchange heterozygosity in a population of *Datura meteloides*. *Gs.*, 46:1097-1110.

Sokal, R. R. 1965. Statistical methods in systematics. *Biol. Rev.*, 40:337-391.

——, and J. H. Camin. 1965. The two taxonomies: areas of agreement and conflict. S.Z., 14:176-195.

——, and I. Huber. 1963. Competition among genotypes in *Tribolium castaneum* at varying densities and gene frequencies (the sooty locus). *A.N.*, 97:169-184.

——, and P. H. A. Sneath. 1963. Principles of Numerical Taxonomy. Freeman, San Francisco and London.

Sokoloff, A. 1964. A dominant synthetic lethal in *Tribolium castaneum* Herbst. *A.N.*, 98:127-128.

——. 1966. Morphological variation in natural and experimental populations of *Drosophila pseudoobscura* and *Drosophila persimilis*. *E.*, 20:49-71.

——, and I. M. Lerner. 1967. Laboratory ecology and mutual predation of *Tribolium* species. *A.N.*, 101:261-276.

——, I. M. Lerner, and F. K. Ho. 1965. Self-elimination of *Tribolium castaneum* following xenocide of *T. confusum*. *A.N.*, 99:399-404.

Solima Simmons, A. 1966. Experiments on random genetic drift and natural selection in *Drosophila pseudoobscura*. *E.*, 20:100-103.

Sonneborn, T. M. 1957. Breeding systems, reproductive methods, and species problems in *Protozoa*. Pp. 155-324 *in:* E. Mayr (Ed.), "The Species Problem," American Association for the Advancement of Science, Washington.

Sorensen, F. 1969. Embryonic genetic load in coastal Douglas fir. *Pseudotsuga menziesii* var. *menziesii*. *A.N.*, 103:389-398.

Spassky, B., Th. Dobzhansky, and W. W. Anderson. 1965. Genetics of natural populations. XXXVI. Epistatic interactions of the components of the genetic load in *Drosophila pseudoobscura*. *Gs.*, 52:623-664.

——, N. Spassky, H. Levene, and Th. Dobzhansky. 1958. Release of genetic variability through recombination. I. *Drosophila pseudoobscura*. *Gs.*, 43:845-867.

Spencer, W. P. 1947a. Genetic drift in a population of *Drosophila immigrans*. *E.*, 1:103-110.

——. 1947b. Mutations in wild populations of *Drosophila*. *Adv. Genetics,* 1:359-402.

Sperlich, D. 1958. Modellversuche zur Selektionswirkung verschiedener chromosomaler Strukturtypen von *Drosophila subobscura* Coll. *Z.i.A.V.*, 89:422-436.

——. 1966a. Equilibria for inversions induced by X-rays in isogenic strains of *Drosophila pseudoobscura*. *Gs.*, 53:835-842.

——. 1966b. Unterschiedliche Paarungsaktivität innerhalb und zwischen verschiedenen geographischen Stämmen von *Drosophila subobscura*. *Z.i.A.V.*, 98: 10-15.

——, and H. Feuerbach. 1966. Ist der chromosomale Strukturpolymorphisms von *Drosophila subobscura* stabil oder flexibel? *Z.i.A.V.*, 98:16-24.

Spickett, S. G., and J. M. Thoday. 1966. Regular responses to selection. 3. Interaction between located polygenes. *G.R.*, 7:96-121.

Spiegelman, S., N. R. Pace, D. R. Mills, R. Levisohn, T. S. Eikhom, M. M. Taylor, R. L. Petersen, and D. H. L. Bishop. 1969. Chemical and mutational studies of a replicating RNA molecule. *P.* XII *I.C.G.*, 3:127-154.

Spiess, E. B. 1957. Relation between frequency and adaptive values of chromosomal arrangements in *Drosophila persimilis*. *E.*, 11:84-93.

——. 1959. Release of genetic variability through recombination. II. *Drosophila persimilis*. *Gs.*, 44:43-58.

——. 1968. Low frequency advantage in mating of *Drosophila pseudoobscura* karyotypes. *A.N.*, 102:363-379.

——, and A. C. Allen. 1961. Release of genetic variability through recombination. VII. Second and third chromosomes of *Drosophila melanogaster*. *Gs.*, 46:1531-1553.

——, and B. Langer. 1964a. Mating speed control by gene arrangement in *Drosophila pseudoobscura* homokaryotypes. *P.N.A.S.*, 51:1015-1019.

——, and B. Langer. 1964b. Mating speed control by gene arrangement carriers in *Drosophila persimilis*. *E.*, 18:430-444.

——, B. Langer, and L. D. Spiess. 1966. Mating control by gene arrangements in *Drosophila pseudoobscura*. *Gs.*, 54:1139-1149.

——, and L. D. Spiess. 1967. Mating propensity, chromosomal polymorphism and dependent conditions in *Drosophila persimilis*. *E.*, 21:672-688.

Spiess, L. D., and E. B. Spiess. 1969. Minority advantage in interpopulational matings of *Drosophila persimilis*. *A.N.*, 103:155-172.

Spieth, H. T. 1966. Mating behavior of *Drosophila ananassae* and *ananassae*-like flies from the Pacific. *U.T.P.*, 6615:133-145.

——. 1968. Evolutionary implications of sexual behavior in *Drosophila*. *E. B.*, 2:157-193.

——, and T. C. Hsu. 1950. The influence of light on the mating behavior of seven species of the *Drosophila melanogaster* group. *E.*, 4:316-325.

Sprague, G. F., W. A. Russell, and L. H. Penny. 1960. Mutations affecting quantitative traits in the selfed progeny of doubled monoploid maize stocks. *Gs.*, 45:855-866.

Stadler, L. J. 1954. The gene. *S.*, 120:1811-1819.

Staiger, H. 1954. Der Chromosomendimorphismus beim Prosobranchier *Purpura lapillus* in Beziehung zur Ökologie der Art. *Ch.*, 6:419-478.

——. 1955. Reciproke Translocationen in natürlichen Populationen von *Purpura lapillus*. *Ch.*, 7:181-197.

Stakman, E. C. 1947. Plant diseases are shifty enemies. *Sci. Progr.*, 5:235-279.

——, W. Q. Loegering, R. C. Cassell, and L. Hines. 1943. Population trends of physiologic races of *Puccinia graminis tritici* in the United States for the period 1930-1941. *Phytopathology*, 33:884-898.

Stalker, H. D. 1954. Parthenogenesis in *Drosophila*. *Gs.*, 39:4-34.

——. 1956. On the evolution of parthenogenesis in *Lonchoptera (Diptera)*. *E.*, 10:345-359.

——. 1960. Chromosomal polymorphism in *Drosophila paramelanica* Patterson. *Gs.*, 45:95-114.

——. 1964a. Chromosomal polymorphism in *Drosophila euronotus*. *Gs.*, 49:669-682.

——. 1964b. The salivary gland chromosomes of *Drosophila nigromelanica*. *Gs.*, 49:883-893.

——. 1965. The salivary gland chromosomes of *Drosophila micromelanica* and *Drosophila melanica*. *Gs.*, 51:487-507.

——. 1966. The phylogenetic relationships of the species in *Drosophila melanica* group. *Gs.*, 53:327-342.

Stansfield, W. D., G. E. Bradford, C. Stormont, and R. J. Blackwell. 1964. Blood groups and their associations with production and reproduction in sheep. *Gs.*, 50:1357-1367.

Stauber, L. A. 1950. The problem of physiological species with special reference to oysters and oyster drills. *Ecology*, 31:109-118.

Stebbins, G. L. 1939. Structural hybridity in *Paeonia californica* and *P. brownii*. *J. Genetics*, 38:1-36.

——. 1950. Variation and Evolution in Plants. Columbia Univ. Press, New York.

——. 1957a. Self-fertilization and population variability in the higher plants. *A.N.*, 91:337-354.

——. 1957b. The hybrid origin of microspecies in the *Elymus glaucus* complex. *Cytologia,* Suppl. Vol.:336-340.

——. 1958a. The inviability, weakness, and sterility of interspecific hybrids. *Adv. Genetics,* 9:147-215.

——. 1958b. Longevity, habitat, and release of genetic variability in the higher plants. *C.S.H.,* 23:365-378.

——. 1969. The Basis of Progressive Evolution. Univ. North Carolina Press, Chapel Hill.

——, and A. Day. 1967. Cytogenetic evidence for long continued stability in the genus *Plantago. E.,* 21:409-428.

——, and E. Yagil. 1966. The morphogenetic effects of the hooded gene in barley. I. The course of development in hooded and awned genotypes. *Gs.,* 54:727-741.

Stebbins, R. C. 1949. Speciation in salamanders of the plethodontid genus *Ensatina. Univ. California Publ. Zool.,* 48:377-526.

——. 1957. Intraspecific sympatry in the lungless salamander *Ensatina eschscholtzi. E.,* 11:265-270.

Stehli, F. G. 1968. Taxonomic diversity gradients in pole location. The recent model. Pp. 163-227 *in:* E. T. Drake (Ed.), "Evolution and Environment," Yale Univ. Press, New Haven.

——, R. G. Douglas, and N. D. Newell. 1969. Generation and maintenance of gradients in taxonomic diversity. *S.,* 164:947-949.

Steinberg, A. G. 1959. The genetics of diabetes. A review. *Ann. New York Acad. Sci.,* 82:197-207.

——. 1962. Progress in the study of genetically determined human gamma globulin types (the *Gm* and *Inv* groups). Pp. 1-33 *in:* A. G. Steinberg and A. G. Bearn (Eds.), "Progress in Medical Genetics," Vol. 2, Grune & Stratton, London.

——. 1967. Genetic variations in human immunoglobulins: the *Gm* and *Inv.* types. Pp. 75-98 *in:* F. J. Greenwalt (Ed.), "Advances in Immunogenetics," Lippincott, Philadelphia.

——. 1969. Globulin polymorphisms in man. *Ann. Rev. Genetics,* 3:25-52.

——, H. K. Bleibtreu, Th. W. Kurczynski, A. O. Martin, and A. M. Kurczynski. 1967. Genetic studies on an inbred human isolate. Pp. 267-289 *in:* J. F. Crow and J. V. Neel (Eds.), "Proceedings of the 3rd International Congress on Human Genetics," Johns Hopkins Press, Baltimore.

Stephens, S. G. 1946. The genetics of "corky." *J. Genetics,* 47:150-161.

——. 1949. The cytogenetics of speciation in *Gossypium.* I. Selective elimination of the donor parent genotype in interspecific backcrosses. *Gs.,* 34:627-637.

——. 1950. The internal mechanism of speciation in *Gossypium. Bot. Rev.,* 16: 115-149.

Stern, C. 1960. Principles of Human Genetics. 2nd Ed. Freeman, San Francisco.

——, G. Carson, M. Kinst, E. Novitski, and D. Uphoff. 1952. The viability of heterozygotes for lethals. *Gs.,* 37:413-450.

——, and C. Tokunaga. 1968. Autonomous pleiotropy in *Drosophila. P.N.A.S.,* 60:1252-1259.

Stevenson, A. C. 1959. The load of hereditary defect in human populations. *Radiation Res.,* Suppl. 1:306-325.

——. 1961. Frequency of congenital and hereditary disease. *Brit. Med. Bull.,* 17:254-259.

——, H. A. Johnston, M. I. P. Stewart, and D. R. Golding. 1966. Congenital Malformations. World Health Organization, Geneva.

Stimpfling, J. H., and M. R. Irwin. 1960. Evolution of cellular antigens in *Columbidae*. *E.*, 14:417-426.

Stone, W. S., W. C. Guest, and F. D. Wilson. 1960. The evolutionary implications of the cytological polymorphism and phylogeny of the *virilis* group of *Drosophila*. *P.N.A.S.*, 46:350-361.

——, F. M. Johnson, K. Kojima, and M. R. Wheeler. 1968. Isozyme variation in island populations of *Drosophila*. I. An analysis of a species of the *nasuta* complex in Samoa and Fiji. *U.T.P.*, 6818:157-170.

——, F. D. Wilson, and V. L. Gerstenberg. 1963. Genetic studies of natural populations of *Drosophila: Drosophila pseudoobscura*, a large dominant population. *Gs.*, 48:1089-1106.

Stormont, C. 1959. On the application of blood groups in animal breeding. *P. X I.C.G.*, I:206-224.

Strauss, B. S. 1964. Chemical mutagens and the genetic code. *Progr. Med. Genetics*, 3:1-48.

Straw, R. M. 1956. Floral isolation in *Penstemon*. *A.N.*, 90:47-53.

Streisinger, G., Y. Okada, J. Emrich, J. Newton, A. Tsugita, E. Terzagi, and M. Inouye. 1966. Frameshift mutations and the genetic code. *C.S.H.*, 31:77-84.

Strickberger, M. W. 1963. Evolution of fitness in experimental populations of *Drosophila pseudoobscura*. *Gs.*, 17:40-55.

——. 1968. Genetics. Macmillan, New York.

Stubbe, H. 1940. Kritische Bemerkungen zu *Antirrhinum rhinantoides* Lotsy. *Biol. Zentral.*, 60:590-597.

——. 1950. Über den Selektionswert von Mutanten. Sitzungsber. *Deutsch. Akad. Wissen. Berlin,* Landwirtshafliche Klasse, No. 1:1-42.

Sturtevant, A. H. 1920-21. Genetic studies on *Drosophila simulans*. *Gs.*, 5:488-500; 6:179-207.

——. 1926. A crossover reducer in *Drosophila melanogaster* due to inversion of a section of the third chromosome. *Biol. Zentral.*, 46:697-702.

——. 1929. The genetics of *Drosophila simulans*. *Carnegie Inst. Washington Publ.* 399:1-62.

——. 1948. The evolution and function of genes. *Amer. Scientist,* 36:225-236.

——, and E. Novitski. 1941. The homologies of the chromosome elements in the genus Drosophila. *Gs.*, 26:517-541.

Sullivan, B., and P. E. Nute. 1968. Structural and functional properties of polymorphic hemoglobins from orangutans. *Gs.*, 58:113-124.

Sullivan, W. 1964. We Are Not Alone. McGraw-Hill, New York.

Sun, S., and H. Rees. 1967. Genotypic control of chromosome behaviour in rye. IX. The effect of selection on the disjunction frequency of interchange associations. *Hy.*, 22:249-254.

Suomalainen, E. 1947a. Parthenogenese und Polyploidie bei Russelkäfern (*Curculionidae*). *Hs.*, 33:425-456.

——. 1947b. On the cytology of the genus *Polygonatum* group *alterniflora*. *Ann. Acad. Sci. Fennicae,* A, 13:1-66.

——. 1958. Über das Vorkommen und spätere Verschwinden von *Epinephele lycaon* Rott. *(Lep. Satyridae)* in Finland. *Ann. Entom. Fennici,* 24:168-181.

——. 1962. Significance of parthenogenesis in the evolution of insects. *Ann. Rev. Entom.*, 7:349-366.

——. 1965. On the chromosomes of the geometrid moth genus *Cidaria*. *Ch.*, 16:166-184.

——. 1966. Achiasmatische Oogenese bei trichopteren. *Ch.*, 18:201-207.

——. 1969. Evolution in parthenogenetic *Curculionidae*. *E.B.*, 3:261-296.

Sved, J. A. 1968. Possible rates of gene substitution in evolution. *A.N.*, 102:283-293.

——, T. E. Reed, and W. F. Bodmer. 1967. The number of balanced polymorphisms that can be maintained in a natural population. *Gs.*, 55:469-481.

Takahashi, E. 1966. Growth and environmental factors in Japan. *Human Biol.*, 38:112-130.

Takenouchi, Y., and Takagi, R. 1967. A chromosome study of two parthenogenetic scolytid beetles. *Annot. Zool. Japon.*, 40:105-110.

Takhtajan, A. 1966. Systema et Phylogenia Magnoliophytorum. Nauka, Moscow-Leningrad (Russian).

Tal, M. 1967. Genetic differentiation and stability of some characters that distinguish *Lycopersicon esculentum* Mill. from *Solanum pennellis* Cor. *E.*, 21:316-333.

Tan, C. C. 1935. Salivary gland chromosomes in the two races of *Drosophila pseudoobscura*. *Gs.*, 20:392-402.

——. 1946. Mosaic dominance in the inheritance of color patterns in the ladybird beetle *Harmonia axyridis*. *Gs.*, 31:195-210.

Tanner, J. M. 1962. Growth at Adolescence. 2nd Ed. Blackwell, Oxford.

Tano, S., and A. B. Burdick. 1965. Female fecundity of *Drosophila melanogaster* second chromosome recessive lethal heterozygotes in homozygous and heterozygous genetic background. *Gs.*, 51:121-135.

Tantawy, A. O., and M. R. El-Helw. 1966. Studies on natural populations of *Drosophila*. V. Correlated response to selection in *Drosophila melanogaster*. *Gs.*, 53:97-110.

Taylor, J. H. 1969. Replication and organization of chromosomes. *P. XII I.C.G.*, 3:177-189.

Taylor, K. M. 1967. The chromosomes of some lower chordates. *Ch.*, 21:181-188.

Teilhard de Chardin, P. 1959. The Phenomenon of Man. Harper & Row, New York.

Temin, R. G. 1966. Homozygous viability and fertility loads in *Drosophila melanogaster*. *Gs.*, 53:27-46.

——, H. U. Meyer, P. S. Dawson, and J. F. Crow. 1969. The influence of epistasis on homozygous viability depression in *Drosophila melanogaster*. *Gs.*, 61:497-519.

Thoday, J. M. 1953. Components of fitness. *Symposium Soc. Exp. Biol.*, 7:96-113.

——. 1959. Effects of disruptive selection. I. Genetic flexibility. *Hy.*, 13:187-203.

——. 1967. New insights into continuous variation. Pp. 339-350 *in:* J. F. Crow and J. V. Neel (Eds.), "Proceedings of the 3rd International Congress on Human Genetics," Johns Hopkins Press, Baltimore.

——, and T. B. Boam. 1959. Effects of disruptive selection. II. Polymorphism and divergence without isolation. *Hy.*, 13:205-218.

——, J. B. Gibson, and S. G. Spickett. 1964. Regular responses to selection. 2. Recombination and accelerated response. *G.R.*, 5:1-19.

Thomas, R. E., and P. A. Roberts. 1966. Comparative frequency of X-ray induced crossover suppressing aberrations recovered from oocytes and sperm of *Drosophila melanogaster*. *Gs.*, 53:855-862.

Thompson, J. B. 1956. Genetic control of chromosome behaviour in rye. II. Disjunction at meiosis in interchange heterozygotes. *Hy.*, 10:99-108.

Thompson, P. E. 1962. Asynapsis and mutability in *Drosophila melanogaster*. *Gs.*, 47:337-349.

Thorpe, W. H. 1930. The biology of the petroleum fly (*Psilofa petrolei* Cog.). *Trans. Entom. Soc. London*, 78:331-343.

——. 1963. Learning and Instinct of Animals. 2nd Ed. Harvard Univ. Press, Cambridge.

Timofeeff-Ressovsky, H. 1934. Über den Einfluss des genotypischen Milieus und der Ausenbedingungen auf die Realisation des Genotyps. *Nachr. Ges. Wiss. Göttingen, Biol.*, N. F. 1:53-106.

——. 1935. Auslösung von Vitalitätsmutationen durch Röntgenbestrahlung bei *Drosophila melanogaster. Nachr. Ges. Wiss. Göttingen, Biol.*, N. F., 1:163-180.

——. 1937. Experimentelle Mutationsforschung in der Vererbungslehre. Theodor Steinkopff, Dresden and Leipzig.

——. 1939. Genetik und Evolution. *Z.i.A.V.*, 76:158-218.

——, and N. W. Timofeeff-Ressovsky. 1927. Genetische Analyse einer freilebenden *Drosophila melanogaster* Population. *Arch. Entwicklungsmech. Organ.*, 109: 70-109.

Tinbergen, N. 1965. Some recent studies of the evolution of sexual behavior. Pp. 1-33 *in:* F. A. Beach, "Sex and Behavior," John Wiley, New York.

——. 1966. Social Behavior in Animals. 2nd Ed. Methuen, London.

Tinkle, D. W. 1965. Population structure and effective size of a lizard population. *E.*, 19:569-573.

Tjio, J. H., and A. Levan. 1956. The chromosome number in man. *Hs.*, 42:1-6.

Tobari, I. 1966. Effects of temperature on the viabilities of homozygotes and heterozygotes for second chromosomes of *Drosophila melanogaster. Gs.*, 54: 783-791.

——, and K. I. Kojima. 1967. Selective modes associated with inversion karyotypes in *Drosophila ananassae.* I. Frequency-dependent selection. *Gs.*, 57:179-188.

Tobias, Ph. V. 1965. Early man in Africa. *S.*, 149:22-33.

Torroja, E. 1964. Genetic loads in irradiated experimental populations of *Drosophila pseudoobscura. Gs.*, 50:1289-1298.

——. 1966. An experiment on the effects of sex-linked lethals in heterozygous condition in *Drosophila melanogaster. A.N.*, 100:77-80.

Tretzel, E. 1955. Intragenerische Isolation und interspecifische Konkurrenz bei Spinnen. *Zeit. Morph. Ökol. Tiere,* 44:43-162.

Trevor, J. C. 1953. Race crossing in man. *Eugenics Lab. Mem.* 36, Univ. London.

Tshetverikov, S. S. 1926 (1959). On certain aspects of the evolutionary process from the standpoint of genetics. *Zhurnal Exp. Biol.*, 1:3-54 (Russian); English translation: *Proc. Amer. Phil. Soc.* 105:167-195.

Turesson, G. 1922. The genotypical response of the plant species to the habitat. *Hs.*, 3:211-350.

——. 1925. The plant species in relation to habitat and climate. *Hs.*, 3:147-236.

——. 1930. The selective effect of climate upon the plant species. *Hs.*, 14:99-152.

Turpin, R., and J. Lejeune. 1965. Les chromosomes humains. Gauthier-Villars, Paris.

Twitty, V. 1959. Migration and speciation in newts. *S.*, 130:1735-1743.

——, D. Grant, and O. Anderson. 1967. Home range in relation to homing in the newt *Taricha rivularis. Copeia,* 1967:649-653.

Unrau, J. 1959. Cytogenetics and wheat breeding. *P.* X *I.C.G.*, 1:129-141.

Upcott, M. 1939. The nature of tetraploidy in *Primula kewensis. J. Genetics,* 39: 79-100.

Uvarov, B. P. 1928. Locusts and Grasshoppers. Imperial Bureau of Entomology, London.

——, and J. G. Thomas. 1942. The probable mechanism of phase variation in the pronotum of locusts. *Proc. Royal Soc.*, A, 17:113-118.

Uzzell, Th. M., and S. M. Goldblatt. 1967. Serum protein of salamanders of the *Ambystoma jeffersonianum* complex, and the origin of the triploid species of this group. *E.*, 21:345-354.
Vandel, A. 1941. Étude des garniture chromosomiques de quelques crustacés isopodes terrestres et d'eau douce européens. *Cytologia,* 12:44-65.
——. 1964. Biospéologie. Gauthier-Villars, Paris.
Vann, E. 1966. The fate of X-ray induced chromosomal rearrangements introduced into laboratory populations of *Drosophila melanogaster. A.N.*, 100:425-449.
Vanyushin, B. F., A. N. Belozersky, N. A. Kokurina, and D. X. Kadirova. 1968. 5-methylcytosine and 6-methylaminopurine in bacterial DNA. *N.*, 218:1066.
Vavilov, N. I. 1926. Studies on the Origin of Cultivated Plants. *Bull. Applied Botany.* Leningrad.
Vernadsky, V. I. 1965. Chemical Structure of the Biosphere of the Earth and Its Environs. Nauka, Moscow (Russian).
Vetukhiv, M. 1953. Viability of hybrids between local populations of *Drosophila pseudoobscura. P.N.A.S.*, 39:30-34.
——. 1954. Integration of the genotype in local populations of three species of *Drosophila. E.*, 8:241-251.
——. 1956. Fecundity of hybrids between geographic populations of *Drosophila pseudoobscura. E.*, 10:139-146.
Vickery, R. K. 1959. Barriers to gene exchange within *Mimulus guttatus (Scrophulariaceae). E.*, 13:300-310.
——. 1964. Barriers to gene exchange between members of the *Mimulus guttatus (Scrophulariaceae). E.*, 18:52-69.
——. 1966. Speciation and isolation in section *Simiolus* of the genus *Mimulus. Taxon,* 15:55-63.
Viemeyer, G. 1958. Reversal of evolution in the genus *Penstemon. A.N.*, 92:129-137.
Vogel, F., H. J. Pettenkofer, and W. Helmbold. 1961. Über die Populationsgenetik der ABO Blutgruppen. *Acta Gen. Stat. Med.*, 10:267-294.
Vogt, M. 1950. Analyse durch Athylurethan bei *Drosophila* induzirter Mutationen. *Z.i.A.V.*, 83:324-340.
Voipio, P. 1950. Evolution at the Population Level with Special Reference to Game Animals and Practical Game Management. Finnish Fdn. for Game Preservation. Helsinki.
Volpe, E. P. 1952. Physiological evidence for natural hybridization of *Bufo americanus* and *Bufo fowleri. E.*, 6:393-406.
Vries, Hugo de. 1901. Die Mutationstheorie. Veit, Leipzig.
Waagen, W. 1869. Die Formenreihe des *Ammonites subradiatus. Benecke geognostischpaläontol. Beitr.*, 2:179-257.
Waddington, C. H. 1953. Genetic assimilation of an acquired character. *E.*, 7:118-126.
——. 1957. The Strategy of the Genes. Allen & Unwin, London.
——. 1960. Evolutionary adaptation. Pp. 381-402 *in:* S. Tax (Ed.), "Evolution after Darwin," Univ. Chicago Press, Chicago.
Wagner, M. 1889. Die Entstehung der Arten durch räumliche Sonderung. Schwalbe, Basel.
Walker, J. C. 1951. Genetics and plant pathology. *In:* L. C. Dunn, "Genetics in the Twentieth Century," Macmillan, New York.
Walker, T. J. 1957. Specificity in the response of female tree crickets to calling songs of the males. *Ann. Entom. Soc. Amer.*, 50:626-636.

Wallace, B. 1953. On coadaptation in *Drosophila*. *A.N.*, 87:343-358.

——. 1954. Genetic divergence of isolated populations of *Drosophila melanogaster*. *P.* IX *I.C.G.*, *Caryologia*, 6 (Suppl.): 761-764.

——. 1955. Interpopulation hybrids in *Drosophila melanogaster*. *E.*, 9:302-316.

——. 1958. The average effect of radiation-induced mutations on viability in *Drosophila melanogaster*. *E.*, 12:532-556.

——. 1962. Temporal changes in the roles of lethal and semilethal chromosomes within populations of *Drosophila melanogaster*. *A.N.*, 96:247-256.

——. 1963a. Genetic diversity, genetic uniformity, and heterosis. *Canad. J. Gen. Cytol.*, 5:239-253.

——. 1963b. Further data on the overdominance of induced mutations. *Gs.*, 48: 633-651.

——. 1965. The viability effects of spontaneous mutations in *Drosophila melanogaster*. *A.N.*, 99:335-348.

——. 1966a. On the dispersal of *Drosophila*. *A.N.*, 100:551-563.

——. 1966b. Distance and allelism of lethals in a tropical population of *Drosophila melanogaster*. *A.N.*, 100:565-578.

——. 1966c. Natural and radiation-induced chromosomal polymorphism in *Drosophila*. *Mutation Res.*, 3:194-200.

——. 1968a. Topics in Population Genetics. Norton, New York.

——. 1968b. Polymorphism, population size, and genetic load. Pp. 87-108 *in:* R. C. Lewontin (Ed.), "Population Biology and Evolution," Syracuse Univ. Press, Syracuse.

——, and Th. Dobzhansky. 1947. Experiments on sexual isolation in *Drosophila*. VIII. Influence of light on the mating behavior of *Drosophila subobscura, D. persimilis*, and *D. pseudoobscura*. *P.N.A.S.*, 32:226-234.

——, and Th. Dobzhansky. 1962. Experimental proof of balanced genetic loads in *Drosophila*. *Gs.*, 47:1027-1042.

——, J. C. King, C. V. Madden, B. Kaufmann, and E. C. McGunnigle. 1953. An analysis of variability arising through recombination. *Gs.*, 38:272-308.

——, and C. Madden. 1953. The frequencies of sub- and supervitals in experimental populations of *Drosophila melanogaster*. *Gs.*, 38:456-470.

——, and C. Madden. 1965. Studies on inbred strains of *Drosophila melanogaster*. *A.N.*, 99:495-510.

——, E. Zouros, and C. B. Krimbas. 1966. Frequencies of second and third chromosome lethals in a tropical population of *Drosophila melanogaster*. *A.N.*, 100: 245-251.

Wallace, M. E. 1965. How homozygous are our inbred lines and closed colony stocks? *Fd. Cosmet. Toxicol.*, 3:165-175.

Walles, B. 1963. Macromolecular physiology of plastids. IV. On amino acid requirements of lethal chloroplast mutants in barley. *Hereditas*, 50:317-344.

Walters, J. L. 1942. Distribution of structural hybrids in *Paeonia californica*. *Amer. J. Bot.*, 29:270-275.

Wasserman, M. 1960. Cytological and phylogenetic relationships in the *repleta* group of the genus *Drosophila*. *P.N.A.S.*, 46:842-859.

——. 1963. Cytology and phylogeny of *Drosophila*. *A.N.*, 97:333-352.

Watanabe, T. K., 1969a. Persistence of a visible mutant in natural populations of *Drosophila melanogaster*. *Japan. J. Genetics*, 44:15-22.

——. 1969b. Frequency of deleterious chromosomes and allelism between lethal genes in Japanese natural populations of *Drosophila melanogaster*. *Japan. J. Genetics*, 44:171-187.

Watson, J. D. 1965. The Molecular Biology of the Gene. Benjamin, New York.

——, and F. C. Crick. 1953. Genetical implication of the structure of deoxyribose nucleic acid. *N.*, 171:964.

Watt, W. B. 1968. Adaptive significance of pigment polymorphisms in *Colias* butterflies. I. Variation of melanin pigment in relation to thermoregulation. *E.*, 22: 437-458.

Weinberg, W. 1908. Über den Nachweis der Vererbung beim Menschen. *Jahreshefte Verein, Naturk. Würtemberg*, 64:368-382.

Welch, D'Alte A. 1938. Distribution and variation of *Achatinella mustellina* Michels in the Waianae Mountains, Oahu. *Bull. Bishop Museum*, 152:1-164.

Weisbrot, D. R. 1963. Studies on differences in the genetic architecture of related species of *Drosophila*. *Gs.*, 48:1121-1139.

——. 1966. Genotypic interactions among competing strains and species of *Drosophila*. *Gs.*, 53:427-435.

Weiss, M. C., and B. Ephrussi. 1966. Studies of interspecific (rat × mouse) somatic hybrids. I. Isolation, growth and evolution of the karyotype. *Gs.*, 54:1095-1109.

——, and H. Green. 1967. Human-mouse hybrid cell lines containing partial complements of human chromosomes and functioning human genes. *P.N.A.S.*, 58: 1104-1111.

Welshons, W. J. 1965. Analysis of a gene in *Drosophila*. *S.*, 150:1122-1129.

——, and L. B. Russell. 1959. The Y chromosome as the bearer of male determining factors in the mouse. *P.N.A.S.*, 45:560-566.

Wheeler, M. R. 1959. A nomenclatorial study of the genus *Drosophila*. *U.T.P.*, 5914:181-205.

White, M. J. D. 1949. A cytological survey of wild populations of *Trimetotripis* and *Circotettix* (*Orthoptera, Acrididae*). I. The chromosomes of twelve species. *Gs.*, 34:537-563.

——. 1954. Animal Cytology and Evolution. 2nd Ed. Cambridge Univ. Press, Cambridge.

——. 1957. Cytogenetics of the grasshopper *Moraba scurra*. II. Heterotic systems and their interaction. *Austral. J. Zool.*, 5:305-337.

——. 1966. Further studies on the cytology and distribution of the Australian parthenogenetic grasshopper *Moraba virgo*. *Rev. Suisse Zool.*, 73:383-398.

——. 1968. Models of speciation. S., 159:1065-1070.

——, H. L. Carson, and Y. Cheney. 1964. Chromosomal races in the Australian grasshopper *Moraba viatica* in a zone of geographic overlap. *E.*, 18:417-429.

——, J. Cheney, and K. H. L. Key. 1963. A parthenogenetic species of grasshopper with complex structure heterozygosity (*Orthoptera, Acridoidea*). *Austral. J. Zool.*, 11:1-19.

——, R. C. Lewontin, and L. E. Andrew. 1963. Cytogenetics of the grasshopper *Moraba scurra*. VII. Geographic variation of adaptive properties of inversions. *E.*, 17:147-162.

——, and N. H. Nickerson. 1951. Structural heterozygosity in a very rare species of grasshopper. *A.N.*, 85:239-246.

Wiener, A. S. 1965. Blood groups of chimpanzees and other nonhuman primates. *Trans. New York Acad. Sci.*, 27:488-504.

Wilkie J. S. 1959. Buffon, Lamarck and Darwin, the originality of Darwin's theory of evolution. Pp. 262-307 *in:* P. R. Bell (Ed.), "Darwin's Biological Work," Cambridge Univ. Press, Cambridge.

Williams, C. B. 1960. The range and pattern of insect abundance. *A.N.*, 94:137-151.

Williams, G. C. 1966. Adaptation and Natural Selection. Princeton Univ. Press, Princeton.

Williamson, D. L., and L. Ehrman. 1967. Induction of hybrid sterility in non-hybrid males of *Drosophila paulistorum*. *Gs.*, 55:131-140.

Wills, C. 1966. The mutational load in two natural populations of *Drosophila pseudoobscura*. *Gs.*, 53:281-294.

——. 1968. Three kinds of genetic variability in yeast populations. *P.N.A.S.*, 61:937-944.

Wilson, A. C., and V. M. Sarich. 1969. A molecular time scale for human evolution. *P.N.A.S.*, 63:1088-1093.

Wilson, E. O., and W. H. Bossert. 1963. Chemical communication among animals. *Recent Progr. Hormone Res.*, 19:673-716.

——, and W. L. Brown. 1953. The subspecies concept and its taxonomic application. S.Z., 2:97-111.

Winge, H. 1965. Interspecific hybridizaton between the six cryptic species of *Drosophila willistoni* group. *Hy.*, 20:9-19.

Witkin, E. M. 1947. Genetics of resistance to radiation in *Escherichia coli*. *Gs.*, 32:221-248.

——. 1966. Radiation-induced mutations and their repair. S., 152:1345-1353.

Wittmann-Liebold, B., and H. G. Wittmann. 1963. Die primäre Proteinstruktur von Stämmen des Tabakmosaikvirus. *Z.i.A.V.*, 9::427-435.

Wolf, B. E. 1968. Adaptiver chromosomaler Polymorphismus und flexible Kontrolle der Rekombination bei *Phryne cincta* (*Diptera, Nematocera*). *Zool. Beitr.*, 14:125-153.

Wolpoff, M. H. 1969. The effect of mutations under conditions of reduced selection. *Social Biol.*, 16:11-23.

Wolstenholme, G. E. W., and C. M. O'Connor (Eds.). 1957. Drug Resistance in Microorganisms. CIBA Foundation Symposium, Little, Brown, Boston.

Woodson, R E. 1964. The geography of flower color in butterfly weed. *E.*, 18:143-163.

Woodworth, C. M., E. R. Leng, and R. W. Jugenheimer. 1952. Fifty generations of selection for protein and oil in corn. *Agron. J.*, 44:60-66.

Woolf, Ch. M. 1965. Albinism among Indians in Arizona and New Mexico. *Amer. J. Human Genetics*, 17:23-35.

——, and F. C. Dukepoo. 1969. Hopi Indians, inbreeding and albinism. S., 164: 30-37.

Woolpy, J. H., and B. E. Ginsburg. 1967. Wolf socialization: a study of temperament in a wild social species. *Amer. Zoologist*, 7:357-363.

Wright, S. 1921. Systems of mating. *Gs.*, 6:111-178.

——. 1931. Evolution in Mendelian populations. *Gs.*, 16:97-159.

——. 1932. The roles of mutation, inbreeding, crossbreeding, and selection in evolution. *P.* VI *I.C.G.*, I:356-366.

——. 1940. Breeding structure of populations in relation to speciation. *A.N.*, 84:232-248.

——. 1943a. Isolation by distance. *Gs.*, 28:114-138.

——. 1943b. An analysis of local variability of flower color in *Linanthus parryae*. *Gs.*, 28:139-156.

——. 1948. On the roles of directed and random changes in gene frequency in the genetics of natural populations. *E.*, 2:279-294.

——. 1949. Adaptation and selection. Pp. 365-389 *in:* G. L. Jepsen, G. G. Simpson, and E. Mayr (Eds.), "Genetics, Paleontology and Evolution," Princeton Univ. Press, Princeton.

——. 1955. Classification of factors of evolution. *C.S.H.*, 20:16-24.

——. 1960. On the appraisal of genetic effects of radiation in Man. Pp. 18-24 *in:* "The Biological Effects of Atomic Radiations," National Academy of Science, Washington.

——. 1964a. Pleiotropy in the evolution of structural reduction and of dominance. *A.N.*, 98:65-69.

——. 1964b. The distribution of self-incompatibility alleles in populations, *E.*, 18:609-619.

——. 1966. Polyallelic random drift in relation to evolution. *P.N.A.S.*, 55:1074-1081.

——, and Th. Dobzhansky. 1946. Genetics of natural populations. XII. Experimental reproduction of some of the changes caused by natural selection in certain populations of *Drosophila pseudoobscura*. *Gs.*, 31:125-150.

——, Th. Dobzhansky, and W. Hovanitz. 1942. Genetics of natural populations. VII. The allelism of lethals in the third chromosome of *Drosophila pseudoobscura*. *Gs.*, 27:373-394.

Wynne-Edwards, V. C. 1962. Animal Dispersion in Relation to Social Behaviour. Oliver & Boyd, Edinburgh and London.

Yanagishima, S. 1961. CuSO4 resistance in *Drosophila melanogaster*. III and IV. *Mem. College Sci. Univ. Kyoto*, B, 28:1-52.

Yang, S. Y., and R. K. Selander. 1968. Hybridization in the grackle *Quiscalus quiscula* in Louisiana. *S.Z.*, 17:107-143.

Yanofsky, Ch., H. Berger, and W. J. Brammar. 1969. *In vivo* studies on the genetic code. *P.* XII *I.C.G.*, 3:155-165.

——, G. R. Drapeau, J. R. Guest, and B. C. Carlton. 1967. The complete amino acid sequence of the tryptophan synthetase. A protein (α subunit) and its colinear relationship with the genetic map of the A gene. *P.N.A.S.*, 57:296-298.

Yarbrough, K., and K. I. Kojima. 1967. The model of selection at the polymorphic esterase 6 locus in cage populations of *Drosophila melanogaster*. *Gs.*, 57:677-686.

Yosida, T. H., and K. Amano. 1965. Autosomal polymorphism in laboratory bred and wild Norway rats, *Rattus norvegicus*, found in Misima. *Ch.*, 16:658-667.

——, A. Nakamura, and T. Fukaya. 1965. Chromosomal polymorphism in *Rattus rattus* (L) collected in Kusudomari and Misima. *Ch.*, 16:70-78.

Zander, C. D. 1962. Untersuchungen über einen arttrenrenden Mechanismus bei lebendgebärenden Zahnkarpfen aus der Tribus *Xiphophorini*. *Mitt. Hamburg Zool. Mus. Inst.*, 60:205-264.

Zaslavsky, V. A. 1966. Isolating mechanism and its role in the ecology of two allied *Chilocorus* species. *Zool. Zhurnal*, 45:203-212. (Russian).

——. 1967. Reproductive self-destruction as an ecological factor. *J. Gen. Biol.*, 28:3-11. (Russian).

Zavadsky, K. M. 1968. Species and Species Formation. Nauka, Leningrad. (Russian).

Zimmerman, K. 1931. Studien über individuelle und geographische Variabilitat paläarktischer *Polistes* und verwandter Vespiden. *Zeit. Morph. Ökol. Tiere*, 22:173-230.

Zuckerkandl, E. 1964. Controller-gene diseases. *J. Mol. Biol.*, 8:128-147.

——. 1965. The evolution of hemoglobin. *Sci. American*, 212(5):110-118.

——, and L. Pauling. 1965. Evolutionary divergence and convergence in proteins. Pp. 97-166 *in:* V. Bryson and H. J. Vogel (Eds.), "Evolving Genes and Proteins," Academic Press, New York.

Zürcher, C. 1963. Der Faktor e^{ug} bei *Drosophila melanogaster*. *Ga.*, 34:1-33.

INDEX OF AUTHORS

SUBJECT INDEX